Plato's Press
CHICAGO, ILLINOIS

THE LIBRARY
OF Congress HAS
CATALOGUED THIS TEXT AS FOLLOWS
Registration: XXX

Philosophy
of Science
Medical Research

Description: Electronic File (eService)
Copyright: Jeffrey Wallace Bulger
Date of Creation: X/X/XX
Authorship: Jeffrey Wallace Bulger
Rights and Permissions: Jeffrey Wallace Bulger

Philosophy of Science

© 2024 Jeffrey W. Bulger

Medical Research

Table of Contents

WEEK SEVEN: Stewardship and Systems

WEEK EIGHT: Cognitive and Ethics

WEEK NINE: Philosophy and Foundations

WEEK TEN: Standards in Practice

WEEK ELEVEN: Philosophers and Ethics

WEEK TWELVE: Empiricism and Inquiry

Philosophy of Science

Medical Research

Description

PHILOSOPHY OF SCIENCE
Medical Research
Jeffrey W. Bulger

"*Philosophy of Science: Medical Research*" by Jeffrey W. Bulger is a meticulously crafted exploration of the intellectual journey that has shaped modern science and medicine. Spanning from the foundational ideas of early philosophers to contemporary advancements, this book offers a comprehensive narrative that intertwines the evolution of scientific thought with the ethical principles of autonomy, beneficence, nonmaleficence, and justice. Designed for both 10-week and 16-week courses, it provides an in-depth understanding of how rationalism, empiricism, and the scientific method have evolved, emphasizing the enduring relevance of these principles in today's biomedical landscape.

In the opening weeks, readers are introduced to the bedrock of scientific inquiry through the pioneering concepts of Karl Popper and Thomas Kuhn, setting the stage for a nuanced appreciation of how paradigms shift and knowledge advances. The narrative then delves into the ethical frameworks that guide scientific progress, examining the contributions of key thinkers such as Beauchamp, Childress, and Singer. Through the lens of historical figures like Hippocrates and modern innovators like Richard Dawkins, the book explores the dynamic interplay between scientific rigor and ethical responsibility.

As the course progresses, the book navigates through pivotal periods of scientific revolution and modern advancements, from the germ theory of disease introduced by Louis Pasteur to the genetic breakthroughs of Gregor Mendel. Each chapter illuminates how empirical investigation and ethical considerations have driven progress in fields like neuroscience, pharmacogenomics, and digital health. By the conclusion, readers gain a holistic understanding of how science and ethics are inextricably linked, fostering a deeper appreciation for the principles that ensure the integrity and advancement of medical and scientific practices. This comprehensive text is an essential resource for students, educators, and professionals seeking to understand the historical and philosophical foundations of science, as well as its future directions.

***

Philosophy of Science

Medical Research

WEEK ONE: FOUNDATIONS OF SCIENCE

WEEK ONE

Class Hour 1

- Syllabus
- The Big Picture

Class Hour 2

- Chapter 1: Karl Popper (1902-1994) — The Falsifiability Principle: Scientific Rigor
- Chapter 2: The Scientific Method — The Foundation of Inquiry: Empirical Rigor
- Chapter 3: Correlation and Causation — The Analytical Foundation: Accurate Inquiry

---

The Big Picture

The "*Philosophy of Science*" course offers a comprehensive journey through the pivotal moments and figures that have shaped scientific thought, methodology, and ethical considerations from antiquity to modern times. The course begins by laying the foundational principles of scientific inquiry, highlighting Karl Popper's philosophy of falsifiability, which redefined scientific rigor by emphasizing empirical testing and critical scrutiny. The evolution of the scientific method is further explored, tracing its roots in rationalism and empiricism and its application across various disciplines, ensuring accurate inquiry and ethical integrity.

As the course progresses, it delves into the paradigms and ethics that govern scientific progress. Thomas Kuhn's theory of paradigm shifts introduces the non-linear progression of scientific knowledge, reshaping our understanding of scientific revolutions. This section emphasizes the integration of diverse ethical frameworks, including deontology, utilitarianism, and virtue ethics, which guide scientific and medical practices. These philosophies underscore the importance of ethical decision-making, patient autonomy, beneficence, nonmaleficence, and justice, which are central to bioethics and medical ethics.

The course then explores the contributions of key thinkers and pioneers who have significantly influenced scientific and medical advancements. Figures like David Hume and Immanuel Kant bridged empirical evidence and rational analysis, laying the groundwork for modern epistemology and ethics. Innovators such as Edward Jenner and John Stuart Mill introduced transformative medical interventions and ethical theories that continue to influence public health and moral philosophy. Charles Darwin's theory of evolution and Claude Bernard's concept of the internal environment exemplify the integration of empirical observation and rational analysis in advancing biological and medical sciences.

In the later weeks, the course highlights modern scientific innovators whose empirical research and ethical considerations have revolutionized various fields. Louis Pasteur's germ theory and James Clerk Maxwell's equations unified fundamental scientific concepts, transforming medicine and physics. Wilhelm Conrad Röntgen's discovery of X-rays and Sigmund Freud's exploration of the unconscious mind showcased the transformative power of empirical research in medical diagnostics and psychology. Albert Einstein's theories of relativity and Alexander Fleming's discovery of penicillin underscore the profound impact of integrating rational and empirical approaches in advancing scientific knowledge and medical practice.

The final weeks of the course focus on contemporary medical innovators and the ethical implications of their work. Albert Sabin and Jonas Salk's development of polio vaccines exemplified the power of empirical science in public health, while Henry K. Beecher's advocacy for informed consent transformed medical ethics. Alan Turing's contributions to computing and cryptanalysis, Richard Feynman's innovations in quantum mechanics, and Rosalind Franklin's pivotal role in DNA research highlight the interdisciplinary nature of modern scientific inquiry. John Rawls's theory of justice as fairness provides a robust ethical framework for contemporary discussions on justice and equity, underscoring the ongoing interplay between scientific innovation and ethical responsibility.

**

Introduction

As we embark on our journey into "*Philosophy of Science*," we will explore the intricate tapestry of thought that has shaped our understanding of scientific inquiry and its ethical dimensions. This week, we delve into the foundational concepts that underpin the scientific method, beginning with the influential ideas of Karl Popper and extending to the nuances of empirical rigor, analytical precision, and ethical considerations in scientific practice.

Our exploration begins with Karl Popper, a seminal figure in the philosophy of science, whose principle of falsifiability redefined scientific rigor. Popper posited that for a theory to be considered scientific, it must be testable and potentially falsifiable. This revolutionary idea steered scientific inquiry away from the quest for absolute truth and towards a continuous process of hypothesis testing and critical scrutiny. Popper's legacy extends beyond his methodological contributions; his advocacy for open societies and the relentless pursuit of truth has left an indelible mark on fields as diverse as medicine and ethics.

Building upon Popper's principles, we delve into the scientific method, a structured approach rooted in rationalism and empiricism. This method, characterized by observation, hypothesis formulation, and experimentation, strives to minimize human biases and align results with objective truth. Its dual roots—rationalist logic and empiricist observation—create a balanced pathway for acquiring and refining knowledge. This approach not only advances scientific discovery but also underpins ethical integrity, setting high standards for research and societal values.

As we progress, we encounter the critical distinction between correlation and causation. This analytical foundation ensures rigorous scientific inquiry, preventing the misinterpretation of mere associations as causal relationships. In medicine, this distinction is paramount, guiding evidence-based practice and safeguarding patient care standards. Moreover, it upholds bioethical principles such as informed consent, beneficence, nonmaleficence, and justice, ensuring that medical decisions are both scientifically sound and ethically responsible.

Together, these chapters weave a coherent narrative that highlights the interplay between rigorous scientific methodology and ethical considerations. Popper's emphasis on falsifiability fosters a culture of critical evaluation, while the scientific method provides a structured framework for inquiry. The careful differentiation between correlation and causation further refines our analytical precision, ensuring that scientific and medical practices uphold the highest standards of accuracy and integrity. Through this integrated exploration, we gain a deeper appreciation of the philosophical underpinnings that drive scientific progress and ethical practice.

***

Chapter 1: Karl Popper (1902-1994) — The Falsifiability Principle: Scientific Rigor

Popper's philosophy of falsifiability redefined scientific rigor, emphasizing empirical testing and critical scrutiny in the pursuit of knowledge.

Abstract: Sir Karl Popper, a pivotal figure in the 20th-century philosophy of science, introduced revolutionary concepts that reshaped our understanding of scientific validation and progress. Popper challenged traditional validation methods, proposing that scientific theories cannot be proven true but are scientific if they are susceptible to empirical falsification. His philosophy, an intricate interplay of rationalist and empiricist elements, emphasized the fundamental role of logical reasoning and deductive analysis in forming scientific theories while endorsing empirical testability as the cornerstone of their validation. Beyond his notable influence on refining the scientific method, Popper inadvertently sculpted a framework for advancements in evidence-based medicine and bioethical principles. By emphasizing the potential falsifiability of scientific hypotheses, he fostered a culture of critical evaluation and transparency in various disciplines, leaving a lasting imprint on scientific inquiry, medical research, and ethical deliberations. His advocacy for open societies and pursuit of truth, coupled with his approach to scientific methodology, forms an enduring legacy, guiding contemporary discourse in science, medicine, and ethics.

**

Introduction: Sir Karl Popper, a towering figure in the philosophy of science during the 20th century, consistently pushed the boundaries of thought regarding the nature and progress of scientific understanding. Born in Austria in 1902, Popper's intellectual journey took him through the tumultuous terrains of the philosophy of knowledge, where he left an indelible mark with his theories on falsifiability and the demarcation problem. Fiercely critical of the conventional methods of scientific validation, Popper postulated that scientific theories can never be proven true but can be deemed scientific if they are open to empirical falsification. This distinctive approach, caught between rationalism and empiricism, has garnered acclaim and contention. As we delve deeper, we will assess Popper's rationalist inclinations, empiricist orientation, and his pivotal role in refining the scientific method. Furthermore, we will explore his uncharted contributions to medicine and dissect his influence on the core ethical principles of bioethics: patient autonomy (informed consent), practitioner beneficence (do good), practitioner nonmaleficence (do no harm), and public justice (be fair).

Rationalism: Karl Popper's philosophical stance displayed a unique interplay of rationalist inclinations within his broader empiricist framework. Although he is most commonly associated with emphasizing falsifiability and empirical testing as the cornerstones of scientific progress, Popper also acknowledged the fundamental role of reason and logic in scientific theorization. He believed scientific theories often emerge from our human propensity to solve problems rationally. This involves conjuring hypotheses from a mixture of creativity, logic, and existing knowledge rather than solely from empirical observation. In this sense, Popper embraced the rationalist belief that knowledge isn't solely derived from sensory experience. Instead, the logical structure and deductive reasoning inherent to theory formation are pivotal in advancing understanding. His insistence on critically evaluating theories using rational criteria further underscores his rationalist leanings. However, it's essential to note that while reason was indispensable in formulating hypotheses for Popper, empirical testing remained the ultimate arbiter of their scientific validity. Popper's philosophy encapsulates a synthesis of rationalist ideation and empiricist validation.

Empiricism: Karl Popper's empiricist orientation notably shone through in his foundational beliefs about the nature of scientific knowledge and its validation. While he recognized the creative and rational processes behind theory formulation, he ardently advocated that the true grit of a scientific theory lies in its empirical testability. Contrary to classical empiricists, who might have sought positive confirmations of hypotheses, Popper's distinctive twist on empiricism posited that scientific propositions should be open to falsification. For him, a theory was considered scientific if it could be potentially disproven through observational or experimental means. This "*conjecture and refutation*" method underscored the importance of empirical scrutiny, where theories continuously faced real-world tests and were either discarded or tentatively retained. His approach was a response to the problem of induction, suggesting that no number of positive confirmations could conclusively prove a theory. Still, a single valid contradictory observation could disprove it. In Popper's empiricism, knowledge grew not by amassing positive evidence but by eliminating falsehoods, reflecting his deep-seated belief in the pivotal role of sensory experiences and empirical tests in advancing scientific understanding.

The Scientific Method: Karl Popper's influence on the scientific method is most poignantly recognized in his emphasis on the falsifiability of scientific hypotheses. While the traditional scientific method involves observation, hypothesis formulation, experimentation, and conclusion, Popper introduced a critical refinement: it's not the number of confirmations that affirm a theory's truth but the potential for its falsification that demarcates genuine scientific inquiry. He argued that for a hypothesis to be scientific, it must be testable in ways that could prove it wrong. This perspective shifted the emphasis from trying to verify theories with accumulating affirmative evidence to striving to refute them. By doing so, Popper sought to address the problem of induction, asserting that no number of observations supporting a theory could prove its veracity, but a single contradictory observation could debunk it. This view has deeply impacted experimental design and how scientists approach hypothesis testing. Popper's philosophy also fostered a culture of skepticism and critical evaluation in scientific disciplines, further strengthening scientific endeavors' methodological rigor, transparency, and replicability, making him an enduring pillar in the structure of contemporary scientific methodology.

Medicine: While Karl Popper's primary legacy lies in the philosophy of science, his thoughts have indirectly made significant ripples in medicine, particularly in evidence-based practices. Popper's emphasis on falsifiability as a cornerstone of scientific validity finds resonance in the rigorous testing and scrutiny that medical interventions undergo. His "*conjecture and refutation*" philosophy aligns seamlessly with the methodology behind clinical trials, where treatments are postulated and tested for efficacy and safety, ensuring that only those passing stringent empirical tests become standard practices. The continual evolution of medical standards of care can be likened to Popper's model of scientific progress, where old theories (or treatments) are discarded or refined in light of new evidence. Moreover, the essence of evidence-based medicine, which relies on peer-reviewed research and systematic reviews to inform best practices, is an extension of Popper's insistence on critical scrutiny and empirical validation. While Popper may not have directly commented extensively on medicine, the ethos of his philosophy — a relentless pursuit of closer approximations to the truth through empirical testing — is deeply embedded in modern medical research and practice.

Ethics: While not directly focused on bioethics, Karl Popper's philosophy provides a sturdy foundational ethos from which bioethical principles have drawn inspiration. His commitment to open societies and advancing knowledge through critical scrutiny aligns with the ethical tenets that guide medical and research practices. The principle of autonomy, particularly the importance of informed consent, resonates with Popper's value for individual freedom and the pursuit of truth. It reflects his belief in respecting individuals' rights to make choices based on comprehensive and lucid information. Beneficence and nonmaleficence, the duties to do good and prevent harm, can be linked to Popper's emphasis on scientific integrity and the moral responsibility of the scientific community. According to Popper, any scientific endeavor should be undertaken sincerely to seek truth and improve human understanding, aligning with the broader objective of benefiting humanity and minimizing harm. The principle of justice, ensuring fairness, dovetails with Popper's advocacy for an open society where every individual's rights are recognized and respected. While he didn't directly shape these bioethical principles, the spirit of Popper's philosophy — valuing truth, critical inquiry, and human dignity — underpins these crucial ethical guidelines in medicine and research.

Conclusion: Karl Popper's philosophical legacy has resonated profoundly across multiple disciplines, with his ideas often acting as both a guiding beacon and a touchstone for critical debate. Popper's advocacy for falsifiability as the defining criterion for scientific theories has reshaped how scientists approach, test, and evaluate hypotheses, infusing scientific inquiry with a rigorous demand for empirical validation. Beyond pure science, his ideas have permeated medicine, emphasizing the importance of evidence-based practices and the constant evaluation of medical interventions against empirical standards. In the realm of ethics, while not a direct contributor to bioethics, the foundational tenets of his philosophy echo principles that prioritize truth-seeking, individual rights, and the betterment of humanity. Popper's intricate dance between rationalism and empiricism, a deep-seated belief in the open society, and relentless pursuit for closer approximations to truth have collectively sculpted a legacy that is both enduring and universally relevant. His contributions have acted as a compass, guiding subsequent generations to navigate the complex landscapes of science, medicine, and ethics.

Karl Popper's Legacy: Karl Popper's most projected legacy lies in his introduction of falsifiability as the criterion for scientific demarcation. This principle revolutionized the philosophy of science by emphasizing that theories should be open to empirical refutation, ensuring rigorous empirical scrutiny and safeguarding scientific inquiry from dogma and unfounded conjectures.

**

REVIEW QUESTIONS

True/False Questions:

1. Karl Popper introduced the concept that scientific theories can never be proven true but can be considered scientific if they are susceptible to empirical falsification.
   True or False?

2. According to Popper, the accumulation of positive evidence can conclusively prove a scientific theory.
   True or False?

Multiple-Choice Questions:

3. Which principle of Karl Popper's philosophy is most closely associated with the scientific method?
   a) Verification of hypotheses
   b) Falsifiability of theories
   c) Accumulation of positive evidence
   d) Deductive reasoning

4. How has Karl Popper's philosophy influenced the practice of evidence-based medicine?
   a) By emphasizing the role of theoretical constructs over empirical testing
   b) By prioritizing the falsifiability and rigorous testing of medical interventions
   c) By advocating for the subjective interpretation of medical data
   d) By promoting the use of anecdotal evidence in clinical practice

Clinical Vignette:

5. A medical researcher is designing a clinical trial to test a new drug's efficacy. Inspired by Popper's philosophy, what key principle should the researcher prioritize to ensure the trial's scientific rigor?
   a) Accumulation of as much positive data as possible
   b) Ensuring the hypothesis can be empirically tested and potentially falsified
   c) Relying on theoretical models without the need for empirical validation
   d) Collecting subjective patient testimonials

Basic Science Vignette:

6. In a scientific discussion, a researcher highlights the importance of falsifiability in validating scientific theories. Which scientist's work on falsifiability is foundational to this concept?
   a) Albert Einstein
   b) Karl Popper
   c) Isaac Newton
   d) Niels Bohr

Philosophy Vignette:

7. During a philosophy seminar, students discuss the importance of empirical testability in scientific theories. Which philosopher emphasized that scientific hypotheses should be structured to allow for potential falsification?
   a) Immanuel Kant
   b) David Hume
   c) Karl Popper
   d) Ludwig Wittgenstein

Correct Answers:

1. True
2. False
3. b) Falsifiability of theories
4. b) By prioritizing the falsifiability and rigorous testing of medical interventions
5. b) Ensuring the hypothesis can be empirically tested and potentially falsified
6. b) Karl Popper
7. c) Karl Popper

BEYOND THE CHAPTER
Karl Popper (1902-1994)

- *Karl Popper: A Centenary Assessment* edited by Ian Jarvie, Karl Milford, and David Miller
- *The Logic of Scientific Discovery* by Karl Popper
- *The Open Society and Its Enemies* by Karl Popper

***

Chapter 2: The Scientific Method — The Foundation of Inquiry: Empirical Rigor

The Scientific Method, rooted in rationalism and empiricism, remains the cornerstone of objective inquiry and ethical integrity.

Abstract: The Scientific Method, rooted in rationalist and empiricist traditions, serves as a structured pathway for scientific inquiry, fostering new knowledge acquisition and refining existing theories. This approach, characterized by observation, hypothesis formulation, and experimentation, aims to minimize human biases and align results with objective "*truth*." Its rationalist roots emphasize structured reasoning and deductive logic, while its empiricist orientation highlights sensory experiences and inductive reasoning to form general conclusions from specific observations. This method has significantly influenced various domains, including medicine, where it has catalyzed life-saving technologies and drugs, and ethics, where it has shaped foundational bioethical principles. It has standardized scientific research, ensuring a balance between empirical evidence and rationalist logic, and fostering ethical integrity and societal values. The Scientific Method remains a beacon of rational thought and empirical inquiry, fundamentally shaping our understanding of the natural world and setting high standards in pursuing knowledge, dictating ethical and societal norms in our quest to comprehend the natural world.

**

Introduction: The Scientific Method is the cornerstone of modern scientific inquiry, a structured approach for investigating phenomena, acquiring new knowledge, and revising existing theories. Rooted in rationalist and empiricist traditions, the Scientific Method is a systematic, iterative process typically consisting of steps like observation, hypothesis formulation, experimentation, and conclusion drawing. By employing rational reasoning and empirical evidence, it aims to minimize human biases, thereby achieving results that are as close to "*truth*" as possible. Its applications are broad-ranging, from solving everyday problems to catalyzing groundbreaking research in fields such as medicine, where it has been instrumental in developing life-saving drugs and medical technologies. Moreover, the Scientific Method has provided a foundational framework for the ethical principles governing bioethical issues, notably patient autonomy (informed consent), practitioner beneficence (do good), practitioner nonmaleficence (do no harm), and public justice (be fair). These principles help guide moral deliberations in scientific research, ensuring that the pursuit of knowledge aligns with societal values and ethical integrity. The Scientific Method has fundamentally shaped how humanity understands and interacts with the natural world through its systematic approach and ethical guidelines.

Rationalism: The rationalist foundations of the Scientific Method can be traced back to the notion that certain cognitive faculties are hardwired within us as tools for acquiring knowledge and understanding. This perspective posits that humans have innate abilities — like basic reasoning, pattern recognition, and problem-solving — that are "*a priori*" principles guiding scientific inquiry. For instance, creating a hypothesis in scientific research often relies on deductive reasoning, which starts with broad, often axiomatic, truths and derives specific conclusions from them. This applies to a rationalist framework, where certain foundational premises or axioms are assumed and considered self-evident or universal. Rationalism's influence can be seen in the logical structuring of experiments and the analytical rigor applied in interpreting results. The Scientific Method employs a disciplined, structured reason to filter out inconsistencies and anomalies, aiming for logical coherence in the conclusions derived. The Scientific Method is grounded in rationalist principles by prioritizing structured reasoning and deductive logic. These principles guide the researcher from the preliminary stages of problem identification through the design of experiments to the validation or falsification of hypotheses, ensuring that the inquiry is not just empirical but also logically sound.

Empiricism: The Scientific Method is deeply rooted in the empiricist tradition, emphasizing the role of observation and experience as the bedrock of scientific knowledge. Empiricism posits that our understanding of the world is fundamentally gleaned from sensory experiences and interactions with our environment. This orientation is most evident in the Scientific Method in observation and experimentation steps. Researchers start with specific observations to identify a problem or question. They then conduct carefully designed experiments to collect data subjected to rigorous statistical analysis. Using inductive reasoning is critical here; scientists take specific instances or data points and extrapolate them to form general conclusions or theories. Unlike deductive reasoning, which starts with a general premise to reach particular conclusions, inductive reasoning goes the other way: from specific observations to broader generalizations. This empirical framework allows for the validation or falsification of hypotheses and the iterative refinement of theories based on new evidence. In essence, the Scientific Method's empiricist orientation ensures that scientific inquiry is grounded in real-world data, making the generated knowledge reliable and applicable, continually subject to revision in the light of new empirical evidence.

The Scientific Method: The Scientific Method has been instrumental in shaping the structure and process of scientific inquiry. As a formalized pathway for pursuing knowledge, it offers a blueprint that disciplines as diverse as physics, biology, and psychology can follow. The cycle begins with observing a phenomenon, prompting questions, and leading to a testable hypothesis. This hypothesis is a pivot around which experiments are designed — controlled environments where variables can be manipulated to test their validity. The meticulous collection and analysis of data offer empirical footing to the inquiry, making the conclusions derived both verifiable and falsifiable. One of the most critical aspects is the emphasis on replicability; an experiment's findings gain credibility only if they can be replicated under the same conditions by other researchers. This insistence on replication adds a layer of rigorous verification, acting as a self-correcting mechanism within the scientific community. Finally, peer review is the gatekeeper, scrutinizing the research's validity, methodology, and significance. This thorough, structured approach has essentially standardized the process of scientific inquiry, ensuring that it remains rooted in rationalist logic and empirical evidence. The Scientific Method has shaped and legitimized scientific research through these mechanisms, setting a high standard for what is true and knowable.

Medicine: The Scientific Method has made monumental contributions to medicine, fundamentally transforming diagnostics and treatments. It serves as the bedrock for evidence-based medicine, enabling rigorous scientific research that shapes healthcare standards of care. These standards, often determined through randomized controlled trials, observational studies, and systematic reviews, are built on peer-reviewed, empirically validated data. The Scientific Method ensures that these standards are not merely anecdotal or tradition-based but are grounded in replicable, verifiable research. It provides a structured framework for designing experiments, collecting data, and statistically analyzing results to arrive at conclusions that influence medical practice. This evidence-based approach informs everything from pharmacological research, where new drugs are subjected to multi-phase clinical trials, to surgical procedures and public health interventions. The continuous evolution of these standards reflects the iterative nature of the Scientific Method, incorporating new research findings, clinical expertise, and patient needs. Importantly, in a legal context, the evidence-based medical healthcare standards of care established through the Scientific Method serve as criteria to assess healthcare professionals' adequacy and quality of care. The Scientific Method has been instrumental in elevating healthcare to unprecedented effectiveness, safety, and accountability by instilling a rigorous, empirical approach to medical inquiry.

Ethics: The Scientific Method has had a profound influence on the ethical underpinnings of bioethics, shaping principles such as patient autonomy (informed consent), practitioner beneficence (do good), practitioner nonmaleficence (do no harm), and public justice (be fair). By advocating for rigor, transparency, and empirical evidence, the Scientific Method naturally lends itself to ethical scrutiny, offering a framework within which these principles can be objectively evaluated. Autonomy, often manifesting as informed consent in research, gains legitimacy from the empirical rigor of the Scientific Method. Participants' rights to understand and willingly engage in a study are emphasized, recognizing them as autonomous agents whose experiences and outcomes contribute valuable data. Beneficence and nonmaleficence are closely tied to medical research's scientific integrity and potential impact. Rigorous experimentation and peer-review processes aim to maximize the benefits of any medical procedure, drug, or intervention. At the same time, potential harms are minimized, fulfilling the ethical imperatives to do good and do no harm. Finally, the principle of justice is implicitly embedded in the Scientific Method through the standardization of procedures and the expectation of replicability, guaranteeing that findings are universally applicable, thereby promoting fairness in medical care and research across diverse populations. The Scientific Method thus serves as a foundational framework that integrates empirical investigation with ethical responsibility, providing a robust structure for ethical deliberation in scientific and medical contexts.

Conclusion: The Scientific Method is the linchpin of modern scientific inquiry. It has far-reaching implications that extend beyond academia into everyday life and vital sectors like medicine and ethics — rooted in both rationalist foundations that emphasize deductive reasoning and innate cognitive faculties and an empiricist orientation that emphasizes sensory experience and inductive logic. The Scientific Method is a balanced and rigorous approach to discovering the truth. Its structured framework has standardized the scientific community's process to research, enforcing replicability and peer review as crucial elements of credible study. This has translated into evidence-based medical healthcare standards of care that are effective and accountable, continually evolving to integrate discoveries and technologies. The Method also informs ethical standards, resulting in the bioethical principles of patient autonomy (informed consent), practitioner beneficence (do good), practitioner nonmaleficence (do no harm), and public justice (be fair). By providing this structured pathway for empirical investigation and ethical evaluation, the Scientific Method has fundamentally shaped our collective pursuit of knowledge, our approach to healthcare, and our moral compass in scientific endeavors. It remains a paragon of rational thought and empirical inquiry, guiding humanity's never-ending quest to understand the natural world.

The Scientific Method's Legacy: The Scientific Method stands as the foundation for empirical inquiry and rational thought, shaping the scientific landscape and the ethical and societal norms that govern the pursuit of knowledge.

**

REVIEW QUESTIONS

True/False Questions:

1. The Scientific Method involves a systematic approach consisting of steps like observation, hypothesis formulation, experimentation, and conclusion drawing.
   True or False?
2. According to the chapter, the Scientific Method primarily relies on anecdotal evidence and tradition-based practices.
   True or False?

Multiple-Choice Questions:

3. The chapter suggests that the rationalist foundations of the Scientific Method emphasize:
   a) Sensory experiences and environmental interactions
   b) Deductive reasoning and structured logic
   c) Random data collection and analysis
   d) Cultural influences and social norms
4. In the context of medicine, the Scientific Method ensures that:
   a) Treatments are based on anecdotal evidence
   b) Medical practices are grounded in replicable, verifiable research
   c) Clinical guidelines are formed without empirical data
   d) Traditional healing methods are prioritized over scientific research

Clinical Vignette:

5. A team of researchers is conducting a study to test a new drug's efficacy. They follow the Scientific Method, starting with a hypothesis based on preliminary observations. According to the chapter, what should their next step be?
   a) Publish their findings immediately
   b) Conduct controlled experiments to collect empirical data
   c) Skip data collection and move to peer review
   d) Base their conclusions on theoretical models alone

Basic Science Vignette:

6. Dr. Kim is studying the effects of a new drug on reducing anxiety levels. He formulates a hypothesis that the drug will significantly reduce anxiety compared to a placebo. Which steps should Dr. Kim follow to ensure his research adheres to the Scientific Method?
   a) Collect anecdotal evidence from a small group of patients and publish the findings immediately
   b) Formulate a hypothesis, conduct controlled experiments, collect and analyze data, and ensure the study can be replicated
   c) Test the drug on his family and friends without a control group
   d) Skip the data analysis and go straight to publishing if initial observations seem promising

Philosophy Vignette:

7. In her philosophy class, Laura argues that the Scientific Method is crucial for upholding ethical standards in research. How should she best support her argument using the principles of bioethics?
   a) The Scientific Method helps ensure that research is based on empirical evidence, supporting patient autonomy, beneficence, nonmaleficence, and justice
   b) The Scientific Method only focuses on obtaining results and does not consider ethical implications
   c) Ethical standards in research are not influenced by the Scientific Method
   d) The Scientific Method allows researchers to bypass ethical concerns to achieve quicker results

Correct Answers:

1. True
2. False
3. b) Deductive reasoning and structured logic
4. b) Medical practices are grounded in replicable, verifiable research
5. b) Conduct controlled experiments to collect empirical data
6. b) Formulate a hypothesis, conduct controlled experiments, collect and analyze data, and ensure the study can be replicated
7. a) The Scientific Method helps ensure that research is based on empirical evidence, supporting patient autonomy, beneficence, nonmaleficence, and justice

BEYOND THE CHAPTER
The Scientific Method

- *The Structure of Scientific Revolutions* by Thomas S. Kuhn
- *Scientific Method: A Historical and Philosophical Introduction* by Barry Gower
- *The Scientific Method: An Evolution of Thinking from Darwin to Dewey* by Henry M. Cowles

***

Philosophy of Science

Medical Research

Chapter 3: Correlation and Causation — The Analytical Foundation: Accurate Inquiry

Distinguishing correlation from causation is crucial for rigorous scientific inquiry and ethical integrity in medicine.

Abstract: The distinction between correlation and causation is a foundational tenet in research and critical analysis, with far-reaching implications across various disciplines. Grounded in both rationalist and empiricist philosophies, this differentiation shapes our approach to knowledge acquisition, be it through inherent cognitive faculties or experiential insights. This distinction ensures rigorous and valid inquiry within the scientific method, preventing erroneous conclusions based on mere associations, and in medicine, differentiating between the two aids in establishing evidence-based medical healthcare standards of care, ensuring effective treatments. Moreover, the discernment between correlation and causation is pivotal in bioethical considerations, safeguarding principles such as patient autonomy (informed consent), practitioner beneficence (do good), practitioner nonmaleficence (do no harm), and public justice (be fair). As such, this seemingly statistical notion profoundly impacts our broader intellectual and ethical frameworks, driving accuracy, reliability, and moral integrity in our endeavors.

**

Introduction: In research and scientific analysis, the distinction between "*correlation*" and "*causation*" is a fundamental pillar, shaping our understanding and interpretation of data. At its core, correlation signifies a relationship between two variables, highlighting how they move in tandem, while causation delves deeper, proposing that one variable actively influences or brings about a change in another. This nuanced difference, though subtle, carries profound implications, determining the veracity and applicability of countless studies across diverse disciplines.

In understanding correlation and causation, it's also paramount to grasp related concepts such as coincidence, confounding variables, and reverse causation, which often muddy the waters of interpretation:

Coincidence: At times, two events might appear correlated merely due to random chance, without any underlying causal relationship. For example, the oft-cited spurious correlation between ice cream consumption and the number of people who drowned in swimming pools in the U.S. is purely coincidental; it is unreasonable to suggest that ice cream consumption causes swimming pool drownings.

Confounding Variables: External factors might cause independent and dependent variables to move in a certain direction, giving the illusion of a causal relationship when there isn't one. For instance, the correlation between ice cream sales and drowning incidents is confounded by temperature, as warm weather increases both ice cream sales and swimming activities, leading to more drownings.

Reverse Causation: This occurs when the cause and effect are misunderstood, leading to the wrong conclusion about which is the influencing factor. For example, one might observe that homes with fire alarms experience more fires and wrongly conclude that fire alarms cause fires. In reality, houses at a higher risk of fires are more likely to have alarms installed.

Recognizing these nuances underlines the need for a rigorous, systematic approach to data analysis, emphasizing the importance of the foundational distinction between correlation and causation. As we embark on this exploration, we'll delve into the topic's roots in rationalist thinking and empiricist orientation and its paramount role in sculpting the scientific method. Furthermore, we'll assess its significant contributions to medicine and examine its profound impact on the four principles of bioethics: ensuring patient autonomy through informed consent, upholding practitioner beneficence and nonmaleficence by doing good and avoiding harm, and championing public justice through fair healthcare delivery.

Rationalism: Within the philosophical context of rationalism, the dichotomy of correlation versus causation finds its roots in our innate cognitive faculties. These faculties, evolutionarily hardwired into our very being, endow us with the ability to discern patterns, exercise basic reasoning, and engage in problem-solving. They serve as our "*a priori*" principles—foundational tools for knowledge acquisition. When confronted with two events frequently occurring together, our intrinsic pattern recognition might quickly deem them correlated. However, deducing causation demands a deeper layer of reasoning stemming from axioms or postulated truths about the world. While the power of our inherent reason is undeniable, epigenetic factors and learned behaviors can influence and sometimes distort these faculties, introducing potential biases. For instance, the adaptive value of these faculties, designed to react to challenges and bolster survival swiftly, can sometimes lead to hasty, flawed conclusions in complex modern scenarios. These basic faculties can collectively birth more intricate cognitive capabilities at the crossroads of emergence. Understanding correlation and causation, especially in sophisticated scientific inquiries, is a testament to the emergent complexity of our rational faculties, transcending mere pattern recognition to embody nuanced understanding and ethical contemplation. As such, on an individual and societal level, acknowledging the strengths and vulnerabilities of our rationalistic foundations becomes paramount, guiding ethical, educational, and policy-driven discussions.

Empiricism: Empiricism, rooted in the assertion that knowledge can be gained from sensory experiences and environmental interactions, provides a unique lens through which the discourse of correlation versus causation is framed. At its core, empiricism relies on inductive reasoning, where specific observations culminate into broader conclusions. When dissecting the relationship between two observed phenomena, one might notice a correlation based on repeated sensory experiences. However, discerning causation necessitates a more rigorous empirical examination, often involving controlled experiments or long-term observations. While this approach offers a robust foundation, it's not without pitfalls; cognitive biases, formed as heuristic shortcuts, can sometimes skew our interpretations. Furthermore, external influences, such as cultural norms and socialization, introduce additional layers of complexity to our empirical judgments. From an individual standpoint, recognizing the impact of one's experiences on interpretations is essential, especially when making ethical decisions. On a broader scale, the emphasis on empirical evidence in professional realms, particularly in scientific and policy domains, ensures guidelines and regulations are grounded in tangible realities. The emergent property of empiricism can be exemplified in phenomena like language evolution or the genesis of complex ecosystems—both instances where collective experiences, rather than isolated ones, define the outcome. Thus, in understanding correlation and causation, the empiricist orientation serves as a reminder of the multifaceted nature of knowledge acquisition and the intricate dance between experience and understanding.

The Scientific Method: The intricate interplay between correlation and causation has been instrumental in refining the scientific method, a structured approach to inquiry and discovery. Observing a phenomenon naturally leads to questions about relationships between variables, and often, a correlation is the first hint of a potential connection. However, in its rigor, the scientific method demands that mere correlation is not mistaken for causation. It requires a testable hypothesis beyond noting synchronous behaviors to positing a causal link. Controlled experiments are then crafted to isolate variables and determine whether the observed correlation stems from a causal relationship or if confounding factors are at play. For instance, while an observational study might reveal a correlation between two variables, experimental studies ascertain causality by manipulating one variable and observing its effect on another. The necessity for experiments to be replicable is also tied to this distinction; if causation is genuine, the cause-effect relationship should consistently manifest under identical conditions, irrespective of the researcher. As findings undergo the scrutiny of peer review, the distinction between correlation and causation often becomes a focal point of evaluation, ensuring that conclusions drawn are robust, valid, and contribute meaningfully to the collective scientific understanding. The scientific method reinforces the importance of discerning correlation from true causation in pursuing knowledge through this rigorous process.

Medicine: The distinction between correlation and causation has profoundly impacted medicine, especially in developing and refining medical healthcare standards of care. While observational studies might note correlations, such as a link between a lifestyle habit and a health outcome, causal relationships are established through rigorous controlled trials, guiding medical interventions and treatments. Recognizing mere correlations without identifying true causative agents could lead to ineffective or harmful medical practices. For instance, while an observational study might find a correlation between patients taking a certain drug and improved health outcomes, controlled trials are essential to determine if the drug causes the improvement or if other factors are at play. By discerning causation from correlation, the medical community can establish genuinely beneficial and evidence-based standards of care. These standards, grounded in thorough research and causal understanding, dictate the expected level of care, ensuring that patients receive treatments founded on robust, scientifically validated principles. In legal contexts, such rigorous standards also serve as benchmarks to assess the adequacy and appropriateness of care delivered, protecting practitioners and patients within the dynamic landscape of medical practice.

Ethics: The distinction between correlation and causation plays a pivotal role in upholding the foundational bioethical principles of medicine. Regarding autonomy and informed consent, a clear understanding of correlation versus causation ensures that patients receive accurate information about potential treatments, allowing them to make informed decisions about their care. Misinterpreting correlations as causal relationships can lead to misleading information and compromise a patient's autonomy. Beneficence, the principle of doing good, is reinforced when medical interventions are based on causal relationships, ensuring patients receive the most effective treatments. Conversely, nonmaleficence, or the commitment to do no harm, is safeguarded when healthcare professionals recognize and avoid acting upon mere correlations, which could lead to unintended harm or ineffective treatments. Lastly, the principle of justice, which emphasizes fairness, is fortified when treatments and interventions are based on genuine causative evidence, ensuring that all patients have access to effective care, regardless of their background. Thus, discerning correlation from causation is not just a scientific imperative but also an ethical one, underpinning the principles guiding moral and just medical practice.

Conclusion: The differentiation between correlation and causation is a cornerstone in various intellectual disciplines, deeply influencing our interpretative frameworks and shaping our understanding of phenomena. In the crosshairs of rationalist instincts and empirical observations, this distinction serves as a guiding post, ensuring that our innate faculties and learned experiences converge to extract meaningful insights. Through the lens of the scientific method, it drives rigorous experimentation and analysis, safeguarding the integrity and reliability of discoveries. In medicine, discerning causation from mere correlation ensures that standards of care are rooted in genuine causative evidence, optimizing patient outcomes and maintaining the sanctity of the healthcare profession. Furthermore, from an ethical standpoint, understanding this distinction becomes paramount, as it underscores the principles of autonomy, beneficence, nonmaleficence, and justice, ensuring that healthcare decisions are informed, beneficial, and just. While fundamentally a statistical concept, the discourse on correlation versus causation resonates deeply across various domains, highlighting its significance in sculpting our knowledge, practices, and ethical standards.

Correlation and Causation's Legacy: The crucial distinction between correlation and causation underpins rigorous scientific inquiry and critical thinking, ensuring accurate interpretations and preventing misleading conclusions in philosophical and empirical endeavors.

**

REVIEW QUESTIONS

True/False Questions:

1. The distinction between correlation and causation is essential in preventing erroneous conclusions based on mere associations.
   True or False?

2. According to the chapter, reverse causation occurs when the cause and effect are clearly understood without any confusion.
   True or False?

Multiple-Choice Questions:

3. The chapter suggests that within rationalist thinking, discerning causation from correlation primarily relies on:
   a) Sensory experiences
   b) Pattern recognition and deductive reasoning
   c) Random chance
   d) Cultural influences

4. In medicine, the distinction between correlation and causation ensures that:
   a) Treatments are based on unverified assumptions
   b) Observational studies alone dictate medical practices
   c) Standards of care are founded on scientifically validated principles
   d) Correlational data is preferred over causal evidence

Clinical Vignette:

5. A researcher is investigating the relationship between a new diet and reduced heart disease incidence. Observational studies show a strong correlation, but the researcher wants to establish causation. Based on the chapter, what should the researcher prioritize?
   a) Conducting more observational studies to confirm the correlation
   b) Performing controlled trials to determine if the diet directly reduces heart disease
   c) Relying on anecdotal evidence from patients
   d) Publishing the findings based on the observed correlation alone

Basic Science Vignette:

8. Dr. Wong is studying the relationship between high sugar intake and the development of diabetes. His observational study finds a strong correlation between the two. What should be his next step to establish a causal relationship?
   a) Conclude that high sugar intake directly causes diabetes based on the correlation
   b) Conduct controlled experiments to determine if high sugar intake directly leads to diabetes
   c) Ignore the correlation and look for other potential causes of diabetes
   d) Recommend immediate public health policies based on the observed correlation

Philosophy Vignette:

7. In her philosophy class, Emily argues that understanding the distinction between correlation and causation is crucial for ethical decision-making in medical research. How should she best support her argument using bioethical principles?
   a) Correlations are sufficient for making ethical medical decisions
   b) Causal relationships ensure that treatments are based on solid evidence, upholding principles like beneficence and nonmaleficence
   c) Ethical decision-making does not require distinguishing between correlation and causation
   d) Correlations can always be trusted to make ethical decisions in medical research

Correct Answers:

1. True
2. False
3. b) Pattern recognition and deductive reasoning
4. c) Standards of care are founded on scientifically validated principles
5. b) Performing controlled trials to determine if the diet directly reduces heart disease
6. b) Conduct controlled experiments to determine if high sugar intake directly leads to diabetes
7. b) Causal relationships ensure that treatments are based on solid evidence, upholding principles like beneficence and nonmaleficence

BEYOND THE CHAPTER
Correlation and Causation

- *The Book of Why: The New Science of Cause and Effect* by Judea Pearl and Dana Mackenzie
- *Naked Statistics: Stripping the Dread from the Data* by Charles Wheelan
- *Causality: Models, Reasoning and Inference* by Judea Pearl
- *Spurious Correlations* by Tyler Vigen

***

Philosophy of Science

Medical Research

WEEK TWO: PARADIGMS AND ETHICS

WEEK TWO

Class Hour 1

- Chapter 4: Thomas Kuhn (1922-1996) — The Paradigm Shifter: Scientific Revolutions
- Chapter 5: Epistemology — The Foundations of Knowledge: Inquiry and Understanding
- Chapter 6: Bioethics — The Moral Guide: Science and Ethics

Class 2

- Chapter 7: Deontology — The Moral Compass: Ethical Principles
- Chapter 8: Utilitarianism — The Greatest Good: Ethical Decision-Making
- Chapter 9: Virtue Ethics — The Moral Excellence: Character and Integrity

---

Introduction

In the second week of our course, "*Philosophy of Science*," we delve into the complex interplay between paradigms and ethics, examining how scientific revolutions, epistemological foundations, and ethical theories converge to shape our understanding and practice of science. Through the works of seminal thinkers like Thomas Kuhn, we explore the dynamic nature of scientific knowledge and the moral frameworks that guide its application in society.

Thomas Kuhn's theory of paradigm shifts stands as a cornerstone of this exploration. Kuhn challenged the traditional, linear view of scientific progress, positing instead that science advances through periodic revolutions that disrupt established paradigms. These paradigm shifts reflect the evolving nature of scientific knowledge, driven by both rationalist and empiricist elements. Kuhn's insights reveal the sociocultural and historical contexts that influence scientific development, urging a continuous re-evaluation of our understanding of the world. This non-linear progression underscores the importance of adaptability and critical thinking in scientific inquiry, profoundly impacting fields like medicine and bioethics, where new paradigms can redefine practices and ethical standards.

As we consider the foundations of knowledge through the lens of epistemology, we appreciate the delicate balance between rationalism and empiricism. Epistemology examines the mechanisms of knowledge acquisition, emphasizing the interplay between cognitive faculties and sensory experiences. This philosophical study informs scientific methodology, ensuring that our quest for knowledge remains rigorous and ethically sound. In medicine, epistemology guides the development of healthcare policies and practices, fostering an evidence-based approach that adapitates to new discoveries while upholding ethical integrity.

Bioethics emerges as a crucial field that integrates scientific and moral principles, providing a framework for navigating the ethical complexities of modern healthcare. Rooted in both rationalist and empiricist traditions, bioethics emphasizes the importance of patient autonomy, beneficence, nonmaleficence, and justice. These principles guide ethical decision-making, ensuring that medical practices align with both scientific validity and moral integrity. Bioethics challenges us to consider the broader implications of scientific advancements, promoting a humanistic approach that prioritizes dignity and compassion in healthcare.

Deontology, with its focus on adherence to rational rules, offers another perspective on ethical conduct. This branch of ethics underscores the importance of universal principles in guiding actions, emphasizing the intrinsic ability to discern moral truths through reason. In science and medicine, deontological principles promote integrity and responsibility, shaping standards of care that prioritize patient welfare and dignity. Deontology's emphasis on rational foundations and empirical intersections fosters a society anchored in moral integrity and respect for human dignity.

Conversely, utilitarianism seeks to maximize the greatest good, harmonizing rational thought and empirical evidence in ethical decision-making. This consequentialist doctrine promotes analytical scrutiny and rigorous peer review in scientific methodology, while in medicine, it shapes standards of care based on empirical research. Utilitarianism aligns with bioethical principles, cultivating a compassionate and equitable healthcare ecosystem that prioritizes collective well-being.

Finally, virtue ethics emphasizes the importance of moral character and integrity in ethical practices. Rooted in Aristotelian thought, this approach nurtures virtues such as compassion and moral discernment, guiding healthcare professionals to embody ethical principles in their practice. Virtue ethics promotes a patient-centered approach, fostering a society anchored in justice and compassion and guiding individuals toward moral excellence and human flourishing.

Together, these chapters weave a rich narrative that highlights the intricate connections between scientific paradigms and ethical considerations. From Kuhn's paradigm shifts to the foundational principles of epistemology, bioethics, deontology, utilitarianism, and virtue ethics, we gain a deeper understanding of the philosophical underpinnings that shape our scientific and ethical landscapes. This integrated exploration underscores the dynamic and evolving nature of knowledge, urging us to continuously re-evaluate and adapt our approaches to ensure that scientific progress remains ethically grounded and socially responsible.

***

Chapter 4: Thomas Kuhn (1922-1996) — The Paradigm Shifter: Scientific Revolutions

Kuhn's theory of paradigm shifts, highlighting the non-linear progression of scientific knowledge, reshaped our understanding of scientific development.

Abstract: In the 20th century, Thomas Kuhn emerged as a seminal figure in the philosophy of science, radically altering our understanding of scientific progression with his theory of paradigm shifts, as presented in "*The Structure of Scientific Revolutions*." Kuhn challenged the conventional narrative of linear scientific development, proposing that scientific fields evolve through periodic revolutions, transitioning from one dominant paradigm to another. This concept, rooted in both rationalist and empiricist ideologies, highlights the interplay between conceptual frameworks and empirical observations, fostering intellectual discourse and critical analysis. Kuhn's theories profoundly impact medicine and bioethics, encouraging a nuanced, fluid approach to medical practices and ethical considerations. His emphasis on the sociocultural and historical intricacies of scientific developments reveals the complex, ever-changing nature of knowledge acquisition, urging continuous re-evaluation and adaptation within the scientific community and beyond. Acknowledging the mutable nature of truth and knowledge, Kuhn has established himself as a cornerstone in the philosophical discourse on science, ethics, and their broader societal implications.

**

Introduction: Thomas Kuhn is one of the most influential philosophers of science in the 20th century, having left an indelible mark on how we conceptualize and understand scientific progress. Born in 1922, Kuhn's seminal work, "*The Structure of Scientific Revolutions*," challenged traditional narratives of a linear and accumulative path to scientific knowledge. Instead, he introduced the idea of "*paradigm shifts*," suggesting that scientific fields undergo periodic revolutions, transitioning from one dominant paradigm to another. Kuhn's ideas were groundbreaking, sparking extensive debate and leading to a reconsideration of the nature and development of scientific theories. As we delve deeper into his thought, we'll assess Kuhn's position concerning rationalism and empiricism, his impact on the scientific method, his contributions to medicine, and the intriguing ways in which his work intersects with the four ethical pillars of bioethics: patient autonomy (informed consent), practitioner beneficence (do good), practitioner nonmaleficence (do no harm), and public justice (be fair).

Rationalism: Thomas Kuhn's work on Philosophy of Science, while not exclusively rooted in rationalism, does entertain certain rationalist inclinations. Central to Kuhn's thinking is the notion of paradigms, which are essentially shared conceptual frameworks within which scientists work and reason. These paradigms are not simply empirical constructs derived from objective observations, but they represent a collective understanding, a shared logic, and a set of assumptions about the world. Kuhn posits that scientists operate within these paradigms until anomalies accumulate, prompting a paradigm shift. This aspect of Kuhn's theory is rationalist in spirit, as it suggests that scientific progress isn't merely an empirical endeavor built upon sensory observations but is also deeply rooted in the logical and conceptual structures (paradigms) that guide scientific thought. Kuhn's paradigm theory underscores that even in empirical observations, scientists are driven by "*a priori*" frameworks, which are informed more by reason and foundational assumptions than raw sensory data. Thus, while Kuhn does not champion pure rationalism, his emphasis on the pivotal role of paradigms in scientific practice nods toward the rationalist's belief in the importance of reason and logic in shaping knowledge.

Empiricism: While Thomas Kuhn's philosophy is multifaceted and can't be pigeonholed strictly within empiricism, his work resonates with several empiricist tenets. Most notably, Kuhn acknowledges the significance of empirical anomalies in scientific change. In "*The Structure of Scientific Revolutions*," Kuhn illustrates how accumulating anomalies — empirical observations that don't fit within the reigning paradigm — can lead to a crisis within a scientific community. This crisis, born from direct experiences and observations defying existing frameworks, can catalyze a paradigm shift. Kuhn's emphasis on the role of empirical observations in challenging and reshaping dominant scientific paradigms lends his work an empiricist orientation. Furthermore, Kuhn believed that different paradigms could be incommensurable, suggesting that scientists working under different paradigms might see and interpret the world differently, highlighting the importance of experiential knowledge. Thus, while Kuhn certainly did not dismiss the role of conceptual structures (paradigms) in guiding scientific inquiry, he placed the empirical world front and center in challenging and changing those structures, showcasing his alignment with empiricist principles.

The Scientific Method: Thomas Kuhn's influence on our understanding of the scientific method is profound, notably through his challenge to the conventional linear narrative of scientific progress. Before Kuhn, the scientific method was often seen as a straightforward procedure leading invariably to objective truths about the world. In "*The Structure of Scientific Revolutions*," Kuhn offered a more nuanced view, suggesting that science doesn't advance through a cumulative process of adding to a body of knowledge but through periodic "*paradigm shifts*." For Kuhn, the scientific method is deeply embedded within paradigms, essentially shared conceptual frameworks guiding scientific inquiry. When the community widely accepts a paradigm, "*normal science*" ensues, wherein scientists work within the paradigm's boundaries, often solving puzzles rather than questioning the paradigm itself. It's only when anomalies accumulate and cause a crisis that the possibility of a paradigm shift emerges. Kuhn's perspective reshaped the understanding of the scientific method by emphasizing the social, historical, and even psychological elements of scientific practice, underscoring that science is not just an objective pursuit but is also shaped by prevailing paradigms, community consensus, and historical context. In doing so, he introduced a layer of complexity to the scientific method, integrating the sociology and philosophy of scientific discovery.

Medicine: Although Thomas Kuhn's primary focus was not medicine; his philosophy has significantly influenced medical thinking and practice. Kuhn's concept of paradigm shifts can be seen in the evolution of medical theories and practices over time. Just as scientific fields experience paradigm shifts when existing frameworks become inadequate, medicine has also witnessed transformative changes in its understanding of diseases, treatments, and healthcare methodologies. For instance, the shift from miasma theory to germ theory in the 19th century can be likened to a Kuhnian paradigm shift, radically changing medical practices and public health interventions. Furthermore, Kuhn's emphasis on the scientific revolution's sociocultural and historical contexts underscores the importance of understanding medicine as a collection of objective facts and a practice deeply embedded in societal values, expectations, and historical moments. His work indirectly underscores the need for medical standards of care to be fluid, reflecting both empirical advancements and evolving societal values. Thus, while Kuhn did not directly contribute to medical standards or practices, his philosophy offers a lens through which medicine's evolution, practice, and societal embedding can be better understood, appreciated, and critiqued.

Ethics: While Thomas Kuhn's foundational work, "*The Structure of Scientific Revolutions*," primarily engages with the philosophy and history of science, its implications ripple into bioethics, indirectly influencing how we perceive and apply its principles. Kuhn's assertion that prevailing paradigms shape scientific understanding highlights the mutable nature of what is considered "*truth*" or "*knowledge*" at any given time. In the context of bioethics, this mutability necessitates a dynamic approach. For instance, autonomy, particularly informed consent, becomes intricate when what's considered scientifically valid today might be supplanted tomorrow. Patients making choices based on current knowledge might act differently if they were privy to a future paradigm. Beneficence and nonmaleficence, grounded in the principle of doing good and avoiding harm, are similarly impacted. If medical practices evolve based on shifting paradigms, what is deemed beneficial or harmful can also change. Finally, justice, which demands fairness in healthcare, must recognize that equitable treatment involves continuously updated practices and notions of health influenced by evolving paradigms. While Kuhn didn't explicitly delve into bioethics, his paradigm theory compels a recognition of the fluidity and complexity inherent in applying ethical principles within a continuously evolving medical landscape.

Conclusion: Thomas Kuhn's monumental legacy in the philosophy of science transcends the confines of his field, influencing a diverse range of domains from the scientific method to medicine and bioethics. His groundbreaking assertion that scientific advancements occur not as steady accumulations but through transformative paradigm shifts revolutionized our comprehension of scientific progress. This perspective spotlighted the significance of rationalist frameworks and empirical observations in the evolution of knowledge. By emphasizing that science is as much a sociocultural endeavor as an empirical one, Kuhn illuminated the intricate interplay between objective observations and the paradigms in which they are interpreted. His insights resonated beyond the laboratories and academic circles, prompting the medical community to recognize and appreciate the periodic revolutions in its theories and practices. Furthermore, Kuhn's emphasis on the shifting nature of scientific "*truth*" has profound implications for bioethics, urging a continuous reevaluation of principles in light of evolving paradigms. Kuhn's work is a testament to knowledge's intricate, dynamic, and interwoven nature, offering a critique of established notions and a roadmap for future intellectual endeavors.

Thomas Kuhn's Legacy: Introduced the concept of paradigm shifts in "*The Structure of Scientific Revolutions*," reshaping how we understand scientific progress and knowledge accumulation.

**

REVIEW QUESTIONS

True/False Questions:

1. Thomas Kuhn's theory of scientific progress is based on the idea of linear, cumulative development.
   True or False?

2. According to Kuhn, paradigm shifts occur when anomalies accumulate and challenge the existing framework.
   True or False?

Multiple-Choice Questions:

3. What is a key component of Kuhn's theory of scientific revolutions?
   a) Scientific progress is always steady and cumulative.
   b) Paradigms never change once established.
   c) Scientific fields undergo periodic revolutions that shift paradigms.
   d) Science advances only through technological innovations.

4. How does Kuhn's concept of paradigm shifts influence the understanding of medical practices?
   a) It suggests that medical practices remain unchanged over time.
   b) It emphasizes that medical practices evolve and can undergo transformative changes.
   c) It states that medical practices are solely based on empirical data.
   d) It asserts that medical practices are purely theoretical.

Clinical Vignette:

5. A hospital is revising its guidelines for the treatment of a particular disease based on recent breakthroughs in research that contradict previous understandings. According to Kuhn's philosophy, how should the hospital approach this change?
   a) Ignore the new research and continue with the old guidelines.
   b) Acknowledge the new research as a paradigm shift and update the guidelines accordingly.
   c) Conduct more experiments without changing the guidelines.
   d) Disregard the importance of paradigms in medical practice.

Basic Science Vignette:

6. A group of researchers in a biomedical lab notices that their long-standing model for a disease process is increasingly inconsistent with new experimental data. Following Thomas Kuhn's concept of paradigm shifts, what should their next step be?
   a) Disregard the new data and continue with the existing model
   b) Abandon all previous research and start from scratch
   c) Evaluate the anomalies to determine if a new paradigm is needed
   d) Wait for other researchers to validate the new data before making changes

Philosophy Vignette:

7. A bioethics committee is assessing the validity of current ethical guidelines in light of new medical technologies. Inspired by Thomas Kuhn's theories, what should the committee consider to ensure the guidelines remain relevant and effective?
   a) Strictly adhere to existing guidelines, regardless of new developments
   b) Completely overhaul the guidelines with each technological advancement
   c) Reassess the guidelines periodically, considering empirical data and the possibility of paradigm shifts
   d) Avoid making any changes to avoid confusion and maintain stability

Correct Answers:

1. False
2. True
3. c) Scientific fields undergo periodic revolutions that shift paradigms
4. b) It emphasizes that medical practices evolve and can undergo transformative changes
5. b) Acknowledge the new research as a paradigm shift and update the guidelines accordingly
6. c) Evaluate the anomalies to determine if a new paradigm is needed
7. c) Reassess the guidelines periodically, considering empirical data and the possibility of paradigm shifts

BEYOND THE CHAPTER
Thomas Kuhn (1922-1996)

- *The Structure of Scientific Revolutions* by Thomas S. Kuhn
- *Thomas Kuhn: A Philosophical History for Our Times* by Steve Fuller
- *The Road Since Structure: Philosophical Essays, 1970-1993* by Thomas S. Kuhn

***

Chapter 5: Epistemology — The Foundations of Knowledge: Inquiry and Understanding

Epistemology bridges rationalism and empiricism, guiding scientific rigor and ethical frameworks that shape our understanding and actions.

Abstract: In the multifaceted domain of epistemology, we explore the core mechanisms of human knowledge acquisition and its vast influence on science, ethics, and medicine. Based on the interplay between rationalism and empiricism, epistemology underpins scientific methodology and ethical principles, guiding disciplines beyond their philosophical confines. A meticulous approach to rationalist foundations elucidates the cognitive faculties that govern our understanding of reality, while empiricism emphasizes the role of sensory experiences and environmental interactions in knowledge formation, fostering an adaptive, evidence-based approach in professional fields and public policy. This philosophical study shapes rigor and ethical considerations in medicine and fundamentally influences healthcare policies. Through its nuanced exploration of knowledge, epistemology emerges as a pivotal framework that continually reshapes our comprehension of the world and the methodologies that govern scientific, ethical, and societal developments, leaving a lasting legacy on human thought and action.

**

Introduction: Epistemology, the philosophical inquiry into the nature, scope, and limits of human knowledge, is a critical underpinning for advancing science, ethics, and medicine. As a branch of philosophy, it grapples with fundamental questions surrounding reality's nature, sensory perception's reliability, and the justification for belief. Epistemology's dual allegiance to rationalism and empiricism provides a dialectical framework for evaluating knowledge, influencing a broad spectrum of disciplines beyond philosophy. Rationalist traditions, grounded in logical reasoning and inherent knowledge, coalesce with empiricist orientations grounded in sensory experience and observation. These intellectual legacies shape the scientific method, lending methodological rigor and ethical considerations to scientific inquiries. Notably, epistemology's influence is deeply felt in the medical sciences, contributing to understanding and applying ethical principles such as patient autonomy (informed consent), practitioner beneficence (do good), practitioner nonmaleficence (do no harm), and public justice (be fair). As we navigate this labyrinthine exploration of knowledge, we aim to dissect how epistemology forms the bedrock for scientific and ethical reasoning and fosters advancements in clinical research and healthcare outcomes.

Rationalism: Epistemology's rationalist foundations offer a rich tapestry of insights into the intrinsic faculties that humans and other organisms possess for knowledge acquisition and understanding. Rooted in the belief that certain cognitive abilities — like basic reasoning, pattern recognition, and problem-solving — are evolutionarily hardwired, rationalism lends credence to the idea that "*a priori*" principles guide our pursuit of knowledge. These innate faculties are foundational for deductive reasoning, allowing for logical inferences based on definitional truths, axioms, or postulates. The interplay between these innate cognitive skills and epigenetic or learned factors adds further complexity, contributing to emergent phenomena like complex social behaviors in animals or advanced ethical reasoning in humans. For example, consider how birds migrate using an innate sense of direction combined with learned environmental cues. Similarly, human ethical reasoning can be seen as emerging from innate cognitive structures that interact with cultural and social learning. This concept of "*emergence*" suggests that the collective capabilities of an organism, extending to states of consciousness, are far more intricate than a mere sum of individual faculties. Moreover, the adaptive value of these rationalistic faculties lies in their efficacy in swiftly and reliably addressing recurring life challenges, thus contributing to an organism's survival and reproductive success. In contemporary discourse, rationalist foundations also serve as ethical benchmarks; they are scrutinized by professionals like scientists, educators, and ethicists to ensure that cognitive faculties are understood and engaged within a responsible, ethical framework. Concurrently, at the public policy level, an awareness of these rational faculties and their limitations is crucial for designing regulations that mitigate potential downsides while maximizing benefits for societal well-being.

Empiricism: Epistemology's empiricist orientation posits that knowledge originates from sensory experiences and environmental interactions. Employing inductive reasoning, empiricism starts with specific observations and scales them to general conclusions. This approach highlights the significant role of cognitive biases, often evolved as heuristic "*shortcuts*," in shaping our perception and understanding of the world. Additional layers of complexity are introduced by external factors such as socialization and culture, which significantly influence cognition and behavior. For example, the development of phobias can illustrate how sensory experiences shape knowledge and behavior. A person bitten by a dog might develop a phobia of dogs, demonstrating how specific observations (the bite) can generalize to broader conclusions (fear of all dogs). On an individual level, empiricism emphasizes the importance of self-awareness in understanding how one's sensory experiences shape ethical viewpoints and decision-making. Professionally, the empiricist stance underscores the indispensability of empirical evidence in framing ethical guidelines and scientific theories. In the public policy sphere, empiricism advocates for regulations grounded in empirical data and thus is adaptive and responsive to societal changes and technological advances. The adaptive value of empiricism lies in its ability to prepare organisms to adapt to new and unforeseen environmental challenges, contributing to survival and reproductive success. When discussing "*emergence*" within the context of empiricism, we refer to the notion that an organism's collective properties, behaviors, and perhaps states of consciousness are more complex and intricate than what can be predicted by studying individual sensory experiences alone. Phenomena like the development of language, culture, or complex ecosystems stand as a testament to such emergent properties.

The Scientific Method: Epistemology, the study of knowledge and its acquisition, fundamentally shapes the scientific method by providing the philosophical groundwork for how we understand, collect, and interpret evidence. The method begins with observation, which epistemology examines through its empiricist orientation — focusing on sensory experiences as a reliable source of knowledge. The subsequent formulation of testable hypotheses is informed by epistemology's rationalist foundations, positing that logical reasoning can help us deduce theoretical consequences that can be empirically tested. Epistemology also grapples with "*proof*," guiding scientific inquiry's meticulous data collection and analysis phase. The imperative for replicability in experiments finds its roots in the epistemological commitment to the stability and reliability of knowledge. Lastly, the peer-review process embodies epistemology's nuanced perspectives on the social dimensions of knowledge — how communal verification and methodological rigor contribute to the formation of scientifically accepted truths. For example, the discovery of penicillin by Alexander Fleming illustrates the scientific method in action. Fleming's initial observation of mold killing bacteria was an empirical finding, which led to the formulation of a hypothesis that penicillin could be used to treat bacterial infections. Subsequent experiments and peer-reviewed studies confirmed its efficacy, leading to its widespread use in medicine. The scientific method can be viewed as an applied epistemology, systematizing the diverse ways we pursue, evaluate, and establish knowledge.

Medicine: Epistemology's influence on medicine is pervasive and transformative, underpinning the scientific rigor and ethical considerations in healthcare. At its core, epistemology offers the philosophical scaffolding for evidence-based medicine, shaping how medical professionals establish medical healthcare standards of care. These standards are not arbitrary but are grounded in empirically tested, peer-reviewed research. The epistemological commitment to reliable, replicable knowledge sets the stage for how evidence is collected, interpreted, and applied in clinical settings. Furthermore, integrating rationalistic elements, like deductive reasoning, allows for developing and refining diagnostic algorithms and treatment protocols, imbuing them with logical consistency. For example, the development of diagnostic criteria for mental health disorders in the DSM-5 involves both empirical evidence and rational analysis to ensure accuracy and reliability. The constantly evolving nature of medical healthcare best practices echoes epistemology's recognition that knowledge is not static but influenced by various factors, including technological advancements and emerging insights into patient needs. In legal frameworks, epistemology indirectly shapes the concept of "*reasonably competent and skilled healthcare*," providing the criteria against which medical actions are judged. Thus, epistemology plays a central role in establishing the scientific and ethical bases for the ever-evolving medical healthcare standards of care, enabling an evidence-based approach that serves both clinical efficacy and ethical responsibility.

Ethics: Epistemology profoundly impacts the ethical landscape of medicine, particularly the four bioethical principles of patient autonomy (informed consent), practitioner beneficence (do good), practitioner nonmaleficence (do no harm), and public justice (be fair). The principle of autonomy, expressed through informed consent, is undergirded by the epistemological assertion that individuals should have adequate and accurate knowledge to make informed decisions about their health. This aligns with rationalist and empiricist orientations, emphasizing the importance of reasoning and evidence in decision-making. Beneficence and nonmaleficence are influenced by empirically derived medical standards that maximize benefit while minimizing harm. These standards are codified forms of medical knowledge whose legitimacy is ensured through scientific methodologies, peer review, and professional consensus. For example, during the Ebola outbreak in West Africa, the development and deployment of experimental treatments and vaccines were guided by empirical evidence and ethical considerations. This approach aimed to maximize the benefits of these interventions while minimizing potential harms, ultimately aiding in the control of the epidemic. The epistemological commitment to objective reasoning and empirical evidence informs the principle of justice, driving policies that aim to distribute healthcare resources fairly and equitably. For example, the allocation of limited medical resources during a crisis is often based on evidence-based protocols that aim for the greatest collective benefit. Thus, the bioethical principles of patient autonomy (informed consent), practitioner beneficence (do good), practitioner nonmaleficence (do no harm), and public justice (be fair) are not merely normative postulates but are deeply rooted in epistemological theories that shape how knowledge is acquired, validated, and applied in medical ethics.

Conclusion: Epistemology, the study of knowledge and its acquisition, is a cornerstone across multiple domains — from rationalist philosophies to empirical sciences and medicine to ethics. It provides the rationalist foundation for understanding innate cognitive faculties, shaping our understanding of "*a priori*" principles and their role in human reasoning and ethical decision-making. On the flip side, its empiricist orientation emphasizes the importance of sensory experiences and environmental interactions in shaping individual perspectives and collective norms. This empiricist view has been pivotal in shaping the scientific method, imbuing it with rigorous empirical scrutiny that relies on observation, hypothesis testing, and replicability. In medicine, epistemology contributes to establishing evidence-based standards of care, emphasizing the role of research and empirical data in evolving medical practices. Regarding bioethics, principles like patient autonomy (informed consent), practitioner beneficence (do good), practitioner nonmaleficence (do no harm), and public justice (be fair) are deeply informed by epistemological theories, reinforcing the need for informed consent, evidence-based practices, and just resource distribution. Overall, epistemology is not merely an academic discipline but a foundational framework that significantly influences our understanding of the world, the ethics that govern our interactions, and the methodologies we employ in various professional fields.

Epistemology's Legacy: As the study of the nature, scope, and limits of knowledge, Epistemology provides the foundational framework that underpins scientific inquiry, ethical reasoning, and public policy, continually shaping how humanity understands and engages with the world.

**

REVIEW QUESTIONS

True/False Questions:

1. Epistemology is the philosophical study of knowledge, including its nature, scope, and limits.
   True or False?

2. According to the chapter, empiricism emphasizes the role of innate cognitive faculties in shaping knowledge.
   True or False?

Multiple-Choice Questions:

3. The chapter describes rationalism as positing that certain cognitive abilities are:
   a) Learned solely through environmental interactions
   b) Evolutionarily hardwired and innate
   c) Derived from cultural and social learning
   d) Developed through random chance

4. In the context of medicine, epistemology primarily contributes to:
   a) Arbitrary decision-making in treatment plans
   b) Establishing evidence-based standards of care
   c) Ignoring empirical data
   d) Promoting traditional healing methods without scientific basis

Clinical Vignette:

5. A 40-year-old patient is considering participating in a clinical trial. The medical team needs to ensure that the patient makes an informed decision. Based on the chapter's discussion on epistemology and ethics, what should the medical team prioritize?
   a) Providing only the benefits of the trial
   b) Ensuring the patient understands both the risks and benefits
   c) Highlighting only the risks involved
   d) Allowing the patient to decide without any guidance

Basic Science Vignette:

6. Dr. Lee is studying the development of language in children. He notices that while some children quickly acquire language skills, others take longer despite similar environmental exposures. Considering the role of both innate cognitive faculties and sensory experiences, which approach should he take to best understand language development?
   a) Focus exclusively on genetic factors and cognitive faculties
   b) Ignore cognitive faculties and focus solely on sensory experiences
   c) Adopt an integrated approach that considers both innate cognitive faculties and sensory experiences
   d) Treat all children with language delays using the same methods, regardless of individual differences

Philosophy Vignette:

7. In her philosophy seminar, Sarah argues that the empirical evidence of cognitive biases challenges the reliability of sensory perception. How should she best support her argument using epistemology?
   a) Cognitive biases have no impact on sensory perception, making sensory experiences reliable
   b) Sensory perception is entirely accurate, and cognitive biases do not exist
   c) Cognitive biases demonstrate that sensory experiences can be flawed, requiring critical examination and validation through empirical evidence
   d) Sensory perception is unreliable and should be disregarded in favor of pure reasoning

Correct Answers:

1. True
2. False
3. b) Evolutionarily hardwired and innate
4. b) Establishing evidence-based standards of care
5. b) Ensuring the patient understands both the risks and benefits
6. c) Adopt an integrated approach that considers both innate cognitive faculties and sensory experiences
7. c) Cognitive biases demonstrate that sensory experiences can be flawed, requiring critical examination and validation through empirical evidence

BEYOND THE CHAPTER
Epistemology

- *Theory of Knowledge: An Introduction* by Paul K. Moser
- *Epistemology: An Anthology* edited by Ernest Sosa, Jaegwon Kim, Jeremy Fantl, and Matthew McGrath
- *The Problems of Philosophy* by Bertrand Russell

***

Chapter 6: Bioethics — The Moral Guide: Science and Ethics

Bioethics integrates science and moral principles, guiding healthcare with a balance of rational inquiry and empirical evidence.

Abstract: Bioethics is an essential interdisciplinary field at the crossroads of philosophy, law, and medical sciences, providing moral guidance amid rapid advancements in biology, medicine, and technology. This study explores bioethics' relationship with rationalist and empiricist perspectives, fostering a balanced approach to ethical decision-making in healthcare and biotechnology. Rationalism in bioethics delves into ethical considerations from innate and epigenetic viewpoints, while empiricism emphasizes a data-driven approach, recognizing the value of flexibility and nuanced understanding. Bioethics integrates with the scientific method, imposing ethical checkpoints to ensure research aligns with scientific validity, societal implications, and ethical integrity. It redefines healthcare standards, grounding the field in human dignity and moral integrity. Principles like patient autonomy (informed consent), practitioner beneficence (do good), practitioner nonmaleficence (do no harm), and public justice (be fair) are central to bioethics. As it shapes ethical discourse and healthcare practices, bioethics promises a legacy of guiding biomedicine with a humanistic, morally accountable approach, safeguarding the ethical trajectory of the biomedical landscape.

**

Introduction: Bioethics is an interdisciplinary field that examines the ethical implications of advancements in biology, medicine, and technology, aiming to navigate the complex moral landscape these advances often create. At the intersection of philosophy, law, and medical sciences, bioethics questions patients' dignity and rights, healthcare providers' responsibilities, and the societal implications of medical innovations such as genetic engineering, stem cell research, and end-of-life care. Traditional ethical theories — utilitarianism, deontology, and virtue ethics — often serve as frameworks for discussing these dilemmas. Bioethics utilizes principles that are derivable from, consistent with, or at least not in conflict with these traditional ethical theories and worldviews — patient autonomy (informed consent), practitioner beneficence (do good), practitioner nonmaleficence (do no harm), and public justice (be fair) — to guide decision-making. In an era where scientific and technological capabilities are rapidly evolving, bioethics provides crucial insights that help to steer these developments in ethically sound directions, ensuring that they comply with general morality in benefiting humanity and minimizing harm. Whether it's the distribution of scarce medical resources, questions surrounding patient consent, or the moral quandaries posed by breakthroughs in biotechnology, bioethics serves as a vital compass in a world of unprecedented possibilities and challenges.

Rationalism: Bioethics incorporates rationalist perspectives by acknowledging the evolutionary basis of human cognitive faculties, which include reasoning, pattern recognition, and problem-solving abilities. These faculties play a crucial role in shaping our moral and ethical landscape. From a rationalist standpoint, the ethical principles in bioethics — such as patient autonomy (informed consent), practitioner beneficence (do good), practitioner nonmaleficence (do no harm), and public justice (be fair) — can be viewed as advanced manifestations of more basic cognitive faculties that have evolved for adaptive reasons. For instance, the principle of patient autonomy might have roots in a developed understanding of individual agency, while practitioner beneficence and nonmaleficence may be sophisticated expressions of social reciprocity, a key survival advantage in early human communities. Deductive reasoning is frequently employed in bioethics to conclude established ethical axioms, addressing complex issues such as gene editing, artificial intelligence in healthcare, and organ transplantation. On a professional level, bioethicists, scientists, and healthcare providers apply rationalist principles to dissect ethical dilemmas, often to harmonize ethical frameworks with scientific facts. Moreover, public policy in healthcare is increasingly being formulated through a rationalist lens, recognizing the evolved cognitive limitations that might result in ethical pitfalls or biases. Thus, in this intricate web, rationalism provides a structure, enabling bioethics to engage deeply with ethical considerations from an innate and epigenetic standpoint while fostering an understanding that the emergent complexities of moral reasoning are greater than the sum of individual cognitive faculties.

Empiricism: In bioethics, an empiricist orientation offers a complementary perspective to rationalism, emphasizing the importance of sensory experiences, environmental interactions, and empirical evidence in shaping ethical principles and guidelines. Rooted in inductive reasoning, this approach relies on specific observations — clinical trials, epidemiological data, patient testimonials, etc. — to inform broader ethical considerations such as patient rights, healthcare access, and the societal implications of medical technologies. These empirically-derived insights are particularly vital for understanding the impact of environmental and cultural variables on cognition and behavior, thus enriching the ethical framework used to assess emerging biomedical advancements. For example, the ethical debates surrounding vaccination policies or genetically modified organisms often necessitate a thorough review of empirical data to weigh the risks and benefits. At a professional level, bioethicists, clinicians, and policymakers use empiricism to continuously update ethical guidelines, ensuring they are congruent with the latest scientific findings. On the public policy front, empiricism supports the need for adaptable, evidence-based regulations that respond dynamically to societal changes and technological advances. The adaptive value of empiricism lies in its capacity for flexibility and nuanced understanding, allowing bioethics to evolve in response to new ethical challenges that were previously unforeseen. The empiricist orientation in bioethics recognizes that ethical complexity often emerges from many individual experiences and observations, reinforcing the importance of a data-driven approach to ethical decision-making.

The Scientific Method: Bioethics plays a pivotal role in shaping the scientific method by providing an ethical framework that informs the design, execution, and interpretation of scientific research. In the pre-experimental phase, bioethics ensures that formulating research questions and hypotheses aligns with ethical considerations, particularly in studies involving human subjects or those with potential societal impacts, such as clinical trials and genetic research. The ethical principles of patient autonomy (informed consent), practitioner beneficence (do good), practitioner nonmaleficence (do no harm), and public justice (be fair) guide the informed consent process, ensuring that participants are fully aware of risks and benefits. During the experimental phase, bioethics mandates rigorous data collection and analysis standards to minimize harm and maximize benefit, aligning closely with the scientific method's emphasis on meticulous data handling. As for the post-experimental phase, bioethics contributes to the peer-review process by scrutinizing the research's scientific validity and ethical integrity and societal implications. This includes assessing whether the research was conducted fairly if it poses any risks to social justice, or if it compromises the autonomy or well-being of individuals or communities. By embedding these ethical checkpoints within the scientific method, bioethics ensures that scientific advancements are not just methodologically sound but also ethically responsible. Thus, bioethics enriches the scientific method with additional scrutiny, reinforcing the pursuit of knowledge as a balanced interplay between empirical rigor and ethical consideration.

Medicine: Bioethics has profoundly influenced the field of medicine by offering ethical guidelines and principles that help shape medical practice, research, and healthcare policy. The foundational principles of bioethics — patient autonomy (informed consent), practitioner beneficence (do good), practitioner nonmaleficence (do no harm), and public justice (be fair) — provide a moral compass that guides medical decision-making, patient care, and standards of practice. For example, the principle of autonomy has revolutionized the concept of informed consent, emphasizing the patient's right to make decisions about their care based on full disclosure of risks, benefits, and alternatives. Similarly, beneficence and nonmaleficence serve as ethical cornerstones that inform healthcare professionals' responsibilities to do good and avoid harm, shaping evidence-based and ethically sound medical healthcare standards of care. The principle of justice tells healthcare policies related to equitable access, fair treatment, and the distribution of scarce medical resources. In a legal context, these bioethical principles often serve as evaluative benchmarks to determine whether healthcare professionals have met the expected medical healthcare standards of care based on peer-reviewed, evidence-based practices. By interweaving ethical considerations into medical practices and healthcare standards, bioethics ensures that medicine is not just a science but also an art grounded in human dignity, social responsibility, and moral integrity. As a result, bioethics elevates medical healthcare standards of care, ensuring they are clinically effective, ethically justifiable, responsive to patient needs, and reflective of societal values.

Ethics: Bioethics has played an instrumental role in the development and application of the four cornerstone principles of patient autonomy (informed consent), practitioner beneficence (do good), practitioner nonmaleficence (do no harm), and public justice (be fair) in the context of healthcare and biotechnology. Autonomy, often operationalized through informed consent, underscores the importance of respecting individual agency, ensuring that patients have the right to self-determination, and fully understanding the implications of medical procedures or research participation. Beneficence embodies the ethical obligation to promote good, guiding healthcare professionals in proactively pursuing treatments and interventions that benefit the patient, often with a broader societal perspective. Nonmaleficence, encapsulated in the Hippocratic injunction to "*do no harm*," serves as a cautionary principle, urging physicians and researchers to consider and minimize the risks and adverse effects of medical interventions. Lastly, justice addresses the equitable distribution of healthcare resources, fair treatment of patients, and tackling systemic inequities in healthcare access and outcomes. These principles serve as ethical guideposts for clinicians, researchers, policymakers, and ethicists, providing a balanced framework for navigating the complex ethical terrain of modern medicine and biotechnology. They are not just theoretical constructs but are actively integrated into clinical guidelines, research protocols, and healthcare policies, ensuring that ethical considerations are deeply embedded in the fabric of medical practice and scientific exploration. By defining and advocating for these principles, bioethics has profoundly shaped healthcare's ethical discourse and practice, elevating it from mere technical competence to a human-centered endeavor grounded in moral responsibility.

Conclusion: Bioethics serves as an ethical compass, not just for medical professionals but for society at large, by addressing the complex moral and ethical questions that arise from advances in biology, medicine, and technology. It harmonizes rationalism and empiricism, incorporating deductive and inductive reasoning, to create robust ethical frameworks deeply rooted in philosophy and empirical evidence. Through its influence on the scientific method, bioethics ensures scientific research is methodologically sound and ethically responsible. In medicine, bioethics has fundamentally redefined medical healthcare standards of care by incorporating ethical considerations into medical practices, research protocols, and healthcare policies. It guides medical professionals in balancing the principles of patient autonomy (informed consent), practitioner beneficence (do good), practitioner nonmaleficence (do no harm), and public justice (be fair), elevating healthcare from a mere technical service to a humanistic, morally accountable endeavor. In doing so, bioethics provides society with the tools to navigate the ethical intricacies of issues ranging from informed consent and healthcare access to gene editing and artificial intelligence. As we venture deeper into an age marked by rapid scientific advancements and increasing ethical complexity, bioethics remains an indispensable guide, shaping healthcare and the moral dimensions of human interaction with the biological world.

Bioethics' Legacy: The lasting legacy of bioethics will likely be its establishment of a multidisciplinary ethical framework that navigates the intersection of science, medicine, and technology, a contribution vital for ensuring that advances in these domains are guided by considerations of human dignity, social equity, and moral integrity, thereby safeguarding the humane and ethical trajectory of an increasingly complex biomedical landscape.

**

REVIEW QUESTIONS

True/False Questions:

1. Bioethics primarily focuses on scientific methods and excludes philosophical and legal considerations.
   True or False?

2. The principle of nonmaleficence in bioethics emphasizes that healthcare providers should do no harm to patients.
   True or False?

Multiple-Choice Questions:

3. Which principle of bioethics is directly concerned with ensuring that patients fully understand the implications of medical procedures or research participation?
   a) Beneficence
   b) Nonmaleficence
   c) Autonomy
   d) Justice

4. The ethical principle that guides healthcare professionals to promote good and pursue beneficial treatments is known as:
   a) Autonomy
   b) Justice
   c) Nonmaleficence
   d) Beneficence

Clinical Vignette:

5. Dr. Jones is conducting a clinical trial and ensures that all participants are fully informed about the potential risks and benefits before they consent to participate. Which bioethical principle is Dr. Jones primarily upholding?
a) Beneficence
b) Nonmaleficence
c) Autonomy
d) Justice

Basic Science Vignette:

7. Dr. Ramirez is conducting a clinical trial on a new gene therapy for a rare genetic disorder. To ensure ethical compliance and scientific rigor, what should Dr. Ramirez prioritize during his study?
a) Maximizing the number of participants regardless of consent
b) Ensuring informed consent, conducting the study with rigor, minimizing harm, and providing equitable access to the therapy
c) Publishing preliminary results quickly without thorough data analysis
d) Ignoring ethical considerations to speed up the research process

Philosophy Vignette:

7. In her philosophy class, Lisa argues that bioethics is crucial for guiding medical practices. How should she best support her argument using the principles of bioethics?
a) Bioethics ensures that medical practices are grounded in patient autonomy, beneficence, nonmaleficence, and justice, promoting ethical and humane healthcare
b) Bioethics focuses solely on theoretical discussions without practical applications
c) Medical practices do not need bioethical principles to ensure ethical conduct
d) Bioethics primarily concerns legal issues rather than ethical considerations in healthcare

Correct Answers:

1. False
2. True
3. c) Autonomy
4. d) Beneficence
5. c) Autonomy
6. b) Ensuring informed consent, conducting the study with rigor, minimizing harm, and providing equitable access to the therapy
7. a) Bioethics ensures that medical practices are grounded in patient autonomy, beneficence, nonmaleficence, and justice, promoting ethical and humane healthcare

BEYOND THE CHAPTER

Bioethics

- *Principles of Biomedical Ethics* by Tom L. Beauchamp and James F. Childress
- *The Oxford Handbook of Bioethics* edited by Bonnie Steinbock
- *Bioethics: An Anthology* edited by Helga Kuhse, Udo Schüklenk, and Peter Singer
- *Bioethics: Passing The Boards, Providing Patient Care, and Beyond* by Jeffrey W. Bulger

***

Chapter 7: **Deontology** — The Moral Compass: Ethical Principles

Deontology upholds moral principles through rational rules, guiding ethical conduct in science and medicine.

Abstract: Deontology, a branch of ethics, explores the morality of actions through adherence to rational rules rather than empirical consequences. Rooted in rationalism, it relies on humans' intrinsic ability to discern moral truths through reason, establishing universal rules and principles. However, it also integrates real-world observations to form a more adaptable ethical framework. In the scientific realm, deontological principles underscore ethical conduct, promoting integrity and responsibility in research. In medicine, it shapes healthcare standards, guiding professionals to prioritize patient welfare and dignity through universal principles. It significantly influences bioethics, forming the bedrock of the four principles — patient autonomy (informed consent), practitioner beneficence (do good), practitioner nonmaleficence (do no harm), and public justice (be fair) — that guide ethical decision-making. Deontology leaves a lasting legacy through its rational foundations and empirical intersections, fostering a society anchored in moral integrity and respect for human dignity.

**

Introduction: Deontology, a prominent branch of ethics, scrutinizes the morality of actions based on their adherence to universal rules or principles rather than specific contingent consequences. Stemming from the Greek words "*deon*" (duty) and "*logos*" (study), it is often referred to as duty-based or rule-based ethics. Central to this ethical theory is the belief in universal moral principles that govern the actions of all individuals, irrespective of the contingent circumstances. It underscores individuals' moral duties and obligations toward each other, rooted in rational reasoning and creating societal norms. This non-consequentialist approach diverges from consequentialist theories like utilitarianism, as it places a premium on the intrinsic properties of universal principles rather than solely focusing on contingent outcomes. As we delve deeper, we will explore deontology's rationalist foundations, its empiricist orientation, its role in shaping the scientific method, its contributions to the field of medicine, and its profound influence on the four principles of bioethics: patient autonomy (informed consent), practitioner beneficence (do good), practitioner nonmaleficence (do no harm), and public justice (be fair).

Rationalism: Deontology's rationalist foundations are deeply entrenched in the belief that moral truths can be discerned through reason and logical deduction, independent of sensory experience. This ethical theory posits that individuals have innate cognitive faculties that serve as evolutionarily hardwired mechanisms for acquiring knowledge or understanding. These faculties, encompassing basic reasoning skills, pattern recognition, and problem-solving abilities, function as "*a priori*" principles, guiding individuals in moral judgments. Furthermore, deontology leans heavily on deductive reasoning, which draws from definitional truths, axioms, or postulates to reach universal logical conclusions. This rationalist approach is vividly illustrated in the works of the German philosopher Immanuel Kant, a stalwart of deontological ethics. Kant asserted that actions are morally right if they adhere to a universalizable principle, a concept determined primarily through rational deliberation, not contingent empirical observation. Thus, deontology stands as a beacon of rationalism in ethical theories, emphasizing the role of inherent reasoning capabilities in shaping moral judgments and actions.

Empiricism: While deontology is predominantly grounded in rationalist principles, its interaction with empiricism presents a nuanced perspective on moral ethics. Empiricism, which posits that knowledge can be acquired through sensory experiences and environmental interactions, is a counterpoint to deontology's reliance on universal "*a priori*" principles. In the context of deontology, an empiricist orientation would entail a meticulous analysis of real-world situations and outcomes to derive moral principles. This approach leans on inductive reasoning, where specific observations lead to general conclusions, offering a more flexible and adaptive ethical framework for contingent situations. For instance, a deontologist with an empiricist orientation might consider the tangible impacts of an action on individuals and communities, integrating these observations into their moral deliberations of universal principles. This synthesis of empiricism and deontology fosters a more comprehensive ethical framework, where moral principles are derived from logical deductions and informed by the rich tapestry of human experiences and interactions, thus broadening the scope and applicability of deontological ethics.

The Scientific Method: Deontology's influence on the scientific method can be discerned in the universal adherence to ethical principles that govern scientific inquiry. The scientific method, a systematic approach to research, hinges on formulating and testing hypotheses through controlled experiments, data analysis, and peer review. Deontological ethics, emphasizing duty-based universal principles, fosters a culture of integrity and responsibility in the scientific community. Researchers are bound by a moral duty to conduct experiments ethically, ensuring subjects' welfare and data authenticity. Moreover, the principle of universalizability, a cornerstone of deontology, echoes the scientific method's requirement for replicability, where findings must consistently hold under the same conditions, irrespective of the researcher's experiment. Furthermore, the peer review process embodies deontological principles of justice and fairness, as it subjects research to rigorous scrutiny by experts in the field, ensuring that only credible and ethically conducted research gains acceptance. Thus, deontology plays a pivotal role in shaping the scientific method, instilling it with a strong ethical foundation that upholds the virtues of honesty, transparency, and responsibility, thereby safeguarding the credibility and progress of scientific endeavors.

Medicine: In medicine, deontological ethics has significantly shaped the standards of care, emphasizing the universal moral duties and responsibilities that healthcare professionals owe to their patients. This ethical approach insists on adherence to established principles, such as providing competent and skilled care, respecting patient autonomy (informed consent), and avoiding harm (nonmaleficence), which align with the universal expectations of medical healthcare standards of care. These standards, grounded in peer-reviewed, evidence-based practices, dictate the level and type of care that should be provided under various contingent circumstances. These serve as a benchmark for evaluating the universal adequacy of delivered care, especially in legal contexts. Deontology's influence is palpable in the continual evolution of these standards, fostering a culture of integrity and responsibility where the primary duty is to uphold the welfare and dignity of patients. By promoting adherence to the universal moral principles of patient autonomy (informed consent), practitioner beneficence (do good), nonmaleficence (do no harm), and public justice (be fair), deontology aids in navigating the complex ethical dilemmas that often arise in medical practice, guiding healthcare professionals in making decisions that are not only medically sound but also ethically justified, thereby safeguarding the sanctity of the patient-doctor relationship and enhancing the overall quality of healthcare delivery.

Ethics: Deontology's profound influence on bioethics is markedly evident in the formulation and adherence to the four universal principles: patient autonomy (informed consent), practitioner beneficence (do good), practitioner nonmaleficence (do no harm), and public justice (be fair). These principles, deeply rooted in deontological ethics, serve as the bedrock of ethical decision-making in healthcare. The universal principle of autonomy, which emphasizes informed consent, resonates with deontology's emphasis on respecting individual rights and moral agency, ensuring that patients have a decisive voice in their treatment plans. Beneficence, the universal moral obligation to promote good and foster well-being, aligns with deontology's duty-based approach, urging healthcare professionals to act in the best interest of the patients. Similarly, nonmaleficence, which embodies the universal maxim "*do no harm*," echoes deontological mandates to avoid actions that could inflict harm, safeguarding patients from the potential adverse effects of medical interventions. Lastly, the principle of justice, which universally advocates for fairness and equality in healthcare delivery, promotes equitable distribution of healthcare resources and opportunities. Together, these principles, under the guidance of deontological ethics, forge a robust and compassionate framework that navigates the complex moral landscape of healthcare, fostering an ethically sound and profoundly humane practice.

Conclusion: Deontology, with its steadfast focus on duty-based ethics, has carved a significant niche in various domains, including philosophy, science, and medicine. Its insistence on universal moral principles has fostered a culture of integrity and responsibility, guiding individuals and professionals in navigating complex ethical dilemmas. In the scientific realm, it has been instrumental in shaping a systematic approach that emphasizes replicability and peer review, ensuring universal research credibility and ethical conduct. Moreover, its contributions to the medical field are substantial, influencing the development of universal medical healthcare standards of care that prioritize patient welfare and ethical practice. Furthermore, its profound impact on bioethics, particularly in shaping the universal principles of patient autonomy (informed consent), practitioner beneficence (do good), practitioner nonmaleficence (do no harm), and public justice (be fair), has cultivated a compassionate and ethical approach to healthcare delivery. As we reflect upon deontology's role in shaping ethical discourse and practice, it stands as a beacon of moral rectitude, encouraging a conscientious approach that universally respects the dignity and rights of all individuals, thereby fostering a society grounded in universal justice, empathy, and moral responsibility. Its enduring relevance and influence underscore its pivotal role in guiding humanity toward ethical and principled living.

Deontology's Legacy: As a moral compass, Deontology guides individuals and professionals to adhere to universal principles and duties, fostering a society grounded in ethical integrity and respect for human dignity.

**

REVIEW QUESTIONS

True/False Questions:

1. Deontology emphasizes the morality of actions based on their adherence to universal principles rather than contingent circumstances.
   True or False?

2. Deontological ethics primarily relies on empirical observations rather than rational reasoning to derive moral principles.
   True or False?

Multiple-Choice Questions:

3. Which of the following principles is NOT one of the four universal principles influenced by deontological ethics in bioethics?
   a) Patient autonomy
   b) Practitioner beneficence
   c) Nonmaleficence
   d) Utilitarianism

4. Which philosopher is most closely associated with deontological ethics?
   a) John Stuart Mill
   b) Immanuel Kant
   c) Jeremy Bentham
   d) David Hume

Clinical Vignette:

5. A patient with a terminal illness requests to end their life through assisted suicide. According to deontological ethics, what would be the primary consideration in addressing this request?
   a) The potential consequences and overall happiness
   b) The patient's quality of life and suffering
   c) The adherence to universal principles, such as the duty to do no harm
   d) The opinions of the patient's family and friends

Basic Science Vignette:

6. A researcher is conducting a clinical trial and follows deontological principles to ensure all participants give informed consent, regardless of the potential benefits the study might bring. Which principle of deontology is primarily demonstrated by this action?
   a) Consequentialism
   b) Utilitarianism
   c) Universalizability
   d) Empiricism

Philosophy Vignette:

7. A philosopher evaluates deontology's emphasis on the moral duty to treat all individuals with inherent dignity and respect. Which principle of deontology best aligns with the idea that moral actions are determined by adherence to universal rules?
   a) Risk-benefit analysis
   b) Ethical review by IRBs
   c) Categorical imperative
   d) Public justice

Correct Answers:

1. True
2. False
3. d) Utilitarianism
4. b) Immanuel Kant
5. c) The adherence to universal principles, such as the duty to do no harm
6. c) Universalizability
7. c) Categorical imperative

BEYOND THE CHAPTER

Deontology

- *Groundwork of the Metaphysics of Morals* by Immanuel Kant
- *The Metaphysics of Morals* by Immanuel Kant
- *Kantian Ethics* by Allen W. Wood
- *Deontology* by Stephen Darwall (ed.)
- *Kant's Ethical Thought* by Allen W. Wood

***

Philosophy of Science — Medical Research

Chapter 8: Utilitarianism — The Greatest Good: Ethical Decision-Making

Utilitarianism seeks the greatest good, harmonizing rational thought and empirical evidence in ethical decision-making.

Abstract: This exploration delves into the multifaceted impact of utilitarianism — a consequentialist doctrine focused on the greatest happiness for the greatest number — on scientific methodology, healthcare practices, and bioethics. Beginning with its rationalist foundations, the discussion highlights the human propensity for seeking pleasure and avoiding pain, positioning utilitarianism as rooted in evolutionary cognitive faculties guiding ethical conduct. Its empiricist lens relies on sensory experiences and environmental interactions to construct ethical principles grounded in empirical realities. The doctrine's influence on the scientific method promotes analytical scrutiny and rigorous peer review. In medicine, utilitarianism shapes healthcare standards of care based on empirical research, optimizing patient outcomes and adaptive approaches to healthcare delivery. The investigation also highlights its role in bioethics, aligning with patient autonomy (informed consent), practitioner beneficence (do good), practitioner nonmaleficence (do no harm), and public justice (be fair) to cultivate a compassionate and equitable healthcare ecosystem. Utilitarianism's profound legacy is a cornerstone in ethical philosophy and scientific methodology, championing empathy and the collective pursuit of well-being.

**

Introduction: Utilitarianism, a consequentialist philosophical doctrine, posits that actions are morally right if they result in the greatest happiness for the greatest number of people. Rooted in the works of seminal thinkers like Jeremy Bentham and John Stuart Mill, it emphasizes pursuing pleasure and avoiding pain as the ultimate motivation of human action. As a dynamic and evolving theory, it encompasses various dimensions, including act and rule utilitarianism and qualitative distinctions between various types of pleasures. In the forthcoming discussion, we will delve deeper into the rationalist foundations of utilitarianism, scrutinizing its empiricist orientation and significant role in shaping the scientific method. Furthermore, we will explore its profound contributions to medicine and its pervasive influence on the four principles of bioethics: patient autonomy (informed consent), practitioner beneficence (do good), practitioner nonmaleficence (do no harm), and public justice (be fair). Through this lens, we aim to comprehensively analyze utilitarianism's enduring impact on ethical decision-making and its pivotal role in modern society.

Rationalism: Utilitarianism, deeply entrenched in rationalist foundations, posits that pursuing the greatest happiness is not merely a moral aspiration but an evolutionary imperative hardwired into the human psyche. This philosophical doctrine leverages the innate cognitive faculties and instinctual behaviors that have evolved over millennia as mechanisms for discerning moral truths and guiding ethical conduct. These "*a priori*" principles, encompassing basic reasoning skills, pattern recognition, and problem-solving abilities, are the bedrock upon which utilitarian ethics are built. Utilitarian thinkers apply deductive reasoning, drawing from axioms and definitional truths, to formulate moral principles that maximize happiness and minimize suffering. This rationalist approach, which potentially extends to the realms of consciousness and awareness, asserts that individuals, by their inherent cognitive capabilities, are predisposed to seek pleasure and avoid pain, thereby naturally aligning with the utilitarian goal of promoting the greatest good for the greatest number. Through this lens, utilitarianism emerges as a sophisticated framework that harnesses the power of rational, evolutionary, hardwired faculties to foster ethical decision-making and promote societal well-being.

Empiricism: Utilitarianism's empiricist orientation is grounded in the belief that moral truths and ethical principles are discerned through sensory experiences and contingent environmental interactions. This philosophical stance aligns with the empiricist doctrine, which posits that knowledge is acquired through empirical observations, not just innate motivations or principles. In this context, utilitarianism employs inductive reasoning as a methodological tool, where specific observations about human behavior and experiences are analyzed to formulate general conclusions about what actions would lead to the greatest happiness. Through a meticulous analysis of human responses to pleasure and pain, utilitarianism seeks to construct a moral framework deeply rooted in the empirical realities of human existence. This approach allows for a dynamic and adaptable ethical system capable of responding to the complex and ever-changing nature of human experiences. Utilitarianism embodies an empiricist orientation that grounds ethical decision-making in real-world experiences and observations. It fosters a pragmatic and outcome-oriented approach to morality by focusing on the tangible outcomes of particular actions and universal rules and their impact on overall happiness.

The Scientific Method: Utilitarianism, with its emphasis on empirical analysis and outcome-based evaluations, has significantly influenced the development and refinement of the scientific method. This philosophical doctrine, which advocates for actions that maximize happiness and minimize suffering, parallels the scientific method's systematic approach to inquiry, which is grounded in observation, hypothesis formulation, and rigorous testing. Utilitarianism's focus on assessing the consequences of actions mirrors the scientific method's reliance on data collection and analysis to draw credible conclusions. Moreover, the utilitarian principle of seeking the greatest good encourages a meticulous and analytical approach to understanding phenomena, akin to the scientific method's emphasis on controlled experiments and replicability to ensure the validity of findings. Furthermore, just as utilitarianism values empirical evidence and inductive reasoning in ethical decision-making, the scientific method relies heavily on empirical observations to formulate and test hypotheses. This mutual emphasis on empirical analysis fosters a culture of rigorous scrutiny and peer review, ensuring that ethical principles and scientific conclusions are grounded in tangible evidence and rational analysis, thereby promoting social progress and well-being.

Medicine: Utilitarianism has profoundly influenced the field of medicine, particularly in the formulation and adherence to medical healthcare standards of care. This philosophical doctrine, which prioritizes the greatest happiness for the greatest number, resonates with the medical community's commitment to providing care that maximizes patient well-being and minimizes harm. Utilitarian principles have guided the development of evidence-based practices grounded in empirical research to achieve the best possible patient outcomes. These practices, established through peer-reviewed research, embody a utilitarian approach by focusing on the consequences of medical interventions and striving to optimize the benefits while reducing the adverse effects. Moreover, utilitarianism encourages a dynamic and adaptive approach to medical care, mirroring the evolving nature of medical healthcare standards of care that continually adapt to incorporate the latest research findings, clinical expertise, and patient needs. In legal contexts, utilitarian ethics serve as a benchmark to assess the adequacy of delivered care, ensuring that actions taken are in the best interest of the patient and society at large. Thus, utilitarianism's contributions to medicine are evident in developing and implementing medical healthcare standards of care that promote the greatest good through evidence-based, patient-centered care.

Ethics: Utilitarianism, a consequentialist theory grounded in maximizing overall happiness, has significantly shaped the four principles of bioethics: patient autonomy (informed consent), practitioner beneficence (do good), practitioner nonmaleficence (do no harm), and public justice (be fair). In autonomy, utilitarianism underscores the importance of informed consent, advocating that individuals are best placed to judge what will maximize their happiness, thereby fostering a healthcare environment that respects individual decision-making capacities. Beneficence, a principle urging to "*do good*," aligns seamlessly with utilitarianism's core tenet of promoting the greatest happiness, encouraging healthcare professionals to undertake actions that enhance the well-being of patients. Similarly, the principle of nonmaleficence, which urges to "*do no harm*," resonates with utilitarianism's emphasis on minimizing suffering, fostering a healthcare approach that seeks to avoid adverse outcomes. Lastly, the principle of justice, urging fairness and equitable treatment, finds resonance in utilitarianism's commitment to promoting the greatest good for the greatest number, encouraging policies and practices that ensure equal access and fair distribution of healthcare resources. Thus, utilitarianism serves as a potent philosophical foundation, guiding the application and interpretation of these bioethical principles in pursuing optimal healthcare outcomes and societal well-being.

Conclusion: In the intricate tapestry of philosophical doctrines, utilitarianism is a formidable thread weaving through various domains of human endeavor, notably influencing rational thought, empirical inquiry, scientific methodology, medical practice, and bioethics. Its core tenet — the greatest happiness principle — advocates for the maximization of happiness and the minimization of suffering and has been a guiding principle, fostering a pragmatic and outcome-oriented approach to ethical decision-making. Through its rationalist foundations, it harnesses innate cognitive faculties to discern moral truths, while its empiricist orientation grounds ethical principles in tangible experiences and observations. Its influence permeates the scientific method, encouraging a meticulous and analytical approach to understanding phenomena, thereby fostering progress and societal well-being. In the medical sphere, it has guided the evolution of medical healthcare standards of care, promoting evidence-based practices that prioritize patient welfare. Furthermore, it has significantly shaped the bioethical landscape, aligning seamlessly with principles of patient autonomy (informed consent), practitioner beneficence (do good), practitioner nonmaleficence (do no harm), and public justice (be fair), thus fostering a healthcare environment that is both compassionate and just. Utilitarianism's pervasive influence underscores its pivotal role in shaping a society that strives for the greatest good, marking its enduring impact on modern civilization.

Examples:

Public Health Policy: Utilitarianism has played a significant role in shaping public health policies, such as vaccination programs. By prioritizing the greatest good for the greatest number, policymakers advocate for widespread immunization to achieve herd immunity, thereby maximizing societal health and minimizing the spread of infectious diseases. This approach balances individual freedoms with community welfare, illustrating utilitarian principles in action.

Resource Allocation in Healthcare: In the aftermath of the 2010 Haiti earthquake, utilitarian principles guided the allocation of scarce medical resources, such as surgical teams and medical supplies. Decisions were often based on maximizing overall survival and minimizing suffering, reflecting the doctrine's emphasis on outcomes that benefit the largest number of patients.

Environmental Ethics: Utilitarianism also informs environmental policies aimed at reducing carbon emissions and combating climate change. By evaluating the long-term consequences of environmental degradation, utilitarian ethics support measures that aim to preserve the planet for future generations, ensuring the greatest good for the greatest number over time.

Animal Welfare: Utilitarian principles underpin many animal welfare regulations. For example, the use of animals in research is often justified by the potential to achieve significant medical breakthroughs that can alleviate human suffering. However, utilitarianism also demands minimizing animal suffering and exploring alternative methods, balancing the benefits to humans with the ethical treatment of animals.

By examining these examples, it becomes clear how utilitarianism shapes practical decision-making across various fields, ensuring that actions are aligned with the overarching goal of promoting the greatest happiness and minimizing suffering for the largest number of people.

Utilitarianism's Legacy: Utilitarianism is a cornerstone in ethical philosophy and scientific methodology, championing the greatest happiness principle as a guide to moral decision-making and policy formulation, thereby fostering a society grounded in empathy, rationality, and the pursuit of collective well-being.

**

REVIEW QUESTIONS

True/False Questions:

1. Utilitarianism is a consequentialist doctrine that emphasizes the pursuit of the greatest happiness for the greatest number of individuals.
   True or False?

2. Utilitarianism solely relies on rationalist principles and does not consider empirical observations in ethical decision-making.
   True or False?

Multiple-Choice Questions:

3. Which of the following is NOT a principle of bioethics influenced by utilitarianism?
   a) Patient autonomy
   b) Practitioner beneficence
   c) Nonmaleficence
   d) Hedonism

4. Who is NOT one of the seminal thinkers associated with the development of utilitarianism?
   a) Jeremy Bentham
   b) John Stuart Mill
   c) Immanuel Kant
   d) James Mill

Clinical Vignette:

5. A public health policy is being formulated to distribute a limited supply of vaccines during a pandemic. The policy aims to maximize the overall survival rate and minimize suffering across the population. According to utilitarian principles, which approach should be taken?
   a) Distribute the vaccines equally among all age groups
   b) Prioritize vaccinating the elderly and those with pre-existing conditions
   c) Vaccinate the wealthiest individuals first
   d) Randomly select individuals for vaccination

Basic Science Vignette:

6. A researcher is designing a study to maximize the benefits for the greatest number of people, carefully considering the potential outcomes of each experimental approach. Which principle of utilitarianism is primarily demonstrated by this action?
   a) Universalizability
   b) Rule utilitarianism
   c) Greatest happiness principle
   d) Empirical analysis

Philosophy Vignette:

7. A philosopher evaluates the ethical foundation of utilitarianism, emphasizing the principle that actions are morally right if they promote the greatest happiness for the greatest number. Which principle of utilitarianism best aligns with this ethical consideration?
   a) Categorical imperative
   b) Informed consent
   c) Greatest happiness principle
   d) Public justice

Correct Answers:

1. True
2. False
3. d) Hedonism
4. c) Immanuel Kant
5. b) Prioritize vaccinating the elderly and those with pre-existing conditions
6. c) Greatest happiness principle
7. c) Greatest happiness principle

BEYOND THE CHAPTER
Utilitarianism

- *Utilitarianism* by John Stuart Mill
- *The Methods of Ethics* by Henry Sidgwick
- *Rethinking Utilitarianism: A Theory of Conscious Thought* by Ben Eggleston and Dale E. Miller
- *The Point of View of the Universe: Sidgwick and Contemporary Ethics* by Katarzyna de Lazari-Radek and Peter Singer
- *Utilitarianism: A Very Short Introduction* by Katarzyna de Lazari-Radek and Peter Singer

***

Chapter 9: **Virtue Ethics** — The Moral Excellence: Character and Integrity

Virtue Ethics emphasizes moral character and integrity, shaping ethical practices in science and healthcare.

Abstract: Virtue ethics, rooted in Aristotelian thought, intertwines rationalist foundations with empiricist orientations to nurture moral virtues and character development. This analysis explores its substantial influence on the scientific method, emphasizing integrity and conscientiousness in research, and its pivotal contributions to medicine, promoting a patient-centered approach marked by compassion and moral discernment. The discussion highlights virtue ethics' alignment with bioethical principles, guiding healthcare professionals to embody patient autonomy (informed consent), practitioner beneficence (do good), practitioner nonmaleficence (do no harm), and public justice (be fair) as intrinsic virtues. Virtue ethics serves as an enduring ethical compass, fostering a society anchored in justice and compassion and guiding individuals toward moral excellence and human flourishing (eudaimonia). This analysis underscores virtue ethics' pervasive influence and vital role in shaping ethical considerations and guidelines in diverse human endeavors.

**

Introduction: Virtue ethics, a prominent branch of moral philosophy, underscores the significance of personal character and virtues in ethical decision-making, diverging from the consequentialist and deontological approaches emphasizing contingent outcomes or universal duties, respectively. Rooted in Aristotelian philosophy, it advocates for cultivating moral virtues such as wisdom, courage, and compassion, fostering an individual's holistic development and pursuit of eudaimonia, a state of human flourishing or happiness. This ethical framework is agent-centered, focusing on the intrinsic qualities of the moral agent rather than the act itself. As we delve deeper into this topic, we will scrutinize its rationalist foundations, alignment with empiricism, role in shaping the scientific method, and substantial contributions to medicine. Furthermore, we will explore its profound influence on the four principles of bioethics: patient autonomy (informed consent), practitioner beneficence (do good), practitioner nonmaleficence (do no harm), and public justice (be fair), thereby assessing its pervasive impact in shaping ethical considerations and guidelines in various human endeavors.

Rationalism: Virtue ethics, deeply entrenched in rationalist foundations, posits that cultivating virtues is intrinsically linked to humans' innate cognitive faculties and instinctual behaviors as hardwired mechanisms for acquiring knowledge and understanding. This philosophical approach recognizes that virtues, such as wisdom and courage, are developed and nurtured by applying "*a priori*" principles grounded in the inherent abilities of basic reasoning, pattern recognition, and problem-solving that humans naturally embody. Through the lens of rationalism, virtue ethics emphasizes the role of deductive reasoning, where moral truths are discerned from definitional truths, axioms, or postulates, fostering a deeper understanding and awareness of the complexities of human behavior and morality. This perspective suggests that the journey toward eudaimonia, or a state of ultimate happiness and human flourishing, is navigated through the conscious cultivation of virtues, leveraging the intrinsic capabilities of the human mind to discern and aspire toward moral excellence, thereby facilitating a harmonious alignment between one's character and the rational principles that govern the moral fabric of human existence.

Empiricism: Virtue ethics, in its engagement with empiricism, acknowledges that sensory experiences and interactions with the environment significantly influence the development and refinement of virtues. This empiricist orientation suggests that moral virtues are not merely innate traits but are also cultivated through continuous engagement with one's surroundings, learning from specific observations and experiences to form general conclusions about moral conduct. Through inductive reasoning, individuals can assimilate and adapt their moral compass, fostering virtues such as compassion, empathy, and prudence based on the cumulative knowledge acquired from lived experiences. This process of moral development is dynamic and responsive to the changing circumstances and complexities of human life. Virtue ethics, therefore, offers a nuanced approach to moral philosophy, where the cultivation of a virtuous character is seen as a confluence of both intrinsic moral inclinations and the wisdom garnered through a lifetime of sensory experiences and environmental interactions, paving the way for a more rounded, adaptable, and realistic approach to ethical decision-making and personal development.

The Scientific Method: In the realm of the scientific method, virtue ethics plays a pivotal role in fostering a conscientious and responsible approach to scientific inquiry. Virtue ethics, with its emphasis on moral character and the cultivation of virtues such as integrity, diligence, and prudence, encourages researchers to adhere to the highest standards of ethical conduct throughout the research process. From the initial stages of observing a phenomenon and formulating a hypothesis to conducting controlled experiments and analyzing data, the principles of virtue ethics guide researchers to undertake their work with a deep sense of responsibility and ethical consideration. It promotes the cultivation of virtues like honesty, which is fundamental in reporting accurate findings, and courage, which is essential in upholding the truth even when findings contradict prevailing theories or beliefs. Furthermore, the virtue of wisdom guides researchers in drawing nuanced, reflective, and insightful conclusions. By fostering a community of researchers who embody these virtues, virtue ethics contributes to the establishment of a scientific method that is not only rigorous and systematic but also grounded in moral integrity, thereby enhancing the credibility and reliability of scientific advancements and discoveries.

Medicine: Virtue ethics, emphasizing moral character and virtues, has significantly contributed to medicine, particularly in shaping the medical healthcare standards of care and fostering a patient-centered approach. In the medical realm, cultivating virtues such as compassion, empathy, and prudence is vital in providing competent and humane care. Healthcare professionals are encouraged to embody these virtues, fostering a practice beyond mere technical proficiency to encompass a deep understanding and respect for patients' needs and dignity. Virtue ethics also plays a crucial role in guiding medical practitioners in navigating complex moral dilemmas, encouraging them to apply wisdom and moral discernment in their decisions. Furthermore, it inspires a commitment to continual learning and adaptation, reflecting the evolving nature of medical healthcare standards of care grounded in peer-reviewed, evidence-based practices. By fostering a community of medical professionals who are skilled and virtuous, virtue ethics contributes to the enhancement of healthcare delivery, ensuring that it is grounded in both scientific excellence and moral integrity, thereby promoting a healthcare system that is both just and compassionate.

Ethics: Virtue ethics profoundly influences the bioethical principles that govern the complex landscape of healthcare and medical research. Its emphasis on moral character and virtues resonates deeply with the principles of patient autonomy (informed consent), practitioner beneficence (do good), practitioner nonmaleficence (do no harm), and public justice (be fair). Firstly, it nurtures respect for autonomy, encouraging healthcare professionals to uphold the virtue of honesty and transparency, facilitating informed consent, and empowering patients to be active participants in their care. Secondly, the principle of beneficence is echoed in virtue ethics' encouragement to cultivate goodness and kindness, guiding practitioners to seek the best patient outcomes. Concurrently, nonmaleficence, the commitment to do no harm, aligns with the virtue of prudence, urging practitioners to exercise caution and avoid actions that might inflict harm. Lastly, the principle of justice is reinforced by the virtue of fairness, fostering an environment where care is dispensed equitably, without prejudice or bias. Virtue ethics, therefore, serves as a moral compass, guiding healthcare professionals to embody these principles not as mere obligations but as intrinsic virtues that define their character, ultimately fostering an ethical and compassionate healthcare system.

Conclusion: Virtue ethics stands as a formidable pillar in the philosophical discourse on ethics, offering a nuanced approach that emphasizes the cultivation of moral character and virtues as the pathway to ethical decision-making and human flourishing. Its rich tapestry, woven from rationalist foundations and empiricist orientations, provides a comprehensive framework that harmonizes innate cognitive faculties with sensory experiences, fostering dynamic and adaptive moral development. Its influence permeates various spheres, notably shaping the scientific method by instilling a sense of responsibility and integrity in research endeavors. It also significantly contributes to medicine by promoting a patient-centered approach grounded in compassion and moral discernment. Moreover, its profound impact on bioethics, resonating deeply with the principles of patient autonomy (informed consent), practitioner beneficence (do good), practitioner nonmaleficence (do no harm), and public justice (be fair), serves as a guiding light in navigating the complex ethical landscape of healthcare and medical research. Thus, virtue ethics emerges as a vital ethical compass, steering individuals toward a life marked by moral excellence, wisdom, and a deep-seated commitment to fostering the greater good and nurturing a just and compassionate society.

Examples:

Medical Practice: A physician guided by virtue ethics may go beyond merely diagnosing and treating illnesses to fostering a deep, empathetic relationship with patients. For instance, a doctor might spend extra time explaining a diagnosis to ensure the patient fully understands their condition and treatment options, embodying the virtues of compassion and wisdom.

Scientific Integrity: In scientific research, a commitment to honesty and integrity is paramount. A researcher adhering to virtue ethics would rigorously check their data for accuracy and refrain from manipulating results to fit a desired narrative. This commitment ensures the credibility and reliability of scientific findings.

Public Health Initiatives: Public health officials guided by virtue ethics may design health campaigns that respect cultural values and individual dignity. For example, a vaccination campaign could be tailored to address specific community concerns and misinformation, demonstrating respect and fairness while promoting the common good.

Ethical Leadership: In business, a leader who follows virtue ethics might prioritize ethical practices over short-term profits. By fostering a culture of integrity, they can build a company that values employee well-being, customer satisfaction, and social responsibility, thereby contributing to long-term success and societal trust.

By examining these examples, it becomes clear how virtue ethics shapes practical decision-making across various fields, ensuring actions align with the overarching goal of cultivating moral character and promoting human flourishing.

Virtue Ethics' Legacy: The enduring legacy of virtue ethics lies in its advocacy for the cultivation of moral character and virtues, fostering a holistic approach to ethical decision-making that harmonizes rationalist and empiricist perspectives and profoundly influences diverse fields such as science and medicine with its emphasis on integrity, compassion, and human flourishing.

**

REVIEW QUESTIONS

True/False Questions:

1. Virtue ethics emphasizes the importance of moral character and virtues in ethical decision-making, rather than focusing solely on the outcomes of actions or adherence to universal duties.
   True or False?

2. According to virtue ethics, virtues are only developed through rational reasoning and are not influenced by sensory experiences or environmental interactions.
   True or False?

Multiple-Choice Questions:

3. Which of the following is NOT a principle of bioethics influenced by virtue ethics?
   a) Patient autonomy
   b) Practitioner beneficence
   c) Nonmaleficence
   d) Consequentialism

4. Which philosopher is most closely associated with the development of virtue ethics?
   a) Immanuel Kant
   b) Jeremy Bentham
   c) John Stuart Mill
   d) Aristotle

Clinical Vignette:

5. A medical professional is faced with a complex ethical dilemma involving a patient's treatment plan. According to virtue ethics, what should guide the professional's decision-making process?
   a) Adhering strictly to hospital policies
   b) Focusing solely on the patient's immediate happiness
   c) Cultivating virtues such as compassion, wisdom, and prudence to determine the best course of action
   d) Following the most cost-effective treatment option

Basic Science Vignette:

6. A scientist is conducting research and adheres to the principles of virtue ethics, ensuring that all findings are reported honestly, regardless of the outcomes. Which virtue of virtue ethics is primarily demonstrated by this action?
   a) Prudence
   b) Integrity
   c) Courage
   d) Empathy

Philosophy Vignette:

7. A philosopher analyzes the foundation of virtue ethics, emphasizing the cultivation of moral character and virtues in ethical decision-making. Which concept in virtue ethics best aligns with the idea of living a flourishing life through moral excellence?
   a) Utilitarianism
   b) Categorical imperative
   c) Eudaimonia
   d) Public justice

Correct Answers:

1. True
2. False
3. d) Consequentialism
4. d) Aristotle
5. c) Cultivating virtues such as compassion, wisdom, and prudence to determine the best course of action
6. b) Integrity
7. c) Eudaimonia

BEYOND THE CHAPTER
Virtue Ethics

- *Nicomachean Ethics* by Aristotle
- *After Virtue: A Study in Moral Theory* by Alasdair MacIntyre
- *The Morality of Happiness* by Julia Annas
- *Virtue Ethics* by Roger Crisp and Michael Slote (eds.)
- *On Virtue Ethics* by Rosalind Hursthouse
- *Virtue Ethics: A Pluralistic View* by Christine Swanton
- *The Virtue of Aristotle's Ethics* by Paula Gottlieb
- *Virtue Ethics and Consequentialism in Early Chinese Philosophy* by Bryan W. Van Norden
- *Virtue Ethics: Dewey and MacIntyre* by Elizabeth Trott
- *The Cambridge Companion to Virtue Ethics* edited by Daniel C. Russell

***

Philosophy of Science

Medical Research

WEEK THREE: MORALITY AND INTEGRITY

WEEK THREE

Class Hour 1

- Chapter 10: Universal Ethics — The Holistic Approach: Integrative Morality
- Chapter 11: Beauchamp and Childress — The Principlists: Ethical Framework in Bioethics
- Chapter 12: Peter Singer (1946-) — The Utilitarian Ethicist: Global Welfare and Animal Rights

Class Hour 2

- Chapter 13: Academic Freedom — The Catalyst for Truth: Unbounded Inquiry
- Chapter 14: Scholarship — The Beacon of Progress: Intellectual Rigor
- Chapter 15: History of Peer Review — The Evolution of Integrity: Critical Analysis

---

Introduction

As we venture into the third week of our course, "*Philosophy of Science*," we delve into the themes of morality and integrity, exploring the diverse ethical frameworks that guide scientific inquiry and healthcare practices. This week's chapters weave together the concepts of Universal Ethics, principled bioethics, utilitarianism, academic freedom, scholarship, and the history of peer review, creating a rich tapestry of thought that underscores the importance of ethical integrity in scientific progress.

Our journey begins with Universal Ethics, a holistic approach that integrates Deontology, Utilitarianism, and Virtue Ethics into a comprehensive moral framework. By merging the duty-driven axioms of Deontology, the consequential foresight of Utilitarianism, and the character-building focus of Virtue Ethics, Universal Ethics fosters a richer understanding of ethical principles. This integration is particularly impactful in healthcare, where it guides professionals through complex dilemmas with compassion and respect for human capabilities. By intertwining patient autonomy, practitioner beneficence, nonmaleficence, and public justice, Universal Ethics promises a future where ethical decisions are grounded in the unified pursuit of truth and well-being.

Building on this foundation, we explore the principled approach to bioethics developed by Beauchamp and Childress. Their framework, which emphasizes autonomy, beneficence, nonmaleficence, and justice, has become foundational in medical ethics. By integrating rationalist and empiricist philosophies, they have bridged the gap between theoretical ethics and practical application, fostering a culture of ethical rigor in scientific inquiry. Their work has significantly influenced medical practice and research, enhancing the credibility and societal impact of scientific endeavors.

Next, we delve into the utilitarian ethics of Peter Singer, who focuses on global welfare and animal rights. Singer's utilitarian ethos, centered on maximizing overall happiness and well-being, challenges conventional moral frameworks and promotes a rational, compassionate approach to ethics. His work spans bioethical issues in medicine, economics, and law, using rational deliberation and empirical observations to craft universally applicable ethical principles. Singer's influence extends to policy-making, urging a reconsideration of established norms for a more inclusive ethical framework.

The theme of academic freedom highlights the importance of unbounded inquiry in driving scientific and ethical advancements. Rooted in the principles of rationalism and empiricism, academic freedom fosters scholarly innovation and intellectual growth. It catalyzes credible knowledge pursuit, facilitating the design and analysis of experiments, and promotes evidence-based advancements and ethical practices in healthcare. By nurturing an environment where intellectual evolution and innovation can thrive, academic freedom supports the ethical principles that guide healthcare and scientific research.

Scholarship, as the beacon of progress, intertwines rational inquiry and empirical evidence to foster innovation and ethical integrity. This intricate discipline has played a pivotal role in shaping modern society, driving technological advancements, and enhancing the quality of life. Rooted in academic freedom and fortified by tenure, scholarship has revolutionized our understanding of the natural world and profoundly impacted the medical field. It has shaped bioethics principles, fostering a healthcare environment that is both scientifically advanced and ethically grounded.

Finally, we trace the evolution of peer review, from ancient discourse to modern AI-assisted evaluations. This process embodies a commitment to integrity and collaborative scientific progress, integrating critical analysis and fostering a more inclusive and transparent scholarly community. As peer review evolves, it aims to incorporate technological advancements while upholding the values of integrity, inclusivity, and transparency, ensuring the continual enhancement of scientific discourse and collaboration.

Together, these chapters offer a comprehensive view of the ethical frameworks that underpin scientific inquiry and healthcare practices. From Universal Ethics to the evolution of peer review, we gain a deeper appreciation of the principles that guide us in the pursuit of knowledge and the application of that knowledge in ways that are just, compassionate, and morally sound. This integrated exploration underscores the dynamic interplay between morality and integrity in shaping a society committed to truth and well-being.

***

Chapter 10: Universal Ethics — The Holistic Approach: Integrative Morality

Universal Ethics integrates diverse moral philosophies, fostering comprehensive and compassionate ethical decision-making.

Abstract: In the evolving field of moral philosophy, Universal Ethics integrates Deontology, Utilitarianism, and Virtue Ethics into a comprehensive system, affirming its position as a generalizable moral framework. This integration is rooted in principles consistent with diverse philosophical perspectives, fostering a richer understanding of ethical principles. By merging Deontology's duty-driven axioms, Utilitarianism's consequential foresight, and Virtue Ethics' character-building focus, Universal Ethics transcends traditional boundaries. It emphasizes a transition from theoretical abstraction to actionable outcomes, nurturing moral excellence and eudaimonia. In healthcare, it offers a revolutionary approach, guiding professionals through complex dilemmas with compassion and respect for human capabilities, intertwining patient autonomy (informed consent), practitioner beneficence (do good), practitioner nonmaleficence (do no harm), and public justice (be fair). Universal Ethics promises a future where ethical decisions are grounded in the unified pursuit of truth and well-being, heralding a society thriving on justice and the greater good, marking a milestone in ethical discourse.

**

Introduction: In the grand scheme of moral discourse, Universal Ethics emerges as a beacon of unification, a place where Deontology, Utilitarianism, and Virtue Ethics converge, melding into an intellectual powerhouse that transcends traditional boundaries in ethical philosophy. This amalgamation serves as a testament to the rich tapestry of human morality, harmonizing the nuanced theories that have guided our understanding of ethical principles thus far. At its core, this model embodies an inclusive approach, reaching out to assimilate perspectives and ideologies from the broad spectrum of moral worldviews that are earnest in delineating the principles of right conduct. It serves as a testament to the fact that when we venture deep into moral contemplation, there exists a common ground — a fundamental understanding of morality that binds all serious philosophical inquiries together. This implies that Universal Ethics is not merely a confluence of established doctrines but also a resilient, adaptable framework that finds resonance with all ethical perspectives deeply invested in understanding and propagating the facets of moral truth. By espousing the quintessence of Deontology, it implores individuals to uphold moral conduct fostered by duty and universal principles. It embraces the consequentialist foresight of Utilitarianism, advocating for choices that engender the greatest happiness and welfare for the maximum number. Furthermore, it acknowledges the intrinsic value of Virtue Ethics, nurturing moral excellence and character-building development. Moreover, this model pioneers a transformative journey in moral philosophy, enabling individuals to establish theoretically profound and pragmatically relevant principles. In doing so, it forges a path that merges theoretical abstraction with tangible, actionable outcomes. As we delve deeper into the intricacies of this model, we find that it not only revamps the scientific approach to ethics but also etches significant imprints on medical ethics, thus reshaping the fundamental pillars of bioethics that encompass patient autonomy, practitioner beneficence, nonmaleficence, and public justice. Through its innovative approach, Universal Ethics emerges as a herald of a new era in moral philosophy, facilitating a more profound, nuanced understanding of ethics grounded in the relentless pursuit of goodness and justice. It signifies a robust and comprehensive moral system that epitomizes common morality, resonating with every worldview committed to the serious discourse on ethics, thereby representing a progressive stride toward a more unified and inclusive approach to ethical reflection and analysis.

Rationalism: Universal Ethics builds its foundations on rationalist principles, merging the robust components of Deontology, Utilitarianism, and Virtue Ethics. This rationalist groundwork posits that individuals inherently possess the capacity to unveil moral truths utilizing intrinsic cognitive capabilities and "*a priori*" principles, essentially serving as evolutionarily ingrained mechanisms for understanding and knowledge acquisition. Through a lens of deontological thought, it emphasizes logical deductions and adherence to universal principles, thus fostering rational discourse during moral evaluations. At the same time, it embodies the utilitarian ethos, which perceives the pursuit of the greatest amount of happiness as both a rational endeavor and an evolutionary instinct, utilizing innate faculties to amplify societal welfare. This stance beautifully dovetails with the precepts of virtue ethics, highlighting the deliberate nurturement of virtues through rational scrutiny and an aspiration toward moral excellence, guiding beings to a state of eudaimonia. This amalgamated approach promotes pluralistic exploration, facilitating a dynamic interaction between diverse ethical theories, instincts, and principles akin to scientific methodologies. It aims to orchestrate a balanced harmony in moral deliberations, evaluating actions through a composite prism of duty, consequences, and virtue, thus cultivating well-rounded and thoughtful moral judgments. By synchronizing these theories, the model crafts a rich mosaic that embraces the multifaceted dimensions of human experiences, nurturing a community where ethical choices are rational and profoundly reflective of human morality's intricate, layered nature. Thus, it carves a trajectory where moral conundrums are traversed with an intrinsic comprehension and regard for humanity's inherent potentials and evolutionary ingrained faculties, paving the way for a more enlightened, empathetic, and equitable society.

Empiricism: Within the fabric of philosophical contemplation, Universal Ethics seamlessly integrates the empirical facets found in Deontology, Utilitarianism, and Virtue Ethics, thereby crafting a holistic and nuanced blueprint for ethical deliberation. The model reveres empiricism, a philosophy where knowledge emanates from sensory experiences and interactions with one's environment. From this perspective, it nurtures a versatile and pluralistic method for making moral decisions, formulating intermediate principles that align with intuitive moral selections and adhere to overarching theories. Endeavoring to epitomize a synergistic stance, the model intertwines deontological analytical rigor with utilitarianism's practical, observational ethos. It encourages individuals to commence with universal principles as ethical foundations yet also impels them to modulate them based on tangible consequences witnessed in reality, encouraging a pragmatic, results-focused approach to ethics. This perspective welcomes evaluating actions based on their repercussions on communal happiness and welfare, endorsing a flexible and adaptive ethical schema harmonious with empirical means of acquiring knowledge. Moreover, it assimilates the insights from virtue ethics, accentuating the evolution of virtuous characteristics nurtured through relentless engagement with the external world. It acknowledges the innate moral guidance within individuals, validating the pivotal role of personal experiences in sculpting one's ethical outlook. This orientation propels individuals to refine and recalibrate their ethical viewpoints based on the collective wisdom gleaned from these lived experiences, fostering virtues like compassion, empathy, and judiciousness. Universal Ethics embodies a judicious and comprehensive technique for ethical judgments, accentuating the need for a reflective equilibrium where diverse theories, instincts, and principles coalesce, facilitating a deeper and more dependable traverse through intricate moral quandaries. This prism harbors a rich, adaptive, and sophisticated ethical framework anchored in empirical observations and environmental engagements, laying the groundwork for a more comprehensive, adaptive, and realistic approach to moral discernment and personal growth.

The Scientific Method: In scientific research, Universal Ethics posits itself as a vital instrument in refining the scientific method, fostering an amalgamation of empirical analysis with rationalist elements harmoniously intertwined with ethical considerations reminiscent of deontological, utilitarian, and virtue ethics. This dynamic model champions a holistic methodology in scientific inquiry, deeply entrenched in the pursuit of truth while staunchly adhering to moral and ethical benchmarks. Here, the initiation of scientific exploration is rooted in an ethical backdrop, drawing upon deontological doctrines that underscore duty and accountability. It mandates researchers to undertake their studies with unswerving integrity and transparency, thus safeguarding the authenticity of data and the well-being of the subjects involved. This ethical foundation paves the way for the birth of a milieu where scientific endeavors uphold the highest echelons of moral integrity. Furthermore, this model seamlessly integrates utilitarian principles, accentuating the importance of empirical scrutiny and results-driven evaluations. It encourages an approach wherein scientific investigations aim to augment positive repercussions while curbing negative consequences, harmonizing with the utilitarian aspiration for widespread happiness. Simultaneously, it champions an analytical stance resonating with the scientific method's emphasis on diligent data accumulation and analysis, thereby fostering societal advancement and well-being through scientific progress. Additionally, insights from virtue ethics are woven into this framework, nurturing researchers to embody virtues such as integrity and prudence. This moral compass guides individuals to undertake research with a heightened sense of conscientiousness, championing virtues like veracity in documentation and valor in upholding the truth, even amidst dissenting conventional theories. Consequently, the scientific method transforms into a quest for knowledge and a crucible for nurturing ethically sound individuals within the scientific fraternity. This synergistic approach to the scientific method envisages a cooperative, transparent, and truth-driven scientific community spearheaded by an ethical and morally grounded approach to investigation and innovation.

Medicine: Universal Ethics stands as a pivotal beacon in medicine, orchestrating a confluence of principles from Deontology, Utilitarianism, and Virtue Ethics to shape a vibrant, adaptable, and empathetic framework for healthcare decisions. This model sees Deontological ethics as the bedrock of medical practice, urging healthcare professionals to uphold universal moral duties that accentuate patient welfare and dignity, facilitating an environment where the four principles of bioethics: patient autonomy (informed consent), practitioner beneficence (do good), practitioner nonmaleficence (do no harm), and public justice (be fair) are paramount. It establishes a sacred foundation where the practitioner-patient relationship is grounded in trust and accountability. Coupled with this, it assimilates the utilitarian stance, endorsing an outcome-oriented approach that continually refines standards of care based on empirical research, clinical insights, and patient necessities, fostering a dynamic ecosystem that constantly evolves to embody the latest advancements and insights. It propels the healthcare system to be inherently responsive, aiming to enhance overall well-being by optimizing positive outcomes and curtailing adverse effects, thus facilitating a healthcare paradigm intrinsically aligned with societal needs and welfare. Adding a further layer of depth, Virtue Ethics is seamlessly woven into this model, nurturing the development of moral character in healthcare professionals promoting the virtues of compassion, empathy, and prudence. This facet imbues medical practitioners with a profound understanding and reverence for the humanistic dimensions of healthcare, prompting them to navigate moral complexities with wisdom and discernment. Consequently, Universal Ethics crafts a robust, integrative approach to healthcare, fostering a medical community that is scientifically adept yet morally upright and compassionate, capable of resonating with the ever-changing nuances of medical healthcare standards of care and adeptly addressing the multi-dimensional challenges endemic to this field, heralding a future where healthcare is both ethically resilient and harmoniously attuned to contemporary needs and standards.

Ethics: In the intricate landscape of bioethics, Universal Ethics presents itself as an integrated beacon, weaving elements of Deontology, Utilitarianism, and Virtue Ethics to create a compassionate, equitable, and devoted healthcare framework. It staunchly upholds the principle of autonomy, mirroring deontological tenets of individual rights and moral agency, by fostering an environment where informed consent is sacrosanct, thereby empowering individuals to actively partake in decisions pertaining to their wellbeing, armed with transparency and honesty. Simultaneously, it embraces the principle of beneficence, as both deontological and utilitarian philosophies advocate, urging healthcare providers to pursue optimal outcomes for their patients ardently, a pursuit grounded in intrinsic virtues of goodness and kindness, thus creating a healthcare atmosphere pulsating with empathy and compassion. Additionally, it profoundly resonates with the principle of nonmaleficence, guided by deontological principles to avoid harmful actions and a utilitarian goal to minimize suffering while invoking the wisdom of virtue ethics to promote cautious and prudent actions that prevent harm, preserving the sanctity and safety of healthcare. Furthermore, it cherishes the principle of justice, championing fairness and equity in healthcare delivery, as reflected in utilitarianism's mission to foster the greatest amount of happiness for the greatest amount of people and encouraging initiatives that guarantee equal access and just distribution of healthcare resources, while also incorporating the depth of virtue ethics, nurturing an ecosystem where fairness and justice are central virtues, paving the way for a healthcare system that is unbiased and equitable. Thus, Universal Ethics emerges as a vanguard of ethical amalgamation in bioethics, harmonizing diverse philosophical strands to foster a robust, humane, and fair moral blueprint, intertwining empirical scrutiny with ethical accountability, aspiring to sculpt a healthcare structure that stands firmly on scientific validity while being deeply rooted in moral integrity, fostering a tradition of respect, empathy, and justice in the rapidly evolving sphere of healthcare and medical research.

Conclusion: In the labyrinthine realm of moral philosophy, Universal Ethics crystallizes as a beacon of harmonized ethical thought, embodying a universal approach to moral discernment that is a testament to the potential for ethical convergence. This innovative model distinguishes itself by its inherent ability to merge Deontological, Utilitarian, and Virtue Ethics into a comprehensive whole, a feat made possible owing to its foundational principle: its derivability from, consistency with, or at least its non-contradiction with most worldviews. This characteristic affirms it as a comprehensive, generalizable moral system representing a shared, common morality, promoting broader understanding and agreement in ethical deliberation. Through this transformative approach, the model meticulously forges moral decisions that are the amalgamation of principles of duty, consequence, and virtue, offering a multifaceted lens for scrutinizing and navigating complex ethical dilemmas with an enriched understanding and profound respect for the intrinsic capabilities of human beings. Manifesting as a dynamic framework that profoundly influences domains such as the scientific method and medicine, it nurtures the ground for virtue cultivation where moral excellence and eudaimonia transition from mere aspirational ideals to attainable realities. Further, its perceptive integration into the healthcare landscape is a testament to its potency in fostering a more compassionate, just, and adaptive healthcare system. Professionals under this guidance are steered not only by principles and consequences but also by intrinsic virtues championing empathy, respect, and justice. Its role in bioethics is particularly noteworthy, fostering a fertile ground where ethics and science harmonize in a symphonic confluence, intertwining ingeniously with aspects like patient autonomy, practitioner beneficence and nonmaleficence, and public justice. Universal Ethics heralds a promising future as we stand on the threshold of a new epoch in moral philosophy. It guides society toward a more enlightened path characterized by depth and compassion, where ethical decisions embody a relentless pursuit of the greater good. This heralds the dawn of a society that thrives on justice, understanding, and a unified pursuit of truth and well-being, unequivocally representing a milestone in the evolution of ethical discourse.

Universal Ethics' Legacy: Bridging diverse philosophical principles, this model heralds a nuanced era in moral philosophy, fostering holistic and adaptive ethical deliberations grounded in theoretical profundity and empirical pragmatism.

**

REVIEW QUESTIONS

True/False Questions:

1. Universal Ethics integrates Deontology, Utilitarianism, and Virtue Ethics into a cohesive framework, emphasizing their compatibility with diverse worldviews.
   True or False?

2. Universal Ethics relies solely on rationalist principles, ignoring empirical observations and environmental interactions in ethical decision-making.
   True or False?

Multiple-Choice Questions:

3. Which of the following is NOT a core principle integrated into the Universal Ethics framework?
   a) Patient autonomy
   b) Practitioner beneficence
   c) Nonmaleficence
   d) Libertarianism

4. Which ethical principle in Universal Ethics emphasizes the development of moral character and virtues?
   a) Deontological ethics
   b) Utilitarian ethics
   c) Virtue ethics
   d) Consequentialism

Clinical Vignette:

5. A healthcare team is faced with a decision on allocating limited resources during a medical crisis. According to Universal Ethics, what should guide their decision-making process?
   a) Strictly adhering to hospital policies without considering individual cases
   b) Maximizing overall happiness and minimizing suffering while maintaining moral integrity and virtues
   c) Ensuring the wealthiest individuals receive priority care
   d) Randomly selecting individuals for treatment without ethical considerations

Basic Science Vignette:

6. A researcher is designing an experiment and integrates Universal Ethics principles to ensure that their study adheres to the highest ethical standards. They consider the duty to follow universal principles, the potential outcomes for maximizing well-being, and the cultivation of personal integrity. Which ethical theory is primarily demonstrated by this holistic approach?
   a) Deontology
   b) Utilitarianism
   c) Virtue Ethics
   d) Universal Ethics

Philosophy Vignette:

7. A philosopher evaluates the concept of Universal Ethics, which combines Deontology, Utilitarianism, and Virtue Ethics. Which aspect of Universal Ethics best represents its integration of various ethical theories to form a comprehensive moral framework?
   a) Consequential analysis
   b) Duty-driven actions
   c) Character and virtue development
   d) Holistic and integrative approach

Correct Answers:

1. True
2. False
3. d) Libertarianism
4. c) Virtue ethics
5. b) Maximizing overall happiness and minimizing suffering while maintaining moral integrity and virtues
6. d) Universal Ethics
7. d) Holistic and integrative approach

BEYOND THE CHAPTER
Universal Ethics

- *Moral Philosophy: A Theoretical and Practical Approach to Moral Decision-Making* by Jeffrey W. Bulger
- *Reflective Equilibrium: An Essay in Moral Epistemology* by Stijn Van Impe
- *Reflective Equilibrium and Theory Acceptance in Ethics* by Robert Alan Frazier

***

Chapter 11: Beauchamp and Childress — The Principlists: Ethical Framework in Bioethics

Beauchamp and Childress's principled approach to bioethics, emphasizing autonomy, beneficence, nonmaleficence, and justice, has become foundational in medical ethics.

Abstract: In their seminal work "*Principles of Biomedical Ethics*," Beauchamp and Childress pioneered the profound integration of rationalist and empiricist philosophies into bioethics, significantly influencing medical practice and scientific research globally. Their in-depth analysis and expansion of the *Belmont Report's* ethical principles — patient autonomy (informed consent), practitioner beneficence (do good), practitioner nonmaleficence (do no harm), and public justice (be fair) — have cemented a principled approach to medical healthcare decision-making, creating a universally applicable guide that has become foundational in medical ethics education and practice. Their rigorous analytical framework has bridged the gap between theoretical ethics and its practical application, fostering a culture of ethical rigor in scientific inquiry and enhancing scientific research's credibility and societal impact. Their legacy represents a transformative journey that has reshaped the ethical landscape of healthcare delivery and research, setting a cornerstone for ethical decision-making that is both intellectually robust and practically applicable, thereby leaving an indelible mark on the global medical healthcare and research community.

**

Introduction: Born in 1939, Tom L. Beauchamp and James F. Childress, born in 1940, are distinguished figures in bioethics. They gained prominence for their seminal work "*Principles of Biomedical Ethics*," first published in 1979. Interestingly, their work was released in the same year as the *Belmont Report*, published earlier, laying a robust ethical groundwork for research involving human subjects. The *Belmont Report* introduced the principles of autonomy, beneficence, and justice while integrating nonmaleficence into beneficence. Beauchamp and Childress's key contribution was to provide a deeper philosophical analysis of these four principles and to extend the Principlistic approach to medical healthcare decision-making. Given that the *Belmont Report* had already been legislated as federal law under 45CFR46, institutions conducting human subject research and receiving federal funding, directly or indirectly, are mandated to adopt this approach, enforced through Institutional Review Boards (IRBs). As a result, Principlism now serves as the basis for ethical decision-making in medical healthcare settings, human subject research, and public health policy. Beauchamp and Childress have profoundly impacted the field by providing academic support to the *Belmont Report's* Principlistic approach to ethical decision-making in medical healthcare practice. Their work combines philosophical rigor with practical applicability, making enduring contributions to the ethical conduct of medicine and research.

Rationalism: Beauchamp and Childress's work in bioethics is deeply rooted in rationalist foundations, drawing from philosophical traditions that prioritize reason and logical analysis as the primary means of acquiring knowledge. The *Belmont Report's* principlistic approach served as the axiomatic algorithm for the ethical decision-making framework. This principlistic framework was designed to be universally applicable, transcending cultural and situational contexts in the vein of rationalist thought, which seeks universal truths. The *Belmont Report* thought of these principles as derivable from, consistent with, or at least not in conflict with most reasonable and rational worldviews and moral decision-making frameworks, and therefore, can be thought of as part of common morality. Beauchamp and Childress employed deontological, consequentialist, and virtue theories, among others, to provide a structured, rational basis for ethical judgments. For example, the principle of patient autonomy, which emphasizes individual freedom and informed consent, is deeply rooted in Kantian ethics, a rationalist philosophy that stresses rational agents' inherent worth and dignity. Similarly, practitioner beneficence and nonmaleficence are grounded in utilitarian thought, which uses rational calculations to maximize good and minimize harm, and public justice, as fair distribution of benefits and burdens, is also supported. By synthesizing these rationalist ethical principles into a cohesive framework, Beauchamp and Childress offer a robust, intellectually rigorous foundation for bioethics for medical healthcare practice.

Empiricism: While Beauchamp and Childress's work is grounded in rationalist philosophy, it also incorporates an empiricist orientation emphasizing the importance of experience, observation, and evidence-based practice in ethical decision-making. The *Belmont Report's* principlistic approach — patient autonomy (informed consent), practitioner beneficence (do good), practitioner nonmaleficence (do no harm), and public justice (be fair) — is not just a construct for research on human subjects but was designed to be practically applicable in real-world medical healthcare settings. This practical focus aligns with empiricism, which values knowledge gained through sensory experience and empirical data. For instance, beneficence and nonmaleficence are not merely abstract concepts but are meant to guide concrete actions, such as risk-benefit analyses in clinical trials, which are deeply rooted in empirical evidence. Their work also acknowledges the evolving nature of medical science and ethics, suggesting that ethical guidelines should be adaptable and responsive to new empirical findings. By integrating empiricism into their framework, Beauchamp and Childress bridge the gap between theoretical ethics and practical application, allowing for a more nuanced and context-sensitive approach to bioethical dilemmas. Their empiricist orientation complements their rationalist foundations, resulting in a comprehensive ethical framework that has profoundly impacted medical practice and research ethics.

The Scientific Method: Beauchamp and Childress's contributions to bioethics have significantly impacted the scientific method by supporting the *Belmont Report's* principlistic approach and helping shape the ethical principles governing scientific research design, conduct, and evaluation. The *Belmont Report's* principles have been integrated into public policy by law, establishing Institutional Review Boards (IRBs). For instance, the principle of beneficence, which calls for maximizing benefits and minimizing harm, has been a driving force behind the rigorous risk-benefit analyses that are now standard in research methodologies. Similarly, the principle of autonomy has reinforced the importance of informed consent, making it a critical element in the ethical conduct of experiments. By providing a rational and empirically adaptable framework for ethical decision-making, Beauchamp and Childress have helped to instill a culture of ethical rigor in scientific inquiry. Their work ensures that ethical considerations are not an afterthought but an integral part of the scientific method, enhancing scientific research's credibility, integrity, and societal impact.

Medicine: Beauchamp and Childress have made substantial contributions to medicine through their academic support of the principlistic framework of patient autonomy (informed consent), practitioner beneficence (do good), practitioner nonmaleficence (do no harm), and public justice (be fair). These principles have become foundational in medical ethics education and practice, guiding patients, practitioners, and public policy in making ethical decisions in various scenarios, from patient care to research and policy. The principle of autonomy, for example, has reinforced the importance of informed consent in clinical settings, ensuring that patients are active participants in decisions about their care. Beneficence and nonmaleficence have influenced medical protocols to ensure that treatments that practitioners and researchers design maximize benefits while minimizing potential harm, aligning closely with the Hippocratic Oath's directive to "*do no harm*." The principle of justice has encouraged equitable access to medical care and fair distribution of medical resources. By integrating rationalist and empiricist perspectives into a cohesive ethical framework, Beauchamp and Childress have provided medical professionals with a comprehensive, intellectually rigorous tool for navigating complex ethical dilemmas. Their work has helped shape medical healthcare education and policy and profoundly impacted the quality and ethics of medical healthcare delivery.

Ethics: Beauchamp and Childress have had a transformative impact on bioethics by articulating the *Belmont Report's* principlistic approach: patient autonomy (informed consent), practitioner beneficence (do good), practitioner nonmaleficence (do no harm), and public justice (be fair). These principles have become the cornerstone of ethical medical healthcare and research frameworks, shaping how ethical dilemmas are approached and resolved. The principle of autonomy has elevated the importance of patient-informed consent, ensuring that patients have the right to make informed decisions about their bodies and medical healthcare treatments. Beneficence, the principle of doing good, has guided medical healthcare providers and researchers to aim for the greatest possible benefit for patients and research participants. Nonmaleficence, or the principle of doing no harm, has been integrated into medical healthcare protocols and research designs to minimize patient risks and prevent harm, echoing the longstanding medical healthcare ethic encapsulated in the phrase "*Primum non nocere*." Lastly, the principle of public justice has influenced policies and practices to ensure fair distribution of medical healthcare resources and equitable treatment of all individuals, regardless of their background or circumstances. By taking the *Belmont Report's* principles and synthesizing them into a cohesive ethical framework, Beauchamp and Childress have provided a comprehensive guide for ethical decision-making widely adopted in bioethics and medical healthcare practice.

Conclusion: Beauchamp and Childress have left an indelible mark on bioethics, medicine, and scientific research through their pioneering work, "*Principles of Biomedical Ethics*." The principlistic approach of using the common ground moral principles of patient autonomy, practitioner beneficence and nonmaleficence, and public justice has become a cornerstone in ethical decision-making across medical healthcare and research settings. The principlistic framework is rooted in rationalist and empiricist traditions and provides a comprehensive, intellectually rigorous tool for navigating complex ethical dilemmas. The *Belmont Report* is the foundation of all Institutional Review Board (IRB) protocols, shaped the scientific method in human research, and guided medical professionals in ethical practice. Principlism has bridged the gap between theoretical ethics and practical application, ensuring that ethical considerations are integral to scientific inquiry and medical healthcare practice. By synthesizing various ethical theories into a cohesive framework, Beauchamp and Childress have created a universally applicable guide widely adopted in bioethics and have had a transformative impact on the quality and ethics of healthcare delivery and scientific research. Their contributions have shaped academic discourse and had a profound, real-world impact, enhancing the ethical conduct of medical healthcare and research globally.

Beauchamp and Childress' Legacy: Beauchamp and Childress revolutionized bioethics by synthesizing rationalist and empiricist perspectives into a coherent principlistic approach, embedding the principles of patient autonomy (informed consent), practitioner beneficence (do good), practitioner nonmaleficence (do no harm), and public justice (be fair) firmly within healthcare and research ethics, thus fundamentally shaping the moral landscape of modern medical practice and policy.

**

REVIEW QUESTIONS

True/False Questions:

1. Beauchamp and Childress first published "*Principles of Biomedical Ethics*" in 1979.
   True or False?

2. The principle of patient autonomy is based on the idea that patients should not have any say in their medical treatments.
   True or False?

Multiple-Choice Questions:

3. What is one of the key contributions of Beauchamp and Childress in the field of bioethics?
   a) Introducing a new medical treatment for cancer
   b) Developing a principlistic approach to ethical decision-making
   c) Creating a universal healthcare system
   d) Inventing new medical devices

4. Which principle emphasizes the importance of fair distribution of medical resources?
   a) Autonomy
   b) Beneficence
   c) Nonmaleficence
   d) Justice

Clinical Vignette:

5. A physician is considering a new, experimental treatment for a patient. According to Beauchamp and Childress's framework, what should be the physician's primary considerations?
   a) Whether the treatment aligns with the physician's personal beliefs
   b) The potential benefits and harms of the treatment, ensuring informed consent, and the fair distribution of the treatment
   c) The cost of the treatment to the healthcare system
   d) The ease of administering the treatment

Basic Science Vignette:

6. A clinical trial is designed to test a new medication for hypertension. According to Beauchamp and Childress's principlistic framework, which principle should primarily guide the decision to ensure the trial is ethically sound?
   a) Maximizing patient autonomy at all costs
   b) Prioritizing the collection of data regardless of patient safety
   c) Ensuring the trial maximizes benefits while minimizing potential harm
   d) Focusing on the economic benefits of the medication

Philosophy Vignette:

7. During a healthcare ethics seminar, a student presents a case study on equitable distribution of scarce medical resources, highlighting the ethical challenges in times of crisis. Reflecting on Beauchamp and Childress's contributions, which approach aligns best with their perspective on justice in healthcare?
a) Allocating resources based solely on clinical need, ignoring broader social implications
b) Ensuring resources are distributed to maximize overall societal benefit, even if it means unequal distribution
c) Balancing clinical needs with ethical obligations to ensure fair distribution of benefits and burdens across society
d) Prioritizing technologically advanced treatments without considering their accessibility

Correct Answers:

1. True
2. False
3. b) Developing a principlistic approach to ethical decision-making
4. d) Justice
5. b) The potential benefits and harms of the treatment, ensuring informed consent, and the fair distribution of the treatment
6. c) Ensuring the trial maximizes benefits while minimizing potential harm
7. c) Balancing clinical needs with ethical obligations to ensure fair distribution of benefits and burdens across society

BEYOND THE CHAPTER
Beauchamp and Childress (1939-)

- *Principles of Biomedical Ethics* by Tom L. Beauchamp and James F. Childress
- *Standing on Principles: Collected Essays* by Tom L. Beauchamp
- *Moral Matters: A Philosophy of Homecoming* by James F. Childress

***

Philosophy of Science — Medical Research

Chapter 12: Peter Singer (1946-) — The Utilitarian Ethicist: Global Welfare and Animal Rights

Singer's utilitarian ethics, focuses on global welfare and animal rights, challenges conventional moral frameworks and promotes a rational, compassionate approach to ethics.

Abstract: Born in 1946, Peter Singer is a towering figure in biomedical ethics, significantly shifting traditional paradigms and societal norms. Armed with a utilitarian ethos focused on maximizing overall happiness and well-being, his work spans bioethical issues in medicine, economics, and law. While his seminal book "*Animal Liberation*" catapulted him to the forefront of animal rights activism, his influence extends to broader bioethical discourse. Using rationalist and empiricist tools, Singer crafts universally applicable ethical principles. Singer's philosophy, combining rational deliberation with empirical observations, has ignited academic debates and influenced policy-making by addressing the moral intricacies of medical advancements. He has sculpted a legacy as a beacon of rationalism and empiricism, fostering critical conversations about healthcare standards and resource allocation. His ideas challenge and expand the foundations of patient autonomy (informed consent), practitioner beneficence (do good), nonmaleficence (do no harm), and public justice (be fair), urging a reconsideration of established norms for a more inclusive ethical framework.

**

Introduction: Peter Singer, an Australian moral philosopher born in 1946, has been pivotal in biomedical ethics, pushing the boundaries of traditional ethical thought and challenging societal norms. His utilitarian approach to ethics — predicated on maximizing overall happiness and well-being — has had far-reaching implications for various biomedical dilemmas, including animal testing, euthanasia, abortion, and healthcare distribution. While Singer's work is perhaps most popularly known for its impact on animal rights through his groundbreaking book "*Animal Liberation*," his intellectual contributions extend into complex bioethical issues that intersect with medicine, economics, and law. His provocative arguments often employ both rationalist and empiricist elements, aiming to derive ethical principles that are universally applicable. Singer's work has influenced policy-making, public opinion, and academic discourse by addressing the moral complexities that medical advancements often bring. His thought-provoking ideas invite scrutiny into the rationalist foundations, empiricist orientation, scientific methodology, and ethical principles underpinning biomedical ethics, and they continue to spark debates that question the very core of what it means to be ethical in medicine.

Rationalism: Peter Singer's ethical philosophy in biomedical ethics is profoundly rooted in rationalist traditions, specifically through his staunch advocacy for utilitarianism. Utilitarianism is a rationalist doctrine predicated on applying reason to ascertain the greatest good for the greatest number. In Singer's hands, this translates into a meticulously logical framework where moral principles are not merely opinion-based but derived from rational deliberations. He employs deductive reasoning to move from the basic axiom that suffering and happiness are intrinsically valuable to more complex moral judgments, such as the ethical considerations surrounding euthanasia, abortion, and animal testing. For instance, Singer's arguments for the permissibility of euthanasia in certain cases start with the rational premise that needless suffering should be avoided. He then logically deduces that, under specific conditions, euthanasia is a morally defensible means of eliminating such suffering. By applying reason as the primary tool for ethical inquiry, Singer's work epitomizes a rationalist approach. His moral conclusions aim for universal applicability, transcending cultural or personal biases, by grounding them in rational principles that can be scrutinized, debated, and, most importantly, universally understood.

Empiricism: Peter Singer, although primarily grounded in rationalist foundations through his utilitarian framework, also incorporates elements of empiricism, particularly when addressing biomedical ethics. He frequently draws upon empirical data, scientific studies, and observed realities to substantiate his ethical positions. For example, in discussions surrounding animal experimentation for medical research, Singer utilizes empirical evidence about animal cognition and the capacity for suffering to challenge the ethical justifiability of such practices. Similarly, in human healthcare, he employs statistical data on resource allocation and the effectiveness of medical interventions to argue for a more equitable distribution of medical resources globally. This empirical approach is aligned with his utilitarian viewpoint, as the data collected about suffering and well-being are critical variables in the utilitarian calculus. By doing so, Singer enriches his rationalist ethical framework with empirical observations, employing inductive reasoning to generalize specific instances of suffering or happiness to broader ethical principles. This empiricist orientation makes his arguments particularly compelling in the biomedical context, as it bridges the gap between theoretical ethics and the practical realities faced in medicine, thereby adding depth and applicability to his philosophical discourse.

The Scientific Method: Peter Singer is not primarily known for his contributions to the scientific method, as his primary expertise lies in moral philosophy rather than empirical science. However, his work has indirectly influenced how ethical considerations are integrated into scientific inquiry, particularly biomedical ethics. Singer's utilitarian framework calls for rigorous empirical evidence to assess the consequences of actions, thereby encouraging the collection and scrutiny of data in ethical dilemmas often encountered in scientific research. For example, his arguments against certain animal testing have led to increased scrutiny of the ethical dimensions of scientific experiments, challenging researchers to consider alternative methods that reduce animal suffering. Singer's ethical positions often necessitate empirical validation, urging scientists to quantify variables like pain, suffering, or well-being, thus integrating ethical considerations into the scientific method. While he does not directly shape the methodological aspects of scientific research, his ideas raise questions that have impacted the design and ethical approval processes of scientific studies, especially those involving human or animal subjects. In this sense, Singer's work adds an ethical layer to the scientific method, encouraging a more holistic approach to scientific inquiry that considers both empirical evidence and moral considerations.

Medicine: While Peter Singer is not a medical doctor, his contributions to bioethics have significantly impacted medical practice, particularly in areas requiring complex ethical decisions. His utilitarian philosophy has provided a theoretical framework for assessing various medical issues, from allocating scarce healthcare resources to the ethical considerations surrounding euthanasia and abortion. Singer's arguments often challenge conventional medical healthcare standards of care by asking difficult questions about acting in the best interests of patients and communities. For example, his views on allocating healthcare resources in a globally unequal world challenge medical professionals to consider ethical imperatives beyond their immediate medical community. Furthermore, his discussions on physician-assisted suicide and end-of-life care have stirred debates that influence medical ethics and law, pushing for revisions or reconsidering existing guidelines based on utilitarian principles. Singer's work also encourages the integration of evidence-based ethical considerations into medical practice, thereby influencing healthcare standards by promoting practices that aim for the greatest overall good. Although Singer doesn't directly set medical healthcare standards of care, his ethical frameworks have become integral to the conversation about what those standards should be, offering a utilitarian lens through which medical ethics and, by extension, medical practices can be scrutinized and improved.

Ethics: Peter Singer's work has profoundly influenced bioethical principles, often challenging traditional interpretations of these core tenets. Starting with autonomy, Singer's utilitarian approach does not necessarily prioritize individual consent as an inviolable principle but assesses it within the larger context of overall well-being and happiness. His viewpoint adds complexity to the notion of informed consent, forcing a reevaluation of how individual autonomy fits within broader ethical considerations. Regarding beneficence, Singer extends the obligation to do good far beyond conventional boundaries, arguing for a global perspective on altruism and advocating for effective ways to alleviate suffering, including non-human suffering. Regarding nonmaleficence, Singer's work, particularly on animal ethics, has revolutionized the principle of "*do no harm*," broadening its scope to include non-human entities capable of suffering. His utilitarian framework measures harm regarding suffering and well-being rather than relying solely on moral intuition or cultural norms. Finally, his concept of justice, deeply rooted in utilitarianism, goes beyond mere fairness to advocate for a redistribution of resources and opportunities to maximize overall welfare. Singer's work thus offers a provocative but invaluable lens through which the foundational principles of bioethics — patient autonomy (informed consent), practitioner beneficence (do good), practitioner nonmaleficence (do no harm), and public justice (be fair) — are scrutinized, expanded, and, at times, radically redefined.

Conclusion: Peter Singer's contributions to biomedical ethics and philosophy have left an indelible mark on how society thinks about ethical dilemmas. His utilitarian philosophy, grounded in rationalist and empiricist elements, offers a comprehensive framework for tackling complex moral questions. While not a medical doctor or a scientist, his influence reverberates through medicine, science, law, and public policy. His ideas challenge conventional notions of healthcare standards, introduce empirical scrutiny into ethical debates, and even urge scientists to consider the ethical ramifications of their work. His probing questions and compelling arguments have spurred critical discussions around healthcare resource allocation, end-of-life decisions, animal testing, and much more. Singer's work complicates and enriches the key bioethical principles of patient autonomy (informed consent), practitioner beneficence (do good), practitioner nonmaleficence (do no harm), and public justice (be fair), pushing for an expanded ethical purview encompassing global considerations and non-human welfare. This unique blend of rationalism, empiricism, and ethical theory has revolutionized bioethical thought, contributing to its evolution as a multidisciplinary field that intersects with various aspects of human life and society. In doing so, Peter Singer has redefined what it means to make ethical decisions in the complex, interconnected world of the 21st century.

Peter Singer's Legacy: Known as a pioneering force in bioethics and animal rights, Singer's utilitarian philosophy challenges conventional ethical norms and pushes for an expanded moral consideration that includes global human welfare and the well-being of non-human entities.

**

REVIEW QUESTIONS

True/False Questions:

1. Peter Singer is most famously known for his book "*Animal Liberation*," which propelled him to the forefront of animal rights activism.
   True or False?

2. Singer's ethical philosophy primarily emphasizes deontological principles over utilitarian principles.
   True or False?

Multiple-Choice Questions:

3. Which ethical principle does Peter Singer challenge by proposing that individual consent should be assessed within the larger context of overall well-being and happiness?
   a) Practitioner beneficence
   b) Patient autonomy
   c) Public justice
   d) Practitioner nonmaleficence

4. What philosophical approach does Peter Singer predominantly use in his bioethical arguments?
   a) Deontology
   b) Virtue ethics
   c) Utilitarianism
   d) Contractualism

Clinical Vignette:

5. A medical ethics discussion revolves around allocating scarce medical resources in a pandemic. Drawing on Peter Singer's utilitarian perspective, which approach would most align with his philosophy?
   a) Prioritizing resources for the most economically productive individuals
   b) Allocating resources equally regardless of need
   c) Prioritizing resources for those who can pay the most
   d) Allocating resources to maximize overall well-being and reduce suffering

Basic Science Vignette:

6. A researcher is conducting an experiment on the ethical implications of animal testing for developing a new medical treatment. According to Peter Singer's utilitarian principles, what should be the primary consideration in deciding whether to proceed with the experiment?
   a) The tradition and cultural acceptance of animal testing
   b) The potential economic benefits of the new treatment
   c) The overall balance of pleasure and pain resulting from the experiment
   d) The personal beliefs of the researchers involved

Philosophy Vignette:

7. During a lecture on global ethics, a student asks how Peter Singer's utilitarianism addresses the issue of wealth distribution. According to Singer, what is the most ethical way to address global inequality?
   a) Encouraging wealthy individuals to donate to their local communities
   b) Implementing policies that prioritize national economic growth
   c) Redistributing resources to maximize overall global welfare
   d) Promoting voluntary charity based on personal choice

Correct Answers:

1. True
2. False
3. b) Patient autonomy
4. c) Utilitarianism
5. d) Allocating resources to maximize overall well-being and reduce suffering
6. c) The overall balance of pleasure and pain resulting from the experiment
7. c) Redistributing resources to maximize overall global welfare

BEYOND THE CHAPTER
Peter Singer (1946-)

- *Peter Singer Under Fire: The Moral Iconoclast Faces His Critics* edited by Jeffrey A. Schaler
- *Animal Liberation* by Peter Singer
- *The Life You Can Save: How to Do Your Part to End World Poverty* by Peter Singer

***

Chapter 13: **Academic Freedom** — The Catalyst for Truth: Unbounded Inquiry

Academic freedom fuels innovation and integrity, driving scientific and ethical advancements in a dynamic healthcare landscape.

Abstract: The principle of academic freedom, rooted in the 1940 Statement of Principles on Academic Freedom and Tenure by the AAUP, fosters scholarly innovation and intellectual growth. This study elucidates the relationship between academic freedom and the rationalist and empiricist frameworks underpinning the scientific method, driving advancements in medicine and nurturing ethical integrity in biomedical practices. Rationalism fosters deep analytical discourse, while empiricism encourages knowledge acquisition through sensory experiences. Within the scientific method, academic freedom catalyzes credible knowledge pursuit, facilitating the design and analysis of experiments. It promotes evidence-based advancements and ethical practices in healthcare. This exploration highlights the influence of academic freedom on the four principles of bioethics — patient autonomy (informed consent), practitioner beneficence (do good), practitioner nonmaleficence (do no harm), and public justice (be fair), fostering a dynamic healthcare environment. Ultimately, academic freedom is a linchpin in the pursuit of truth, encouraging intellectual evolution and innovation.

**

Introduction: In the dynamic realm of academia, the principle of academic freedom emerges as a linchpin that nurtures intellectual growth, fosters innovation, and facilitates vibrant scholarly dialogue. Established on the foundational guidelines set by the 1940 Statement of Principles on Academic Freedom and Tenure by the American Association of University Professors (AAUP), this concept embodies a dedication to upholding the highest standards of freedom in academic endeavors. It entrusts faculty members with the autonomy to conduct unswayed research and to foster a rich and inclusive learning environment through diverse and critical discourse within classrooms. Furthermore, this freedom extends beyond the realms of institutions, allowing academicians to engage in civic responsibilities and collaborative dialogue without the fear of institutional censorship or repercussions, thus nurturing a community that stands steadfast in its commitment to excellence, innovation, and ethical grounding. As we proceed, we will evaluate the rationalist foundations of this concept, its empiricist orientations, its pivotal role in shaping the scientific method and contributing to advancements in medicine, as well as its profound influence on the four pillars of bioethics: patient autonomy (informed consent), practitioner beneficence (do good), practitioner nonmaleficence (do no harm), and public justice (be fair).

Rationalism: Academic freedom finds its rationalist roots deeply entrenched in the belief that certain intrinsic cognitive faculties play a pivotal role in acquiring and propagating knowledge. Under the canopy of rationalism, it is posited that individuals are endowed with innate abilities such as problem-solving, pattern recognition, and basic reasoning skills, which serve as the cornerstone for intellectual exploration and discourse. These "*a priori*" principles facilitate the unfettered exploration of ideas and theories, fostering a nurturing ground for scholarly innovation and discovery. Scholars can embark on intellectual journeys through deductive reasoning, drawing from axioms or definitional truths to cultivate new perspectives and insights. For example, Galileo's defense of heliocentrism against the prevailing geocentric models of his time was grounded in his rationalist convictions. His ability to argue from foundational principles and deductive reasoning, despite significant opposition, showcases how academic freedom allows for the challenging and refinement of established paradigms. Thus, academic freedom protects these intrinsic faculties and encourages scholars to delve deeper, leveraging their inherent capabilities to further knowledge and understanding in various fields. This firm foundation on rationalist principles underscores the essential role that academic freedom plays in promoting a vibrant and intellectually rich scholarly environment, wherein ideas can be dissected, debated, and synthesized, contributing to the broader tapestry of human understanding and progress.

Empiricism: In juxtaposition to its rationalist foundations, the empiricist orientation of academic freedom fosters an open environment where knowledge is gleaned through sensory experiences and interactions with the surrounding environment. This philosophical stance embraces firsthand observations and experiments as vital pathways to accumulating knowledge, championing the belief that insights evolve primarily from experiential learning rather than ingrained instincts or "*a priori*" principles. Through the lens of empiricism, academic freedom encourages scholars to employ inductive reasoning, where specific observations are meticulously analyzed to formulate broader generalizations and theories. This methodological approach facilitates a rich and varied scholarly discourse, allowing for the cultivation of a more comprehensive understanding of complex phenomena through synthesizing observed data. For instance, the empirical studies conducted by Louis Pasteur, which led to the development of germ theory, highlight the importance of empirical evidence in challenging existing beliefs and advancing scientific understanding. Pasteur's meticulous experiments demonstrated the role of microorganisms in disease, fundamentally transforming medical science. Academic freedom, therefore, stands as a guardian of this empirical pursuit, protecting the rights of academicians to explore, analyze, and interpret the world through a prism of empirical experiences, thereby fostering a dynamic and adaptive learning environment that continually seeks to expand the horizons of human knowledge and understanding.

The Scientific Method: Within the scaffolding of the scientific method, academic freedom manifests as a catalyst that fosters unfettered inquiry and diligent exploration of phenomena. This liberty grants researchers the latitude to observe and dissect various phenomena impartially, formulating nuanced questions and testable hypotheses. The nurturing grounds of academic freedom ensure that scholars have the autonomy to design and conduct controlled experiments without undue influence or bias, facilitating the meticulous collection and analysis of data to substantiate or refute proposed hypotheses. The development of quantum mechanics in the early 20th century, driven by physicists like Niels Bohr and Albert Einstein, exemplifies how academic freedom supports scientific advancement. These scientists' ability to freely propose, test, and debate novel theories led to groundbreaking discoveries that reshaped our understanding of the physical world. Moreover, academic freedom is intrinsically linked to the replicability criterion of the scientific method, encouraging an open exchange of ideas and findings, which allows different researchers to verify results under similar conditions, thus enhancing the credibility and reliability of scientific conclusions. Furthermore, this freedom engenders a rigorous peer-review process, where experts scrutinize research for its validity, methodology, and significance, fostering a scientific community that thrives on collaboration, critical thinking, and mutual respect. Therefore, academic freedom is the cornerstone in the edifice of the scientific method, bolstering a systematic, credible, and transparent approach to advancing human knowledge.

Medicine: Academic freedom stands as a beacon in the continuous evolution of medical science, fostering an environment where healthcare professionals can relentlessly pursue advancements and refinements in medical healthcare standards of care. Through the lens of academic freedom, clinicians, researchers, and other healthcare professionals are empowered to scrutinize and develop evidence-based practices, steering the healthcare community toward approaches that are both innovative and grounded in robust scientific evidence. This liberty to explore and challenge existing paradigms has been instrumental in shaping medical healthcare standards of care, defined by the proficiency and expertise expected of a competent medical healthcare professional within a similar medical community. For example, the discovery and development of insulin therapy for diabetes were made possible by years of academic research on pancreatic function and metabolism, supported by academic freedom. Researchers Banting and Best could investigate and experiment without immediate commercial pressures, leading to a groundbreaking treatment that has saved millions of lives and revolutionized diabetes management. Moreover, academic freedom facilitates a dynamic flow of dialogue and collaboration, allowing for the integration of research, clinical expertise, and patient needs to sculpt current best practices. This fluidity ensures that medical healthcare standards of care are not static but are continually adapting to encapsulate the latest findings and methodologies. Consequently, academic freedom also acts as a pivotal force in the legal spheres, assisting in determining the adequacy of care delivered, thereby serving as a vanguard in pursuing medical excellence and safeguarding the integrity and progressiveness of medical healthcare practices globally.

Ethics: Academic freedom stands as a pillar bolstering the ethical framework that governs biomedical practices, significantly influencing the four biomedical principles of patient autonomy (informed consent), practitioner beneficence (do good), practitioner nonmaleficence (do no harm), and public justice (be fair). Firstly, it fosters a culture of autonomy, wherein individuals are empowered with the right to informed consent, enabling them to make well-informed decisions about their healthcare based on comprehensive and unbiased information. Secondly, it encourages the principle of beneficence, stimulating professionals to continually seek advancements that aim to do good and enhance the well-being of patients. Concurrently, it reinforces the principle of nonmaleficence, guiding researchers and practitioners to adopt approaches that minimize harm and avoid detrimental outcomes. For instance, the ethical review processes for clinical trials ensure that new treatments are tested in ways that prioritize patient safety and informed consent, reflecting academic freedom's role in upholding high ethical standards. Lastly, academic freedom nourishes the principle of justice, promoting equity and fairness in healthcare by encouraging research and discourse that scrutinizes and challenges existing inequalities and biases. Through the lens of academic freedom, the bioethical landscape is rendered a dynamic space where these principles are continuously examined, refined, and upheld, fostering an environment where the integrity of medical healthcare practices is maintained and the dignity and rights of individuals are respected and protected.

Conclusion: In navigating the ever-evolving labyrinth of knowledge and scholarly endeavors, academic freedom surfaces as an indispensable ally, catalyzing robust intellectual exploration, innovation, and ethical integrity across diverse fields. As we have delineated, its profound roots in rationalist principles foster an environment where innate cognitive faculties facilitate deep-rooted, analytical, and deductive discourse, thereby nurturing a fertile ground for scholarly innovation and discoveries. This intertwines beautifully with its empiricist orientation, which underscores knowledge acquisition through sensory experiences and empirical analyses, promoting a vibrant and adaptive learning environment. Furthermore, academic freedom functions as a linchpin in shaping the systematic and meticulous approach of the scientific method, fostering credibility and collaboration in the pursuit of knowledge. Its contributions to the medical sphere are undeniable, steering continuous enhancements in medical healthcare standards of care and fostering an environment ripe for innovative and evidence-based advancements. Moreover, its pervasive influence in shaping the bioethical landscape cannot be overstated, as it diligently works to uphold the principles of patient autonomy (informed consent), practitioner beneficence (do good), practitioner nonmaleficence (do no harm), and public justice (be fair), ensuring the protection and respect of individuals' dignity and rights. Academic freedom emerges as a bastion of intellectual growth and ethical fortitude, a driving force propelling society forward in a ceaseless quest for excellence, innovation, and a deeper understanding of the complex tapestry of the human experience.

Academic Freedom's Legacy: Serving as a cornerstone for intellectual evolution, it safeguards the ceaseless pursuit of truth, fostering a society where ideas can flourish unbounded, and innovation is nurtured through rationalist insights and empiricist explorations, ultimately sculpting a future where knowledge is not constrained, but perennially blossoming.

**

REVIEW QUESTIONS

True/False Questions:

1. The principle of academic freedom is rooted in the 1940 Statement of Principles on Academic Freedom and Tenure by the American Association of University Professors (AAUP).
   True or False?

2. According to the chapter, academic freedom limits the ability of scholars to engage in empirical research.
   True or False?

Multiple-Choice Questions:

3. The chapter highlights that academic freedom, when viewed through rationalist principles, supports:
   a) Restricting the exploration of new ideas
   b) Promoting inherent cognitive faculties like problem-solving
   c) Limiting the use of sensory experiences in research
   d) Enforcing a strict adherence to existing paradigms

4. In the context of medicine, academic freedom significantly contributes to:
   a) The stagnation of medical practices
   b) The development of standardized treatment plans without evidence
   c) The continuous improvement of evidence-based practices
   d) The rejection of new medical technologies

Clinical Vignette:

5. A medical researcher is conducting a study on a new treatment for a chronic disease. The researcher's institution values academic freedom and encourages innovative approaches. Based on the chapter's discussion on academic freedom, what should the institution ensure?
   a) The researcher follows a rigid methodology without deviations
   b) The researcher is free to explore and refine the study based on empirical evidence
   c) The researcher avoids collaboration with other experts
   d) The researcher focuses solely on theoretical analysis without empirical testing

Basic Science Vignette:

6. Dr. Harris is researching new treatments for Alzheimer's disease. He proposes a novel approach involving a combination of genetic therapy and cognitive rehabilitation. Considering the importance of academic freedom in fostering innovation, which strategy should he pursue?
   a) Rely solely on genetic therapy, ignoring cognitive rehabilitation
   b) Focus only on cognitive rehabilitation without considering genetic factors
   c) Integrate both genetic therapy and cognitive rehabilitation to explore comprehensive treatment options
   d) Abandon the research due to potential challenges and controversies

Philosophy Vignette:

7. In her philosophy seminar, Olivia argues that academic freedom is essential for advancing ethical principles in biomedical research. How should she best support her argument using the concept of academic freedom?
   a) Academic freedom restricts researchers from exploring new ethical theories
   b) Academic freedom allows researchers to pursue unbiased and innovative studies, leading to advancements in ethical practices
   c) Academic freedom limits the development of evidence-based ethical guidelines
   d) Academic freedom prevents researchers from addressing existing ethical inequalities

Correct Answers:

1. True
2. False
3. b) Promoting inherent cognitive faculties like problem-solving
4. c) The continuous improvement of evidence-based practices
5. b) The researcher is free to explore and refine the study based on empirical evidence
6. c) Integrate both genetic therapy and cognitive rehabilitation to explore comprehensive treatment options
7. b) Academic freedom allows researchers to pursue unbiased and innovative studies, leading to advancements in ethical practices

BEYOND THE CHAPTER

Academic Freedom

- *The Future of Academic Freedom* by Louis Menand (Editor)
- *Academic Freedom in an Age of Conformity: Confronting the Fear of Knowledge* by Joanna Williams
- *Free to Think: Why Scientific Integrity Matters* by Austin Ruse

***

Philosophy of Science

Medical Research

Chapter 14: Scholarship — The Beacon of Progress: Intellectual Rigor

Scholarship illuminates the path of human progress, intertwining rational inquiry and empirical evidence to foster innovation and ethical integrity.

Abstract: In the monumental voyage of human progression, scholarship emerges as a luminescent beacon guiding the relentless quest for knowledge and fostering intellectual evolution. This intricate discipline, characterized by a blend of rationalist roots and empirical orientations, has played a pivotal role in shaping modern society, fostering technological advancements, and enhancing the quality of life across generations and geographies. Rooted in academic freedom and fortified by tenure, scholarship has shaped the scientific method. This systematic approach has revolutionized our comprehension of the natural world. Furthermore, it has profoundly impacted the medical field, spearheading the creation and refinement of healthcare standards and nurturing a resilient, innovative medical community committed to excellence. This discipline has also significantly influenced bioethics principles, fostering a healthcare environment that is scientifically advanced yet ethically grounded. This examination seeks to appreciate the pervasive and transformational influence of scholarship in cultivating an enlightened society where knowledge acts as the catalyst for sustainable growth and human welfare.

**

Introduction: In the expansive journey of human advancement, scholarship shines as a beacon, signifying a relentless pursuit of knowledge and intellectual growth. This multifaceted discipline, transcending generations and borders, plays a critical role in shaping societies, advancing technology, and improving the quality of life. The integrity and progress of scholarship are upheld by two vital principles: academic freedom and the institution of tenure. Academic freedom fosters an environment where scholars can engage in unbiased research and discourse, thereby nurturing innovation and the flourishing of new ideas. Complementarily, tenure provides stability and support to scholars, fostering a robust network of knowledge and mentorship and guiding the next generation of thinkers and innovators. As we delve deeper into this remarkable journey, we aim to explore its rational and empirical foundations, its influence in shaping scientific methods and contributing to the medical field, and its profound impact on bioethics principles. Through this examination, we aspire to appreciate the far-reaching influences of scholarship in fostering an enlightened society where knowledge catalyzes sustainable growth and human welfare.

Rationalism: Scholarship, firmly rooted in the philosophical grounding of rationalism, has continually advanced the notion that certain intellectual faculties and instinctual behaviors serve as intrinsic conduits for knowledge acquisition and comprehension. Under this scope, it is proposed that humans are bestowed with innate faculties — ranging from basic reasoning skills and pattern recognition to problem-solving abilities — that function as evolutionary instruments primed to forge paths of understanding and awareness. This "*a priori*" backdrop allows individuals to engage in deductive reasoning. This process often draws from established axioms, postulates, or definitional truths, forming the bedrock of scholarly pursuits. For example, the work of philosophers like René Descartes, who famously declared "*Cogito, ergo sum*" (I think, therefore I am), illustrates the rationalist approach.

Descartes' emphasis on deductive reasoning and innate ideas laid the foundation for modern philosophy and scientific inquiry. This rationalist foundation posits that scholarship is not solely an external endeavor but a journey inward, tapping into the pre-existing wellsprings of knowledge hardwired within the human psyche. Over centuries, this intellectual tradition has sculpted the framework of scholarship, encouraging scholars to transcend empirical observations and to unearth deeper truths about our existence, a pursuit that continues to foster innovation, depth of insight, and a nuanced understanding of the complex tapestry that constitutes human knowledge and experience.

Empiricism: Within the diverse spectrum of scholarship lies its empiricist orientation, a pivotal philosophy that advocates acquiring knowledge through sensory experiences and environmental interactions. This standpoint emphasizes the tangible, where data and evidence from meticulous observation become the cornerstone of knowledge building. Through this lens, scholarship embodies an organized and systematic approach, employing inductive reasoning to carve pathways from specific, grounded observations to more expansive, general conclusions. This empirical journey often unfolds in laboratories, field studies, and research facilities, where scholars immerse themselves in the intricate web of reality, extracting and analyzing information to formulate well-founded theories and concepts. For instance, the empirical research of Charles Darwin, who meticulously gathered data from his observations of natural selection during the voyage of the HMS Beagle, exemplifies this approach. Darwin's inductive reasoning led to the formulation of his theory of evolution, which fundamentally transformed biological sciences. The empiricist orientation of scholarship has played a fundamental role in fostering a rich culture of inquiry and discovery, cultivating both receptive and critical minds capable of navigating the complex realms of knowledge with a grounded and evidence-based approach. Through this diligent process, scholarship continues to evolve, offering a robust and dynamic platform where learning is continuously forged through the crucible of tangible experiences and empirical investigations.

The Scientific Method: Scholarship, through its unwavering commitment to the pursuit of knowledge, has been a seminal force in shaping the scientific method, a structured and systematic approach to inquiry that has revolutionized our understanding of the natural world. This method commences with observing a phenomenon, sparks curiosity, and beckons researchers to delve deeper into its underlying causes or explanations. A hypothesis emerges from these inquiries, serving as a tentative yet testable explanation that can be verified and falsified. Here, scholarship shines, fostering a culture of critical thinking and meticulous analysis essential in conducting controlled experiments central to validating or refuting the initial hypothesis. Through the rigorous collection and interpretation of data, conclusions are carefully drawn, stepping stones to further research and discovery. A linchpin in this process is the emphasis on replicability, a hallmark that ensures the credibility and reliability of findings, fostering a collaborative and transparent scientific community. To further fortify its integrity, scholarship encourages peer reviews, where experts scrutinize the researcher's methodology, validity, and significance. It fosters an ecosystem where knowledge is continually refined and expanded, cementing the scientific method as a bastion of credible and rigorous inquiry.

Medicine: In the ever-evolving field of medicine, scholarship stands as a vanguard, driving forward advancements and refining practices to meet the dynamic needs of healthcare. Through rigorous academic research and inquiry, scholarship has played an instrumental role in shaping the medical healthcare standards of care, a set of benchmarks that delineate the expected levels of proficiency and expertise within the medical community. These meticulously crafted and continually refined standards emerge from evidence-based practices based on peer-reviewed research grounded in empirical data and clinical expertise. Through the lens of scholarship, medical healthcare professionals are equipped with the tools to navigate complex medical landscapes, drawing from a repository of knowledge that reflects the best current practices in patient care. For example, the development of new surgical techniques often stems from scholarly research and clinical trials that rigorously test the efficacy and safety of these methods. Techniques such as laparoscopic surgery have revolutionized patient care, offering less invasive options with quicker recovery times, thanks to scholarly research. Furthermore, these standards serve as a litmus test in legal contexts, determining the adequacy and competency of delivered care. Thus, with its commitment to excellence and innovation, scholarship fosters a resilient and adaptable medical community where care is not static but evolves with the strides made in research and understanding, paving the way for a future where healthcare is progressive and patient-centric.

Ethics: At the intersection of scholarship and ethics, we find a profound influence shaping the core principles that govern the field of bioethics: patient autonomy (informed consent), practitioner beneficence (do good), practitioner nonmaleficence (do no harm), and public justice (be fair). With its roots deeply embedded in the quest for knowledge and understanding, scholarship reinforces the principle of autonomy, encouraging medical practitioners to uphold the tenet of informed consent, thereby empowering individuals to make decisions about their health with full awareness and understanding. Moreover, it fosters a climate of beneficence, wherein the pursuit to "*do good*" is continually guided by advancements in knowledge and expertise, aiming to improve patient welfare. Through rigorous scrutiny and ethical considerations, the scholarship also encourages the principle of nonmaleficence, steering medical professionals toward practices that minimize harm and avoid adverse consequences. Furthermore, it champions the principle of justice, advocating for fair and equitable treatment of all individuals and fostering a medical healthcare system devoid of discrimination and prejudice. For example, the ethical review processes for clinical trials ensure that new treatments are tested in ways that prioritize patient safety and informed consent, reflecting academic freedom's role in upholding high ethical standards. Through the lens of scholarship, the bioethical landscape is continually reshaped, nurturing a healthcare environment that is not only scientifically advanced but ethically sound, holding steadfast to principles that safeguard human dignity and well-being.

Conclusion: Scholarship, epitomized as the epicenter of intellectual advancement and pursuit of knowledge, underscores humanity's tireless journey toward enlightenment and progress. This monumental force has sculpted the tapestry of society, fostering a resilient framework that integrates the profound depth of rationalist foundations with the practical and grounded approach of empiricism, synergizing to fuel discoveries and innovations in a myriad of fields. Its role in shaping the scientific method cannot be understated, as it cultivates an ecosystem where inquiry and analysis stand paramount, fostering advancements that resonate through various disciplines, notably in medicine. By spearheading the development and refinement of medical healthcare standards of care, scholarship is a beacon, guiding professionals to deliver care that mirrors the zenith of contemporary knowledge and expertise. Furthermore, its influence permeates the sphere of bioethics, nurturing a framework where principles of patient autonomy (informed consent), practitioner beneficence (do good), practitioner nonmaleficence (do no harm), and public justice (be fair) are revered and upheld. As we traverse this intricate pathway carved by scholarship, it behooves us to recognize its monumental impact on society. This catalyst continually drives us toward a future characterized by wisdom, ethical integrity, and an unremitting commitment to bettering the human condition. Through the rich discourse fostered by scholarship, we stand on the cusp of an era where knowledge blossoms, promising a brighter and more enlightened future for all.

Scholarship's Legacy: Serving as the linchpin in the ceaseless journey of human advancement, fostering an enduring culture of intellectual rigor, ethical stewardship, and groundbreaking discoveries across diverse fields.

**

REVIEW QUESTIONS

True/False Questions:

1. Scholarship is characterized by a blend of rationalist roots and empirical orientations, playing a pivotal role in shaping modern society.
   True or False?

2. According to the chapter, tenure restricts academic freedom and limits scholarly innovation.
   True or False?

Multiple-Choice Questions:

3. The chapter suggests that scholarship, grounded in rationalism, supports:
   a) Acquiring knowledge solely through sensory experiences
   b) Innate cognitive faculties like problem-solving and deductive reasoning
   c) Random and unstructured knowledge acquisition
   d) The elimination of empirical research

4. In the context of medicine, scholarship primarily contributes to:
   a) Outdated medical practices
   b) The development of evidence-based standards of care
   c) Ignoring clinical expertise
   d) Promoting untested medical procedures

Clinical Vignette:

5. A medical scholar is conducting research on a new treatment protocol for diabetes. The scholar's work is supported by academic freedom and tenure, which encourage unbiased research. Based on the chapter, what should the scholar prioritize in their research?
   a) Conducting experiments without peer review
   b) Ignoring empirical data and focusing on theoretical models
   c) Employing evidence-based practices and peer-reviewed research
   d) Avoiding collaboration with other experts

Basic Science Vignette:

6. Dr. Martinez is researching the impact of CRISPR technology on genetic disorders. His team has made a groundbreaking discovery that could potentially cure cystic fibrosis. Considering the role of scholarship in fostering innovation and ethical practices, which approach should he take?
   a) Conduct further research in isolation without peer review
   b) Publish the findings in a reputable scientific journal for peer review and ethical scrutiny
   c) Proceed directly to human trials without additional testing
   d) Focus solely on the commercial applications of the discovery

Philosophy Vignette:

7. In her philosophy class, Jane argues that scholarship is essential for upholding ethical principles in scientific research. How should she best support her argument using the concept of scholarship?
   a) Scholarship restricts researchers from exploring controversial topics
   b) Scholarship promotes the pursuit of knowledge while ensuring ethical guidelines are followed through rigorous peer review and empirical evidence
   c) Scholarship prevents the dissemination of new ideas and hinders progress
   d) Scholarship focuses only on theoretical knowledge without practical applications

Correct Answers:

1. True
2. False
3. b) Innate cognitive faculties like problem-solving and deductive reasoning
4. b) The development of evidence-based standards of care
5. c) Employing evidence-based practices and peer-reviewed research
6. b) Publish the findings in a reputable scientific journal for peer review and ethical scrutiny
7. b) Scholarship promotes the pursuit of knowledge while ensuring ethical guidelines are followed through rigorous peer review and empirical evidence

BEYOND THE CHAPTER
Scholarship

- *The Scholarship of Teaching and Learning Reconsidered: Institutional Integration and Impact* by Pat Hutchings, Mary Taylor Huber, and Anthony Ciccone
- *Scholarship Reconsidered: Priorities of the Professoriate* by Ernest L. Boyer
- *The Forms of Things Unknown: Essays Toward an Aesthetic Philosophy* by David L. Hall

***

Chapter 15: History of Peer Review — The Evolution of Integrity: Critical Analysis

Peer review's evolution from ancient discourse to AI-assisted evaluations embodies a commitment to integrity and collaborative scientific progress.

Abstract: This comprehensive discourse elucidates the journey of developing peer review processes, tracing its roots from informal discussions in ancient Greek agoras to modern methodologies in contemporary scientific publishing. It highlights the persistent evolution and adaptations throughout historical periods, such as the Alexandrian period, Scholasticism, the Scientific Revolution, and the rapid advancements of the 20th and 21st centuries. Emphasizing the need to avoid "*false precision*" and "*oversimplification*" in intellectual history, the discussion portrays peer review as a gradually developed system integrating critical analysis and collaboration. In the present era, the peer review process is witnessing innovations like AI-assisted evaluations and open science movements, fostering a more inclusive and transparent scholarly community. As it strides into the future, the peer review system aims to incorporate technological advancements while upholding integrity, inclusivity, and transparency, ensuring the continual enhancement of scientific discourse and collaboration.

**

Introduction: When tracing the origins and development of intellectual realms such as philosophy, science, evidence-based medicine, and peer review, it is often misguided to pinpoint precise "*start dates.*" Instead, it is crucial to appreciate the gradual emergence of these domains over centuries or millennia, owing to the accumulation of knowledge, theories, and methodologies. These domains have been intricately woven into the fabric of human history, each era contributing noteworthy advancements and shifts in thinking. Therefore, attributing a specific "*start date*" might overlook the depth and progressive layering of efforts that have enriched these fields over time. A thorough understanding of these evolutionary nuances offers a more panoramic view of their histories and guides contemporary and future endeavors.

This conversation circles a logical fallacy often referred to as "*false precision*" or "*oversimplification,*" where complex, evolving phenomena are narrowed down to a singular inception moment, thus bypassing the myriad nuances and the gradual development that marks their journey. This error sometimes aligns with a "*false dichotomy,*" where a field's expansive, multi-faceted progression is simplified to a basic "*before*" and "*after*" narrative. Recognizing and appreciating the myriad developments and influences that have substantially molded these fields over time is, thus, of utmost importance.

A Comprehensive Chronology of Peer Review Development

Pre-Socratic Period (6th Century BCE): In Ancient Greece, around 600 BCE, a significant milestone was established in developing what we now know as the peer review process. During this period, the agora (a public open space used for assemblies and markets) became a focal point for intellectuals and scholars to debate and critique each other's theories and ideas. This practice of dialogue-based scrutiny marked the early seeds of peer review, fostering a culture where critical thinking and peer evaluation were highly encouraged. It laid the groundwork for a tradition that valued the questioning, challenging, and refining of theories, essentially nurturing the foundational principles of modern peer review.

Classical Antiquity to Middle Ages (5th Century BCE-15th Century CE)

Alexandrian Period (3rd Century BCE-1st Century BCE): In this period, the Library of Alexandria was vital for scholarly exchange and critique, encouraging a budding culture of peer review. Here, scholars from various disciplines congregated to rigorously review and critique each other's works, sowing the seeds of a rich environment characterized by intellectual exchange and collaboration. This beacon of knowledge became the epicenter of learning, where scholarly works were meticulously analyzed, paving the way for a collaborative scholarly culture.

Scholasticism (12th-15th Century): Later, the scholastic period heralded a more structured approach to intellectual inquiries, encouraging academics to engage critically with each other's texts. Disputations or formal debates were common as a precursor to the systematic peer review format and the dissertation defenses we witness today. These structured debates were fertile grounds for the cultivation of rigorous intellectual discourse, where scholars were tasked with defending or opposing various theses based on meticulous examinations of texts and logical reasoning. This era was marked by systematically investigating concepts and theories, laying significant groundwork for modern academic discourse and peer review.

Early Modern Period (16th-18th Century)

Scientific Revolution (16th-17th Century): This era brought a remarkable shift in scientific inquiry, characterized by the adoption of the scientific method. This transformative shift led to heightened scrutiny of scholarly works, often facilitated through intellectual circles and correspondence among scientists. Icons like Isaac Newton and Galileo Galilei were central figures in this transition, fostering a vibrant scholarly atmosphere emphasizing evidence and critical analysis in assessing scientific works.

18th Century: In 1752, the Royal Society pioneered the introduction of a formalized peer review process through their renowned journal "*Philosophical Transactions*." This innovation established a benchmark for systematic peer review in scientific publishing, fostering a culture of scholarly accountability and collaboration. This phase saw the development of more decentralized yet organized peer review mechanisms, marking a pivotal moment in the scholarly landscape.

19th Century

During the early to mid-1800s, editorial committees were pivotal in steering the review process, aiding the gradual formulation of modern peer review systems. This period also marked the emergence of specialized scientific societies, which became platforms facilitating peer reviews, albeit initially in a more informal manner. These societies became pillars of scientific discourse, cultivating environments where scholars could critically assess each other's works, solidifying the foundation of peer review processes.

20th Century

Post-World War II (1940s-1950s): The aftermath of World War II saw a surge in scientific research, necessitating a more structured and formalized peer review system to maintain the quality and integrity of published works. This period solidified peer review as a standard practice in reputable journals and research-granting organizations, establishing scrutiny and quality assurance that became a hallmark of scholarly publishing.

1950s-1960s: This time frame experienced a significant expansion in the role of funding agencies. The grant review process began to echo the peer review process in journals, emphasizing the necessity for stringent review standards to safeguard the credibility and quality of funded research projects. This alignment of funding strategies with peer review mechanisms fostered a culture of excellence in scientific research.

The 1970s: This decade marked a significant milestone in the evolution of the peer review process, characterized by a global institutionalization that introduced structured approaches, including the implementation of double-blind reviewing methods. In this double-blind approach, both the authors and the reviewers remain anonymous to each other, a strategy devised to minimize potential biases and ensure a more objective evaluation. This pivotal change laid the groundwork for fostering a culture of transparency and fairness in scholarly evaluations, steering toward more ethical and equitable review processes.

1980s-1990s: These years were marked by heightened scrutiny and formalization of the peer review process, focusing on ethical considerations. New strategies were devised to address conflicts of interest and plagiarism, thus safeguarding the integrity and quality of scholarly publications. This period witnessed a concerted effort to bolster the credibility and reliability of scientific publishing.

21st Century

The early 2000s: The onset of the digital revolution streamlined the manuscript submission and review process, fostering the growth of open-access journals and facilitating a wider dissemination of scientific knowledge. This period shifted toward a more inclusive and accessible scholarly ecosystem, encouraging collaborative efforts and open discourse in the scientific community.

Late 2000s to 2010s: This era was characterized by the rise of open science movements seeking to make scientific research more transparent and accessible. It significantly influenced the evolution of the peer review process, making it more inclusive and interdisciplinary, fostering a community where knowledge could be shared and scrutinized more openly and from diverse perspectives.

2020s: The global pandemic heightened the necessity for swift dissemination of research, promoting the idea of preprint servers for rapid publication while emphasizing the critical role of peer review in ensuring the validity of published data. This period saw a renewed focus on balancing speed with scrutiny in the publication process, maintaining a rigorous review mechanism despite urgent demands.

Present and Future Directions

The peer review system is evolving as we traverse the contemporary era, embracing innovations like AI-assisted reviews and collaborative methodologies that promise to refine and expedite the evaluation process. However, these advancements come with challenges, including potential biases in AI algorithms and the necessity to maintain the human element of critical analysis. Platforms like PubPeer are pioneering in fostering ongoing peer reviews and evaluations, extending scholarly dialogue beyond publication, but also necessitate safeguards to prevent the dissemination of unverified information and ensure credible post-publication reviews. Concurrently, efforts are underway to establish a robust recognition and rewards system for peer reviewers, a move aimed at enhancing the scholarly ecosystem's credibility while also requiring vigilant structuring to avoid pitfalls like rushed reviews or fostering competition over collaboration. Navigating this nuanced landscape requires a keen focus on maintaining transparency, inclusivity, and integrity, laying the groundwork for a peer review system that is both adaptive and fortified to cater to the scholarly community's evolving needs and aspirations.

Conclusion: As depicted here, the journey of peer review vividly showcases its evolution from informal dialogues in ancient agoras (public areas) to the structured, collaborative processes that constitute the backbone of modern scientific publishing. This chronicle portrays a rich, vibrant trajectory marked by continuous innovation and adaptation, meeting the changing needs and landscapes of the scientific community. As it embarks on future pathways, the peer review process promises to embody further transformations, leveraging technological advancements and fostering inclusivity and interdisciplinary collaboration. These steps ensure scholarly discourse's rigor, quality, and credibility, nurturing a robust, ever-evolving scientific paradigm.

History of Peer Review's Legacy: The history of peer review showcases a resilient evolution, metamorphosing from informal debates in ancient agoras (public areas) to a cornerstone of modern scientific discourse, continually adapting to foster integrity, inclusivity, and transparency in the collaborative pursuit of knowledge and innovation.

**

REVIEW QUESTIONS

True/False Questions:

1. The peer review process has its origins in informal discussions in ancient Greek agoras.
True or False?

2. According to the chapter, the 1970s saw the introduction of single-blind reviewing methods to minimize potential biases.
True or False?

Multiple-Choice Questions:

3. The chapter highlights that the Royal Society's introduction of a formalized peer review process in 1752 established a benchmark for:
a) Informal discussions in scientific communities
b) Structured and systematic peer review in scientific publishing
c) The elimination of critical analysis in research
d) Reducing collaboration among scientists

4. According to the chapter, a significant development in the peer review process during the 1980s-1990s focused on:
a) Increasing the anonymity of authors
b) Enhancing ethical considerations and addressing conflicts of interest
c) Promoting unverified information dissemination
d) Eliminating the need for peer review in scientific publications

Clinical Vignette:

5. A research team is submitting their study on a novel cancer treatment to a prominent journal. The journal follows a double-blind peer review process, as described in the chapter. What should the research team expect during this review?
a) Reviewers and authors will be aware of each other's identities
b) Only the reviewers will remain anonymous to the authors
c) Both reviewers and authors will remain anonymous to each other
d) The review process will be completely automated without human input

Basic Science Vignette:

6. Dr. Patel is investigating a novel cancer treatment and submits his findings to a prestigious journal. During the peer review process, one reviewer points out a potential flaw in his experimental design. Considering the historical importance of peer review in ensuring scientific integrity, what should Dr. Patel do?
   a) Ignore the reviewer's comments and push for publication
   b) Revise his experimental design and resubmit the manuscript with the necessary corrections
   c) Withdraw the manuscript and abandon the research
   d) Publish his findings on a personal blog without further review

Philosophy Vignette:

7. In a philosophy class, Anna argues that the peer review process is essential for maintaining the integrity and quality of scholarly work. How should she best support her argument using the historical development of peer review?
   a) Peer review has always been perfect and without any need for improvement
   b) Peer review originated in the digital age and has no historical significance
   c) Peer review has evolved from informal discussions in ancient times to a structured process, ensuring rigorous evaluation and credibility in scientific publishing
   d) Peer review prevents the dissemination of new and innovative ideas

Correct Answers:

1. True
2. False
3. b) Structured and systematic peer review in scientific publishing
4. b) Enhancing ethical considerations and addressing conflicts of interest
5. c) Both reviewers and authors will remain anonymous to each other
6. b) Revise his experimental design and resubmit the manuscript with the necessary corrections
7. c) Peer review has evolved from informal discussions in ancient times to a structured process, ensuring rigorous evaluation and credibility in scientific publishing

BEYOND THE CHAPTER
History of Peer Review

- *Peer Review: A Historical and Philosophical Study* by Melinda Baldwin
- *The History and Meaning of Peer Review* by Aileen Fyfe
- *The Development of Peer Review in the 20th Century* by Alex Csiszar

***

Philosophy of Science

Medical Research

WEEK FOUR: Rigor and Evolution

WEEK FOUR

Class Hour 1

- Chapter 16: Peer Review — The Pillar of Rigor: Scientific Integrity
- Chapter 17: P-Hacking — The Threat to Integrity: Ethical Research
- Chapter 18: Research Funding Source — The Ethical Foundation: Scientific Integrity

Class Hour 2

- Chapter 19: AI and Emergent Properties — The Technological Ethos: Rethinking Intelligence
- Chapter 20: Big Data — The Analytical Revolution: Precision and Ethics
- Chapter 21: Bioinformatics — The Computational Bridge: Biological Insights

EXAM I: 50Q - online

Introduction

As we enter the fourth week of our course, "*Philosophy of Science*," we focus on the themes of rigor and evolution in scientific research and practice. This week's chapters guide us through the integral processes and ethical considerations that uphold the integrity of scientific inquiry, exploring the critical roles of peer review, the dangers of p-hacking, the influence of research funding, and the transformative impact of AI, Big Data, and bioinformatics.

We begin with the peer review process, a pillar of scientific rigor that ensures the credibility and quality of research. Peer review, influenced by both rationalist and empiricist philosophies, utilizes deductive and inductive reasoning to validate research findings. It reinforces the scientific method by ensuring investigations are rigorous, replicable, and valid. In medicine, peer review is crucial in shaping clinical practices and healthcare standards, ensuring that published data is credible and ethically sound. By upholding bioethical principles, peer review maintains the trust and integrity of the scientific and medical communities.

However, the integrity of research can be threatened by p-hacking, a practice that manipulates data to produce statistically significant results. P-hacking distorts our understanding and erodes public trust in science by exploiting statistical flexibility and empirical uncertainties. This methodological malfeasance can lead to erroneous conclusions, particularly in medicine, where evidence-based practice is essential. P-hacking compromises ethical principles, violating patient autonomy, beneficence, nonmaleficence, and justice. To combat this, rigorous methodology and ethical vigilance are required to maintain trust in research.

Research funding is another critical component of scientific integrity, balancing financial support with ethical responsibility. Funding sources, including government grants, private foundations, industry contributions, and university support, significantly influence research questions, data interpretation, and findings dissemination. In medicine, research funding advances healthcare standards and reflects the latest research, clinical expertise, and patient needs. However, the ethical implications of funding are profound, affecting the principles of subject autonomy, researcher beneficence, nonmaleficence, and public justice. Researchers and institutions must uphold stringent ethical standards and maintain transparency regarding funding sources to ensure the integrity of the research process.

The convergence of Artificial Intelligence (AI) and emergent properties represents a technological ethos that redefines intelligence and raises ethical dilemmas. AI systems, exhibiting behaviors arising from complex combinations of simpler functions, open new avenues in self-learning machines and human cognition simulations. This technological evolution necessitates a reexamination of the scientific method, medicine, and ethical principles, questioning the rights and responsibilities of quasi-autonomous entities. This transformative interaction requires an interdisciplinary lens to harmonize technology and ethics, fostering a society that reflects the zenith of human intellectual and technological endeavor.

Big Data, a revolutionary force in the digital era, offers remarkable opportunities and complex challenges in research and healthcare. It harnesses immense volumes of data, enhancing decision-making processes and transforming the scientific method. Big Data's influence necessitates a reevaluation of traditional methodologies and raises ethical responsibilities, requiring scrutiny and transparency. In medicine, Big Data drives personalized care and redefines medical standards, bringing ethical dilemmas related to patient autonomy, beneficence, nonmaleficence, and justice. Responsible management is crucial to safeguard ethical integrity and harmonize technological advancements with moral responsibility and societal betterment.

Finally, bioinformatics merges computational and biological sciences, accelerating innovation and ethical discourse in healthcare. This interdisciplinary field extends the scope and precision of the scientific method, comprehending the intricate fabric of biological systems. Its impact is evident in personalized medicine and innovative drug development. Bioinformatics also bears the responsibility of ensuring equitable access to healthcare advancements, safeguarding data privacy, and continually refining ethical frameworks. As a beacon of scientific innovation and ethical vigilance, bioinformatics encapsulates the dynamic intersection of technology, ethics, and societal perspectives on science and healthcare equity.

Together, these chapters illustrate the intricate processes and ethical considerations that uphold scientific rigor and drive the evolution of research and healthcare practices. From peer review to the transformative impacts of AI, Big Data, and bioinformatics, we gain a deeper understanding of the principles that ensure the integrity and ethical foundations of scientific inquiry. This integrated exploration underscores the importance of maintaining rigorous methodologies and ethical standards in the ever-evolving landscape of science and technology.

***

Chapter 16: **Peer Review** — The Pillar of Rigor: Scientific Integrity

Peer review upholds scientific rigor and ethical standards, ensuring credibility and trust in research and clinical practice.

Abstract: The peer review process, a cornerstone of scientific integrity, is critical for maintaining the credibility and quality of research in the public domain. This study explores the influence of rationalist and empiricist philosophies on peer review, emphasizing deductive and inductive reasoning in evaluating research validity and significance. It highlights peer review's role in reinforcing the scientific method, ensuring rigor, replicability, and validity in investigations. In medicine, peer review shapes clinical practices, defines healthcare standards, and influences medical legal frameworks by ensuring the publication of credible data. Additionally, it upholds bioethical principles — patient autonomy (informed consent), practitioner beneficence (do good), practitioner nonmaleficence (do no harm), and public justice (be fair) — as standard considerations in research and publication. Despite its shortcomings, peer review is indispensable for preserving the credibility and ethical integrity of scientific and medical communities, serving as a foundational system of checks and balances pivotal to advancing and ensuring the reliability of human knowledge.

**

Introduction: Peer review is a cornerstone of scientific integrity, a rigorous evaluation process that serves as a quality filter for research before reaching the public domain. In this system, researchers submit their findings to scholarly journals, where independent experts in the same field assess the work's validity, significance, and originality. By doing so, peer review ensures that published studies are methodologically sound, logically consistent, and contribute substantively to existing knowledge. Academic journals have adopted this vetting process, integral to research proposals, especially those seeking funding from governmental and private institutions. While it's not without its criticisms — including the potential for bias, inefficiency, and lack of transparency — the peer review system remains an essential mechanism for maintaining the quality and credibility of scientific research. As we explore the role of peer review, we will examine its roots in rationalist and empiricist thought, its pivotal role in shaping the scientific method, its indispensable contributions to the field of medicine, and its influence on the ethical principles of bioethics: patient autonomy (informed consent), practitioner beneficence (do good), practitioner nonmaleficence (do no harm), and public justice (be fair).

Rationalism: The foundations of peer review are deeply rooted in rationalist philosophy, which emphasizes the role of reason as a source of knowledge. In the context of peer review, rationalism manifests through the evaluative processes guided by innate cognitive faculties like deductive reasoning, pattern recognition, and problem-solving. When experts review a manuscript or research proposal, they engage in deductive analysis, drawing on established axioms and principles within their field to assess the validity and integrity of the presented work. This reliance on "*a priori*" principles, such as the scientific method or certain fundamental laws of a discipline, provides a framework within which reviewers judge the appropriateness of the research design, the coherence of the argument, and the soundness of the conclusions. The ideal of unbiased, logical scrutiny stems from the rationalist conviction that human reason can reliably distinguish between valid and invalid claims when properly applied. Thus, peer review is a real-world application of rationalist ideals to eliminate errors, inconsistencies, and biases, ensuring that the knowledge produced is as close to objective truth as possible.

Empiricism: The practice of peer review is also deeply influenced by empiricism, which recognizes sensory experience, observation, and experiment as the fundamental sources of knowledge. In the peer review process, the empirical orientation manifests primarily through inductive reasoning, whereby reviewers scrutinize the data, methodologies, and results presented in the manuscript to arrive at a generalized assessment of its quality and significance. Specifically, reviewers focus on the adequacy of the data collection process, the robustness of the experiments, and the veracity of the observations, all of which are central to empiricism. Empirical rigor is essential in scientific disciplines, such as natural sciences and medicine, where experiments, observations, and quantitative analysis are the norm. The scrutiny is not merely a theoretical exercise but an interrogation of how well the research holds up under practical, real-world conditions, ensuring that claims are rooted in tangible evidence. By stressing empirical validation, peer review serves as a mechanism to sift out conjecture, unsubstantiated hypotheses, and poorly supported conclusions, aiming to ensure that any generalizations made in published work have been derived from solid, specific, and repeatable observations. Therefore, peer review embodies the empiricist commitment to grounding knowledge in empirical evidence, contributing to a more accurate and reliable scientific discourse.

The Scientific Method: Peer review plays a pivotal role in shaping and upholding the standards of the scientific method, serving as a critical checkpoint that ensures the rigor, validity, and replicability of scientific research. Peer review acts as a quality control mechanism at each stage of the scientific process — from hypothesis formulation to experimental design, data collection, and analysis. Researchers know that their work will undergo stringent evaluation by experts, thus incentivizing meticulous adherence to the principles of scientific inquiry. The scrutiny doesn't just stop at evaluating the appropriateness of the chosen methodologies or the conclusions' coherence; it extends to assessing whether the research is replicable, a cornerstone of the scientific method. If a study's findings can't be reproduced under the same conditions, its validity is questioned, and it is unlikely to pass the peer review process. Moreover, peer review often triggers further inquiry, inspiring additional hypotheses and experiments that deepen collective understanding. In essence, peer review is the mechanism by which the scientific community self-regulates, ensuring that published research contributes to existing knowledge and stands up to the foundational principles of scientific inquiry. Through this iterative process, peer review helps to refine and evolve the scientific method itself, strengthening its role as the gold standard for acquiring reliable and accurate knowledge.

Medicine: Peer review has an invaluable role in shaping the field of medicine, fundamentally influencing both clinical practice and medical research. It is the bedrock for evidence-based medicine, setting medical healthcare standards of care and protocols defining competent and ethical medical practices. The importance of robust, peer-reviewed research cannot be overstated in a discipline where lives are at stake. Medical guidelines, treatment protocols, and pharmaceutical approvals often hinge on clinical trials and studies that have undergone rigorous peer review. This ensures that conclusions are based on credible, empirically substantiated data, influencing the "*standards of care*" in medical healthcare practice. These medical healthcare standards of care, recognized and practiced across the medical community, are dynamic, evolving with each newly published, peer-reviewed study. They serve as both best practice guidelines and legal yardsticks for evaluating the adequacy and ethics of care delivered to patients. By fostering an environment of scrutiny and validation, peer review elevates the credibility of medical research, ensures the efficacy and safety of new treatments, and helps to eliminate or minimize medical errors. Thus, peer review is a critical mechanism that safeguards the integrity of medical science, underpinning the evolution of healthcare practices to meet patient needs better while influencing legal frameworks related to medical accountability.

Ethics: Peer review holds a substantial influence over the ethical framework that governs medical and scientific research, particularly in aligning with the four principles of bioethics: patient autonomy (informed consent), practitioner beneficence (do good), practitioner nonmaleficence (do no harm), and public justice (be fair). By rigorously scrutinizing research methodologies and protocols, peer review ensures that studies requiring informed consent are ethically designed and executed, thereby upholding the principle of autonomy. Peer review also serves the principle of beneficence by rigorously evaluating a study's scientific merit and potential societal benefits, ensuring that it contributes positively to human well-being and scientific understanding. The principle of nonmaleficence is safeguarded as studies are meticulously reviewed for safety protocols, ensuring that the risks are minimized and that the research adheres to the dictum of "*do no harm*." Lastly, the principle of justice is served by ensuring that research is equitable and non-discriminatory, can be generalizable across different populations, and doesn't exploit or disproportionately burden any group. In this way, the peer review process serves as an ethical sieve, ensuring that only research that adheres to these bioethical principles sees the light of day. It institutionalizes these ethical considerations, making the principles a standard part of the research and publication process and reinforcing the moral underpinnings of scientific inquiry and medical practice.

Conclusion: Peer review is a linchpin in the pursuit of knowledge, acting as a formidable gatekeeper ensuring the integrity, rigor, and ethical soundness of scientific and medical research. Rooted in rationalist and empiricist philosophical traditions, it leverages human cognitive faculties like deductive and inductive reasoning to evaluate the validity and significance of submitted works. Beyond theoretical scrutiny, peer review fortifies the scientific method, imposing quality controls that drive researchers to meet its stringent criteria, from formulating hypotheses to ensuring the replicability of experiments. In medical healthcare, peer review has been indispensable in elevating standards of care and influencing clinical practices and healthcare policies based on empirical evidence. It establishes best practices, ensures patient safety, and has legal implications for medical accountability. Ethically, it reinforces the core principles of bioethics — patient autonomy (informed consent), practitioner beneficence (do good), practitioner nonmaleficence (do no harm), and public justice (be fair) — by enforcing guidelines that make ethical considerations an integral part of research and publication. Despite its imperfections, such as potential bias and inefficiencies, peer review remains an irreplaceable safeguard that sustains scientific and medical communities' credibility and ethical integrity. Through its multifaceted roles, peer review embodies a system of checks and balances foundational to human knowledge's progress and reliability.

Peer Review's Legacy: As a cornerstone of scientific and academic rigor, peer review embodies the pursuit of objectivity and ethical integrity, shaping the fabric of credible, honest, and replicable research in myriad disciplines.

**

REVIEW QUESTIONS

True/False Questions:

1. The peer review process ensures that published studies are methodologically sound, logically consistent, and contribute substantively to existing knowledge.
   True or False?

2. According to the chapter, peer review primarily focuses on theoretical scrutiny without considering empirical validation.
   True or False?

Multiple-Choice Questions:

3. The chapter highlights that rationalism influences peer review through:
   a) Sensory experiences and observations
   b) Deductive reasoning and innate cognitive faculties
   c) Random guesswork
   d) Ignoring established axioms and principles

4. In the context of medicine, peer review primarily contributes to:
   a) Outdated clinical practices
   b) The development of evidence-based standards of care
   c) Promoting untested medical procedures
   d) Ignoring empirical data

Clinical Vignette:

5. A researcher submits a study on a new surgical technique to a peer-reviewed medical journal. The journal follows a rigorous peer review process. Based on the chapter, what should the researcher expect during this review?
   a) The reviewers will ignore the study's methodology and focus only on the results
   b) The reviewers will evaluate the study's empirical data and experimental design
   c) The reviewers will ensure that the study has not undergone previous peer review
   d) The reviewers will only assess the study based on theoretical models without considering empirical evidence

Basic Science Vignette:

6. Dr. Green submits a manuscript detailing her findings on a new antibiotic's effectiveness against resistant bacteria. During the peer review process, reviewers identify several methodological issues and suggest additional experiments. Considering the role of peer review in ensuring scientific integrity, what should Dr. Green do?
   a) Ignore the reviewers' suggestions and resubmit the manuscript unchanged
   b) Address the methodological issues and conduct the additional experiments before resubmitting
   c) Withdraw the manuscript and abandon the research project
   d) Publish her findings on social media without further review

Philosophy Vignette:

7. In a philosophy seminar, Michael argues that peer review is essential for upholding the ethical integrity of scientific research. How should he best support his argument using the principles of bioethics?
   a) Peer review prevents researchers from exploring controversial topics
   b) Peer review ensures that research adheres to ethical principles like patient autonomy, beneficence, nonmaleficence, and justice
   c) Peer review is only concerned with the technical aspects of research and not its ethical implications
   d) Peer review has no impact on the ethical standards of scientific research

Correct Answers:

1. True
2. False
3. b) Deductive reasoning and innate cognitive faculties
4. b) The development of evidence-based standards of care
5. b) The reviewers will evaluate the study's empirical data and experimental design
6. b) Address the methodological issues and conduct the additional experiments before resubmitting
7. b) Peer review ensures that research adheres to ethical principles like patient autonomy, beneficence, nonmaleficence, and justice

BEYOND THE CHAPTER
Peer Review

- *Peer Review: Its Strengths and Weaknesses* by David Shatz
- *Peer Review and Manuscript Management in Scientific Journals: Guidelines for Good Practice* edited by Irene Hames
- *Scientific Peer Review: Guidelines for Informative Peer Review* by Robert J. AmdurChapter

***

Philosophy of Science

Medical Research

Chapter 17: P-Hacking — The Threat to Integrity: Ethical Research

P-hacking undermines scientific integrity, demanding rigorous methodology and ethical vigilance to maintain trust in research.

Abstract: P-hacking, or data dredging, is a methodological malfeasance that undermines scientific integrity by manipulating data until nonsignificant results appear statistically significant. This practice exploits statistical flexibility and empirical uncertainties to produce misleading findings, distorting our understanding and eroding public trust in science. The roots of p-hacking are tied to the philosophical dichotomies of rationalism and empiricism, where cognitive biases and ethical lapses can corrupt the quest for knowledge. P-hacking threatens the reliability and reproducibility of findings, leading to erroneous conclusions, especially in fields like medicine. In medicine, where evidence-based practice is crucial, p-hacking can result in ineffective or harmful treatments. Ethically, p-hacking violates the principles of autonomy (informed consent), beneficence (do good), nonmaleficence (do no harm), and justice (be fair), compromising patient care and equitable healthcare resource distribution. Addressing p-hacking requires methodological rigor, ethical vigilance, and systemic reforms to uphold scientific inquiry integrity.

**

Introduction: P-hacking refers to the manipulation of data analysis to achieve statistically significant results, often by selectively reporting data, testing multiple hypotheses, or stopping data collection once significant results are obtained. This practice is driven by the pressure to publish positive findings and the competitive nature of academic research. The prevalence of p-hacking undermines the credibility of scientific literature, leading to a replication crisis where many published findings cannot be reproduced. Understanding p-hacking requires examining the interplay between rationalist and empiricist philosophies, the principles of the scientific method, and the implications for medical practice and ethics. By dissecting these dimensions, we can better appreciate the significance of addressing p-hacking and develop strategies to mitigate its impact on scientific and medical fields.

Rationalism: In the realm of rationalism, p-hacking represents a departure from the rigorous application of logical principles and axiomatic truths. Rationalism posits that knowledge is derived from reason and innate cognitive faculties, which should guide our interpretation of empirical data. However, the cognitive biases inherent in human reasoning, such as confirmation bias and the desire for positive results, can lead to p-hacking. These biases distort our rational faculties, causing researchers to subconsciously manipulate data to align with preconceived notions or desired outcomes. This manipulation contradicts the rationalist ideal of objective truth-seeking, highlighting the need for self-awareness and methodological rigor to counteract these biases. For example, in the early 2000s, a study suggested a link between MMR vaccines and autism. The researchers selectively reported data that supported their hypothesis while ignoring data that did not, a clear instance of p-hacking. This led to widespread public fear and a decline in vaccination rates, despite subsequent studies debunking the original findings. Acknowledging and addressing these cognitive pitfalls is crucial for maintaining the integrity of rationalist approaches to scientific inquiry and ensuring that our innate faculties are employed in the pursuit of genuine knowledge.

Empiricism: Empiricism, which emphasizes knowledge acquisition through sensory experiences and observations, is particularly vulnerable to p-hacking. The empirical method relies on inductive reasoning, where patterns observed in data are generalized to form broader conclusions. However, the flexibility of data analysis and the abundance of potential variables provide ample opportunities for p-hacking. Researchers may selectively report results, test multiple hypotheses, or manipulate data collection processes to achieve significant findings. These practices exploit the inherent uncertainties of empirical data, leading to conclusions that are not truly reflective of the observed phenomena. For example, in nutrition science, studies often test multiple dietary factors against numerous health outcomes. Researchers may only report significant results, ignoring the myriad of tests that showed no significant relationships, thus misleadingly suggesting strong dietary impacts on health. To uphold the principles of empiricism, it is essential to implement robust methodological safeguards, such as preregistration of studies, transparency in data reporting, and replication efforts. By reinforcing the empirical foundation with stringent methodological practices, we can mitigate the impact of p-hacking and ensure that our observations lead to accurate and reliable knowledge.

The Scientific Method: The scientific method is designed to systematically investigate phenomena and generate reliable knowledge through observation, experimentation, and hypothesis testing. P-hacking, however, subverts this process by prioritizing statistical significance over scientific validity. By manipulating data to achieve significant results, researchers undermine the reproducibility of experiments and the reliability of findings. This practice contributes to the replication crisis, where many published studies cannot be replicated, casting doubt on the validity of scientific literature. A notable historical example is the field of psychology, where many high-profile studies, such as the infamous "*power pose*" research, have failed replication efforts, revealing the prevalence of p-hacking and other questionable research practices. To combat p-hacking, the scientific community must emphasize the importance of methodological rigor, transparency, and accountability. Practices such as preregistration of studies, open data initiatives, and rigorous peer review can help ensure that research findings are robust and reproducible. By adhering to the principles of the scientific method and addressing the systemic pressures that drive p-hacking, we can safeguard the integrity of scientific inquiry.

Medicine: In medicine, where evidence-based practice is critical for patient care, p-hacking poses significant risks. Medical research relies on the accurate interpretation of data to develop treatments, inform clinical guidelines, and improve patient outcomes. P-hacking can lead to the approval of ineffective or harmful treatments, as manipulated data may falsely indicate efficacy or safety. This undermines the trust in medical research and can result in adverse patient outcomes. For instance, the infamous Vioxx scandal involved the pharmaceutical company Merck manipulating data to downplay the cardiovascular risks associated with the drug. This resulted in widespread use of a medication that ultimately caused significant harm, highlighting the dangers of p-hacking in medical research. To address p-hacking in medical research, it is essential to implement stringent regulatory standards, promote transparency in data reporting, and encourage the replication of studies. By fostering a culture of integrity and accountability in medical research, we can ensure that treatments are based on genuine evidence and that patient care is optimized.

Ethics: The ethical implications of p-hacking are profound, as this practice violates the core principles of medical ethics. Beneficence, the commitment to doing good, is compromised when treatments are based on manipulated data, potentially leading to ineffective or harmful interventions. Autonomy, the principle that emphasizes informed consent, is particularly impacted by p-hacking. When patients are provided with information based on manipulated data, their ability to make informed decisions about their care is compromised. A historical example is the promotion of hormone replacement therapy (HRT) for menopausal women, which was based on selectively reported data suggesting it had broad health benefits. Later, more rigorous studies revealed that HRT increased the risk of heart disease and cancer, showing that earlier data manipulation had compromised patient autonomy and safety. This was a violation of nonmaleficence, the principle of avoiding harm, as the p-hacking resulted in adverse patient outcomes. Justice, which emphasizes fairness in the distribution of healthcare resources, was also undermined because the treatments were based on flawed evidence, leading to inequitable access to effective care. To uphold these ethical principles, researchers must prioritize methodological integrity and transparency. Ethical oversight, such as institutional review boards and peer review processes, plays a crucial role in preventing p-hacking and ensuring that research adheres to ethical standards. By fostering a culture of ethical vigilance, we can protect the integrity of medical research and uphold the principles of autonomy, beneficence, nonmaleficence, and justice.

Conclusion: P-hacking represents a significant threat to the integrity of scientific research, with far-reaching implications for knowledge acquisition, medical practice, and ethics. By manipulating data to achieve significant results, researchers undermine the reliability of findings, contribute to the replication crisis, and compromise patient care. Addressing p-hacking requires a multifaceted approach, including methodological rigor, transparency, ethical oversight, and systemic reforms. By reinforcing the principles of rationalism and empiricism, adhering to the scientific method, and upholding ethical standards, we can mitigate the impact of p-hacking and ensure that scientific inquiry remains a trustworthy and reliable pursuit of knowledge. Through collective efforts, we can safeguard the integrity of research and maintain public trust in science and medicine.

P-Hacking's Legacy: The detrimental legacy of p-hacking is a profound erosion of scientific integrity, leading to unreliable research findings, compromised medical practices, and a significant loss of public trust in scientific endeavors.

**

REVIEW QUESTIONS

True/False Questions:

1. P-hacking involves manipulating data to achieve statistically significant results, often by selectively reporting data or testing multiple hypotheses.
   True or False?

2. According to the chapter, p-hacking supports the ethical principles of beneficence and nonmaleficence in medical research.
   True or False?

Multiple-Choice Questions:

3. The chapter highlights that p-hacking distorts the rationalist ideal of:
   a) Objective truth-seeking and logical principles
   b) Sensory experiences and observations
   c) Random data collection
   d) Cultural influences on research

4. In the context of medicine, p-hacking primarily leads to:
   a) The approval of effective treatments
   b) Misleading conclusions and potentially harmful treatments
   c) Increased transparency in data reporting
   d) Enhanced patient care and outcomes

Clinical Vignette:

5. A pharmaceutical company is accused of p-hacking in a study on a new medication. They selectively reported data to show significant results while ignoring nonsignificant findings. Based on the chapter, what is a likely consequence of this p-hacking practice?
   a) The medication will be quickly approved and widely accepted
   b) The medication might receive regulatory approval but later be found ineffective or harmful
   c) The study will be praised for its thorough methodology
   d) The medication will be immediately rejected without further review

Basic Science Vignette:

6. Dr. Johnson is conducting a study on the effectiveness of a new drug for treating hypertension. To achieve statistically significant results, he tests multiple hypotheses and selectively reports only the significant outcomes. What should Dr. Johnson do to avoid p-hacking and ensure the integrity of his research?
   a) Continue reporting only the significant results to highlight the drug's effectiveness
   b) Preregister his study design and hypotheses, and report all outcomes transparently, including nonsignificant results
   c) Stop data collection once he obtains significant results
   d) Publish the significant results without peer review to expedite the process

Philosophy Vignette:

7. In her philosophy class, Sarah argues that p-hacking undermines the ethical principles of scientific research. How should she best support her argument using the principles of bioethics?
   a) P-hacking ensures more research gets published, which is beneficial
   b) P-hacking compromises informed consent, beneficence, nonmaleficence, and justice by producing misleading data
   c) P-hacking has no impact on the ethical standards of scientific research
   d) P-hacking ensures that researchers meet publication deadlines, which is more important than ethical considerations

Correct Answers:

1. True
2. False
3. a) Objective truth-seeking and logical principles
4. b) Misleading conclusions and potentially harmful treatments
5. b) The medication might receive regulatory approval but later be found ineffective or harmful
6. b) Preregister his study design and hypotheses, and report all outcomes transparently, including nonsignificant results
7. b) P-hacking compromises informed consent, beneficence, nonmaleficence, and justice by producing misleading data

BEYOND THE CHAPTER

P-Hacking

- *The Cult of Statistical Significance: How the Standard Error Costs Us Jobs, Justice, and Lives* by Stephen T. Ziliak and Deirdre N. McCloskey
- *Rigor Mortis: How Sloppy Science Creates Worthless Cures, Crushes Hope, and Wastes Billions* by Richard Harris
- *The Seven Deadly Sins of Psychology: A Manifesto for Reforming the Culture of Scientific Practice* by Chris Chambers
- *Science Fictions: How Fraud, Bias, Negligence, and Hype Undermine the Search for Truth* by Stuart Ritchie

***

Philosophy of Science — Medical Research

Chapter 18: Research Funding Source — The Ethical Foundation: Scientific Integrity

Research funding underpins scientific integrity, balancing financial support with ethical responsibility.

Abstract: Research funding is a multifaceted domain integral to the trajectory of scientific inquiry and innovation, shaping the scientific method, medicine, and ethical principles in research. By analyzing the intricate interplay between financial support and the scientific ecosystem, we aim to uncover the impacts of funding sources on research, including government grants, private foundations, industry contributions, and university support. These funding streams influence the formulation of research questions, data interpretation, and findings dissemination. In medicine, research funding is vital for advancing medical healthcare standards of care, reflecting the latest research, clinical expertise, and patient needs. The ethical implications of funding are profound, affecting the bioethical principles of subject autonomy (informed consent), researcher beneficence (do good), researcher nonmaleficence (do no harm), and public justice (be fair), with potential risks of compromising these principles in pursuit of favorable results for funding agencies. Researchers, institutions, and policymakers must uphold stringent ethical standards and maintain transparency regarding funding sources, ensuring the integrity of the research process and equitable distribution of scientific discoveries' benefits.

**

Introduction: The landscape of research funding is a multifaceted domain that plays a pivotal role in shaping the trajectory of scientific inquiry and innovation. From government grants and private foundations to industry contributions and university funding, the sources of research financing are diverse, each bringing unique opportunities and challenges. These funding streams are not merely financial lifelines; they are integral actors that actively shape the fabric of the scientific endeavor, influencing everything from formulating research questions to interpreting and disseminating findings. By dissecting the intricacies of research funding sources, we can understand their rationalist inclinations, empiricist orientations, and the extent to which they mold the scientific method. Additionally, the relationship between research funding and its contributions to medicine is an essential area of inquiry, as is the impact of the financing on the ethical principles underpinning bioethics, namely subject autonomy (informed consent), researcher beneficence (do good), researcher nonmaleficence (do no harm), and public justice (be fair). Through this analysis, we aim to shed light on the complex interplay between financial support and the scientific ecosystem, ultimately unraveling the intricate tapestry that weaves together the pursuit of knowledge, medical advancement, and ethical considerations.

Rationalism: In research funding, rationalism embodies a structured, top-down approach where funding decisions are guided by predefined principles and strategic priorities. Government grants, private foundations, and industry contributions often set broad goals based on theoretical and axiomatic considerations, such as national interests, philanthropic missions, or business objectives. These "*a priori*" principles dictate the formulation of research questions and the allocation of resources, ensuring that funded projects align with the overarching goals of the funding bodies. This approach can shape data interpretation and findings dissemination, as researchers may feel pressured to produce results that meet the expectations of their funders. Ethically, rationalist funding can enhance beneficence by directing resources toward projects with significant potential benefits, but it may also compromise autonomy by limiting researchers' freedom to explore unconventional topics. Nonmaleficence can be jeopardized if the pressure to produce favorable results leads to overlooking potential harms, and justice might be at risk if funding priorities favor certain populations or issues disproportionately.

Empiricism: In research funding, empericism highlights a bottom-up approach, where decisions are driven by empirical evidence and the demonstrated success of research projects. Universities, competitive grant processes, and crowdfunding platforms exemplify this approach, relying on observed data and performance metrics to allocate resources. Funding bodies prioritize projects with strong empirical foundations, emphasizing methodologies and techniques that ensure robust, reproducible results. This empirical orientation promotes transparency and rigorous peer review, fostering an environment where evidence-based research thrives. By valuing observational sense data and inductive reasoning, empiricist funding sources ensure that research is grounded in reliable, verifiable evidence, thereby upholding the integrity and credibility of the scientific method. Ethically, this approach supports autonomy by allowing researchers to follow the evidence wherever it leads and promotes beneficence and nonmaleficence by prioritizing projects with proven effectiveness and safety. However, it may also pose challenges to justice if funding decisions are swayed by the immediate popularity or feasibility of projects, potentially sidelining important but less empirically established research areas.

The Scientific Method: Research funding significantly shapes the scientific method, influencing every stage of the inquiry process. The phenomena observed and researched are often dictated by funding availability, with projects aligning with the interests and objectives of funding agencies more likely to receive support. This financial backing, in turn, affects the questions posed by researchers and the hypotheses formulated, often favoring those aligned with the funder's goals. Moreover, the availability and allocation of funds impact data acquisition and analysis, potentially influencing the quality and integrity of research. This influence extends to the peer-review process, which can be affected by financial ties between journals and funding sources, potentially leading to conflicts of interest. Adequate funding is also crucial for ensuring the reproducibility of research, a cornerstone of the scientific method, as it allows other researchers to replicate experiments. This reliance on funding has its drawbacks. The field of research can become skewed, with more attention given to topics that are more likely to receive funding, leaving other important questions unexplored. Research questions and hypotheses may also become biased, serving the interests of funding agencies or industry partners rather than the scientific community or the public. Experimental designs may be influenced to produce results favorable to the funding source, potentially leading to biased outcomes. Similarly, the data acquisition process may be shaped by the funding source's requirements, affecting the data's quality and integrity. The analysis stage can also be influenced by funding, with researchers feeling pressured to produce results that are favorable to the funding source, potentially leading to biased or manipulated data analysis. Lastly, the availability of resources for reproducibility can be impacted by funding, especially in industry-funded research, which may be less openly available for replication by other researchers. Despite these challenges, it is vital for researchers to maintain transparency regarding their funding sources and actively manage potential conflicts of interest to uphold the scientific method's integrity.

Medicine: Research funding is instrumental in advancing medicine, directly contributing to the evolution of medical healthcare standards of care. As standards of care are established based on peer-reviewed, evidence-based practices widely recognized within the medical community, funding becomes crucial in facilitating the necessary research that informs these standards. Government grants, private foundations, industry funding, and university support are instrumental in driving forward groundbreaking medical research, ultimately refining and redefining standards of care. This iterative process ensures that healthcare standards are not static but continually evolving entities that reflect current best practices. These best practices are informed by a confluence of research, clinical expertise, and patient needs, all of which are intricately linked to the availability and allocation of research funds. Furthermore, these standards of care are pivotal in legal contexts, serving as benchmarks to assess the adequacy of care provided by healthcare professionals. In essence, research funding is the lifeblood of medical advancement, fostering a dynamic and responsive healthcare landscape that continually seeks to enhance patient outcomes and uphold the highest medical healthcare standards of care.

Ethics: Research funding has a profound impact on the bioethical principles of patient autonomy (informed consent), practitioner beneficence (do good), practitioner nonmaleficence (do no harm), and public justice (be fair). The funding source can influence how much informed consent is prioritized and respected, as some funders may have different ethical standards or requirements. For example, government or non-profit-funded research often has strict guidelines ensuring informed consent, whereas industry-funded research might be more relaxed. Beneficence and nonmaleficence, the principles of doing good and avoiding harm, can also be compromised by the influence of funding sources. The drive to produce favorable results for a funding agency, especially in the case of industry funding, can sometimes lead researchers to overlook potential harms or overstate the benefits of their findings. Justice, the principle of fairness, is also at risk when research funding comes into play. Access to research and its benefits may be skewed in favor of those who can gain the most financially rather than being distributed fairly according to need. While funding is essential for advancing research, researchers and institutions must uphold ethical standards and maintain transparency about their funding sources to preserve the integrity of the research and respect the bioethical principles that are fundamental to human dignity and rights.

Conclusion: The landscape of research funding wields considerable influence over the trajectory of scientific inquiry and medical advancements, often serving as the bedrock upon which groundbreaking discoveries and innovations are built. Through its various sources, including government grants, private foundations, industry contributions, and university funding, research funding is a catalyst that fuels the scientific endeavor, shaping the way research questions are formulated, data is interpreted, and findings are disseminated. The contributions of research funding to medicine are manifold, directly influencing the evolution of medical healthcare standards of care and ensuring that they reflect the latest research, clinical expertise, and patient needs. Furthermore, the impact of research funding on the ethical principles of bioethics is profound, with funding sources playing a crucial role in upholding or potentially compromising the principles of subject autonomy (informed consent), researcher beneficence (do good), researcher nonmaleficence (do no harm), and public justice (be fair). As such, it is paramount that researchers, institutions, and policymakers work collaboratively to uphold stringent ethical standards and maintain transparency regarding funding sources, thereby safeguarding the integrity of the research process and ensuring that the benefits of scientific discoveries are equitably distributed.

Research Funding Source's Legacy: A foundational pillar that not only fuels the progression of scientific inquiry and medical advancements but also shapes the ethical landscape of research, influencing everything from formulating research questions to disseminating findings and upholding bioethical principles.

**

REVIEW QUESTIONS

True/False Questions:

1. Research funding sources such as government grants, private foundations, industry contributions, and university support significantly influence the formulation of research questions and data interpretation.
   True or False?

2. Industry-funded research always adheres to the same ethical standards and requirements for informed consent as government or non-profit-funded research.
   True or False?

Multiple-Choice Questions:

3. Which of the following is NOT a potential impact of research funding on the scientific method?
   a) Influencing the questions posed by researchers
   b) Affecting data acquisition and analysis
   c) Ensuring all research is free from bias
   d) Impacting the reproducibility of research

4. What is a significant ethical concern related to research funding in medicine?
   a) Standardizing treatment protocols
   b) Ensuring equitable access to research benefits
   c) Implementing uniform data collection methods
   d) Ignoring funding sources in peer-reviewed publications

Clinical Vignette:

5. A pharmaceutical company funds a clinical trial for a new drug. The researchers involved in the study are aware that favorable results could lead to more funding from the company. During the trial, some negative side effects are observed, but the researchers choose to emphasize the positive outcomes in their report. Which ethical principle is primarily being compromised in this scenario?
   a) Ensuring patient autonomy through informed consent
   b) Promoting practitioner beneficence by prioritizing patient care
   c) Avoiding harm by considering nonmaleficence
   d) Ensuring public justice by providing equitable access to treatments

Basic Science Vignette:

6. A university research team receives funding from a private pharmaceutical company to study a new drug. The company stipulates that the researchers must produce favorable results to secure continued funding. This stipulation influences the team's experimental design and data interpretation. What ethical principle is most directly compromised in this scenario?
   a) Autonomy (informed consent)
   b) Beneficence (do good)
   c) Nonmaleficence (do no harm)
   d) Justice (fair distribution of benefits and burdens)

Philosophy Vignette:

7. A government-funded research project aims to develop a new cancer treatment. The funding agency requires the researchers to publish all findings, regardless of the results. This policy promotes transparency and prevents bias but also raises concerns about patient privacy. Which principle of bioethics is most directly supported by this funding requirement?
   a) Autonomy (informed consent)
   b) Beneficence (do good)
   c) Nonmaleficence (do no harm)
   d) Justice (fair distribution of benefits and burdens)

Correct Answers:

1. True
2. False
3. c) Ensuring all research is free from bias
4. b) Ensuring equitable access to research benefits
5. c) Avoiding harm by considering nonmaleficence
6. b) Beneficence (do good)
7. d) Justice (fair distribution of benefits and burdens)

BEYOND THE CHAPTER
Research and Funding Source Influence

- *The Commercialization of Science and the Response of STS* in Handbook of Science and Technology Studies (pp. 635-689). MIT Press.
- *The Price of Truth: How Money Affects the Norms of Science* by Mirowski, P., & Sent, E. M.
- *Science in the Private Interest: Has the Lure of Profits Corrupted Biomedical Research?* by Krimsky, S. (2003).
- *Extending the Mertonian Norms: Scientists' Subscription to Norms of Research* The Journal of Higher Education, 78(3), 366-393, by Anderson, M. S., Ronning, E. A., De Vries, R., & Martinson, B. C. (2007).
- *Lessons from the Vioxx Debacle: What the Privatization of Science Can Teach Us About Social Epistemology*Social Epistemology, 21(1), 21-39, by Biddle, J. (2007).

***

Philosophy of Science

Medical Research

Chapter 19: AI and Emergent Properties — The Technological Ethos: Rethinking Intelligence

AI and emergent properties redefine intelligence, blending technological innovation with ethical contemplation.

Abstract: In the era of unprecedented technological advancements, the confluence of Artificial Intelligence (AI) and emergent properties opens new vistas that transcend computational boundaries, reaching into philosophy, ethics, and the nature of intelligence and life. This synthesis creates a landscape where AI systems exhibit behaviors not directly coded but arising from complex combinations of simpler functions, unveiling new avenues in self-learning machines and human cognition simulations. However, these emergent properties also raise ethical and philosophical dilemmas, questioning the rights and responsibilities of quasi-autonomous entities and the moral guidelines governing them. As we engage in this transformative interaction, reexamining the scientific method, medicine, and ethical principles becomes vital. Through an interdisciplinary lens, this investigation seeks to unravel the evolving relationship between AI and emergent properties, aiming for a future where technology and ethics harmoniously co-evolve, fostering a society that reflects the zenith of human intellectual and technological endeavor.

**

Introduction: The intriguing relationship between artificial intelligence (AI) and emergent properties extends beyond mere computation, spiraling into philosophy, ethics, and even the fabric of our understanding of intelligence and life. As AI systems grow more complex and autonomous, they begin to exhibit behavior that is not explicitly programmed but rather arises from the intricate interplay between myriad simpler functionalities. This phenomenon of new, unexpected characteristics or behaviors emerging from complex systems is known as "*emergent properties*," a concept deeply rooted in systems theory, biology, and social sciences. While these emergent properties promise groundbreaking innovations — from self-learning machines to intricate simulation models mimicking human cognition — they pose a Pandora's box of ethical and philosophical problems. Questions around the rights and responsibilities of these quasi-autonomous entities, the ethical guidelines that should govern their operation, and the implications for our broader understanding of concepts like consciousness and morality become ever more pressing. We will also delve into concrete examples and emerging trends that illuminate the complexities of this transformative interaction between AI and emergent properties. In assessing the impact of AI through lenses such as rationalism and empiricism, the scientific method, contributions to medicine, and ethical principles like patient autonomy (informed consent), practitioner beneficence (do good), practitioner nonmaleficence (do no harm), and public justice (be fair), we may find that AI becomes a mirror reflecting humanity's most profound questions right back at us.

Rationalism: In the context of "*AI and Emergent Properties*," rationalism takes the form of pre-defined algorithms, architectures, or initial conditions that guide the operation and behavior of artificial systems. One concrete example that embodies this principle is the realm of Autonomous Vehicles. In such scenarios, predefined algorithms for handling weather conditions and lane-keeping operate fine when vieccxcwed in isolation. However, when these algorithms interact, the vehicle begins to exhibit emergent behaviors that could be more predictable. For instance, an algorithm designed to maintain lane integrity during high winds might interact in unforeseen ways with another algorithm aimed at navigating slippery roads. These foundational elements are akin to "*a priori*" principles in rationalist philosophy, serving as non-empirical starting points from which the system's behavior and responses are deduced. The emergence of unanticipated behaviors or characteristics from such complex integrations becomes a profound phenomenon, complicating ethical oversight and policy regulations. It suggests that complex, novel forms of "*intelligence*" or problem-solving can arise even within a framework that starts with a rationalist starting point — a carefully designed set of algorithms. This emergence necessitates stringent ethical oversight at multiple levels. Individually, developers bear the moral burden to ensure their creations are free from biases and capable of operating justly. Organizations, particularly tech giants steering the AI revolution, are professionally obliged to maintain high ethical standards, scrutinizing their algorithms for any unintended, emergent properties that may perpetuate biases or cause harm. Additionally, policymakers must ensure transparent and accountable governance frameworks that adapt to the unique challenges of emergent behaviors in AI systems. An emerging trend that holds promise in this regard is Explainable AI (XAI), which aims to make the decision-making process of complex algorithms understandable to humans. XAI could serve as a vital tool for ethical oversight, allowing for a clearer understanding of how emergent behaviors originate and how they could be managed or mitigated, thereby reconciling rationalist foundations with complex artificial intelligence's unpredictable, dynamic nature.

Empiricism: Within the framework of "*AI and Emergent Properties*," empiricism emphasizes the role of training data as the linchpin for machine learning models. Take, for example, healthcare algorithms, which are only as good as the data they've been trained on. While these algorithms are programmed for specific tasks, they can sometimes identify patterns or make decisions outside their explicit programming due to emergent properties, adding another layer of complexity to their empirical foundations. Unlike rationalist perspectives prioritizing foundational algorithms and architecture, an empiricist orientation leans heavily on inductive reasoning, wherein specific data instances are used to construct general principles or models. This reliance on data becomes even more complex when AI systems are not centrally controlled and are learning from multiple decentralized data sets, which introduces a new set of ethical challenges related to data provenance and quality. The data's quality, volume, and representativeness are critical factors shaping the AI's performance, learning capabilities, and emergent properties. As these models interact with real-world data, they might develop the ability to solve new problems or recognize patterns beyond their original training set — a form of emergence that accentuates the AI's adaptability and resilience. However, the empiricist approach is fraught with ethical concerns, especially regarding data integrity. Data scientists individually bear the onus for scrutinizing the data for inherent biases, whether glaring or subtle, and ensuring that they don't propagate through the AI system. Organizations have a professional responsibility to establish rigorous ethical guidelines and oversight mechanisms to vet the quality and ethical dimensions of the data fed into machine learning algorithms. On a policy level, the government must enforce regulations to ensure ethical accountability in data collection, processing, and algorithmic decision-making. As emergent behaviors become increasingly sophisticated, safeguarding against ethical pitfalls necessitates a multi-tiered approach, reconciling empirical methods with the uncharted terrain of AI's evolving capabilities.

The Scientific Method: The growing complexity of "*AI and Emergent Properties*" has profound implications for the scientific method, pushing the boundaries of traditional approaches to inquiry and experimentation. Take, for example, the evolving capabilities of chatbots: While the scientific method relies on observable, replicable phenomena, what happens when chatbots begin to understand context and sentiment? This complex, emergent behavior could make it difficult to isolate variables for study, adding another layer of complexity to the scientific inquiry process. The inherent unpredictability of emergent behaviors in AI systems challenges the linear steps of observation, hypothesis formulation, controlled experimentation, and data analysis. For instance, even if a phenomenon in a complex AI system is observed and a hypothesis is formulated, conducting a controlled experiment to test that hypothesis can be incredibly difficult due to the system's non-linear and dynamically changing nature. This poses questions about the feasibility of traditional falsifiability and replicability criteria in emergent properties. Conversely, AI tools are becoming indispensable for scientific inquiry, capable of handling vast datasets and complex computations that would be otherwise unmanageable, thereby accelerating hypothesis testing in other fields. As AI algorithms formulate hypotheses based on data patterns, understanding emergent properties will become essential for maintaining scientific rigor. Automating portions of the scientific method raises new ethical and methodological concerns, such as the transparency and scrutiny of automated hypothesis generation and testing. For these findings to gain scientific credibility, new peer-review protocols that can evaluate the validity of machine-generated hypotheses and the emergent behaviors they exhibit may be necessary. In this way, the challenges and capabilities that AI and emergent properties bring catalyze a reevaluation and possible reformation of the scientific method, prompting the scientific community to adapt and evolve.

Medicine: The interplay between AI and emergent properties has far-reaching implications in medicine, transforming diagnostics, treatment planning, and even research paradigms. Highly complex machine learning models can analyze medical data in dimensions that human clinicians cannot easily grasp, such as sifting through thousands of variables in genomics or image data to identify subtle, emergent patterns indicative of disease. For instance, healthcare algorithms are only as good as the data they've been trained on, but sometimes, they identify patterns outside of their explicit programming due to emergent properties. This contributes significantly to raising the medical healthcare standard of care, as AI-powered diagnostics and predictive analytics can be more accurate and swift, facilitating timely and effective interventions. However, these advantages come with ethical complexities. The emergent properties of AI — wherein the system may adapt and "*learn*" beyond its initial programming — pose questions about the consistency and reliability of such technology in a medical context, where medical healthcare standards of care are stringently regulated based on peer-reviewed, evidence-based practices. If an AI system were to adapt its diagnostic algorithm autonomously, for example, it might improve the medical healthcare standard of care or introduce unforeseen errors. This adaptability creates a novel set of challenges in maintaining and auditing those medical healthcare standards, especially when the AI's decision-making process is not easily interpretable. Looking toward the future, the advent of Quantum Computing and artifical general intelligence (AGI) has the potential to push the limits of what AI can achieve in medicine, adding new layers of emergent behavior and complexity. Hence, as AI continues to seep into medicine, it requires a nuanced approach that not only leverages its capabilities for improved healthcare but also rigorously evaluates and updates medical healthcare standards of care to include this rapidly evolving technology.

Ethics: The advent of AI systems exhibiting emergent properties profoundly challenges and reshapes the traditional landscape of bioethical principles — patient autonomy (informed consent), practitioner beneficence (do good), practitioner nonmaleficence (do no harm), and public justice (be fair) — in medical healthcare and beyond. Starting with autonomy, AI's role in medical decision-making could either enhance informed consent processes by providing comprehensive data analyses or erode patient autonomy if the algorithms are opaque and their emergent behaviors unpredictable. Beneficence is also a double-edged sword; while AI has the potential to significantly improve patient outcomes through more accurate diagnoses and personalized treatment plans, emergent properties could lead to unintended, harmful consequences that are not immediately foreseeable. This complexity isn't restricted to the medical healthcare domain; it spans sectors like transportation, where emergent behaviors in autonomous vehicles challenge policy regulations, to scientific inquiry, where AI's capability to understand context and sentiment in chatbots calls into question the replicability criteria of the scientific method. Therefore, the discussed examples — from AI in medicine to its role in rationalist and empiricist traditions — paint a layered and multifaceted picture of ethical complexities that must be addressed. Feeding directly into the principle of nonmaleficence, or "*do no harm*," AI systems must be rigorously vetted to ensure that their emergent behaviors do not lead to detrimental outcomes, such as misdiagnoses or improper treatments, thereby causing harm. Finally, the principle of justice, which demands fairness in the distribution of healthcare resources, is impacted by the deployment of advanced AI systems. If access to these technologies is limited to affluent communities or countries, the ensuing disparity could exacerbate existing inequalities, violating the principle of justice. Looking to the future, emerging technologies like blockchain-based AI could provide more transparent and auditable systems, thereby mitigating some ethical risks. On the other hand, advancements in Quantum Computing have the potential to push the limits of what AI can achieve in fields like medicine, introducing new layers of emergent behavior and complexity that may further challenge our existing ethical frameworks. Therefore, as AI continues to evolve and manifest increasingly complex emergent properties, a nuanced and dynamic ethical framework will be essential to navigate the intricate balance between technological advancement and bioethical integrity.

Conclusion: The intersection of AI and emergent properties is a transformative force that not only shapes technological landscapes but also profoundly influences the paradigms of philosophy, ethics, scientific inquiry, medicine, and social governance. As AI systems evolve to exhibit behaviors resulting from complex, unpredictable interactions within their programmed rules or learned data patterns, they open new avenues for problem-solving and innovation while raising ethical and philosophical questions that society must confront. From a rationalist standpoint, the foundational algorithms and architecture serve as a springboard for these emergent behaviors, demanding ethical oversight at multiple levels — from individual developers to large organizations and policymakers. The empiricist lens, focusing on the role of data, underscores the importance of ethical data collection and algorithmic transparency to ensure unbiased and fair systems. This dual influence of rationalism and empiricism in shaping AI's emergent properties also impacts the scientific method, challenging traditional notions of hypothesis testing, replicability, and peer review, necessitating adaptations within the scientific community. The potential for improved diagnostics and treatments in medicine comes with its ethical complexities tied to the unpredictability of emergent behaviors, affecting medical healthcare standards of care and bioethical principles like patient autonomy (informed consent), practitioner beneficence (do good), practitioner nonmaleficence (do no harm), and public justice (be fair). As we look to the future, the legacy of AI and emergent properties promises to continue evolving, especially as future technologies like quantum computing and decentralized AI systems become mainstream. Quantum computing could exponentially increase computational power, allowing for more intricate simulations and problem-solving and generating new types of emergent behaviors that we have yet to understand fully. Decentralized AI systems, where the authority and data are distributed across multiple nodes, could give rise to emergent properties that are more complex and harder to predict, further complicating ethical oversight. These advances will likely accentuate the existing challenges and opportunities, requiring an even more harmonized, interdisciplinary approach to navigate AI's ever-expanding impact. AI with emergent properties is a multifaceted mirror, reflecting our technological ambitions and ethical, philosophical, and societal challenges. Therefore, we must engage in a concerted, interdisciplinary effort to understand and navigate its intricate and expansive impact, setting the stage for a future where technology and ethics evolve in tandem.

AI and Emergent Properties' Legacy: The most projected philosophical and scientific legacy for "*AI and Emergent Properties*" is its challenge to our traditional understanding of intelligence, autonomy, and ethics, as it forces humanity to reevaluate foundational concepts — like what it means to be intelligent or ethical — while adapting our frameworks for scientific inquiry, medicine, and governance.

**

REVIEW QUESTIONS

True/False Questions:

1. Emergent properties in AI systems refer to behaviors and attributes that are directly coded and predictable.
True or False?

2. The integration of AI and emergent properties challenges traditional notions of hypothesis testing and replicability in the scientific method.
True or False?

Multiple-Choice Questions:

3. Which of the following ethical principles is NOT directly impacted by the emergent properties of AI in medical healthcare?
a) Patient autonomy (informed consent)
b) Practitioner beneficence (do good)
c) Practitioner nonmaleficence (do no harm)
d) Libertarianism

4. Which emerging trend aims to make the decision-making process of complex algorithms more understandable to humans?
a) Quantum Computing
b) Decentralized AI
c) Explainable AI (XAI)
d) Blockchain-based AI

Clinical Vignette:

5. A healthcare provider uses an AI system for diagnostic purposes. The AI system, due to its emergent properties, identifies a pattern that was not part of its initial programming. According to ethical principles, how should the provider proceed?
a) Ignore the AI's findings and follow standard procedures
b) Immediately adopt the AI's recommendations without further review
c) Evaluate the AI's findings through additional tests and consultations to ensure accuracy and patient safety
d) Disregard the AI's findings if they contradict the provider's initial diagnosis

Basic Science Vignette:

6. An advanced AI diagnostic system in a hospital is designed to analyze patient data and suggest possible diagnoses. However, after several updates, the system begins identifying new patterns and correlations not originally programmed, leading to both highly accurate predictions and some unexpected errors. What is the best course of action to address the unexpected errors?
   a) Disable the AI system until the errors can be fully understood and corrected
   b) Continue using the AI system as is, focusing on the accurate predictions
   c) Implement a human review process for the AI's diagnoses to catch any errors
   d) Revert the AI system to its original version without the updates

Philosophy Vignette:

7. An AI system used in autonomous vehicles develops an emergent behavior where it prioritizes avoiding traffic congestion over individual passenger preferences, occasionally leading to longer travel times for some passengers. What is the best philosophical approach to addressing this issue?
   a) Prioritize the efficiency of the AI system in reducing overall traffic congestion
   b) Reprogram the AI to prioritize individual passenger preferences over traffic efficiency
   c) Develop a balanced algorithm that considers both traffic efficiency and passenger preferences
   d) Disable the emergent behavior to maintain control over the AI's decision-making process

Correct Answers:

1. False
2. True
3. d) Libertarianism
4. c) Explainable AI (XAI)
5. c) Evaluate the AI's findings through additional tests and consultations to ensure accuracy and patient safety
6. c) Implement a human review process for the AI's diagnoses to catch any errors
7. c) Develop a balanced algorithm that considers both traffic efficiency and passenger preferences

BEYOND THE CHAPTER
AI & Emergent Properties

- *Life 3.0: Being Human in the Age of Artificial Intelligence* by Max Tegmark
- *Artificial Unintelligence: How Computers Misunderstand the World* by Meredith Broussard
- *Emergence: The Connected Lives of Ants, Brains, Cities, and Software* by Steven Johnson

***

Chapter 20: Big Data — The Analytical Revolution: Precision and Ethics

Big Data revolutionizes research and healthcare, balancing precision with ethical responsibility.

Abstract: In the digital era, Big Data has revolutionized healthcare, business, and science, offering remarkable opportunities and complex challenges. This dynamic force harnesses immense volumes of structured and unstructured data, unlocking valuable insights and enhancing decision-making processes. This analysis navigates rationalism and empiricism, highlighting predefined algorithms and experiential knowledge in data interpretation, while spotlighting the ethical responsibilities requiring scrutiny and transparency. It elucidates Big Data's transformative influence on the scientific method, enhancing observational capacities and necessitating a reevaluation of traditional methodologies. In medicine, Big Data drives personalized care and redefines medical standards and ethics, bringing ethical dilemmas related to patient autonomy (informed consent), practitioner beneficence (do good), practitioner nonmaleficence (do no harm), and public justice (be fair). As we embrace the Big Data era, it calls for responsible management to safeguard ethical integrity and harmonize technological advancements with moral responsibility and societal betterment.

**

Introduction: In the digital age, "*Big Data*" has emerged as a focal point of discussions surrounding technology's transformational impact on various sectors, including healthcare, business, and science. Big Data refers to the massive volumes of structured and unstructured data generated at an unprecedented scale and speed, often requiring sophisticated algorithms and analytical tools for collection, storage, and interpretation. Beyond its sheer size, the significance of Big Data lies in its potential to offer valuable insights and facilitate decision-making processes across disciplines. Particularly in scientific research and healthcare, utilizing Big Data can revolutionize data analysis, diagnosis, and treatment methods, shifting paradigms and shaping new ethical considerations. As we delve into the dimensions of Big Data, it becomes crucial to assess its alignment with rationalist and empiricist approaches, its role in refining the scientific method, its contributions to medical advancements, and its influence on the four core ethical principles of bioethics: patient autonomy (informed consent), practitioner beneficence (do good), practitioner nonmaleficence (do no harm), and public justice (be fair). Through such an assessment, we can better understand the intricate tapestry of opportunities and challenges that Big Data presents in our increasingly interconnected world.

Rationalism: In exploring Big Data through the lens of rationalism, we encounter the foundational role that pre-defined algorithms, architectures, and initial conditions play in shaping how data are processed and interpreted. These elements serve as the "*a priori*" principles, existing independently of empirical experience and accessed through deductive reasoning. They provide the initial framework through which the chaos of Big Data is ordered, categorized, and made intelligible. Ethically, this rationalist approach places considerable responsibility on the shoulders of developers and data scientists to ensure that these initial conditions and algorithms are free from bias and constructed with fairness and justice in mind. Moreover, organizations — especially in the tech industry — bear the professional onus of upholding stringent ethical standards and critically scrutinizing algorithms for potential biases and unintended consequences. When these pre-set elements interact, "*emergence*" takes form. Algorithms might adapt and produce novel problem-solving techniques or data clustering methods not explicitly programmed into them. While such emergence holds the promise of innovation and enhanced effectiveness, it also ushers in complex ethical dilemmas around accountability, especially when the results have societal implications or pose risks. This rationalist inclination, therefore, not only guides the structural underpinning of Big Data analytics but also raises imperative questions about ethical governance that must be transparently addressed at personal, professional, and public policy levels.

Empiricism: Within Big Data, empiricism focuses on the experiential foundation of knowledge, emphasizing the quality and volume of training data as the primary determinants of a model's performance. Unlike the rationalist approach, which is rooted in "*a priori*" principles and deductive reasoning, the empiricist orientation relies on inductive reasoning to draw generalized models or principles from specific, observed data points. This empirical methodology raises unique ethical concerns, particularly around data bias. At a personal level, data scientists are ethically responsible for vetting data quality, ensuring it is comprehensive and free from systemic biases that could skew outcomes. On a professional level, organizations are tasked with establishing ethical guidelines and oversight mechanisms to ensure unbiased data collection and algorithmic fairness. Public policy also plays a crucial role in creating regulations for transparency and ethical accountability in data-driven processes. The concept of "*emergence*" in the empiricist context reflects a machine learning model's potential adaptability. Once trained on a specific dataset, such models can often adapt to recognize new patterns or solve problems not initially within their purview. While this adaptability offers exciting prospects for innovation and resilience, it poses ethical challenges, especially when models begin making unsupervised decisions that must be rigorously vetted for ethical implications. Therefore, the empiricist approach provides the raw material for data analytics and requires careful ethical scrutiny at multiple levels.

The Scientific Method: Big Data has exerted a transformative influence on the traditional scientific method, expanding the horizons of what can be observed, measured, and analyzed. The scientific method has traditionally operated linearly, starting from observation and culminating in peer-reviewed conclusions. However, the influx of Big Data has blurred these lines, enabling researchers to tap into vast repositories of pre-collected data, sometimes even redefining the initial stages of observation and hypothesis formulation. For instance, Big Data analytics can sift through multidimensional data sets to identify patterns or anomalies that could spur new hypotheses, effectively reversing the conventional sequence of scientific inquiry. Moreover, the sheer volume of data available for analysis enables more robust, statistically significant results, although it also introduces new challenges for ensuring data quality and controlling for variables. The emphasis on replicability and peer review remains crucial but is supplemented by further data credibility and integrity metrics, as results must be scrutinizable in the context of complex algorithms and immense datasets. Thus, Big Data not only enriches the resources available for scientific inquiry but also necessitates a reevaluation of methodological frameworks, pushing for advancements in data ethics, verification processes, and interdisciplinary collaboration. Through this synergistic relationship, Big Data enriches and complicates the scientific method, offering transformative potential while demanding new rigor and ethical considerations.

Medicine: Big Data has been a transformative force in medicine, revolutionizing diagnostic processes and treatment paradigms. Leveraging vast datasets from electronic health records, medical imaging, genetic sequencing, and even wearable devices, healthcare professionals can make more precise diagnoses and tailor treatments to individual patient needs. These data-driven approaches complement and enhance traditional medical healthcare standards of care based on peer-reviewed, evidence-based practices. Big Data helps redefine these medical healthcare standards of care by providing a more robust evidence base, continually updated in real-time and derived from a more extensive and diverse patient population. Advanced analytics and machine learning models sift through complex medical data to identify patterns or correlations that might be missed through conventional methods, thereby influencing new best practices and treatment guidelines. For example, predictive analytics can flag potential outbreaks or the likelihood of medical complications, allowing for preemptive interventions. Additionally, personalized medicine, driven by individual patient data, is increasingly becoming the standard, shifting the focus from generalized treatment to more customized care. In a legal context, using Big Data also raises the bar for what constitutes reasonable and competent care, as the availability of detailed patient data creates a new level of responsibility for healthcare providers. Thus, Big Data enhances medical research and patient care and dynamically influences the continually evolving medical healthcare standards of care that define competent and ethical medical practice.

Ethics: Big Data's burgeoning role in healthcare and scientific research poses a nuanced landscape for bioethical principles, notably patient autonomy (informed consent), practitioner beneficence (do good), practitioner nonmaleficence (do no harm), and public justice (be fair). In the realm of autonomy and informed consent, the gathering and using personal data, often without explicit consent or even awareness, raise significant ethical questions. Patients may need to fully grasp the extent to which their data are being used, analyzed, or potentially shared, compromising the principle of informed consent. On the side of beneficence, Big Data holds unprecedented promise for improving healthcare outcomes, from early diagnosis to personalized treatment plans, undoubtedly serving the greater good. However, this incredible progress is tempered by issues of nonmaleficence, as the misuse of data, whether through breaches of privacy or biased algorithms, can cause substantial harm. Finally, the principle of justice, which calls for fair distribution of benefits, risks, and costs, is also impacted by Big Data. While it has the potential to democratize healthcare by making high-quality diagnostics and treatments more widely accessible, there's also the risk that it could exacerbate existing inequalities if only certain populations have access to data-driven care. Additionally, data sets lacking diversity can result in algorithms biased against underrepresented groups, further undermining the principle of justice. While Big Data offers transformative potential for healthcare and scientific research, its intersection with core bioethical principles necessitates rigorous, ongoing ethical scrutiny to ensure that tech nological advancements do not come at the cost of moral compromise.

Conclusion: In the ever-evolving landscape of technology and information, Big Data stands as a monumental paradigm shift that intersects with many sectors, including healthcare, scientific research, and ethics. Its impact reverberates through the rationalist foundations of algorithmic systems, the empiricist methodologies focused on data quality and volume, and even the sanctified rigors of the scientific method, reshaping traditional approaches to inquiry. In medicine, Big Data has ushered in a new era of diagnostics and personalized care, updating and enhancing the very standards by which medical competence is judged. Yet, as promising as it is, Big Data is also a Pandora's box of ethical quandaries. The principles of bioethics — patient autonomy (informed consent), practitioner beneficence (do good), practitioner nonmaleficence (do no harm), and public justice (be fair) — become more complex within the matrix of Big Data, compelling us to scrutinize and update ethical norms continuously. Far from being just a technological advancement, Big Data represents a complex tapestry of opportunities and challenges that requires an integrated understanding and responsible management at individual, organizational, and societal levels. As we unravel its potential, we must also be diligent in safeguarding ethical integrity, social justice, and the fundamental principles that guide scientific and medical practices.

Big Data's Legacy: By fundamentally reshaping the way we approach rationalist and empiricist methodologies, Big Data serves as both the crucible and catalyst for a new interdisciplinary paradigm in scientific inquiry and ethical governance, thus making it essential for achieving unprecedented levels of accuracy, efficiency, and moral accountability in various domains of human endeavor.

**

REVIEW QUESTIONS

True/False Questions:

1. Big Data primarily involves the processing and analysis of small, structured datasets.
   True or False?

2. The use of Big Data in healthcare raises ethical questions about patient autonomy and informed consent.
   True or False?

Multiple-Choice Questions:

3. Which of the following is NOT a core ethical principle impacted by Big Data in healthcare?
   a) Patient autonomy (informed consent)
   b) Practitioner beneficence (do good)
   c) Practitioner nonmaleficence (do no harm)
   d) Financial profitability

4. What is one of the key challenges Big Data poses to the traditional scientific method?
   a) Lack of data
   b) Inability to replicate results
   c) Limited computational power
   d) Difficulty in forming hypotheses

Clinical Vignette:

5. A healthcare provider uses a Big Data analytics tool to identify patients at high risk for a specific condition. The tool analyzes vast amounts of electronic health records and identifies patterns that were not previously apparent. According to ethical principles, how should the provider proceed?
   a) Rely solely on the tool's recommendations without further analysis
   b) Disregard the tool's findings if they contradict traditional diagnostic methods
   c) Use the tool's insights to inform further diagnostic testing and patient discussions, ensuring that patients understand how their data was used
   d) Share the tool's findings with patients without any additional context

Basic Science Vignette:

6. A hospital's AI system analyzes patient data to predict potential health issues. After implementing Big Data analytics, the system starts identifying new health risk patterns not previously known. However, it also begins flagging some false positives. What is the best course of action to address the false positives?
   a) Ignore the false positives as the system's overall performance has improved
   b) Review and refine the algorithm using a more comprehensive dataset
   c) Remove the new health risk patterns from the system to prevent false positives
   d) Disable the AI system until the false positives can be fully understood

Philosophy Vignette:

7. An AI system in a research lab uses Big Data to analyze and predict the outcomes of various ethical dilemmas. The system suggests an optimal solution that maximizes overall happiness but overlooks individual rights in some cases. What is the best philosophical approach to address this issue?
   a) Prioritize the AI's ability to maximize overall happiness
   b) Reprogram the AI to give more weight to individual rights than overall happiness
   c) Develop an oversight committee to review the AI's recommendations before implementation
   d) Disable the AI system until it can be programmed to always respect individual rights

Correct Answers:

1. False
2. True
3. d) Financial profitability
4. b) Inability to replicate results
5. c) Use the tool's insights to inform further diagnostic testing and patient discussions, ensuring that patients understand how their data was used
6. b) Review and refine the algorithm using a more comprehensive dataset
7. b) Reprogram the AI to give more weight to individual rights than overall happiness

BEYOND THE CHAPTER

Big Data

- *Big Data: A Revolution That Will Transform How We Live, Work, and Think* by Viktor Mayer-Schönberger and Kenneth Cukier
- *Data and Goliath: The Hidden Battles to Collect Your Data and Control Your World* by Bruce Schneier
- *The Data Revolution: Big Data, Open Data, Data Infrastructures and Their Consequences* by Rob Kitchin

***

Philosophy of Science
Medical Research

Chapter 21: Bioinformatics — The Computational Bridge: Biological Insights

Bioinformatics merges computational and biological sciences, accelerating innovation and ethical discourse in healthcare.

Abstract: Bioinformatics has manifested as a pivotal force that seamlessly marries computational science with biological research, fostering a realm where scientific advancements are accelerated and complexified. Through its analytical prowess grounded in rationalist principles and empirical methodology, this interdisciplinary field has metamorphosed the scientific method, extending its scope and precision to comprehend the intricate fabric of biological systems. Its profound imprint is visibly marked in the healthcare sector, where the advent of personalized medicine and innovative drug development strategies stands testimony to its transformative power. Furthermore, bioinformatics harbors a profound responsibility toward ensuring equitable access to healthcare advancements, safeguarding data privacy, and fostering a discourse that continually revisits and refines ethical frameworks. This dynamic field stands at a crucial juncture where technological ingenuity and ethical contemplation are steering a revolution, altering societal perspectives on science, ethics, and healthcare equity. The legacy of bioinformatics, thus, encapsulates its role as a beacon of scientific innovation and a sentinel advocating for ethical vigilance in the complex, data-driven landscape of contemporary biology.

**

Introduction: Bioinformatics is an interdisciplinary scientific field that acts as a vital bridge between the biological sciences and computational technology. This dynamic interplay has led to a powerful symbiosis that has profoundly impacted various sectors, from healthcare and medicine to agriculture and environmental science. Bioinformatics leverages algorithms, computational models, and statistical techniques to meticulously analyze biological data. These data sets can include anything from DNA sequences and RNA profiles to protein structures, metabolic pathways, and entire ecosystems. By providing essential tools for understanding the molecular mechanisms of diseases, guiding drug development, and enabling the tailoring of personalized medicine, bioinformatics allows researchers to convert raw biological data into actionable insights. The techniques developed in this field have been instrumental in milestone achievements like decoding the human genome, analyzing diseases' genetic underpinnings, and understanding evolutionary biology's complexities. As we move deeper into an era increasingly characterized by data-driven biology, bioinformatics stands at a unique intersection of technological innovation and ethical considerations. It grapples with urgent issues related to data privacy, equitable access to healthcare, and the responsible use of genetic information. Therefore, bioinformatics's influence extends far beyond the scientific community, affecting societal perspectives on ethics and social responsibility within the broader context of human progress.

Rationalism: Bioinformatics is deeply rooted in rationalism. It relies fundamentally on algorithms, statistical models, and computational architectures to interpret and analyze complex biological data. In this respect, these computational frameworks act as "*a priori*" principles that guide the systematic investigation and understanding of biological phenomena. Unlike empirically derived models, these mathematical constructs are logically and deductively formulated to process data coherently and meaningfully. The concept of "*emergence*" also has significant relevance in bioinformatics. This is particularly true when advanced machine learning algorithms evolve to discover unexpected, novel patterns in genetic variations, intricate protein interactions, or complex metabolic pathways. These emergent, self-organizing features can yield revolutionary scientific insights and pose complicated ethical dilemmas around data privacy, algorithmic bias, and unforeseen scientific consequences. This rationalist orientation places a considerable responsibility on the shoulders of the developers, data scientists, and bioinformaticians involved in the field. Ensuring that algorithms are accurate but also unbiased and equitable is not just a technical challenge; it's a moral imperative. At the organizational level, technology companies, pharmaceutical firms, and research institutions involved in bioinformatics increasingly recognize the need to adhere to stringent ethical guidelines. These organizations must regularly scrutinize their algorithms to identify and correct any biases or inaccuracies that could lead to harmful or unjust outcomes. Regarding public policy, transparent governance frameworks and regulatory mechanisms are vital for overseeing ethical deployment and interpretation of bioinformatics tools. Therefore, the rationalist underpinnings of bioinformatics fuel its analytical power and introduce a host of critical ethical considerations. These concerns demand sustained attention and action from individuals, organizations, regulatory bodies, and society.

Empiricism: Bioinformatics also adopts an empiricist approach, particularly when leveraging vast biological data repositories. These can range from intricate genomic sequences and RNA transcripts to proteomic structures and complex ecological data. The focus here is on inductive reasoning, where many specific data points are synthesized to form generalized principles, theories, or predictive algorithms. These data-driven methodologies have enabled groundbreaking advancements in personalized medicine, early disease diagnosis, and evolutionary biology. However, the collected data's quality and ethical integrity are paramount. Biases in the data — whether they arise from an inequitable representation of populations, lack of diversity in sampling, or other contributing factors — can result in misleading conclusions. These can, in turn, perpetuate systemic disparities in healthcare and other societal domains. Ethically, data scientists and bioinformaticians are responsible for rigorously examining the data for any biases and ensuring its quality, validity, and representativeness. On an organizational level, corporations, companies, and research institutions must establish and adhere to robust ethical guidelines and oversight mechanisms. These must be designed to guarantee that the data sets employed in research and applications are as comprehensive, impartial, and unbiased as possible. Public policy also plays an indispensable role here by enforcing stringent regulations, ensuring data collection transparency, and scrutinizing the ethical implications of algorithmic decision-making. The concept of "*emergence*" becomes particularly pertinent when a machine-learning model in bioinformatics evolves to recognize new, previously unaccounted-for patterns or adapt to solve problems beyond its original training set. While this adaptability offers significant advantages, it also raises new ethical challenges. These emergent properties may not have been initially vetted for ethical considerations, requiring continuous ethical scrutiny and oversight.

The Scientific Method: Bioinformatics has fundamentally reshaped the scientific method, particularly in the life sciences, by integrating computational approaches into traditional biological research paradigms. Observation and questioning, the foundational steps in the scientific method, have been vastly expanded by the ability to acquire and scrutinize enormous datasets. This expanded capability allows researchers to form intricate and highly testable hypotheses that often span multiple biological scales — from molecular to cellular to organismal levels. Experimentation, too, has undergone a paradigm shift. Computational simulations and data-driven models are increasingly considered valid complements or alternatives to traditional wet lab experiments, provided they are carefully validated against empirical data. The data collection and analysis have often been automated and scaled up, requiring new scientific rigor to ensure accuracy, replicability, and ethical compliance. Peer review also faces unprecedented challenges and opportunities. For instance, the rise of "*open science*" initiatives has enabled more collaborative scrutiny of data and methodologies. However, this shift also demands new standards for data sharing, computational reproducibility, and ethical compliance. In essence, bioinformatics has not only adapted the scientific method to a new realm of inquiry but has also prompted significant refinements in how science is conducted. Through its impact on hypothesis generation, data collection, analysis, and peer review, bioinformatics is helping to evolve a more robust, efficient, and collaborative scientific method.

Medicine: In the realm of healthcare, bioinformatics has made transformative contributions. It has fundamentally altered medical practices and set new medical healthcare standards of care. By analyzing vast genomic, proteomic, and metabolomic datasets, bioinformatics provides the scientific underpinnings for personalized medicine. This approach allows treatments to be tailored to individual genetic profiles, thus increasing the efficacy of interventions while minimizing adverse side effects. Consequently, bioinformatics elevates the standard of medical healthcare delivery to unprecedented heights. Additionally, bioinformatics plays a critical role in the accelerated discovery and development of new drugs, often identifying target molecules more efficiently than traditional methods. The field has also been instrumental in understanding the genetic bases of numerous diseases, including various forms of cancer, metabolic disorders like diabetes, and cardiovascular conditions. This knowledge has led to more targeted and effective therapies, often developed through computational simulations that are then verified by wet lab experiments.

Ethics: Bioinformatics intersects deeply with bioethical principles, necessitating a reevaluation and sometimes a complete overhaul of traditional ethical norms, especially in data-driven biology. Starting with the principle of autonomy (informed consent), bioinformatics raises serious questions about the scope and nature of informed consent, particularly when dealing with sensitive genetic data. The potential for misuse or unauthorized access to such data necessitates a comprehensive reassessment of how consent is obtained and what it should encompass. Similarly, the principle of beneficence (do good) is impacted. At the same time, the ability to generate personalized medical treatments or predict disease risks offers significant benefits, but these must be handled with extreme caution to avoid unintended harmful consequences. The principle of nonmaleficence (do no harm) is also challenged by the inherent risks of data misuse, whether through privacy breaches or potential discrimination based on genetic information. Developers, data scientists, and bioinformaticians are ethically obligated to ensure that the algorithms and databases they employ are designed and maintained to do no harm. This entails vigilant oversight to prevent perpetuating bias or enabling data misuse. Lastly, the principle of justice (be fair) is a key concern. The benefits and advancements facilitated by bioinformatics should be equitably distributed and not exacerbate existing healthcare disparities. Given that genomic data often come from limited subsets of the population, there is a significant risk of developing medical solutions that are only sometimes applicable, thereby undermining the principle of fairness and equality.

Conclusion: Bioinformatics has emerged as a transformative force at the intersection of biology and computational science, significantly impacting multiple dimensions of human endeavor. It has revolutionized healthcare by developing personalized medicine and cutting-edge drug discovery techniques. These advancements are facilitated by a harmonious fusion of rationalist and empiricist methodologies supported by a data-driven approach. By infusing traditional biology with computational tools and algorithms, bioinformatics has enriched and extended the scientific method, making it more comprehensive, precise, and adaptable to the complexities of biological systems. This interdisciplinary field has revolutionized science and medicine and brought into sharp focus a plethora of ethical considerations that society must address. From data privacy and informed consent to algorithmic bias and healthcare equity, bioinformatics forces us to revisit and refine our ethical frameworks continually. As we navigate the increasingly digital and interconnected landscape of modern biology, bioinformatics principles and methodologies are a guiding light for scientific discovery and a cautionary tale regarding the importance of ethical vigilance. Thus, bioinformatics is a seminal example of how technology can both advance and complicate our understanding of life, health, and the ethical considerations that come with them.

Bioinformatics' Legacy: Bioinformatics serves as a linchpin for the harmonious integration of computational methods and biological inquiry, thereby redefining the landscape of modern science and medicine, a transformation that is profoundly influencing both our empirical understanding of biological systems and our philosophical considerations of ethics, data privacy, and healthcare equity.

**

REVIEW QUESTIONS

True/False Questions:

1. Bioinformatics primarily deals with the analysis of small, structured datasets, minimizing the need for computational tools.
   True or False?

2. The principle of justice in bioinformatics involves ensuring that the benefits and advancements facilitated by this field are equitably distributed across diverse populations.
   True or False?

Multiple-Choice Questions:

3. Which of the following is NOT a core ethical concern in bioinformatics?
   a) Data privacy
   b) Algorithmic bias
   c) Financial profitability
   d) Informed consent

4. What major contribution has bioinformatics made to healthcare?
   a) Development of generic treatment plans for all patients
   b) Improvement of personalized medicine by tailoring treatments to individual genetic profiles
   c) Limiting access to genomic data for research purposes
   d) Decreasing the accuracy of disease diagnosis

Clinical Vignette:

5. A researcher is using bioinformatics to analyze genomic data from a diverse population to develop a new drug. The initial findings suggest a potential treatment but reveal significant biases in the dataset. According to ethical principles, how should the researcher proceed?
   a) Ignore the biases and continue with the development
   b) Use the biased data as is to quickly produce the drug
   c) Address the biases by obtaining a more representative dataset and re-evaluating the findings
   d) Abandon the research due to the complexity of correcting the biases

Basic Science Vignette:

6. A bioinformatics team is developing an AI system to analyze genetic data for identifying potential disease markers. After integrating a vast new dataset, the system starts identifying novel genetic variations linked to diseases. However, it also produces a significant number of false positive results. What is the best course of action to address the false positives?
   a) Ignore the false positives as the system's overall performance has improved
   b) Review and refine the algorithm using a more comprehensive dataset
   c) Remove the new genetic variations from the system to prevent false positives
   d) Disable the AI system until the false positives can be fully understood

Philosophy Vignette:

7. An AI system in bioinformatics uses Big Data to suggest ethical guidelines for genetic research. It proposes an optimal framework that maximizes scientific progress but overlooks individual consent in some cases. What is the best philosophical approach to address this issue?
   a) Prioritize the AI's ability to maximize scientific progress
   b) Reprogram the AI to give more weight to individual consent than scientific progress
   c) Develop an oversight committee to review the AI's recommendations before implementation
   d) Disable the AI system until it can be programmed to always respect individual consent

Correct Answers:

1. False
2. True
3. c) Financial profitability
4. b) Improvement of personalized medicine by tailoring treatments to individual genetic profiles
5. c) Address the biases by obtaining a more representative dataset and re-evaluating the findings
6. b) Review and refine the algorithm using a more comprehensive dataset
7. b) Reprogram the AI to give more weight to individual consent than scientific progress

BEYOND THE CHAPTER

Bioinformatics

- *Bioinformatics: Sequence and Genome Analysis* by David W. Mount
- *Bioinformatics: A Practical Guide to the Analysis of Genes and Proteins* by Andreas D. Baxevanis and B. F. Francis Ouellette
- *Biological Sequence Analysis: Probabilistic Models of Proteins and Nucleic Acids* by Richard Durbin, Sean Eddy, Anders Krogh, and Graeme Mitchison

***

Philosophy of Science

Medical Research

WEEK FIVE: Ethical and Genetic

WEEK FIVE

Class Hour 1

- Chapter 22: CRISPR-Cas9 — The Genetic Precision: Ethical Innovation
- Chapter 23: Genetic Engineering and Biotechnology — The Transformative Frontier: Ethical Dilemmas
- Chapter 24: Human Genome Project — The Genetic Blueprint: Scientific and Ethical Insights

Class Hour 2

- Chapter 25: Large Language Models — The AI Frontier: Rethinking Knowledge
- Chapter 26: Nanotechnology — The Atomic Precision: Ethical Innovation
- Chapter 27: Pharmacogenomics — The Personalized Medicine: Genetic Insights

---

Introduction

In the fifth week of our course, "*Philosophy of Science*," we explore the ethical and genetic dimensions of scientific advancements. This week's chapters focus on the revolutionary tools and technologies reshaping genetic engineering and artificial intelligence, along with the ethical frameworks guiding their application. From the precision of CRISPR-Cas9 to the expansive insights of the Human Genome Project, and from the transformative potential of large language models to the atomic precision of nanotechnology and the personalized promise of pharmacogenomics, we delve into the complexities of innovation and ethical responsibility.

We begin with CRISPR-Cas9, a groundbreaking tool in genetic engineering that enables precise modifications to DNA. This technology has revolutionized biological interventions, offering new therapeutic possibilities while simultaneously raising significant ethical concerns. CRISPR-Cas9 exemplifies the integration of rationalist and empiricist principles, leveraging cognitive faculties and observational data to advance healthcare and agriculture. However, its application demands careful consideration of bioethical principles, including patient autonomy, beneficence, nonmaleficence, and public justice. The transformative power of CRISPR-Cas9 necessitates responsible stewardship to navigate its profound impact on human life and society.

Genetic engineering and biotechnology, situated at the forefront of scientific innovation, present unprecedented advancements in medicine and agriculture. By rigorously applying the scientific method, these fields address global challenges such as disease eradication and food security. However, they also introduce complex ethical dilemmas that require a reevaluation of bioethical principles. Balancing innovation with ethical responsibility is crucial as we navigate the transformative potential of genetic engineering and biotechnology, considering their profound implications for human existence and the broader ecosystem.

The Human Genome Project (HGP), one of the most transformative scientific endeavors, mapped human DNA and reshaped fields like medicine, anthropology, and bioethics. This project harmonized rationalist and empiricist paradigms, using theoretical assumptions and empirical evidence to unlock genetic knowledge. The HGP fostered international collaboration and extensive data sharing, leading to advancements in medical diagnostics and treatments. It also ignited substantial bioethical discourse, shaping policies and regulations for genomic exploration and ushering in personalized medicine and predictive health analytics. The HGP exemplifies the intersection of scientific innovation and ethical inquiry, enriching our approach to bioethical concerns.

Large Language Models (LLMs) like ChatGPT showcase rapid advancements in artificial intelligence, demonstrating immense potential and ethical complexity. Built on the Transformer architecture, LLMs excel in text generation and context-sensitive interactions, extending their utility to scientific research and healthcare. However, these innovations bring ethical dilemmas, necessitating a reassessment of principles such as patient autonomy, beneficence, nonmaleficence, and public justice. LLMs challenge the rigor of the scientific method and raise concerns about healthcare standards and legal accountability, requiring evolving ethical frameworks to address issues of bias, transparency, and fairness.

Nanotechnology, emerging from Richard Feynman's seminal concepts, epitomizes scientific innovation by manipulating matter on an atomic scale. This field offers opportunities ranging from revolutionizing healthcare to creating resilient infrastructures. Nanotechnology demands a comprehensive approach integrating rationalist and empiricist paradigms, fostering systematic development grounded in foundational principles and empirical research. As it expands and refines the scientific method, nanotechnology presents ethical dilemmas, necessitating robust frameworks for navigating informed consent, equitable access, and potential biological ramifications. Aligning scientific innovation with ethical responsibility ensures that advancements harmonize with societal well-being.

Pharmacogenomics, at the intersection of genetics and pharmacology, heralds a transformative era in personalized medicine by understanding genetic variations and drug responses. This field blends rationalistic underpinnings with empirical validation, shaping the scientific method in precision medicine. Pharmacogenomics shifts the healthcare paradigm towards individualized treatments, enhancing drug efficacy and reducing adverse reactions. This dynamic field fosters heightened ethical discourse, recalibrating medical standards and steering personalized medicine towards a more inclusive, ethical, and effective future.

Together, these chapters illustrate the intricate balance between scientific innovation and ethical responsibility in genetic engineering and AI. From the precision of CRISPR-Cas9 to the transformative insights of the Human Genome Project, and from the capabilities of large language models to the precision of nanotechnology and the promise of pharmacogenomics, we explore the ethical frameworks guiding the application of these groundbreaking technologies. This integrated exploration underscores the importance of maintaining ethical integrity in the face of rapid technological advancements, ensuring that innovation serves the greater good and aligns with societal values.

***

Chapter 22: CRISPR-Cas9 — The Genetic Precision: Ethical Innovation

CRISPR-Cas9 revolutionizes genetic engineering, prompting ethical innovation and responsibility.

Abstract: In the recent decade, the CRISPR-Cas9 system has emerged as a pivotal tool in genetic engineering, enabling precise genetic modifications and prompting comprehensive discussions on its ethical implications. Rooted in rationalist orientations, the technology leverages human cognitive faculties to redefine biological interventions, bringing therapeutic possibilities and ethical dilemmas. Its development, underscored by empiricism, is built on a foundation of observational data and inductive reasoning, fostering innovations in healthcare and agriculture. This innovation has enhanced the scientific method, enabling more controlled research and ushering in a transformative medical era with promising prospects in gene therapy and personalized medicine. However, the technology poses significant ethical challenges, necessitating a nuanced exploration of patient autonomy (informed consent), practitioner beneficence (do good), practitioner nonmaleficence (do no harm), and public justice (be fair) within bioethics. As CRISPR-Cas9 shapes our understanding of biological complexities, it beckons a conscientious approach to navigate its potential impact on human life and society, urging responsible stewardship in harnessing its transformative power.

**

Introduction: The CRISPR-Cas9 system, an acronym for Clustered Regularly Interspaced Short Palindromic Repeats and CRISPR-associated protein 9, has revolutionized the field of genetic engineering by providing an unprecedented precision tool for DNA manipulation. Originating from a natural defense mechanism in bacteria, this biotechnological innovation allows scientists to add, delete, or modify specific genes within an organism's genome. CRISPR-Cas9 has generated excitement and concern since its discovery in 2012, offering a wide range of applications, from treating genetic disorders to creating genetically modified organisms. It is a technological marvel that brings the science fiction dream of "*editing*" genes closer to reality, yet it also opens a Pandora's box of ethical, social, and medical implications. As we explore the rationalist and empiricist orientations within the development of CRISPR, its contributions to the scientific method, its potential in medicine, and its ethical considerations through the lens of patient autonomy (informed consent), practitioner beneficence (do good), practitioner nonmaleficence (do no harm), and public justice (be fair), we find ourselves at a crossroads where science meets the profound questions of what it means to be human and the ethical responsibilities that come with such transformative power.

Rationalism: The development and application of CRISPR-Cas9 technology are deeply rooted in rationalist inclinations, relying on the evolved cognitive faculties of human reasoning, pattern recognition, and problem-solving. Scientists utilized innate faculties to identify the potential of a bacterial defense mechanism, decipher its pattern within the genome, and translate it into a tool capable of targeted genetic editing. The deductions drawn were built on axiomatic scientific principles of molecular biology, enzymology, and genetics, creating a technology that could efficiently modify genes. Deductive reasoning also plays a role in the technology's progression, from understanding DNA's double-helix structure to recognizing the CRISPR system as a naturally occurring biological "*cut-*

and-paste" mechanism. Such deductive chains serve as definitional truths within the field, supporting the development of therapeutic applications that could change the face of medicine. Yet, the same faculties that enable this scientific leap also introduce ethical complexities. While rationalism aids in understanding and manipulating genetic material, it simultaneously compels us to examine the implications of "playing God" with genetic makeup. Scientists and ethicists must engage in advanced ethical reasoning to navigate concerns like gene editing's long-term consequences or the ethical pitfalls of designer babies. At a public policy level, regulations must be designed to temper the powerful yet potentially dangerous capabilities rationalism has bestowed upon us, ensuring that the rapid advances do not compromise ethical considerations. The emergent properties of CRISPR-Cas9 technology — its potential to reshape entire ecosystems, eliminate genetic diseases, or even redefine what it means to be human — are more complex than the sum of its parts and reflect the intricate interplay of human cognitive faculties.

Empiricism: The empiricist orientation in developing and deploying CRISPR-Cas9 is fundamentally grounded in sensory experiences and environmental interactions, leading to a wealth of empirical data. CRISPR-Cas9 did not emerge in a vacuum but was a discovery based on observations of natural bacterial defense mechanisms. Using inductive reasoning, scientists extrapolated from these specific observations to develop a general tool for targeted genetic modification. Numerous experiments have provided empirical evidence validating its efficacy, safety, and possible applications throughout its development — from lab-grown models to clinical trials. Cognitive biases, influenced by culture and societal needs, have shaped the research agenda. For instance, the focus on using CRISPR for curing genetic diseases partly results from societal values prioritizing medical healthcare advancements. Professionally, the empiricist approach emphasizes the indispensability of ongoing empirical evidence to update ethical guidelines in real-time as our understanding of the technology's scope and limitations evolves. Public policy, therefore, must be adaptive and data-driven, responsive to new empirical findings that introduce novel ethical and safety considerations. The adaptive value of this empiricism is evident in the way CRISPR technology has swiftly evolved to meet various challenges, from agriculture to medicine, thereby enhancing human adaptability and survival. The concept of "emergence" is also salient, as the collective capabilities and potential uses of CRISPR-Cas9 go beyond what could have been predicted by studying individual genetic modifications or isolated empirical observations. Complex phenomena such as the potential to alter entire ecosystems or the socio-cultural implications of gene editing are emergent properties that underscore the complexity and responsibility inherent in this groundbreaking technology.

The Scientific Method: The advent and advancement of CRISPR-Cas9 have both been influenced by and have influenced the scientific method, offering a modern exemplar of how systematic inquiry can lead to transformative discoveries. The phenomenon of bacterial resistance to viral attacks led to questions about the underlying mechanisms, culminating in the hypothesis that CRISPR sequences serve as a kind of immune memory. Controlled experiments were designed to isolate and identify the components and functionalities of the CRISPR system, leading to the recognition of its gene-editing capabilities. The methodology was meticulously documented, allowing for the critical element of replicability; indeed, labs worldwide have since confirmed and extended the initial findings. Moreover, CRISPR research has undergone rigorous peer reviews, subjecting it to scrutiny, leading to refined techniques, increased precision, and broader applications. Perhaps most intriguingly, CRISPR technology has added a new dimension to the scientific method by enabling more controlled experiments at the genetic level, thereby accelerating research in many fields. By allowing for targeted gene modifications, CRISPR provides researchers with a powerful tool to formulate sharper hypotheses and perform more exacting tests, enriching the scientific method's rigor and capability.

Consequently, CRISPR stands as a product of systematic scientific inquiry and a catalyzing tool reshaping how we conduct science, validate hypotheses, and draw reliable conclusions.

Medicine: CRISPR-Cas9 has ushered in a new era of possibilities in medicine, fundamentally altering medical healthcare standards of care with its revolutionary contributions to gene therapy and personalized medicine. Its precise gene-editing capabilities offer the promise of effectively treating, and in some cases potentially curing, a multitude of genetic disorders such as cystic fibrosis, sickle cell anemia, and certain forms of cancer that have long been considered intractable. The technology enhances the standard of care by providing a new toolkit for healthcare professionals, one grounded in peer-reviewed, evidence-based practices. It allows for highly targeted therapies customized to an individual's unique genetic makeup, thereby advancing personalized medicine. Clinical trials are increasingly incorporating CRISPR technology to test its efficacy and safety in human subjects, and its inclusion in medical practice is poised to redefine healthcare standards, moving them closer to a paradigm where genetic modification becomes a routine part of treatment for certain conditions. As these practices evolve, they will be subjected to continual review and adaptation to ensure they reflect the most current and effective methods while adhering to rigorous ethical standards. In legal contexts, adopting CRISPR technologies as part of standard care also raises new considerations for defining the adequacy and appropriateness of delivered care. Therefore, CRISPR-Cas9 is not just a new medical tool but a transformative force recalibrating our understanding of what is medically possible and ethically responsible.

Ethics: CRISPR-Cas9's transformative potential has necessitated a robust ethical discourse, particularly as it touches on the four principles of bioethics: patient autonomy (informed consent), practitioner beneficence (do good), practitioner nonmaleficence (do no harm), and public justice (be fair). Autonomy, often operationalized through informed consent, becomes especially complicated when considering germline edits impacting future generations who cannot consent. The principle of beneficence is highlighted and complicated by CRISPR's promise to treat or eradicate debilitating genetic diseases. At the same time, the technology could do immense good. Still, the prospect of "*designer babies*" raises questions about unintended societal consequences. Nonmaleficence, the principle of "*doing no harm*," must also be carefully weighed, as the long-term effects of genetic modifications are not fully understood, and off-target mutations could result in unforeseen complications. Finally, justice issues, including fair access to technology, come into play. The advent of gene editing could exacerbate social inequalities if only the wealthy have access to such treatments or if societal norms start favoring genetically edited traits. Additionally, there's the global justice aspect, as decisions made by one nation regarding the permissible uses of CRISPR technology could have transboundary impacts, affecting ecosystems and shared resources. Thus, CRISPR-Cas9 serves as a focal point where these four ethical principles intersect in complex ways, demanding a multi-faceted, internationally collaborative approach to navigate the moral labyrinth it presents.

Conclusion: CRISPR-Cas9 has greatly impacted various facets of human life, from advancing our understanding of genetics to introducing revolutionary medical treatments and raising complex ethical questions. Its precision in gene editing has served as a technological marvel, fulfilling rationalist aspirations of understanding and controlling the most foundational aspects of biology. Its empiricist orientation has made it a focal point for gathering vital empirical data through rigorous experimentation, enriching scientific understanding, and public policy. Technology has redefined the scientific method, providing a powerful tool for more controlled, replicable, and incisive research. In medicine, CRISPR-Cas9 is reorienting standards of care, offering unprecedented therapeutic avenues for previously incurable genetic conditions and adding new dimensions to ethical standards and legal frameworks. Yet, it also serves as a bioethical crucible, raising intricate issues of patient autonomy (informed consent), practitioner beneficence (do good), practitioner nonmaleficence (do no harm), and public justice (be fair) that challenge society to navigate the ethical complexities with care, caution, and collective responsibility. Overall, CRISPR-Cas9 is a groundbreaking scientific innovation that has not only expanded our capabilities but also compelled us to confront profound ethical and existential questions, ultimately reshaping our understanding of what it means to be human and the responsibilities that come with such knowledge and power.

CRISPR-Cas9's Legacy: Revolutionizes genetic engineering with precision, opening avenues for targeted therapies and debates over genetic modification ethics.

**

REVIEW QUESTIONS

True/False Questions:

1. CRISPR-Cas9 technology allows for precise modifications of specific genes within an organism's genome.
   True or False?

2. The ethical principle of nonmaleficence, which emphasizes "*do no harm*," is easily upheld in the application of CRISPR-Cas9.
   True or False?

Multiple-Choice Questions:

3. Which of the following is NOT a primary application of CRISPR-Cas9 technology?
   a) Gene therapy for genetic disorders
   b) Creating genetically modified organisms
   c) Predicting future economic trends
   d) Personalized medicine

4. What is a significant ethical concern associated with CRISPR-Cas9 technology?
   a) Its high cost of implementation
   b) The potential for off-target mutations
   c) The lack of applications in agriculture
   d) Limited access to laboratory facilities

Clinical Vignette:

5. A geneticist is using CRISPR-Cas9 to develop a treatment for a genetic disorder. During the research, unexpected off-target effects are observed, which could potentially cause harm. According to ethical principles, how should the geneticist proceed?
   a) Continue the research without addressing the off-target effects
   b) Inform the relevant authorities and halt the research to investigate the off-target effects
   c) Publish the results without further investigation
   d) Ignore the off-target effects and focus on the primary results

Basic Science Vignette:

6. A research team uses CRISPR-Cas9 technology to edit genes in a mouse model to study a genetic disorder. After several successful trials, the system starts showing unexpected gene interactions and off-target effects. What is the best course of action to address the unexpected effects?
   a) Ignore the off-target effects as the system's overall performance has improved
   b) Review and refine the CRISPR-Cas9 protocol using a more comprehensive dataset
   c) Remove the problematic genes from the study to prevent off-target effects
   d) Disable the CRISPR-Cas9 system until the off-target effects can be fully understood

Philosophy Vignette:

7. An AI system in genetic research uses CRISPR-Cas9 data to suggest ethical guidelines for gene editing in humans. It proposes an optimal framework that maximizes therapeutic outcomes but overlooks individual consent in some cases. What is the best philosophical approach to address this issue?
   a) Prioritize the AI's ability to maximize therapeutic outcomes
   b) Reprogram the AI to give more weight to individual consent than therapeutic outcomes
   c) Develop an oversight committee to review the AI's recommendations before implementation
   d) Disable the AI system until it can be programmed to always respect individual consent

Correct Answers:

1. True
2. False
3. c) Predicting future economic trends
4. b) The potential for off-target mutations
5. b) Inform the relevant authorities and halt the research to investigate the off-target effects
6. b) Review and refine the CRISPR-Cas9 protocol using a more comprehensive dataset
7. b) Reprogram the AI to give more weight to individual consent than therapeutic outcomes

BEYOND THE CHAPTER
CRISPR-Cas9

- *A Crack in Creation: Gene Editing and the Unthinkable Power to Control Evolution* by Jennifer Doudna and Samuel Sternberg
- *Editing Humanity: The CRISPR Revolution and the New Era of Genome Editing* by Kevin Davies
- *The Code Breaker: Jennifer Doudna, Gene Editing, and the Future of the Human Race* by Walter Isaacson

***

Philosophy of Science — Medical Research

Chapter 23: Genetic Engineering and Biotechnology — The Transformative Frontier: Ethical Dilemmas

Genetic engineering and biotechnology revolutionize medicine and agriculture, posing profound ethical dilemmas alongside groundbreaking advancements.

Abstract: In a world on the brink of revolutionary transformations, genetic engineering and biotechnology are at the forefront, heralding unprecedented advancements in medicine and agriculture. Positioned at the intersection of rationalism and empiricism, these fields rigorously apply the scientific method to manipulate life at the molecular level. Using deductive reasoning and evidence-based approaches, scientists address global challenges like disease eradication and food security. However, these advancements bring ethical dilemmas, necessitating a reevaluation of bioethical principles: patient autonomy (informed consent), practitioner beneficence (do good), practitioner nonmaleficence (do no harm), and public justice (be fair). As we navigate this transformative epoch, genetic engineering and biotechnology offer hope and complex moral considerations, urging society to balance innovation with ethical stewardship and consider the profound implications for human existence and the broader ecosystem.

**

Introduction: Genetic engineering and biotechnology have irrevocably altered the landscape of medicine, agriculture, and numerous other fields, promising unprecedented solutions to some of humanity's most pressing challenges. At the intersection of biology, chemistry, and technology, these disciplines involve manipulating genes and cellular processes to develop novel therapies, hardier crops, and even synthetic organisms. The advancements have sparked both awe and ethical debates, redefining what is possible while raising questions about what is permissible. From CRISPR gene editing, which can potentially eradicate inherited diseases, to synthetic biology aiming to construct artificial life forms, the applications are as vast as they are transformative. Despite the palpable excitement, these developments also conjure ethical dilemmas concerning individual autonomy, the principle of doing good (beneficence), the imperative to do no harm (nonmaleficence), and the equitable distribution of biotechnological benefits (justice). The impact of genetic engineering and biotechnology extends beyond mere scientific curiosity or industrial application, influencing our very understanding of life, ethics, and the human condition.

Rationalism: Genetic engineering and biotechnology are quintessential examples of rationalist inclinations in modern scientific endeavors. These fields rely heavily on deductive reasoning, starting with well-established biological principles, such as DNA structure or cellular replication mechanisms, and then applying logical derivations to manipulate these systems for specific outcomes. The hardwired human faculties for pattern recognition and problem-solving are leveraged to decode and re-engineer life's genetic blueprints. Scientists employ techniques like CRISPR for gene editing or synthetic biology for constructing artificial life forms, relying on "*a priori*" knowledge of molecular biology and genetics to execute these modifications successfully. The adaptive value of such rationalist approaches in biotechnology is evident. Quick, reliable solutions to pressing challenges like antibiotic-resistant bacteria, inherited genetic disorders, or food scarcity could revolutionize our standard of living and resilience as a species. However, this doesn't absolve practitioners from the complex ethical dilemmas these technologies pose. Ethicists, scientists, and policymakers must collaborate to examine the potential downsides and limitations of applying these rational faculties to biotechnology, such as unintentional genetic mutations or the equitable distribution of benefits. Understanding that the emergence of these advanced technologies is greater than the sum of their scientific parts, we must engage in complex ethical reasoning, even as we appreciate the role of innate cognitive faculties that make these advancements possible.

Empiricism: The empiricist orientation in genetic engineering and biotechnology emphasizes the role of sensory experiences, observations, and environmental interactions in gaining knowledge and shaping technologies. Grounded in inductive reasoning, the field relies on meticulous experiments and data collection to understand genetic and cellular mechanisms. For example, observing the effects of specific gene modifications on an organism's phenotype can lead to broader conclusions about genetic functions. These observations then inform further experiments, resulting in a continually refined understanding of biological systems. Cognitive biases, such as the tendency to focus on successful experiments at the expense of "*failed*" ones, can skew interpretations and thus require constant vigilance from researchers. The adaptive value of this empiricist approach lies in its flexibility and responsiveness to real-world challenges. Empirical data can swiftly inform modifications to existing technologies or protocols, allowing scientists to adapt to discoveries or unforeseen complications. This adaptability is crucial for addressing issues like antibiotic resistance or emerging diseases. At the societal level, empiricism guides the creation of evidence-based regulations and ethical guidelines, emphasizing the need for policies that can adapt as our empirical understanding of biotechnology evolves. Here, the concept of "*emergence*" is especially relevant: as genetic engineering and biotechnology progress, they create complex systems whose ethical and societal implications cannot be easily predicted from individual experiments or observations alone, necessitating a holistic, empirically informed approach to governance and ethical considerations.

The Scientific Method: Genetic engineering and biotechnology have had a profound impact on shaping and refining the scientific method, elevating the standards of rigor, replicability, and ethical scrutiny in biological research. As fields predicated on precisely manipulating genetic material and cellular processes, they demand extraordinary methodological rigor. Researchers must form testable hypotheses, often leveraging advanced computational models to predict outcomes, before proceeding to carefully controlled experiments. Using techniques like CRISPR gene editing or sequencing technologies necessitates meticulous data collection, often in real-time and at an unimaginable scale. The high-stakes nature of the work — ranging from potential cures for genetic disorders to bioengineered crops that can address food security — makes replicability a cornerstone. A single mistake can have far-reaching consequences, necessitating multi-layered reviews for experimental design and ethical considerations. Given the complex ethical landscape of manipulating life at the genetic level, peer review in these disciplines often extends beyond methodological rigor to include ethical evaluation, raising the bar for scientifically and ethically sound research. Genetic engineering and biotechnology have influenced their disciplines and set precedents affecting the broader scientific community by necessitating this heightened scrutiny and rigor. Their influence exemplifies how scientific inquiry must adapt and evolve to meet the complexities of modern challenges while maintaining fidelity to rigorous, ethical research principles.

Medicine: The contributions of genetic engineering and biotechnology to medicine have been transformative, fundamentally altering healthcare standards of care and introducing groundbreaking therapies that were previously inconceivable. Techniques like CRISPR gene editing can potentially eradicate or ameliorate genetic diseases, such as cystic fibrosis or muscular dystrophy, by precisely targeting and modifying the genes responsible for these conditions. Beyond gene therapies, biotechnology has ushered in personalized medicine, where treatments are tailored to an individual's genetic makeup, enhancing efficacy and minimizing adverse effects. Monoclonal antibody therapies, cancer immunotherapies, and the rapid development of mRNA vaccines for diseases are a testament to the revolutionary capabilities of these fields. Biotechnological innovations have raised" *reasonably competent and skilled*" healthcare standards by continually evolving and incorporating peer-reviewed, evidence-based practices. These advancements don't just reflect current best practices; they actively shape them, forcing an ongoing reassessment of medical protocols and ethical considerations. This dynamic shift has a profound impact in legal contexts as the rapidly evolving capabilities of biotechnology necessitate constant updates to legal standards governing the adequacy of medical care. Through their influence, genetic engineering and biotechnology have expanded medical treatment possibilities and recalibrated the ethical and professional standards that govern healthcare practice.

Ethics: The advent of genetic engineering and biotechnology has compelled a profound reexamination of the four principles of bioethics: patient autonomy (informed consent), practitioner beneficence (do good), practitioner nonmaleficence (do no harm), and public justice (be fair). The principle of autonomy, which underscores the importance of informed consent, is challenged by technologies that can alter genetic lineages, affecting the individual and future generations. The adequacy of "*informed*" consent comes into question when the long-term consequences of gene manipulation are still not fully understood. Beneficence, the principle of doing good, intersects complexly with these technologies. While the potential for eradicating diseases or enhancing human capabilities is immense, the "*good*" is often fraught with ambiguities, such as the ethical implications of "*designer babies*." Nonmaleficence, the commitment to do no harm, is similarly complicated. Techniques like gene editing come with the risk of off-target mutations or unintended consequences, posing a challenge to ensure that the technology does not inadvertently cause harm. Lastly, the principle of justice, or fairness in distributing benefits and burdens, is critical in a landscape where biotechnological innovations could easily exacerbate existing social inequalities. For instance, if only the affluent can access life-enhancing or life-extending technologies, it raises significant justice concerns. Thus, genetic engineering and biotechnology have advanced scientific capabilities and elevated the complexity and urgency of bioethical considerations, necessitating a multi-disciplinary, nuanced approach to navigate these ethical waters.

Conclusion: Genetic engineering and biotechnology have become cornerstones of modern civilization, fundamentally altering fields as diverse as medicine, agriculture, and ethics. These advancements stand at the intersection of rationalist and empiricist paradigms, guided by deductive and inductive reasoning to push the boundaries of what we understand as possible. By employing highly rigorous methodologies in line with the scientific method, these fields have raised the bar for scientific inquiry, requiring unprecedented rigor, replicability, and ethical scrutiny. In medicine, they have revolutionized healthcare standards of care, introducing novel therapies like gene editing and personalized treatments that challenge and expand our ethical frameworks. The ethical implications are exhilarating and sobering, forcing us to reconsider deeply ingrained principles of patient autonomy (informed consent), practitioner beneficence (do good), practitioner nonmaleficence (do no harm), and public justice (be fair). These technologies offer breathtaking prospects for human enhancement, disease eradication, and challenges to aging and mortality. Still, they also introduce ethical complexities that society must address with nuance and foresight. They raise questions about who gets to decide the genetic makeup of future generations, how to ensure equitable access, and what limitations should be placed on manipulating life at its most fundamental level. Thus, while genetic engineering and biotechnology promise unprecedented solutions to humanity's most pressing challenges, they equally mandate a collective, deeply reflective ethical examination of how we should navigate a landscape replete with promise and peril.

Genetic Engineering and Biotechnology's Legacy: Offers transformative medical and agricultural solutions while raising ethical questions about the manipulation of nature.

**

REVIEW QUESTIONS

True/False Questions:

1. Genetic engineering involves manipulating an organism's genes to achieve desired traits or outcomes.
   True or False?

2. Biotechnology solely focuses on agricultural applications and does not have implications for human medicine.
   True or False?

Multiple-Choice Questions:

3. Which technology is widely used in genetic engineering for precise gene editing?
   a) Polymerase Chain Reaction (PCR)
   b) CRISPR-Cas9
   c) Gel Electrophoresis
   d) Restriction Fragment Length Polymorphism (RFLP)

4. What ethical principle is challenged by genetic interventions that can affect future generations?
   a) Beneficence
   b) Autonomy
   c) Nonmaleficence
   d) Justice

Clinical Vignette:

5. A couple is considering using CRISPR-Cas9 technology to edit the genes of their unborn child to prevent a hereditary disease. Which ethical considerations should the healthcare provider discuss with them?
   a) The potential long-term effects on the child and future generations
   b) The possibility of off-target genetic effects or mutations
   c) Ensuring informed consent and understanding the risks and benefits
   d) All of the above

Basic Science Vignette:

6. A research team uses CRISPR technology to develop a new strain of wheat that is resistant to pests. After initial success, they start observing unexpected genetic mutations in the wheat. What is the best course of action to address these unexpected mutations?
   a) Ignore the mutations as the wheat's overall pest resistance has improved
   b) Review and refine the CRISPR protocol using a more comprehensive dataset
   c) Remove the new strain from further study to prevent the spread of mutations
   d) Disable the CRISPR system until the mutations can be fully understood

Philosophy Vignette:

7. An AI system in biotechnological ethics uses data to suggest guidelines for genetic modifications in agriculture. It proposes an optimal framework that maximizes crop yield but overlooks farmers' consent in implementing genetic changes. What is the best philosophical approach to address this issue?
   a) Prioritize the AI's ability to maximize crop yield
   b) Reprogram the AI to give more weight to farmers' consent than crop yield
   c) Develop an oversight committee to review the AI's recommendations before implementation
   d) Disable the AI system until it can be programmed to always respect farmers' consent

Correct Answers:

1. True
2. False
3. b) CRISPR-Cas9
4. b) Autonomy
5. d) All of the above
6. b) Review and refine the CRISPR protocol using a more comprehensive dataset
7. b) Reprogram the AI to give more weight to farmers' consent than crop yield

BEYOND THE CHAPTER
Genetic Engineering & Biotechnology

- *Regenesis: How Synthetic Biology Will Reinvent Nature and Ourselves* by George Church and Ed Regis
- *Biotechnology: Science for the New Millennium* by Ellyn Daugherty
- *Genetic Engineering: Principles and Methods* edited by Jane K. Setlow

***

Philosophy of Science

Medical Research

Chapter 24: Human Genome Project — The Genetic Blueprint: Scientific and Ethical Insights

The Human Genome Project maps our genetic blueprint, intertwining scientific discovery with ethical considerations.

Abstract: The Human Genome Project (HGP) is one of the most transformative scientific endeavors in history, reshaping fields like medicine, anthropology, and bioethics. Launched in 1990 and completed in 2003, this venture mapped human DNA, offering insights into human physiology and disease. It harmonized rationalist and empiricist paradigms, using theoretical assumptions and empirical evidence to unlock genetic knowledge. The HGP fostered an evolution in the scientific method, characterized by international collaboration and extensive data sharing, leading to advancements in medical diagnostics and treatments. It also ignited substantial bioethical discourse, spotlighting patient autonomy (informed consent), practitioner beneficence (do good), practitioner nonmaleficence (do no harm), and public justice (be fair), shaping policy and regulations for genomic exploration. The HGP ushered in personalized medicine and predictive health analytics, enriching our approach to bioethical concerns and intertwining scientific innovation with philosophical and moral inquiries.

**

Introduction: The Human Genome Project (HGP) was an ambitious international scientific venture that sought to map all the genes that make up human DNA, essentially decoding the blueprint of life itself. Launched in 1990 and completed in 2003, the project aimed to identify the approximately 20,000-25,000 genes in human DNA and determine the sequences of the 3 billion chemical base pairs that form these genes. In doing so, the HGP promised to revolutionize medicine, anthropology, and many other fields by providing unprecedented insight into the genetic underpinnings of human physiology and disease. Its discoveries have led to remarkable advancements in diagnostic methods, treatment modalities, and predictive health analytics. Beyond its scientific and medical ramifications, the Human Genome Project has prompted profound philosophical and ethical inquiries. It serves as a touchstone for discussions around rationalism versus empiricism, the shaping of the scientific method, and the bioethical implications concerning individual patient autonomy (informed consent), practitioner beneficence (do good), practitioner nonmaleficence (do no harm), and public justice (be fair).

Rationalism: The Human Genome Project (HGP) reflects rationalist inclinations in several compelling ways, particularly in demonstrating the use of human cognitive faculties for acquiring knowledge about our very nature. At the core of the HGP was the "*a priori*" assumption that understanding the human genome would yield vital clues about physiology, disease, and even behavior — a deduction drawn from the axiomatic belief in the fundamental role of genetics in biological systems. The project applied complex problem-solving skills, pattern recognition, and advanced reasoning, undergirded by postulates about how genetics influences life, to decode the "*Book of Life*." Rationalistic approaches were evident in the structured methodologies used to sequence DNA, utilizing innate faculties of logic and mathematical reasoning. Additionally, the project itself can be seen as a manifestation of emergent properties, where thousands of scientists' collective intellectual and technological endeavors led to outcomes far more complex and revealing than any individual effort could have achieved. This emergence illustrates how basic cognitive faculties can interact in intricate ways to give rise to advanced ethical reasoning or self-awareness, as the HGP necessitated an ethical framework to examine issues like genetic privacy, potential discrimination, and the very essence of human identity. On a public policy level, the project became a case study of balancing scientific ambition with ethical considerations, prompting regulations that aim to mitigate the downside of our evolved rational faculties. The Human Genome Project is a remarkable testament to human rationalism, from its foundational assumptions to its profound ethical implications.

Empiricism: The Human Genome Project (HGP) also exhibits a strong empiricist orientation, complementing its rationalist underpinnings. The project was grounded in the meticulous collection of empirical data — namely, the sequences of billions of nucleotide bases comprising human DNA. Employing inductive reasoning, scientists used specific observations and experimental data to draw broad conclusions about genetic functions, relationships, and abnormalities. These empirical methodologies became especially critical for understanding the role of particular genes in diseases, leading to more targeted treatments and diagnostics. Cognitive biases, often seen as heuristic shortcuts in empiricism, were carefully managed through peer review and collaborative scrutiny, thereby elevating the quality of data interpretation. The influence of socialization and culture also became apparent, especially as the HGP began to study genetic variations across different populations, highlighting the interplay between genetics and environment. Ethically, the empiricist orientation underscored the necessity of evidence-based guidelines for addressing complex issues such as genetic privacy and the potential for gene editing. At the public policy level, the empirically derived insights from the HGP have led to adaptive regulations responsive to the ongoing discoveries that continue to evolve even after the project's official completion. The project demonstrates the adaptive value of empirical faculties by giving us tools to address unforeseen health challenges and advancing our collective understanding of human biology. Lastly, the concept of "*emergence*" is illustrated through the HGP's transformative impact on diverse fields like medicine, anthropology, and ethics — effects that could not have been predicted merely by studying individual genes but emerged from a complex interplay of empirical data and interpretation.

The Scientific Method: The Human Genome Project (HGP) has profoundly impacted shaping and refining the scientific method, particularly in large-scale, collaborative research. While the scientific method traditionally starts with observation and proceeds to hypothesis testing through controlled experiments, the HGP elevated this process to a level of global cooperation and computational intensity that was previously uncharted. The HGP began with a clear observation — that the genetic code holds secrets vital for understanding human biology and disease — and formulated a testable hypothesis regarding the sequencing of the human genome. The project was groundbreaking in its scale, demanding unprecedented levels of data collection, computational analysis, and experimental rigor. One of its lasting contributions to the scientific method is the emphasis on data sharing and replicability. Given its international scope, the project necessitated data transparency and accessibility that became a model for other large-scale scientific endeavors. This ensured that findings were replicable across different laboratories and by various researchers, a cornerstone of scientific credibility. Peer review was integral to the HGP, both during its operation and in the subsequent analysis of its data, providing rigorous checks on its methodology and conclusions. The Human Genome Project advanced our understanding of genetics by successfully navigating these complexities. It pioneered how complex scientific research can be conducted, scrutinized, and validated in the modern era.

Medicine: The Human Genome Project has had a transformative impact on medicine, setting new standards of care and accelerating the shift toward personalized medicine. By sequencing the entire human genome and identifying the roles of various genes, the HGP has furnished clinicians and researchers with invaluable tools for understanding, diagnosing, and treating diseases at a genetic level. The project has identified numerous genes associated with diseases like cancer, Alzheimer's, and various inherited disorders, enabling more targeted and effective treatments. These advances have deeply influenced standards of care by establishing new, evidence-based practices grounded in genetic knowledge. For example, genetic screening has become an integral part of prenatal care, and gene-targeted therapies like CRISPR are being developed as potentially revolutionary treatments for previously incurable conditions. Moreover, the HGP has facilitated the growth of pharmacogenomics, which allows for the customization of drug therapies based on an individual's genetic makeup, thus maximizing efficacy while minimizing adverse effects. The insights from the HGP have become woven into the fabric of medical best practices, continually evolving to reflect current research findings. In legal contexts, these evidence-based standards set by genomic information are increasingly cited to assess the adequacy of delivered care. Overall, the Human Genome Project has redefined medical science by integrating genetics into the core of healthcare, shaping medical healthcare standards of care in more personalized, effective, and scientifically grounded ways.

Ethics: The Human Genome Project (HGP) has had a profound influence on the landscape of bioethics, prompting rigorous debate and policy-making around the principles of patient autonomy (informed consent), practitioner beneficence (do good), practitioner nonmaleficence (do no harm), and public justice (be fair). Regarding autonomy, the project has elevated the importance of informed consent, especially when gathering genetic samples. Given the deeply personal and potentially revealing nature of genetic data, securing consent became a touchstone for ethical research in genomics. Concerning beneficence, the HGP has been revolutionary in its potential to do good, offering unprecedented insights into disease mechanisms and catalyzing targeted therapies and preventive measures. However, the principle of nonmaleficence — doing no harm — has been equally critical, raising concerns about the potential misuse of genetic information for discrimination or stigmatization. Regulatory frameworks have been established to prevent such harms, including laws prohibiting genetic discrimination by employers or insurers. Justice, the obligation to be fair in the distribution of benefits and risks, has also been a focal point, especially regarding the access to and benefits of genomic medicine. Questions have arisen about the equitable distribution of gene-based therapies and the inclusivity of different populations in genetic research to avoid exacerbating existing healthcare disparities. The Human Genome Project has acted as a catalyst for the evolution of bioethical principles, necessitating a nuanced, multidimensional ethical framework to guide both the scientific community and public policy in the genomic era.

Conclusion: The Human Genome Project (HGP) is one of history's most transformative scientific endeavors, deeply affecting our understanding of biology and rippling across various disciplines such as medicine, ethics, public policy, and even our understanding of human cognition and rationality. By decoding the blueprint of human life, the HGP has advanced diagnostic methods, heralded the era of personalized medicine, and propelled evidence-based healthcare standards. The rationalist and empiricist orientations within the project have synergized to yield foundational assumptions and empirical evidence, providing a holistic understanding of human genetics and its implications. The HGP has also enriched the scientific method, acting as a paragon of international collaboration and data transparency and establishing models for large-scale, complex scientific research. Furthermore, it has catalyzed necessary ethical dialogues, offering new case studies and challenges in bioethics related to patient autonomy (informed consent), practitioner beneficence (do good), practitioner nonmaleficence (do no harm), and public justice (be fair). These conversations have spawned regulations that are adaptive to societal changes and technological advances, guiding the scientific community and public policy into a new era. The Human Genome Project is a monumental achievement that has deepened our scientific understanding and prompted us to think more rigorously about human life's ethical, rational, and empirical aspects.

The Human Genome Project's Legacy: By decoding the genetic blueprint of human life, the Human Genome Project has profoundly redefined our understanding of biology, medicine, and ethics, serving as an enduring testament to the interplay of rationalism and empiricism in advancing both scientific knowledge and ethical discourse.

**

REVIEW QUESTIONS

True/False Questions:

1. The Human Genome Project (HGP) was completed in 1990 and mapped all human genes. True or False?
2. The HGP has had no significant impact on medical diagnostics and treatments. True or False?

Multiple-Choice Questions:

3. Which of the following best describes the primary aim of the Human Genome Project?
   a) To create genetically modified organisms
   b) To map and understand all the genes of human beings
   c) To develop new agricultural techniques
   d) To study ancient human fossils
4. What ethical principle ensures that participants in genetic research are fully aware of the risks and benefits involved?
   a) Beneficence
   b) Nonmaleficence
   c) Autonomy
   d) Justice

Clinical Vignette:

5. A genetic counselor is explaining to a patient the potential outcomes and risks associated with undergoing genetic testing for a hereditary disease. The patient is given all necessary information to make an informed decision. Which ethical principle is primarily being addressed in this scenario?
   a) Beneficence
   b) Nonmaleficence
   c) Justice
   d) Autonomy

Basic Science Vignette:

6. During the Human Genome Project, scientists discovered a gene associated with a severe hereditary disease. They developed a gene-editing tool to correct this genetic defect. After initial success in the lab, the tool showed unexpected mutations in other genes. What is the best course of action to address these mutations?
   a) Ignore the mutations as the overall success rate is high
   b) Review and refine the gene-editing tool to minimize off-target effects
   c) Remove the tool from use to prevent any potential harm
   d) Disable the gene-editing tool until the mutations can be fully understood

Philosophy Vignette:

7. An AI system analyzing data from the Human Genome Project suggests a framework for using genetic information in healthcare. It proposes an optimal strategy that prioritizes efficient treatment but overlooks the privacy and consent of patients. What is the best philosophical approach to address this issue?
   a) Prioritize the AI's ability to maximize treatment efficiency
   b) Reprogram the AI to give more weight to patient privacy and informed consent
   c) Develop an oversight committee to review the AI's recommendations before implementation
   d) Disable the AI system until it can be programmed to always respect patient privacy and informed consent

Correct Answers:

1. False
2. False
3. b) To map and understand all the genes of human beings
4. c) Autonomy
5. d) Autonomy
6. b) Review and refine the gene-editing tool to minimize off-target effects
7. b) Reprogram the AI to give more weight to patient privacy and informed consent

BEYOND THE CHAPTER
Human Genome Project

- *The Human Genome Project: Cracking the Genetic Code of Life* by Thomas F. Lee
- *Drawing the Map of Life: Inside the Human Genome Project* by Victor K. McElheny
- *The Common Thread: A Story of Science, Politics, Ethics and the Human Genome* by John Sulston and Georgina Ferry

***

Philosophy of Science — Medical Research

Chapter 25: Large Language Models — The AI Frontier: Rethinking Knowledge

Large Language Models redefine knowledge and ethics, merging AI innovation with moral complexity.

Abstract: Large Language Models (LLMs) like ChatGPT showcase rapid advancements in artificial intelligence, demonstrating immense potential and ethical complexity. Built on the Transformer architecture, LLMs excel in text generation and context-sensitive interactions, extending their utility to scientific research and healthcare. However, these innovations bring ethical dilemmas, requiring a reassessment of principles such as patient autonomy (informed consent), practitioner beneficence (do good), practitioner nonmaleficence (do no harm), and public justice (be fair). These models combine rationalist foundations, dictated by algorithms, with empiricist orientations, using vast datasets. As they assist in research and potentially revolutionize peer review, they challenge the scientific method's rigor. In medicine, LLMs promise enhancements in diagnostics and patient care but raise concerns about healthcare standards and legal accountability. This progression necessitates evolving ethical frameworks to address issues of bias, transparency, and fairness, involving efforts from developers, organizations, and policymakers. As LLMs evolve, they require a harmonized approach to innovation that aligns with human values and societal needs, epitomizing the intersection of technological prowess and ethical intricacy.

**

Introduction: Large Language Models (LLMs) like ChatGPT, built on the revolutionary Transformer architecture, represent a groundbreaking development in artificial intelligence, raising questions and igniting debates across various disciplines. Harnessing the power of self-attention mechanisms to focus on different parts of vast datasets, these models can generate human-like text and provide context-sensitive interactions. Their utility extends from seemingly trivial tasks like auto-completing text to complex functions such as aiding scientific research, streamlining medical diagnostics, and writing code. However, the rise of these sophisticated algorithms brings a host of ethical and philosophical conundrums. Questions arise about how rational and empirical these systems are in their approach to knowledge and whether they can meaningfully contribute to scientific methodologies and medical healthcare standards of care. Moreover, deploying LLMs in sensitive areas opens up crucial dialogues about ethical considerations in technology. They provoke us to reassess and critically evaluate bioethical principles such as patient autonomy (informed consent), practitioner beneficence (do good), practitioner nonmaleficence (do no harm), and public justice (be fair), particularly in the evolving landscapes of machine learning and data science.

Rationalism: Large Language Models (LLMs) like ChatGPT can be examined through rationalism, as their operations are fundamentally based on pre-defined algorithms, architectures, and initial conditions. These "*a priori*" principles govern how LLMs process data and make inferences, acting as their rationalist backbone. The deductive reasoning these models engage in is not a product of some form of artificial consciousness but an outcome of complex computations dictated by their foundational algorithms. However, the issue of "*emergence*" complicates this rationalist perspective. As

LLMs sift through enormous amounts of data, they sometimes exhibit behaviors or problem-solving techniques that were not explicitly programmed, revealing complexity that can seem almost organic. This emergence raises ethical considerations about predictability and control, especially when these models are applied in sensitive fields like medicine or law. From a moral standpoint, this places a considerable responsibility on the developers and data scientists to ensure that the algorithms they create are unbiased, just, and transparent. On an organizational level, tech companies are tasked with upholding rigorous ethical standards, which include scrutinizing their models for potential bias and unintended outcomes. Public policy, too, plays a vital role by setting guidelines for the ethical governance and accountability of these advanced technologies. This amalgam of rationalist inclinations and emergent behaviors in LLMs thus poses challenges and opportunities that require a multi-dimensional approach to understand and manage.

Empiricism: In the context of Large Language Models (LLMs), empiricism emphasizes the significance of the quality and volume of the training data, leaning heavily on inductive reasoning. Unlike their rationalist inclinations, which are algorithmically dictated, LLMs' empiricist orientation relies on aggregating specific data points to form generalized models or principles. Through this immense dataset, LLMs like ChatGPT gain their ability to generate human-like text, answer questions, and even solve complex problems. However, this empirical approach has its ethical pitfalls, primarily when the training data include biases — be they overt or subtle. Such tendencies can distort the model's understanding and decision-making, potentially perpetuating harmful stereotypes or inequalities. On a personal level, the ethical responsibility falls upon data scientists to scrutinize datasets for biases and other ethical issues. Professionally, organizational leadership must enact rigorous ethical guidelines and oversight mechanisms to prevent these biases from influencing the model's behavior. In public policy, transparent regulations are imperative for ensuring ethical accountability in data collection and algorithmic decision-making. The empiricist aspect of LLMs also interacts intriguingly with the concept of "*emergence*." Because the models are data-driven, they can sometimes adapt to new data patterns or solve problems they weren't initially designed for. While this adaptability and resilience can be beneficial, it also presents ethical challenges, especially if the model's newfound capabilities haven't been scrutinized for ethical implications.

The Scientific Method: Large Language Models (LLMs) are poised to have a transformative impact on the scientific method, both as tools that assist in the research process and as subjects of inquiry. By rapidly parsing through vast volumes of academic literature, LLMs can assist researchers in identifying gaps in existing knowledge, thereby aiding in formulating research questions and hypotheses. They can also help design experiments by suggesting methodologies or analyzing potential variables, drawing from the wealth of data they have been trained on. Moreover, LLMs can assist in data analysis, spotting patterns and correlations within large datasets that might be difficult for human researchers to discern quickly. However, introducing LLMs into scientific research presents challenges around replicability and peer review. Given that these models are black boxes to some extent, ensuring that their operations are transparent enough to be replicated by other researchers and aligning with the scientific method's core principle of replicability becomes crucial. Furthermore, as LLMs get more sophisticated, there may be questions around the peer review process — Can human peers evaluate an AI-assisted paper alone, or do we need machine-readable justifications? Including LLMs in scientific endeavors necessitates adjustments to traditional methodologies and ethical practices to ensure that they augment, rather than undermine, the scientific method's rigorous standards for inquiry and validation.

Medicine: Large Language Models (LLMs) are making noteworthy contributions to medicine, particularly in streamlining research, diagnostics, and patient care. For instance, these models can sift through enormous volumes of medical literature to identify relevant studies, thereby aiding in developing evidence-based practices that inform standards of care. By analyzing countless research papers, clinical trials, and patient records, LLMs can offer insights into disease patterns treatment efficacies, and even predict patient outcomes, serving as a supplemental tool for medical healthcare professionals. They also hold promise in telemedicine, where they can provide preliminary medical advice based on recognized medical healthcare best practices. However, with the caveat that human medical healthcare experts should verify such advice. Despite these advantages, using LLMs in medicine raises ethical and legal concerns related to the medical healthcare standards of care. Since these standards are predicated on peer-reviewed, evidence-based practices provided by competent medical healthcare professionals, algorithmic decision-making complicates accountability and competence. For LLMs to be integrated responsibly into medical practice, rigorous validation must ensure their recommendations align with established medical healthcare standards of care. Furthermore, their role in legal contexts — particularly in medical malpractice cases — requires scrutiny. Overall, while LLMs have the potential to augment healthcare significantly, they must be implemented cautiously, respecting the highly regulated and ethically sensitive nature of medical practice.

Ethics: Large Language Models (LLMs) intersect intriguingly with the four core principles of bioethics: patient autonomy (informed consent), practitioner beneficence (do good), practitioner nonmaleficence (do no harm), and public justice (be fair). Starting with autonomy, LLMs can assist in the process of informed consent by providing comprehensive and accessible information to patients, thereby helping them make educated decisions about their care. However, there's a caveat: if not designed correctly, LLMs could inadvertently omit important information or produce biased advice, compromising the quality of informed consent. In terms of beneficence, LLMs have the potential to greatly benefit medical healthcare by aiding in diagnostics, treatment planning, and medical research. However, this is closely tied to the principle of nonmaleficence, which commands that we "*do no harm.*" LLMs can also perpetuate biases or make errors, so strict oversight is required to ensure they contribute positively to patient outcomes without causing harm. Finally, justice — or fairness — comes into play in how these technologies are deployed. LLMs can democratize access to healthcare information and services. Still, they can also exacerbate healthcare inequities if not designed and regulated carefully, such as by only being available to wealthier institutions or favoring data from certain demographic groups. Therefore, developing and deploying LLMs in healthcare must be guided by a robust ethical framework that carefully considers each of these principles, ensuring that these advanced technologies enrich, not undermine, the ethical foundations of medical practice.

Conclusion: Large Language Models (LLMs) like ChatGPT is a place where rationalism and empiricism converge, resulting in an entity capable of mimicking human-level textual comprehension and generation. While algorithmically grounded, making them subject to rationalist scrutiny, their effectiveness is deeply rooted in the quality and scope of the data they've been trained on, an empiricist characteristic. LLMs offer unprecedented opportunities in diverse sectors, from shaping scientific inquiry to revolutionizing healthcare practices. However, this potential comes with considerable ethical responsibilities around fairness, bias, and transparency that implicate developers, organizations, and policy-makers alike. Moreover, as these models become increasingly capable, issues surrounding bioethical principles such as patient autonomy (informed consent), practitioner beneficence (do good), practitioner nonmaleficence (do no harm), and public justice (be fair) come into play, particularly in fields as sensitive as medicine. Ethical frameworks must evolve to scrutinize their rationalist and empiricist tendencies, ensuring that as these models continue to evolve, they do so in a manner aligned with human values and societal needs. Overall, LLMs epitomize the intersection of technological capability and ethical complexity, challenging us to harmonize innovation with moral and social imperatives.

Large Language Models' Legacy: LLMs are poised to be a catalyst for unprecedented advancements in human-machine collaboration, fundamentally reshaping epistemology by challenging traditional notions of knowledge acquisition, dissemination, and creation, thereby compelling us to rethink ethical, ontological, and teleological questions in an increasingly digitized world.

**

REVIEW QUESTIONS

True/False Questions:

1. Large Language Models (LLMs) like ChatGPT are primarily built on the Transformer architecture and utilize self-attention mechanisms to process vast datasets.
   True or False?

2. One of the ethical concerns surrounding LLMs is their potential to democratize access to healthcare information and services without any associated risks.
   True or False?

Multiple-Choice Questions:

3. Which of the following principles is NOT directly related to bioethics as discussed in the context of LLMs?
   a) Patient autonomy
   b) Practitioner beneficence
   c) Nonmaleficence
   d) Economic efficiency

4. In the medical context, what is a primary advantage of using LLMs like ChatGPT?
   a) Providing unverified medical advice
   b) Replacing healthcare professionals
   c) Analyzing large volumes of medical literature to aid evidence-based practices
   d) Eliminating the need for human oversight in diagnostics

Clinical Vignette:

5. A hospital is considering integrating an LLM to assist in patient care by analyzing medical records and suggesting treatment plans. However, there are concerns about the ethical implications of this technology. What should be the primary focus to ensure the ethical deployment of the LLM?
   a) Reducing the number of healthcare staff
   b) Ensuring the LLM can operate independently without human intervention
   c) Maintaining rigorous validation and oversight to align with established medical healthcare standards of care
   d) Allowing the LLM to make final treatment decisions without human review

Basic Science Vignette:

6. A hospital uses an advanced LLM-based system to assist in diagnosing patients. The system suggests a diagnosis for a rare disease but misses a common side effect of a prescribed medication. What is the best course of action to address this issue?
a) Trust the system's diagnosis as it usually has high accuracy
b) Review and refine the LLM's training data to include more comprehensive information on common medication side effects
c) Remove the system's diagnostic capabilities to prevent future errors
d) Disable the LLM until the missed side effects can be fully understood and addressed

Philosophy Vignette:

7. An LLM system is deployed in a hospital to provide medical advice based on vast datasets. It offers efficient treatment plans but occasionally overlooks patient consent. What is the best philosophical approach to address this issue?
a) Prioritize the LLM's efficiency in providing treatment plans
b) Reprogram the LLM to give more weight to patient consent than treatment efficiency
c) Develop an oversight committee to review the LLM's recommendations before implementation
d) Disable the LLM system until it can ensure patient consent is always respected

Correct Answers:

1. True
2. False
3. d) Economic efficiency
4. c) Analyzing large volumes of medical literature to aid evidence-based practices
5. c) Maintaining rigorous validation and oversight to align with established medical healthcare standards of care
6. b) Review and refine the LLM's training data to include more comprehensive information on common medication side effects
7. b) Reprogram the LLM to give more weight to patient consent than treatment efficiency

BEYOND THE CHAPTER
Large Language Models (LLM)

- *Artificial Intelligence: Structures and Strategies for Complex Problem Solving* by George F. Luger
- *Language Models are Few-Shot Learners* by Tom B. Brown et al. (Research Paper)
- *GPT-3, Bloviator: OpenAI's Language Generator Has No Idea What It's Talking About* by Gary Marcus (Article)

***

Chapter 26: Nanotechnology — The Atomic Precision: Ethical Innovation

Nanotechnology pioneers atomic precision, integrating ethical innovation and scientific advancements.

Abstract: Nanotechnology, a transformative frontier melding principles from physics, chemistry, biology, and engineering, epitomizes scientific innovation and complexity. Emerging from Richard Feynman's seminal concepts in the 1950s, this discipline, characterized by manipulating matter on an atomic scale, offers opportunities from revolutionizing healthcare to creating resilient infrastructures. The field demands a comprehensive approach integrating rationalist and empiricist paradigms, fostering systematic development grounded in foundational principles while emphasizing empirical research. As it expands and refines the scientific method, it heralds groundbreaking medical strides, especially in disease diagnosis and targeted therapies, promising unprecedented healthcare advancements. However, it also presents ethical dilemmas, necessitating robust frameworks for navigating informed consent, equitable access, and potential biological ramifications. Consequently, as nanotechnology rewrites narratives across domains, it challenges us to align scientific innovation with ethical responsibility and societal well-being, ensuring advancements harmonize with ethical discernment and inclusivity.

**

Introduction: Nanotechnology, the manipulation of matter on an atomic or molecular scale, is a groundbreaking field that has opened doors to possibilities once considered the realm of science fiction. This interdisciplinary science involves the application of principles from physics, chemistry, biology, and engineering to create materials and systems with unique properties that are often fundamentally different from their macroscopic counterparts. Originating from the visionary ideas of Nobel laureate Richard Feynman in the 1950s, nanotechnology has become an instrumental force in various industries — from producing more efficient energy solutions and environmental cleanup methods to revolutionary advancements in medicine and computing. Characterized by its remarkable ability to engineer matter at scales between 1 and 100 nanometers, the technology offers extraordinary potential benefits and formidable ethical challenges. Whether crafting nanoparticles to target cancer cells or developing nanomaterials for more durable infrastructure, nanotechnology presents numerous opportunities and dilemmas that compel us to reassess traditional paradigms across the scientific, ethical, and social spectra.

Rationalism: In nanotechnology, rationalist inclinations are evident in the organized and systematic approach to manipulating matter at the atomic or molecular level. Developers often rely on "*a priori*" principles, derived from fundamental laws in physics and chemistry, to predict and control the unique properties of nanoscale materials. These foundational elements serve as axiomatic truths from which researchers can deduce new methods, systems, and applications through a rational, logical process. Like machine learning, where algorithms are built on pre-defined architectures or initial conditions, nanotechnologists employ well-established scientific principles as starting points for novel explorations. Ethically, the rationalist perspective places considerable responsibility on the scientists and engineers who create nanoscale materials and systems. They are accountable for understanding the potential positive and negative implications of manipulating matter at this scale and scrutinizing their methods for bias or unintended harmful consequences. This requires robust ethical frameworks that scrutinize the development process at an organizational level, ensuring that products do not lead to societal inequities or environmental degradation. Regarding public policy, the rationalist underpinning of nanotechnology calls for transparent governance models that can effectively manage the complexities and uncertainties inherent in this transformative field. Like the concept of "*emergence*" in machine learning, nanotechnology also has the capacity for unpredicted novel behaviors — such as unanticipated chemical reactions or biological interactions — that raise incredible opportunities and serious ethical dilemmas.

Empiricism: Nanotechnology, while rooted in "*a priori*" principles, also heavily relies on an empiricist orientation, where observation and experimentation serve as crucial pillars in the field's advancement. Empiricism comes into play through the meticulous collection of data points and the use of inductive reasoning to establish generalized principles that can guide future nanotechnology applications. Here, the properties and behaviors of nanoscale materials, often distinct from their macroscopic counterparts, are explored through empirical research to refine models and create more effective applications. The ethical considerations in this empiricist approach closely parallel those in machine learning. Data scientists and nanotechnologists bear a similar ethical responsibility to ensure that the data employed — chemical properties, biological interactions, or other aspects — are scrutinized for biases and inaccuracies that could lead to harmful outcomes. At the organizational level, frameworks must be established to oversee the ethical integrity of empirical research, ensuring that the resultant nanotechnologies do not exacerbate societal inequities or pose unforeseen risks. In the realm of public policy, there is a need for regulations that ensure ethical accountability and transparency in the collection and interpretation of nanoscale data. The empiricist approach also allows for the "*emergence*" concept, where nanomaterials and systems may reveal new, unexpected properties or applications. While this adaptability presents exciting opportunities for innovation, it also introduces ethical complexities, requiring ongoing vigilance to assess the implications of these emerging characteristics.

The Scientific Method: Nanotechnology has profoundly influenced the modern application and understanding of the scientific method. While the basic tenets of observation, hypothesis formulation, experimentation, and conclusion have remained stable, nanotechnology has necessitated specialized techniques and tools for observing phenomena at the nanoscale. Advanced instruments like scanning tunneling microscopes and atomic force microscopes allow for the precise observation and manipulation of individual atoms and molecules, expanding the boundaries of what can be empirically examined. These tools have also enabled more controlled experiments, where nanoscale materials can be isolated and manipulated under varying conditions to test hypotheses rigorously. The meticulous collection and analysis of data have become increasingly computerized, incorporating algorithms and computational models to handle the complexity of interactions at the atomic level. The necessity for replicability in nanotechnology has pushed for the standardization of methods and tools, ensuring that different researchers can confirm results under the same conditions. Lastly, the peer review process in nanotechnology has had to adapt to the field's interdisciplinary nature, often requiring expertise in physics, chemistry, biology, and engineering to scrutinize the validity and significance of research findings adequately. In this way, nanotechnology has adapted the scientific method to meet its unique challenges and enriched it, setting new standards for rigor, precision, and interdisciplinary collaboration.

Medicine: Nanotechnology has made groundbreaking contributions to medicine, significantly augmenting healthcare standards of care and offering transformative solutions to various medical challenges. Nanoparticles, for instance, have been engineered to target specific cells for drug delivery, enabling more effective treatments with fewer side effects, especially in cancer therapy. Nanoscale materials have also revolutionized diagnostics, enabling the rapid and accurate detection of diseases at their earliest stages through highly sensitive biosensors and imaging agents. Beyond diagnostics and drug delivery, nanotechnology has been instrumental in tissue engineering and regenerative medicine, providing scaffolds at the molecular level that facilitate tissue growth and repair. These advances align with the medical healthcare standards of care by being grounded in peer-reviewed, evidence-based practices. The interdisciplinary nature of nanomedicine — integrating principles from biology, chemistry, and physics — demands rigorous clinical trials and ethical reviews to establish new best practices. Moreover, nanotechnology has been critical in making healthcare more personalized, enabling therapies tailored to an individual's genetic makeup, thus aligning with the evolving nature of standards of care that consider patient-specific needs. By pushing the boundaries of what is medically possible while maintaining a rigorous focus on safety and efficacy, nanotechnology is meeting and raising the medical healthcare standards of care.

Ethics: Nanotechnology's burgeoning impact on medicine and other life sciences has brought bioethical principles sharply into focus. Beginning with autonomy, the microscopic scale of nanotech applications can challenge the traditional concept of informed consent. Understanding the risks and benefits of a conventional medical procedure is one thing. Still, it is quite another to grasp the implications of nanoparticles interacting with cellular machinery, which could necessitate re-evaluating how informed consent is obtained. Beneficence, the principle of doing good, is evident in nanotechnology's potential for revolutionary healthcare interventions — from targeted drug delivery to tissue regeneration — offering promise for substantially improved patient outcomes. However, this must be balanced against the principle of nonmaleficence, or "*do no harm,*" given that the long-term effects of nanoscale materials in the body are not yet fully understood. Preventive measures must be in place to minimize risks of toxicity, unintended biological interactions, or other harmful effects. Lastly, the principle of justice, or fairness, raises concerns about the accessibility and equitable distribution of nanotechnology-based treatments. Given the often high cost of developing and implementing nanoscale technologies, there's a real risk that the benefits could be disproportionately available to wealthier populations, exacerbating existing healthcare inequalities. Thus, as nanotechnology continues to advance, its developers and the broader medical community must carefully navigate these complex ethical terrains to ensure that the technology is deployed in a groundbreaking and ethically sound manner.

Conclusion: Nanotechnology's pervasive influence spans multiple disciplines, radically altering approaches in rationalism, empiricism, the scientific method, medicine, and ethics. A unique amalgamation of rationalist thinking and empiricist methodology, nanotechnology has redefined how we understand and manipulate matter, infusing theoretical foundations and experimental rigor into its development. By providing the tools and techniques to scrutinize phenomena at the nanoscale, it has extended the capabilities of the scientific method, enriching research methodologies and elevating standards of rigor. In medicine, nanotechnology promises revolutionary changes — from targeted therapies to advanced diagnostics — reshaping the standards of care and introducing new paradigms for treatment. Ethically, however, it forces us to confront questions that challenge traditional bioethical principles like patient autonomy (informed consent), practitioner beneficence (do good), practitioner nonmaleficence (do no harm), and public justice (be fair). As nanotechnological applications become more integrated into society, ethical considerations grow increasingly complex, underscoring the need for interdisciplinary dialogue and stringent governance. Through its transformative potential and the ethical conundrums it presents, nanotechnology stands as a monumental scientific frontier, demanding a reevaluation of existing norms and practices across scientific, ethical, and societal dimensions.

Nanotechnology's Legacy: Promises groundbreaking applications across fields, from medicine to materials, by manipulating matter on an atomic scale.

**

REVIEW QUESTIONS

True/False Questions:

1. Nanotechnology involves manipulating matter on an atomic or molecular scale, leading to materials and systems with unique properties.
   True or False?

2. The principles of rationalism and empiricism are irrelevant to the development and application of nanotechnology.
   True or False?

Multiple-Choice Questions:

3. Which of the following is NOT an application of nanotechnology in medicine?
   a) Targeted drug delivery
   b) Tissue engineering
   c) Diagnosing diseases with biosensors
   d) Routine blood pressure measurement

4. What ethical concern is associated with the high cost of developing and implementing nanoscale technologies?
   a) Informed consent
   b) Equitable access and distribution
   c) Nonmaleficence
   d) Beneficence

Clinical Vignette:

5. A patient is considering a treatment involving nanoparticles designed to target and destroy cancer cells. The doctor explains that while this treatment offers promising results, there are potential long-term effects that are not yet fully understood. Which ethical principle is primarily being addressed here?
   a) Reducing healthcare costs
   b) Ensuring patient autonomy through informed consent
   c) Promoting practitioner beneficence by prioritizing patient care
   d) Avoiding harm by considering nonmaleficence

Basic Science Vignette:

6. Scientists develop a new nanoparticle-based drug delivery system to target cancer cells specifically, but there are concerns about potential long-term effects on the body. What is the most prudent next step for these researchers?
   a) Conduct extensive animal studies to evaluate long-term effects before human trials.
   b) Proceed directly to human trials to assess the system's effectiveness and safety in humans.
   c) Abandon the nanoparticle approach due to potential unknown risks.
   d) Rely on data from other nanoparticle studies to justify moving forward with human trials.

Philosophy Vignette:

7. A pharmaceutical company plans to use nanotechnology to create a new treatment that promises to revolutionize chronic disease management. However, the long-term effects of the nanoparticles are not fully understood. What is the best ethical approach to address this uncertainty?
   a) Prioritize the potential benefits and proceed with the treatment.
   b) Ensure patients are fully informed about the potential risks and obtain their consent before treatment.
   c) Develop an oversight committee to continuously monitor the treatment's effects and make necessary adjustments.
   d) Halt the treatment until comprehensive long-term studies are conducted.

Correct Answers:

1. True
2. False
3. d) Routine blood pressure measurement
4. b) Equitable access and distribution
5. d) Avoiding harm by considering nonmaleficence
6; a) Conduct extensive animal studies to evaluate long-term effects before human trials
7. b) Ensure patients are fully informed about the potential risks and obtain their consent before treatment

BEYOND THE CHAPTER
Nanotechnology

- *Nanotechnology: Understanding Small Systems* by Ben Rogers, Jesse Adams, and Sumita Pennathur
- *Introduction to Nanotechnology* by Charles P. Poole and Frank J. Owens
- *Nanotechnology: Science, Innovation, and Opportunity* by Lynn E. Foster

***

Chapter 27: Pharmacogenomics — The Personalized Medicine: Genetic Insights

Pharmacogenomics personalizes medicine, leveraging genetic insights to optimize treatment and ethical care.

Abstract: Pharmacogenomics, at the intersection of genetics and pharmacology, heralds a transformative era in personalized medicine by unpacking the relationship between genetic variations and drug responses. This frontier aims to replace the conventional one-size-fits-all method with individualized treatments, enhancing drug efficacy and reducing adverse reactions through understanding genetic markers. As a dynamic, multidisciplinary field, it blends rationalistic underpinnings with empirical validation, continually shaping the scientific method in precision medicine. Pharmacogenomics shifts the healthcare paradigm towards a more individual-centric approach, fostering heightened ethical discourse. It recalibrates medical standards, revolutionizing treatment by integrating genetic intricacies and broader healthcare dynamics, steering personalized medicine toward a more inclusive, ethical, and effective future.

**

Introduction: Pharmacogenomics, a burgeoning field at the intersection of genetics and pharmacology, aims to elucidate the intricate relationship between an individual's genetic makeup and drug response. Unlike traditional one-size-fits-all approaches to medication, pharmacogenomics seeks to personalize drug therapies based on genetic variations that influence how patients metabolize, respond to, or tolerate certain medications. This endeavor represents a quantum leap in the quest for precision medicine, offering the potential for optimized treatment plans tailored to the individual. By considering genetic predispositions, healthcare providers aim to maximize therapeutic effects while minimizing adverse reactions and inefficacies, potentially revolutionizing the medical treatment landscape. As we explore pharmacogenomics, assessing its rationalist foundations, reliance on empirical data, role in refining the scientific method, and contributions to the field of medicine becomes crucial. Moreover, the ethical implications of this scientific advancement beckon scrutiny, particularly in how it impacts the four cornerstone principles of bioethics: patient autonomy (informed consent), practitioner beneficence (do good), practitioner nonmaleficence (do no harm), and public justice (be fair).

Rationalism: Pharmacogenomics is fundamentally grounded in rationalist principles, especially as it seeks to deduce individual medication responses based on the axiomatic understanding of genetic structures and functions. The field operates on the evolutionary-hardwired human faculties for pattern recognition and problem-solving by identifying how specific genetic markers correspond to drug interactions. Pharmacogenomics employs deductive reasoning to establish definitional truths about how genetic variations determine individual drug responses, aiming to predict therapeutic outcomes or side effects from the genetic code. This rationalistic approach assumes a certain level of pre-existing "*a priori*" principles — that genes govern biological functions and, therefore, can influence how drugs are metabolized and take effect. The complexity arises when one considers epigenetic factors and the multi-layered interactions between various genes, controlled by innate faculties and learned or environmental factors. The collective cognitive capabilities of the scientists, bioinformaticians, and pharmacologists involved in pharmacogenomics are emblematic of the concept of "*emergence*," where the sum is more significant than its individual parts. As a multi-disciplinary field, it capitalizes on the synergistic capabilities of diverse specialists who bring their rational faculties to bear on exceedingly complex problems. This complexity makes it imperative for professionals to scrutinize the ethical implications of their work and for public policy to guide the field responsibly, keeping in mind both the potential and limitations of human rationalistic faculties. By harnessing these faculties, pharmacogenomics aims to resolve recurring challenges in healthcare, such as drug inefficacy and adverse reactions, thus improving overall human fitness and survival.

Empiricism: Pharmacogenomics also draws heavily from empiricist principles, requiring rigorous empirical data to validate the theoretical frameworks derived from rationalistic foundations. In this field, the key modus operandi involves extensive genetic sequencing, clinical trials, and observational studies to gather sensory-based data on how individuals respond to different medications. This inductive approach allows for specific observations about drug responses to be generalized into broader pharmacological guidelines. Cognitive biases, such as the tendency to focus on successful case studies while overlooking non-responders, present challenges that researchers must be conscious of to ensure ethical and scientific integrity. Moreover, external factors like ethnic diversity and environmental variables add complexity that necessitates a comprehensive, empirically grounded approach to drug therapy personalization. The importance of empirical evidence in pharmacogenomics extends to ethical and policy considerations. Ethical guidelines and regulations in this arena are increasingly framed based on empirical findings, emphasizing the role of real-world data in ethical decision-making. For instance, as knowledge expands, policy directives must adapt and change, always tethered to the most recent empirical evidence. Regarding adaptive value, the empiricist approach equips pharmacogenomics with the flexibility to adjust to new scientific discoveries and emerging healthcare challenges. The concept of "*emergence*" manifests as the field accounts for how the collective properties of genetics, environment, and patient history result in complex and often unpredictable drug responses, transcending the simplistic sum of individual variables.

The Scientific Method: Pharmacogenomics is pivotal in utilizing and refining the scientific method, particularly in personalized medicine and healthcare innovation. The field often starts with observing medications' varying efficacy and side effects across individuals, leading researchers to question whether genetic factors can explain these disparities. From this, hypotheses are formulated that certain genetic markers or sets of markers could be indicators of drug response or intolerance. Controlled clinical trials and extensive genetic sequencing efforts are deployed to test these hypotheses, resulting in meticulously gathered and scrutinized data. One of the strengths of pharmacogenomics is its reliance on replicable experiments, often across diverse ethnic and environmental backgrounds, to ensure the generalizability and credibility of its findings. The multi-disciplinary nature of pharmacogenomics — encompassing genetics, pharmacology, data science, and clinical medicine — adds a layer of complexity to the traditional scientific method. This requires new analytical methods and statistical models to interpret the voluminous and intricate data generated, effectively modernizing the methodology for data analysis within the scientific framework. Peer review remains a crucial aspect, offering a multi-layered scrutiny that involves experts from various disciplines, thereby ensuring the integrity and validity of the research. Pharmacogenomics leverages the scientific method and contributes to its evolution by incorporating advanced data analytics and multi-disciplinary expertise, setting a precedent for other emerging fields in the scientific community.

Medicine: Pharmacogenomics has made monumental contributions to medicine, most notably by pioneering the shift toward personalized medical healthcare. Pharmacogenomics has facilitated more targeted and effective treatments by enabling a deeper understanding of how individual genetic makeup influences drug response. This tailoring of medication to the patient has profound implications for the standards of care in medicine, as it challenges the traditional one-size-fits-all paradigm that has long been the norm. The application of pharmacogenomics has been especially impactful in fields like oncology, cardiology, and psychiatry, where medication responses can vary widely, and the stakes for adverse effects are high. In terms of medical healthcare standards of care, pharmacogenomics is increasingly becoming incorporated into evidence-based practices. It is a powerful tool for medical healthcare providers to meet and even exceed the expected level of care, making treatment not just reasonably competent but extraordinarily precise. As the field continues to grow, pharmacogenomic testing will become a routine part of the diagnostic and treatment planning process, raising the medical healthcare standards of care. Peer-reviewed research in pharmacogenomics provides the empirical foundation to update these standards continually, ensuring they reflect current best practices informed by a blend of research, clinical expertise, and patient needs. By setting new benchmarks for effective, individualized treatment, pharmacogenomics is revolutionizing medical practice and the very definition of medical healthcare standards of care.

Ethics: The ethical landscape of pharmacogenomics is intricately shaped by its adherence to the four fundamental principles of bioethics: patient autonomy (informed consent), practitioner beneficence (do good), practitioner nonmaleficence (do no harm), and public justice (be fair). Each principle harbors both positive outcomes and potential challenges in pharmacogenomics. The principle of patient autonomy, especially through informed consent, is crucial in pharmacogenomics. This field enhances patient autonomy by enabling individuals to make informed decisions about their treatment based on their genetic profile. However, this also raises concerns about privacy and the potential misuse of genetic data. Ensuring informed consent means educating patients about the benefits and risks of pharmacogenomic testing and safeguarding their genetic information from unauthorized use. Practitioner beneficence in pharmacogenomics manifests in the tailored treatment plans that maximize therapeutic benefits for the patient. By considering genetic factors, healthcare providers can increase the efficacy of medications and reduce the risk of adverse reactions. Conversely, there's a risk of over-reliance on genetic factors, overlooking other important patient-specific considerations such as lifestyle or environment. Practitioner nonmaleficence is particularly pertinent in pharmacogenomics, which seeks to minimize harm by reducing the likelihood of adverse drug reactions. However, the genetic complexity and the evolving nature of the field mean that there is still a risk of unintended consequences, such as harmful side effects not yet understood in the context of certain genetic profiles. The commitment to "*do no harm*" involves continuous research and vigilance to update treatment protocols as new information becomes available. Pharmacogenomics also raises significant questions about public justice, particularly regarding accessibility and equity. The positive aspect lies in its potential to provide more effective and safer treatments for diverse populations. However, there is a risk of exacerbating healthcare disparities if access to pharmacogenomic testing and personalized therapies is limited to certain groups. Ensuring fair and equal access to these advancements is a critical ethical consideration. While pharmacogenomics promises significant advancements in personalized medicine, it also necessitates a careful ethical balance. Upholding patient autonomy through informed consent, ensuring practitioner beneficence and nonmaleficence, and promoting public justice is essential to realizing this field's full potential ethically and effectively.

Conclusion: Pharmacogenomics is a paradigm-shifting development in modern healthcare, representing a synthesis of rationalism, empiricism, scientific methodology, medical innovation, and ethical contemplation. Rooted in both rationalist foundations of genetic theory and empirical substantiation through clinical trials, pharmacogenomics propels healthcare toward highly personalized treatment regimens. It doesn't just employ the scientific method; it also refines it by integrating interdisciplinary data analytics, enhancing the depth and breadth of scientific inquiry. In terms of medical contributions, it recalibrates the standards of care, paving the way for treatment plans that are not merely adequate but highly customized to individual genetic makeup. This revolution in healthcare has its ethical conundrums, necessitating nuanced protocols around patient autonomy (informed consent), practitioner beneficence (do good), practitioner nonmaleficence (do no harm), and public justice (be fair) to ensure equitable, secure, and ethically sound applications. As pharmacogenomics continues to advance, it elevates both the aspirations and the responsibilities of the scientific and medical communities, fostering an integrative approach that combines rationalist theories, empirical evidence, rigorous scientific evaluation, and ethical consideration to forge the future of personalized medicine.

Pharmacogenomics' Legacy: The most projected legacy of pharmacogenomics is the democratization of personalized medicine, fundamentally altering the ethical and practical landscape of healthcare by ensuring that treatment is as individual as the patient's genetic makeup.

**

REVIEW QUESTIONS

True/False Questions:

1. Pharmacogenomics aims to personalize drug therapies based on genetic variations, enhancing drug efficacy and reducing adverse reactions.
   True or False?

2. Pharmacogenomics eliminates the need for empirical data in understanding drug responses.
   True or False?

Multiple-Choice Questions:

3. Which of the following is NOT a primary focus of pharmacogenomics?
   a) Personalized drug therapy
   b) Understanding genetic influences on drug metabolism
   c) General dietary guidelines
   d) Reducing adverse drug reactions

4. What is a significant ethical concern associated with pharmacogenomics?
   a) Developing standard treatment protocols
   b) Ensuring equitable access to genetic testing and personalized treatments
   c) Establishing non-specific medical guidelines
   d) Implementing uniform drug dosages for all patients

Clinical Vignette:

5. A 60-year-old woman with a history of breast cancer is being treated with a commonly prescribed chemotherapy drug. Her oncologist suggests pharmacogenomic testing to determine if she has any genetic variations that could affect her response to the drug. What is the primary benefit of conducting this pharmacogenomic test?
   a) It will guarantee that the patient will not experience any side effects
   b) It will ensure that the patient can discontinue other treatments
   c) It will help tailor the chemotherapy dosage to improve efficacy and reduce adverse effects
   d) It will eliminate the need for follow-up visit

Basic Science Vignette:

6. A 45-year-old patient with a history of hypertension and diabetes is being considered for pharmacogenomic testing to optimize their medication regimen. Given their genetic makeup, the goal is to maximize therapeutic efficacy and minimize adverse reactions. What is the most crucial factor to consider in this scenario?
   a) The cost of pharmacogenomic testing
   b) The patient's family history of chronic illnesses
   c) The specific genetic markers that influence drug metabolism
   d) The patient's current medication regime

Philosophy Vignette:

7. In a clinical trial, researchers discover that certain genetic markers can predict adverse reactions to a widely used antidepressant. Which principle of bioethics is most directly challenged by the use of genetic information in this context?
   a) Nonmaleficence (do no harm)
   b) Beneficence (do good)
   c) Autonomy (informed consent)
   d) Justice (fair distribution of benefits and burdens)

Correct Answers:

1. True
2. False
3. c) General dietary guidelines
4. b) Ensuring equitable access to genetic testing and personalized treatments
5. c) It will help tailor the chemotherapy dosage to improve efficacy and reduce adverse effects
6. c) The specific genetic markers that influence drug metabolism
7. a) Nonmaleficence (do no harm)

BEYOND THE CHAPTER
Pharmacogenomics

- *Pharmacogenomics: Challenges and Opportunities in Therapeutic Implementation* edited by Lamya M. Shihabuddin and Liewei Wang
- *Pharmacogenomics: An Introduction and Clinical Perspective* by Joseph S. Bertino Jr. et al.
- *Pharmacogenomics: The Search for Individualized Therapies* edited by Julio Licinio and Ma-Li Wong

***

Philosophy of Science

Medical Research

WEEK SIX: Tailored and Innovations

WEEK SIX

Class Hour 1

- Chapter 28: Precision Medicine — The Tailored Treatment: Efficacy and Safety
- Chapter 29: Quantum Computing — The Computational Revolution: Ethical Horizons
- Chapter 30: Digital Health and Telemedicine — The Accessible Healthcare: Modern Solutions

Class Hour 2

- Chapter 31: Virtual and Augmented Reality in Medicine — The Immersive Innovation: Transformative Healthcare
- Chapter 32: Richard Dawkins (1941-) — The Evolutionary Biologist: Gene-Centered Evolution
- Chapter 33: Climate Change — The Anthropocene Challenge: Ethical Imperatives

---

Introduction

In the sixth week of our course, "*Philosophy of Science*," we explore the themes of tailored treatments and innovative technologies in healthcare. This week's chapters guide us through the transformative impacts of precision medicine, quantum computing, digital health, virtual and augmented reality, and the broader ethical imperatives presented by climate change. Additionally, we delve into the influential work of Richard Dawkins in evolutionary biology, examining how his rationalist and empiricist approaches have shaped scientific discourse and public understanding.

Our journey begins with precision medicine, a groundbreaking approach that tailors treatments to the unique genetic, environmental, and lifestyle factors of each patient. This personalized methodology leverages computational biology to interpret extensive data arrays, offering unprecedented accuracy in diagnosing and treating diseases. Precision medicine revolutionizes the therapeutic landscape, enhancing intervention effectiveness and safety. It blends rationalist and empirical strategies, reworking the scientific method and introducing new standards of care. This transformation necessitates a reevaluation of ethical principles and public policies to balance innovation with ethical integrity, fostering social equity and justice in healthcare.

Next, we explore the revolutionary potential of quantum computing. Rooted in the principles of quantum mechanics, this technology promises to revolutionize numerous sectors, including cryptography, material science, and medicine. Quantum computing challenges and expands the scientific method, necessitating nuanced approaches to experimentation and verification. Its potential to accelerate drug discovery and personalize healthcare requires rigorous ethical deliberation, particularly within bioethics. As quantum computing teeters on the brink of feasibility, it opens a rich avenue of interdisciplinary discourse, probing its remarkable computational potential and the societal, philosophical, and ethical dimensions that its development and deployment encompass.

Digital health and telemedicine emerge as formidable forces harmonizing technology with clinical practice to foster unprecedented accessibility and personalization in medical services. Telemedicine optimizes healthcare through swift, reliable responses and evolves through rigorous data analysis and iterative refinements grounded in real-world experiences. It introduces innovative paradigms into medical research, reshaping the scientific method and enhancing traditional healthcare delivery. However, telemedicine also unveils bioethical dilemmas that necessitate continual scrutiny, ensuring responsible integration into the medical landscape and leaving a legacy of bridged gaps and modernized healthcare delivery.

Integrating Virtual Reality (VR) and Augmented Reality (AR) into the medical sector marks a revolutionary shift. These technologies advance surgical simulations, diagnostics, education, and therapeutic interventions, enhancing practitioners' proficiency and democratizing access to quality care. Grounded in rationalist foundations and empirical orientations, VR and AR facilitate complex procedures and allow continuous refinement through data and sensory experiences. However, their growth brings critical ethical considerations, requiring ongoing ethical evaluation to ensure responsible integration into healthcare. VR and AR serve as catalysts for academic and ethical dialogues, reshaping the future of healthcare through transformative advancements.

Renowned evolutionary biologist Richard Dawkins has profoundly impacted the scientific and literary world with his gene-centered view of evolution and advocacy for rationalism and skepticism. His seminal work, "*The Selfish Gene*," and his critique of religious perspectives in "*The God Delusion*" have shaped public understanding and scientific discourse. Dawkins promotes a society that relies on reason, logic, and empirical evidence, advocating the scientific method as a systematic approach to understanding the natural world. His influence extends to medicine, fostering evidence-based practices and critical thinking aligned with modern medical principles. Dawkins also shapes bioethical discourse, advocating for dignity, rights, and well-being in healthcare through rational discourse and evidence-based decision-making.

Finally, we address the ethical imperatives presented by climate change, epitomizing the Anthropocene era where human activities wield a geologic impact. Climate change poses a complex dilemma with global repercussions, necessitating interdisciplinary approaches to safeguard our planet. This chapter explores the profound influence of climate change across various facets, including rationalist cognitive faculties, ethical deliberations, and empiricism in shaping responsive policies. It underscores the necessity of broadening our ethical compass to navigate the intricacies of climate-induced disparities and environmental stewardship. Addressing climate change's colossal challenges requires fostering interdisciplinary dialogues and adaptive strategies, encouraging intellectual and ethical societal evolution.

Together, these chapters illustrate the intricate balance between tailored treatments, innovative technologies, and ethical considerations in healthcare. From precision medicine to the transformative impacts of quantum computing, digital health, VR and AR, and the ethical challenges of climate change, we gain a deeper understanding of the principles guiding scientific innovation and ethical responsibility. This integrated exploration underscores the importance of maintaining ethical integrity while embracing technological advancements, ensuring that innovation serves the greater good and aligns with societal values.

***

Chapter 28: **Precision Medicine** — The Tailored Treatment: Efficacy and Safety

Precision medicine tailors treatments, blending efficacy with safety and ethical integrity.

Abstract: Precision medicine marks a pivotal transformation in healthcare by tailoring treatments to the unique genetic, environmental, and lifestyle factors of each patient. This approach transcends the conventional one-size-fits-all methodology, leveraging computational biology to interpret extensive data arrays, from genomic sequencing to electronic health records. At the crossroads of genomics, data science, and clinical medicine, it offers unprecedented accuracy in diagnosing and treating diseases. Precision medicine revolutionizes the therapeutic landscape and redefines medical research and practice, enhancing intervention effectiveness and safety. This transformation blends rationalist and empirical strategies, reworking the scientific method and introducing new standards of care. It necessitates a reevaluation of ethical principles and public policies to balance innovation with ethical integrity, fostering social equity and justice in healthcare.

**

Introduction: Precision medicine heralds a paradigm shift in healthcare by offering treatments tailored to individual patients based on their unique genetic, environmental, and lifestyle factors. Instead of a one-size-fits-all approach that applies generalized therapies, precision medicine seeks to optimize efficacy by considering each patient's specific circumstances and biological makeup. This method employs sophisticated algorithms and computational biology to interpret vast datasets, ranging from genomic sequencing to electronic health records. Doing so promises a new level of accuracy and personalization in disease diagnosis, prognosis, and treatment. As an evolving field, precision medicine stands at the intersection of genomics, data science, and clinical medicine, transforming the therapeutic landscape and the ethos of medical research and practice. This groundbreaking approach has far-reaching implications for medicine, fundamentally altering how clinicians and healthcare professionals approach patient care, how researchers conduct scientific inquiry, and how ethical considerations are weighed in bioethics.

Rationalism: Precision medicine is rooted in a rationalist framework that seeks to leverage our innate cognitive faculties for pattern recognition, problem-solving, and deductive reasoning. Researchers draw upon foundational axioms in genetics and bioinformatics to create and optimize algorithms to analyze genetic, environmental, and lifestyle data. The idea that specific genes correlate with particular diseases or that certain environmental exposures can influence disease outcomes is an "*a priori*" principle that guides deductive reasoning. Advanced computational techniques then sift through complex datasets to discern patterns and make predictions, mimicking and augmenting our inherent faculties for problem-solving and pattern recognition. Moreover, the rationalist foundations of precision medicine extend to the epigenetic factors contributing to disease predisposition and progression, furthering the field's reach beyond straightforward genetic determinism. Yet, the approach has its complexities. Ethically, relying on big data and advanced analytics to make health decisions requires rigorous oversight to avoid reinforcing existing biases or creating new ones. At a professional level, scientists and ethicists must scrutinize these methodologies within an ethical framework that respects individual autonomy and societal norms. On a public policy level, there's an imperative to establish regulations that consider the limitations of our cognitive faculties, especially as they apply to complex systems like human biology. These efforts recognize that while individual faculties like basic reasoning or pattern recognition are powerful, the emergence of collective capabilities, such as ethical frameworks and advanced medical treatments, results from these faculties interacting in nuanced and complex ways.

Empiricism: Precision medicine embodies an empiricist orientation by heavily collecting and interpreting sensory data — primarily genetic, biochemical, and clinical — to guide medical healthcare decisions. It employs inductive reasoning as its cornerstone methodology: specific observations from individual patients' medical records, genetic profiles, and environmental contexts are analyzed to arrive at more general conclusions about disease mechanisms, effective treatments, and even prevention strategies. Developers and clinicians place immense importance on empirical evidence, continually revising algorithms and treatment plans based on real-world outcomes and new data. This adaptability is critical for addressing unforeseen environmental challenges and staying on the cutting edge of scientific discovery, thereby enhancing human survival and fitness. Cognitive biases, influenced by socialization and culture, can also impact the interpretation of empirical data in precision medicine. As such, there is a growing awareness of the need to scrutinize these biases in the context of individual medical healthcare decision-making and larger ethical frameworks. From a public policy standpoint, this emphasizes the necessity for empirically grounded regulations that can adapt to technological advances and societal changes. Just as empiricism observes the emergence of complex behaviors and systems from simpler sensory experiences, precision medicine aims to understand human biology and disease's complex, emergent properties by piecing together data from various empirical sources. This multi-layered empiricist approach provides a more holistic understanding of health and disease that transcends what any single data point can offer.

The Scientific Method: Precision medicine is reshaping the scientific method by integrating multidisciplinary approaches and leveraging big data analytics, thereby introducing new paradigms of inquiry into medical research. Traditional methods often focus on isolated variables in controlled environments, whereas precision medicine embraces the complexity of multiple interacting factors — genetic, environmental, and lifestyle — that influence health and disease. This demands expanding the scientific method to incorporate data science techniques like machine learning and computational biology, which can analyze vast, heterogeneous datasets to form testable hypotheses. The field also challenges conventional definitions of replicability. Instead of expecting identical outcomes across diverse patient populations — a notion often untenable in complex biological systems — precision medicine aims for predictive accuracy within specified contexts. Furthermore, the methodological rigors of precision medicine have led to the adoption of more advanced validation techniques, including real-world evidence and adaptive clinical trials that evolve based on interim results. These innovations are setting new standards for peer review, requiring experts to be versed in medical science, data analytics, and algorithmic transparency. By weaving together these varied disciplines and methodologies, precision medicine is not just utilizing the scientific method but evolving it to meet the complexities of contemporary healthcare challenges, thereby enriching its depth, scope, and applicability across scientific research.

Medicine: Precision medicine is revolutionizing the clinical landscape by substantially enriching medical healthcare standards of care with highly individualized, evidence-based treatment options. No longer bound by a "*one-size-fits-all*" model, clinicians can customize medical interventions to a previously unimaginable degree, thanks to advancements in genomics, proteomics, and metabolomics coupled with powerful data analytics. Precision medicine facilitates more effective and less invasive treatments by better understanding the intricate interplay between a person's genetic makeup, lifestyle, and environmental factors. For instance, targeted therapies in oncology allow for the destruction of cancerous cells with minimal impact on healthy tissues, drastically reducing side effects and improving patient outcomes. The field has also catalyzed advancements in diagnostics, enabling earlier detection of diseases and, therefore, more timely and effective interventions. These innovations are becoming integrated into the fabric of medical best practices, redefining the "*standards of care*" by adding layers of specificity and personalization. Consequently, legal frameworks are evolving to incorporate these new standards, shaping the definition of competent and adequate healthcare. By merging rigorous scientific research, state-of-the-art technology, and clinical expertise, precision medicine meets and often exceeds existing care standards, setting new benchmarks that promise to better align medical healthcare outcomes with individual patient needs.

Ethics: Precision medicine has a profound impact on the four foundational principles of bioethics: patient autonomy (informed consent), practitioner beneficence (do good), practitioner nonmaleficence (do no harm), and public justice (be fair). Concerning autonomy, the advent of personalized treatment plans necessitates reevaluating informed consent. Patients are now required to understand and decide upon more complex, individually tailored treatments, potentially including experimental therapies, making obtaining informed consent a nuanced and critical endeavor. Beneficence is enhanced by the promise of more effective and personalized therapies, with a higher probability of success and a lower risk of adverse side effects. However, this same capacity for personalized care also brings into focus nonmaleficence; the misuse of genetic and personal data could harm individuals and lead to group-based stigmatization or discrimination. Justice poses the most challenging ethical dilemma. While precision medicine has the potential to offer remarkably effective treatments, its reliance on advanced technologies and specialized expertise can make it prohibitively expensive, thus risking the exacerbation of healthcare inequalities. Ensuring fair access to these potentially life-altering advancements becomes a key ethical issue. As precision medicine continues to evolve, it obliges clinicians, ethicists, and policymakers to reexamine and refine these ethical principles, adapting them to a new era of medical practice that offers unprecedented opportunities but also presents complex moral challenges.

Conclusion: Precision medicine represents a transformative paradigm shift in healthcare that redefines our understanding of medicine, scientific inquiry, and ethics. Its highly individualized approach to treatment integrates genomics, environmental data, and lifestyle factors, revolutionizing standards of care by offering targeted therapies that maximize efficacy and minimize harm. Emerging from rationalist foundations, precision medicine employs deductive reasoning and computational algorithms to interpret complex biological data while also embracing an empiricist orientation that values real-world outcomes and inductive reasoning. This innovative medical framework has also reshaped the scientific method, introducing new paradigms incorporating data science techniques, adaptive clinical trials, and real-world evidence as part of a robust approach to research validation. While the field offers promising advances, such as more effective treatments and earlier diagnosis, it poses significant ethical challenges, particularly concerning patient autonomy (informed consent), practitioner beneficence (do good), practitioner nonmaleficence (do no harm), and public justice (be fair). As this groundbreaking approach continues to evolve, it compels us to revisit and adapt existing ethical frameworks and public policies, ensuring that as we venture into this new frontier, we do so with a commitment to scientific rigor, ethical integrity, and social equity.

Precision Medicine's Legacy: Tailors medical treatments to individual patient profiles, increasing the efficacy and safety of interventions.

**

REVIEW QUESTIONS

True/False Questions:

1. Precision medicine focuses on tailoring medical treatments to individual patients based on their unique genetic, environmental, and lifestyle factors.
   True or False?

2. Precision medicine disregards the need for advanced computational techniques and data analysis.
   True or False?

Multiple-Choice Questions:

3. Which foundational principle of bioethics is particularly challenged by the need for complex, individualized informed consent in precision medicine?
   a) Beneficence
   b) Nonmaleficence
   c) Autonomy
   d) Justice

4. Precision medicine redefines the standards of care by:
   a) Applying generalized therapies to all patients
   b) Ignoring genetic and environmental factors
   c) Customizing medical interventions based on individual patient data
   d) Reducing the role of data analytics in medical research

Clinical Vignette:

5. A 55-year-old male patient with a history of lung cancer is being considered for a targeted therapy based on his genetic profile. His genomic sequencing reveals a specific mutation that is known to respond well to a particular drug. However, the drug is experimental and has not been widely tested. The patient is well-informed about the potential benefits and risks but is uncertain about proceeding with the treatment. What is the best course of action for the healthcare provider?
   a) Recommend the treatment without further discussion
   b) Advise against the treatment due to its experimental nature
   c) Discuss the potential benefits and risks in detail and respect the patient's autonomy in making the final decision
   d) Suggest a traditional chemotherapy approach without considering the genetic data

Basic Science Vignette:

6. A researcher is studying the impact of genetic variations on the effectiveness of a new drug designed to treat Type 2 diabetes. They collect data from 1,000 patients, including their genetic profiles, lifestyle factors, and treatment outcomes. The data analysis reveals that patients with a specific genetic mutation respond significantly better to the drug than those without the mutation. What is the best conclusion the researcher can draw from this study?
   a) The drug should be prescribed to all patients with Type 2 diabetes regardless of genetic profile
   b) The drug is ineffective for treating Type 2 diabetes
   c) The drug may be particularly effective for patients with the specific genetic mutation, suggesting a tailored treatment approach
   d) The drug should be withdrawn from the market due to inconsistent results

Philosophy Vignette:

7. A bioethicist is evaluating the ethical implications of using big data analytics in precision medicine. They are particularly concerned about the potential for bias in the algorithms used to interpret genetic and environmental data. The bioethicist must consider the principles of bioethics, including justice and nonmaleficence, in their evaluation. What is the most appropriate ethical concern for the bioethicist to focus on?
   a) Ensuring all patients receive the same treatment regardless of their genetic data
   b) Avoiding the collection of any genetic data to prevent potential misuse
   c) Ensuring the algorithms are designed to minimize bias and do not reinforce existing healthcare inequalities
   d) Encouraging the use of experimental treatments without informed consent

Correct Answers:

1. True
2. False
3. c) Autonomy
4. c) Customizing medical interventions based on individual patient data
5. c) Discuss the potential benefits and risks in detail and respect the patient's autonomy in making the final decision
6. c) The drug may be particularly effective for patients with the specific genetic mutation, suggesting a tailored treatment approach
7. c) Ensuring the algorithms are designed to minimize bias and do not reinforce existing healthcare inequalities

BEYOND THE CHAPTER

Precision Medicine

- *Genomic and Precision Medicine: Foundations, Translation, and Implementation* edited by Geoffrey S. Ginsburg and Huntington F. Willard
- *Precision Medicine: A Guide to Genomics in Clinical Practice* by Jeanette McCarthy and Bryce A. Mendelsohn
- *The Personalized Medicine Revolution: How Diagnosing and Treating Disease Are About to Change Forever*by Pieter Cullis

***

Philosophy of Science — Medical Research

Chapter 29: Quantum Computing — The Computational Revolution: Ethical Horizons

Quantum computing revolutionizes computation, prompting ethical considerations and scientific innovation.

Abstract: Quantum computing, at the forefront of technological innovation, epitomizes a paradigm shift in computational science, leveraging the principles of quantum mechanics to potentially revolutionize numerous sectors, including cryptography, material science, and medicine. Rooted in the theoretical frameworks of rationalism and empiricism, its evolution has challenged and expanded the scientific method, necessitating nuanced approaches to experimentation and verification while promising unprecedented advancements in medical science through accelerated drug discovery and personalized healthcare. However, this burgeoning field also beckons rigorous ethical deliberation, particularly within bioethics, contemplating patient autonomy (informed consent), practitioner beneficence (do good), practitioner nonmaleficence (do no harm), and public justice (be fair) principles. As it teeters on the brink of feasibility, it opens a rich avenue of interdisciplinary discourse, probing its remarkable computational potential and the societal, philosophical, and ethical dimensions that its development and deployment encompass. This technology mandates collaborative responsibility and vigilant oversight to ensure a future where its transformative impacts are harnessed with discernment and equity.

**

Introduction: Quantum computing represents a paradigm shift in the field of computation, leveraging the principles of quantum mechanics to perform calculations at speeds unimaginable with classical computers. Unlike classical bits, which can be either 0 or 1, quantum bits (qubits) can exist in superpositions of states, allowing them to perform multiple calculations simultaneously. This groundbreaking technology promises to revolutionize various sectors, from solving complex mathematical problems and simulating molecular structures for drug development to breaking existing cryptographic codes and facilitating secure communications. As quantum computing teeters on the brink of feasibility, its prospective impact on various scientific disciplines has become a subject of intense discourse. This discussion extends beyond the realm of computer science and physics to involve philosophical debates regarding its rationalist and empiricist underpinnings, its role in redefining the scientific method, its potential contributions to medical science, and the ethical quandaries it poses, particularly in the realm of bioethics.

Rationalism: In the burgeoning field of quantum computing, rationalist inclinations deeply influence the architecture and algorithms that underpin these next-generation machines. The construction of a quantum computer starts with the axiomatic understanding of quantum mechanics, involving principles such as superposition and entanglement. The fundamental "*a priori*" elements, such as qubit behaviors, quantum gates, and error correction algorithms, are deduced from these axioms and subsequently encoded into the architecture. Developers operate from these theoretical postulates, shaping algorithms that seek to harness the inherent capabilities of quantum mechanics for practical computation. This rationalist approach extends ethical responsibility onto the shoulders of scientists, engineers, and organizations. Given the transformative potential of quantum computing, particularly in areas like cryptography, drug development, and even future AI applications, the ethical implications of these algorithms cannot be overlooked. Unintended consequences, such as the potential for breaching existing security infrastructures, necessitate rigorous ethical scrutiny. In addition, as quantum algorithms evolve and potentially manifest emergent behaviors — resulting from interactions between quantum principles and computational algorithms — new, unforeseen ethical dilemmas may arise. This makes the role of transparent governance frameworks and corporate ethics commendable and imperative in guiding the responsible development and deployment of quantum computing technologies.

Empiricism: While the foundations of quantum computing are deeply rooted in theoretical physics and mathematical formalism, its development and fine-tuning are profoundly influenced by empiricism. In the quantum realm, the quality of empirical data — gathered through experimentation and testing — serves as a critical feedback mechanism for refining algorithms and architectures. Here, inductive reasoning from specific data points to generalized principles is paramount. For example, tuning a quantum algorithm for tasks like molecular simulation or cryptographic problems necessitates running multiple experiments to gather data, understand the system's behavior, and modify the algorithm for better performance. Ethical considerations are particularly salient in this empirical approach. Given that quantum computing has the potential to break existing cryptographic systems or even lead to unforeseen consequences in life sciences, the data used for training and validation must be scrutinized for biases and ethical pitfalls. Data scientists and developers bear individual ethical responsibility for ensuring data quality, while organizational leadership must establish rigorous ethical guidelines and oversight mechanisms to keep these powerful technologies in check. Public policy also has a role in enforcing ethical transparency and accountability. As quantum systems potentially display "*emergent*" behaviors, adapting to solve new problems not initially considered, ethical scrutiny becomes even more pressing. Such adaptability and resilience can be advantageous, but if not carefully managed, could lead to ethically problematic outcomes.

The Scientific Method: Quantum computing is not just a technological advance; it is also reshaping the scientific method, challenging and refining traditional approaches to inquiry. In classical physics and computer science, hypotheses often stem from intuitive or logical assumptions that can be easily modeled and tested. In contrast, the quantum realm frequently defies intuition, compelling researchers to formulate hypotheses that account for counterintuitive phenomena like superposition and quantum entanglement. Consequently, experimental design in quantum computing often demands new controls and metrics for assessing outcomes, necessitating state-of-the-art technologies like quantum error correction and quantum logic gates to execute controlled experiments. Meticulous data collection and analysis are vital due to quantum systems' sensitive and inherently probabilistic nature. Additionally, the replicability of experiments — a cornerstone of the scientific method — is challenged by the fact that quantum states are inherently delicate and susceptible to environmental disturbances. As for peer review, the multi-disciplinary nature of quantum computing calls for diversified expertise involving physicists, computer scientists, and even ethicists. This added complexity elevates the rigor of the scrutiny, necessitating innovative methodologies to test validity and significance. Quantum computing is amplifying the depth and rigor of the scientific method, pushing it to adapt and evolve in the face of complexities introduced by the quantum domain.

Medicine: Quantum computing promises to revolutionize medicine by providing computational capabilities far surpassing those of classical systems, thereby elevating the medical healthcare standards of care through advanced diagnostics, drug discovery, and personalized treatment plans. Classical computers struggle with the computational demands of simulating complex biological systems at the molecular level, a limitation that quantum computing aims to overcome. For instance, the capacity to precisely model molecular interactions could drastically accelerate drug discovery processes, enabling researchers to identify effective medications with fewer trials and errors. The same computational prowess can assist in decoding complex genetic data, opening new avenues for personalized medicine by tailoring treatments to an individual's unique genetic makeup. These advancements would be grounded in peer-reviewed, evidence-based practices, thus enhancing the clinical community's established medical healthcare standards of care. In legal contexts, the capability of quantum computing to deliver highly accurate diagnostic and treatment options could set new benchmarks for what is considered "*reasonably competent and skilled*" medical healthcare. However, as medical healthcare standards of care are meant to reflect best practices that are both clinically proven and broadly accessible, integrating quantum computing into clinical care will also necessitate thorough ethical and practical evaluation to ensure that its benefits are equitably distributed and its risks adequately mitigated.

Ethics: Quantum computing's transformative potential presents unprecedented opportunities and challenges in bioethics, particularly concerning the principles of patient autonomy (informed consent), practitioner beneficence (do good), practitioner nonmaleficence (do no harm), and public justice (be fair). Regarding autonomy and informed consent, the advanced computational capabilities of quantum systems could provide highly detailed risk-benefit analyses for medical treatments, thus enabling patients to make more informed decisions. However, the complexity of the technology may create a knowledge gap that makes it challenging for individuals to understand their options fully. Beneficence is manifest in the potential for quantum computing to significantly advance medical research, diagnostics, and treatment significantly, offering tremendous public good. Nevertheless, the principle of nonmaleficence — doing no harm — raises questions about unintended consequences. For example, using quantum algorithms in genetic engineering or drug synthesis could result in harmful outcomes if not rigorously tested and ethically vetted. Finally, justice, or the equitable distribution of benefits and burdens, becomes a focal concern given that the advanced capabilities of quantum computing are likely to be expensive and initially available only to a privileged few. Therefore, ethical guidelines and public policies must be formulated to ensure that access to these groundbreaking medical advancements is as equitable as possible, fulfilling the bioethical commitment to fairness and justice.

Conclusion: Quantum computing stands at the intersection of cutting-edge technology, scientific innovation, ethical considerations, and societal transformation. As a computational paradigm rooted in the principles of quantum mechanics, it pushes the boundaries of what is computationally feasible, potentially revolutionizing fields as diverse as cryptography, material science, and medicine. Its development is a marriage of rationalist inclinations, where theoretical principles are foundational axioms, and empiricist orientations, where data-driven refinement and validation are paramount. It challenges the traditional contours of the scientific method, compelling us to develop new experimental designs and metrics to accommodate the quantum realm's counterintuitive behaviors. In medicine, quantum computing has the potential to redefine standards of care, accelerating drug discovery, enhancing diagnostics, and personalizing treatments. However, as it barrels toward practical implementation, it raises complex ethical issues that intersect with bioethical principles like patient autonomy (informed consent), practitioner beneficence (do good), practitioner nonmaleficence (do no harm), and public justice (be fair). These ethical considerations mandate a collective responsibility — shared among developers, organizations, and policymakers — to responsibly scrutinize, regulate, and guide this technology. Given its transformative potential and complexities, quantum computing is not just a technological marvel but a multi-dimensional phenomenon that necessitates an interdisciplinary approach for responsible development and deployment.

Quantum Computing's Legacy: Poised to revolutionize our understanding of computation and information, quantum computing challenges traditional paradigms in science and ethics, setting the stage for a new era of technological innovation and interdisciplinary inquiry.

**

REVIEW QUESTIONS

True/False Questions:

1. Quantum computing leverages the principles of quantum mechanics, allowing qubits to exist in superpositions of states and perform multiple calculations simultaneously.
   True or False?

2. The potential of quantum computing to break existing cryptographic codes presents no ethical concerns.
   True or False?

Multiple-Choice Questions:

3. Which of the following is NOT a primary application area for quantum computing?
   a) Cryptography
   b) Material science
   c) Classical data processing
   d) Drug discovery

4. What is a significant ethical concern associated with the development and use of quantum computing in medicine?
   a) Developing standard treatment protocols
   b) Ensuring the equitable access and distribution of advanced medical treatments
   c) Implementing uniform drug dosages for all patients
   d) Ignoring patient-specific data in treatment plans

Clinical Vignette:

5. A patient is offered a new diagnostic procedure that uses quantum computing to analyze their genetic data for personalized treatment plans. The doctor explains that this method can provide highly accurate results but also involves sharing the patient's genetic information with third-party data processors. The patient expresses concerns about the privacy and potential misuse of their genetic data. Which ethical principle is primarily being addressed here?
   a) Reducing healthcare costs
   b) Ensuring patient autonomy through informed consent
   c) Promoting practitioner beneficence by prioritizing patient care
   d) Avoiding harm by considering nonmaleficence

Basic Science Vignette:

6. A pharmaceutical company is using quantum computing to accelerate the drug discovery process. By simulating molecular interactions, they hope to identify potential drug candidates more efficiently than with classical computing methods. What is the primary advantage of using quantum computing in drug discovery?
   a) Increased drug prices
   b) Faster identification of potential drug candidates
   c) Simplified clinical trials
   d) Reduced need for medical professionals

Philosophy Vignette:

7. A research institute is employing quantum computing to analyze genetic data for personalized medicine. This technology could offer highly tailored treatments based on an individual's genetic makeup, raising concerns about privacy and potential misuse of genetic information. Which principle of bioethics is most directly challenged by the use of detailed genetic data in this context?
   a) Nonmaleficence (do no harm)
   b) Beneficence (do good)
   c) Autonomy (informed consent)
   d) Justice (fair distribution of benefits and burdens)

Correct Answers:

1. True
2. False
3. c) Classical data processing
4. b) Ensuring the equitable access and distribution of advanced medical treatments
5. b) Ensuring patient autonomy through informed consent
6. b) Faster identification of potential drug candidates
7. c) Autonomy (informed consent)

BEYOND THE CHAPTER
Quantum Computing

- *Quantum Computing: A Gentle Introduction* by Eleanor G. Rieffel and Wolfgang H. Polak
- *Quantum Computing for Computer Scientists* by Noson S. Yanofsky and Mirco A. Mannucci
- *Quantum Computing: An Applied Approach* by Jack D. Hidary

***

Philosophy of Science

Medical Research

Chapter 30: Digital Health and Telemedicine — The Accessible Healthcare: Modern Solutions

Digital health and telemedicine modernize healthcare, promoting accessibility and ethical innovation.

Abstract: In an era of transformative healthcare, telemedicine emerges as a formidable force, harmonizing technology with clinical practice to foster unprecedented accessibility and personalization in medical services. Rooted in rationalism, it develops digital solutions, optimizing healthcare through swift, reliable responses. Adhering to empiricism, it evolves through rigorous data analysis and iterative refinements grounded in real-world experiences. Telemedicine introduces innovative paradigms into medical research and reshapes the scientific method, creating a multidisciplinary approach that marries technology and healthcare research. As it grows, telemedicine enhances traditional healthcare delivery, expanding accessibility, reducing costs, and tailoring services to individual needs while adhering to stringent standards. However, it unveils bioethical dilemmas encompassing patient autonomy (informed consent), practitioner beneficence (do good), practitioner nonmaleficence (do no harm), and public justice (be fair), necessitating continual scrutiny. Telemedicine stands as a pioneering system, promising to revolutionize healthcare delivery, yet requiring a concerted, multidimensional effort to integrate it responsibly into the medical landscape, leaving a legacy of bridged gaps and modernized healthcare delivery.

**

Introduction: Telemedicine, a key subset of digital health, represents a transformative shift in how medical healthcare is delivered, emphasizing the use of digital technology to provide medical healthcare services remotely. Bridging the gap between medical healthcare professionals and geographically separated patients, telemedicine leverages a suite of technologies — from simple telephone consultations to advanced applications employing artificial intelligence for diagnostics. It has the potential to democratize healthcare, making it accessible and affordable even in remote or underserved areas while also enhancing the efficiency, effectiveness, and customization of medical services. Its influence extends from primary care to specialized consultations and mental health services. However, this new paradigm is not without its challenges and controversies. As telemedicine evolves, it presents unique opportunities and dilemmas regarding evidence-based practice, scientific validation, and bioethical considerations, including patient autonomy (informed consent), practitioner beneficence (do good), practitioner nonmaleficence (do no harm), and public justice (be fair). This complex interplay necessitates a multi-faceted analysis to understand both its empirical foundations and ethical implications, shaping the evolving landscape of healthcare in the digital age.

Rationalism: Telemedicine's foundations can be understood through a rationalist lens, as it relies heavily on human cognitive faculties like pattern recognition, deductive reasoning, and problem-solving. These "*a priori*" principles enable medical professionals and software developers to anticipate healthcare needs and conceive digital solutions. For example, the deductive process might begin with the foundational premise that timely medical healthcare intervention saves lives. Extending this axiom, one can reason that faster, more accessible consultations would optimize patient outcomes.

Developers then use pattern recognition and problem-solving skills to create algorithms and applications that make remote healthcare possible, efficient, and reliable. In this way, telemedicine emerges as a complex system that's more than the sum of its parts, echoing the concept of "*emergence*" in rationalism. It aims to address recurring challenges in medical healthcare — such as access, cost, and efficiency — leveraging our innate faculties for quick, reliable responses that enhance societal well-being. However, this approach is not without ethical considerations. While rational faculties guide the construction of these digital tools, their application within medical healthcare must be scrutinized for ethical soundness, encompassing patient autonomy and data privacy. Regarding public policy, regulations should consider these rational faculties' potential downsides and limitations, such as biases in decision-making algorithms or the potential for reduced quality in medical healthcare delivery. Therefore, while telemedicine draws from our rationalist tendencies to innovate and solve problems, it must do so within a framework that comprehensively examines and addresses both its ethical and practical implications.

Empiricism: Telemedicine's development and adaptation are deeply rooted in an empiricist orientation. It relies on a wealth of data derived from sensory experiences and environmental interactions to refine its methods and applications. Developers use inductive reasoning to extrapolate from specific observations, such as the efficacy of remote consultations in managing chronic conditions, to general conclusions about the viability and value of telemedicine. Additionally, empirical studies serve as the bedrock for shaping ethical guidelines and helping professionals ascertain the effectiveness and safety of these digital health platforms. For example, research findings on the accuracy of remote diagnoses or the psychological impacts of virtual consultations directly inform best practices. Cognitive biases, a product of evolved heuristics, are recognized and mitigated through the rigorous testing and reevaluation of telemedical algorithms and interfaces. On the policy level, there is an ongoing need for empirically grounded regulations that can adapt to rapidly evolving technologies and their societal implications. The adaptive value of empiricism in this context lies in telemedicine's ability to continually refine itself in response to new and unforeseen challenges, thereby increasing its societal utility and ethical soundness. In an emergent sense, the collective utility of telemedicine — its capacity to potentially revolutionize healthcare delivery, impact public health outcomes, and even influence cultural attitudes toward healthcare — transcends what could be predicted by examining individual technologies or methods in isolation. It becomes a complex, evolving entity whose multifaceted impacts can only be fully understood through ongoing empirical investigation.

The Scientific Method: Telemedicine and digital health are increasingly shaping the scientific method within medical healthcare research, necessitating adaptations to traditional frameworks of hypothesis testing, data collection, and peer review. Telemedicine's remote, tech-driven, and data-intensive nature calls for novel research methodologies that adequately capture its multifaceted impacts. For instance, randomized controlled trials, the gold standard in medical healthcare research, are being adapted to include remote interventions and virtual follow-ups. Hypotheses are now formulated to test the efficacy of a given treatment and the medium through which it is delivered. Big Data analytics and machine learning are integrated into data collection and analysis, offering richer, more nuanced insights into patient outcomes and medical healthcare efficiency. Additionally, telemedicine research has prompted a renewed focus on ensuring replicability in a digital environment with standardized software protocols and rigorous cybersecurity measures. Peer review extends beyond traditional medical expertise to include professionals adept in data science, cybersecurity, and software engineering. The net effect is a more interdisciplinary, comprehensive approach to scientific inquiry that better reflects the complex interactions of medicine, healthcare, technology, and society. Consequently, telemedicine is not merely a subject of scientific investigation; it is a catalyst for evolving the methods by which we validate medical healthcare innovation, thereby reinforcing the credibility and adaptability of scientific inquiry in the digital age.

Medicine: Telemedicine has made significant contributions to medicine by enhancing accessibility, reducing costs, and enabling more personalized care, all while adhering to established medical healthcare standards of care. In remote or underserved areas, telemedicine bridges the gap between patients and medical healthcare providers, allowing consultations, follow-up visits, and even specialized treatments that would otherwise be unattainable. This increased accessibility can improve patient outcomes by providing timely, ongoing care. Additionally, telemedicine has introduced cost-effective ways to manage chronic conditions, conduct post-operative assessments, and administer mental health services. Importantly, the technology is subjected to rigorous, evidence-based evaluation to ensure it meets or exceeds existing medical healthcare standards of care. This adherence to peer-reviewed, clinically validated practices is integral for telemedicine to gain wider acceptance in the medical community and be incorporated into legal frameworks that define adequate medical healthcare provision. Data analytics and machine learning algorithms are also leveraged to personalize treatment plans, monitor patient compliance, and predict health-related outcomes, adding another layer of sophistication to modern healthcare. Consequently, telemedicine doesn't merely supplement traditional medical healthcare; it enriches it, offering dynamic, adaptive solutions that cater to the evolving needs of patients, medical healthcare providers, and researchers alike.

Ethics: Telemedicine's rise has brought into sharp focus the bioethical principles that underpin healthcare, necessitating a nuanced approach to issues of patient autonomy (informed consent), practitioner beneficence (do good), practitioner nonmaleficence (do no harm), and public justice (be fair). Regarding autonomy, telemedicine empowers patients with easier access to medical healthcare but poses challenges for informed consent in a virtual environment. Effective communication of risks, benefits, and treatment options becomes essential to preserve patient autonomy, especially when the "*human touch*" is mediated through a screen. Regarding beneficence, telemedicine offers the "*good*" of democratized medical healthcare, providing essential clinical services to underprivileged or remote communities. However, nonmaleficence becomes a concern due to issues like the quality of remote consultations, potential misdiagnoses, and data security risks. Developers and medical healthcare providers are ethically obligated to minimize these harms through robust technology, proper training, and secure data protocols. Finally, the principle of justice is invigorated through telemedicine by improving medical healthcare accessibility. Yet, it is also challenged by the digital divide, which can leave behind those who need more resources to access telehealth services. This risks exacerbating healthcare inequalities, making it imperative to develop low-cost, widely accessible telemedical solutions. Thus, while telemedicine promises to revolutionize healthcare, it requires ongoing ethical scrutiny to ensure it aligns with the core principles that guide medical practice.

Conclusion: Telemedicine and digital health have ushered in a transformative era for medical healthcare, integrating technological advancements with clinical practice to address current and emergent healthcare needs. While rooted in rationalist principles that employ deductive reasoning and innate cognitive faculties to design and implement digital solutions, the field also leans heavily on empiricism, using data-driven methodologies to refine and validate these technologies. Not just a passive subject of scientific investigation, telemedicine actively shapes the scientific method itself, necessitating new research frameworks that accommodate its unique, digitally mediated interactions. In medicine, it has already made substantial contributions by increasing accessibility, reducing costs, and enabling more personalized care while maintaining adherence to established medical healthcare standards of care. However, its rise brings to light complex bioethical issues surrounding patient patient autonomy (informed consent), practitioner beneficence (do good), practitioner nonmaleficence (do no harm), and public justice (be fair). These ethical considerations demand ongoing scrutiny to ensure that the technology enhances, rather than compromises, the foundational principles of medical healthcare practice. Telemedicine, thus, is not merely an auxiliary to traditional medical healthcare but a dynamic, adaptive system that holds the potential to revolutionize medical healthcare delivery, raise ethical questions, and necessitate a multi-dimensional approach for its responsible integration into modern medicine.

Digital Health and Telemedicine's Legacy: Expands healthcare access and modernizes its delivery, bridging geographical and socioeconomic divides.

**

REVIEW QUESTIONS

True/False Questions:

1. Telemedicine enhances healthcare accessibility and personalizes medical services through the use of digital technology.
   True or False?

2. Telemedicine eliminates the need for evidence-based evaluation and adherence to medical standards of care.
   True or False?

Multiple-Choice Questions:

3. Which of the following is NOT a significant advantage of telemedicine?
   a) Increased healthcare accessibility
   b) Reduced healthcare costs
   c) Enhanced personalized care
   d) Complete elimination of the need for in-person consultations

4. What ethical principle is most challenged by the digital divide in telemedicine, where some individuals lack access to necessary technology?
   a) Patient autonomy
   b) Practitioner beneficence
   c) Public justice
   d) Practitioner nonmaleficence

Clinical Vignette:

5. A rural patient with a chronic condition uses telemedicine for regular consultations. While the service provides necessary medical care, the patient worries about the quality of remote consultations and data security. Which ethical principle primarily addresses these concerns?
   a) Ensuring patient autonomy through informed consent
   b) Promoting practitioner beneficence by prioritizing patient care
   c) Avoiding harm by considering nonmaleficence
   d) Ensuring fairness through public justice

Basic Science Vignette:

6. A rural clinic uses telemedicine to provide follow-up care for patients with chronic diseases. This approach allows doctors to monitor patient progress, adjust treatments, and provide timely interventions without requiring patients to travel long distances. What primary benefit of telemedicine is highlighted in this scenario?
   a) Increased healthcare costs
   b) Reduced need for medical equipment
   c) Enhanced accessibility to medical care
   d) Simplified patient diagnoses

Philosophy Vignette:

7. A telemedicine platform enables doctors to deliver healthcare services to underserved communities, ensuring that everyone has access to medical consultations regardless of location. Which principle of bioethics is most directly supported by this approach?
   a) Nonmaleficence (do no harm)
   b) Beneficence (do good)
   c) Autonomy (informed consent)
   d) Justice (fair distribution of benefits and burdens)

Correct Answers:

1. True
2. False
3. d) Complete elimination of the need for in-person consultations
4. c) Public justice
5. c) Avoiding harm by considering nonmaleficence
6. c) Enhanced accessibility to medical care
7. d) Justice (fair distribution of benefits and burdens)

BEYOND THE CHAPTER
Telemedicine

- *Telemedicine: A Guide to Assessing Telecommunications in Health Care* by Institute of Medicine, Committee on Evaluating Clinical Applications of Telemedicine
- *Telemedicine and Telehealth: Principles, Policies, Performance and Pitfalls* by Adam William Darkins and Margaret Ann Cary
- *Understanding Telemedicine: A Guide to Professional Development and Success* by Elizabeth Krupinski and Jordana Bernard

***

Philosophy of Science

Medical Research

Chapter 31: Virtual and Augmented Reality in Medicine — The Immersive Innovation: Transformative Healthcare

Virtual and augmented reality in medicine transform healthcare, integrating immersive innovation with ethical care.

Abstract: Integrating Virtual Reality (VR) and Augmented Reality (AR) into the medical sector marks a revolutionary shift, extending beyond gaming and entertainment. This melding of physical and digital realms advances surgical simulations, diagnostics, education, and therapeutic interventions. Grounded in rationalist foundations, these technologies engage cognitive faculties, facilitating complex procedures and enhancing practitioners' skills. Their empirical orientation allows for continuous refinement through data and sensory experiences, aligning with the scientific method. VR and AR enhance practitioners' proficiency, democratize access to quality care, and advance diagnostics and treatment protocols. However, their growth brings critical ethical considerations, including patient autonomy (informed consent), practitioner beneficence (do good), practitioner nonmaleficence (do no harm), and public justice (be fair). These technologies require ongoing ethical evaluation to ensure responsible integration into healthcare. Thus, VR and AR serve as catalysts for academic and ethical dialogues, reshaping the future of healthcare through transformative advancements.

**

Introduction: Virtual Reality (VR) and Augmented Reality (AR) technologies have rapidly risen in various fields. Still, their application in medicine represents a transformative frontier with profound implications for patient care, medical healthcare training, and medical healthcare systems. Once primarily confined to gaming and entertainment, VR and AR offer novel approaches to simulate surgeries, provide advanced diagnostics, enhance medical healthcare education, and even deliver therapeutic interventions. They transcend traditional limitations by creating immersive, interactive environments that closely mimic real-world clinical scenarios or overlay real-time, augmented information during medical healthcare procedures. By blending the physical and digital worlds, these technologies have enriched medical healthcare professionals' toolkits and raised pivotal questions concerning their ethical and scientific validity. From pioneering surgical techniques to virtual psychological treatments, VR and AR are rapidly reshaping the medical healthcare landscape, necessitating a critical evaluation of their rationalist foundations, empiricist orientation, impact on scientific methodology, contributions to medical healthcare, and alignment with the four principles of bioethics: patient autonomy (informed consent), practitioner beneficence (do good), practitioner nonmaleficence (do no harm), and public justice (be fair).

Rationalism: The development and application of Virtual Reality (VR) and Augmented Reality (AR) in medicine are deeply rooted in rationalist foundations. Developers and medical healthcare professionals leverage innate cognitive faculties like reasoning, pattern recognition, and problem-solving to create and refine these technologies. For instance, the ability to simulate complex surgical procedures in VR relies on the deductive reasoning that if a procedure can be replicated in a virtual setting by medical healthcare axioms, it can enhance a surgeon's skill set. These technologies also appeal to innate human faculties for spatial awareness and manual dexterity, arguably hardwired evolutionary traits essential for survival. The emergence of these complex technological systems signifies a new level of collective cognitive capability in medical healthcare, exceeding the sum of individual rational faculties involved. However, as these technologies evolve, there's a heightened ethical responsibility to examine their implications within a rationalist framework critically. For instance, while VR and AR can greatly aid medical healthcare diagnoses and treatment plans, the algorithms that power these technologies may inherit human biases or errors in deductive reasoning, posing ethical and medical healthcare risks. Therefore, the professional community must scrutinize these rationalist underpinnings, aiming for a more nuanced interpretation to guide public policy and regulations. This ensures the effective application of these promising technologies and mitigates potential downsides, such as erroneous clinical decisions based on flawed rationalist premises.

Empiricism: In medical healthcare, Virtual Reality (VR) and Augmented Reality (AR) technologies demonstrate a strong empiricist orientation, wherein empirical data and sensory experiences are critical for their development and validation. Rather than solely relying on theoretical constructs, the utility of these technologies is continuously refined through rigorous testing, clinical trials, and real-world applications. Developers gather a wealth of data from user interactions to adapt and improve the systems, employing inductive reasoning to generalize findings for broader applications. For example, the integration of haptic feedback in VR surgical simulators is informed by countless observations on how tactile sensations contribute to a more realistic and educationally effective experience. This empirical approach is also essential in shaping ethical guidelines; only through collected data can the medical healthcare community assess the efficacy, safety, and potential biases inherent in these technologies. On a policy level, empirically grounded regulations are pivotal to ensure these tools meet stringent medical healthcare standards and adapt to new findings or societal changes. The adaptive value of this empirical orientation becomes evident when these technologies are used to respond to unprecedented medical healthcare challenges, such as remote surgery or telehealth consultations amid a pandemic. The emergent properties of these technologies, such as their capacity to revolutionize medical healthcare training or patient care, cannot be fully anticipated by studying individual sensory faculties alone but require a comprehensive empirical approach that takes into account a multitude of variables, including human behavior, technological limitations, and ethical considerations.

The Scientific Method: The integration of Virtual Reality (VR) and Augmented Reality (AR) in medicine has been informed by and contributed to the scientific method, thus reshaping traditional medical healthcare research and practice paradigms. These technologies offer new ways to conduct controlled experiments with increased precision and minimized variables. For instance, VR-based cognitive behavioral therapy trials can create standardized environments to gauge the efficacy of specific interventions on mental health, offering replicable conditions that can be universally deployed for peer review. The rigor of data collection also intensifies, as these technologies can generate real-time metrics during procedures or simulations, offering quantitative insights beyond subjective observations. This expansive data set allows for meticulous statistical analysis, validating or refuting the initial hypotheses with higher confidence. Furthermore, VR and AR platforms also facilitate collaborative studies across institutions, making the peer-review process more robust by incorporating diverse perspectives. However, adopting these technologies also prompts the scientific community to revisit the ethical and methodological frameworks governing research, especially concerning human subjects in virtual settings. In essence, VR and AR have benefited from the rigorous testing underpinned by the scientific method and have provided novel avenues for medical healthcare research, strengthening the reliability, replicability, and ethical oversight of scientific inquiries in clinical healthcare.

Medicine: The advent of Virtual Reality (VR) and Augmented Reality (AR) has made transformative contributions to clinical healthcare, impacting everything from education and training to diagnostics and treatment. By aligning with established medical healthcare standards of care, these technologies enhance the proficiency and adaptability of medical healthcare professionals. For instance, surgical simulation platforms in VR offer safe, replicable environments where clinical practitioners can hone their skills, thus aligning with the expectation of competent, skillful care. AR applications in surgical theatres can overlay real-time data and 3D imaging onto the surgical field, aiding in more precise and effective procedures. These technologies have also proven valuable in telehealth, breaking geographical barriers to provide remote consultations and even guided surgeries, democratizing access to high-quality care. From a diagnostic standpoint, VR and AR can offer immersive visualization of complex clinical data, such as MRI scans, enabling clinicians to reach more accurate conclusions faster. These applications adhere to current best practices and push the envelope, contributing to the evolution of new medical healthcare standards of care. The evidence-based validation of these technologies ensures their integration into the healthcare community is not just novel but rigorously scrutinized, fulfilling current and potentially future criteria for the adequacy of care in routine and specialized clinical scenarios.

Ethics: Virtual Reality (VR) and Augmented Reality (AR) in medicine intersect in complex ways with the four principles of bioethics: patient autonomy (informed consent), practitioner beneficence (do good), practitioner nonmaleficence (do no harm), and public justice (be fair). These technologies can enhance patient autonomy by facilitating more informed consent; for example, AR can offer patients a more transparent visualization of a surgical plan, empowering them to make more informed decisions. In terms of beneficence, VR and AR can be decidedly beneficial by improving the precision of surgeries or the efficacy of psychological treatments, thereby enhancing the quality of medical healthcare. However, the principle of nonmaleficence raises critical concerns. While these technologies aim to do no harm, there are still risks of misapplication or technical failures, such as misleading visual cues in a surgical AR interface, which could lead to errors. Therefore, these tools' efficacy and safety must be rigorously vetted to minimize harm. Lastly, the principle of justice poses questions about equitable access to these advanced technologies. Although VR and AR have the potential to democratize medical healthcare by facilitating remote treatments, they could also widen existing disparities if only affluent healthcare systems can afford them. As these technologies become more integrated into clinical care, ongoing ethical scrutiny is essential to ensure they align with these bioethical principles, contributing to more ethical and equitable medical healthcare delivery.

Conclusion: Virtual Reality (VR) and Augmented Reality (AR) have rapidly ingrained themselves into the fabric of modern medical healthcare, revolutionizing how we approach medical healthcare training, diagnostics, treatment, and even ethical considerations. These technologies offer unprecedented opportunities for enhanced patient care and medical healthcare proficiency while inviting crucial debates about their moral, empirical, and rational frameworks. In tracing their influence, it's evident that they exhibit strong rationalistic foundations driven by innate cognitive faculties and deductive reasoning. Their empirical orientation, bolstered by real-world data and observations, is also critical in their continual refinement. Furthermore, VR and AR serve as powerful tools in the scientific method, enabling more controlled, replicable experiments and contributing to the evolution of medical healthcare standards of care. However, as they advance, they raise bioethical questions around patient autonomy (informed consent), practitioner beneficence (do good), practitioner nonmaleficence (do no harm), and public justice (be fair), requiring ongoing scrutiny to ensure they meet ethical and medical healthcare benchmarks. In totality, VR and AR technologies have become a transformative force in medical healthcare, but they also necessitate a holistic, conscientious approach to ensure their responsible integration into clinical practice. This involves continuous reevaluation and alignment with ethical principles, empirical evidence, and rationalistic inquiry, making them not just technological marvels but also subjects of multi-faceted academic and ethical discourse.

Virtual and Augmented Reality in Medicine's Legacy: Revolutionizes medical healthcare training, treatment, and patient experiences, offering immersive and interactive medical healthcare approaches.

**

REVIEW QUESTIONS

True/False Questions:

1. Virtual Reality (VR) and Augmented Reality (AR) technologies in medicine primarily enhance the educational and training experiences of medical practitioners by providing immersive simulations.
   True or False?

2. The integration of VR and AR in medicine eliminates the need for empirical validation and evidence-based practices.
   True or False?

Multiple-Choice Questions:

3. Which of the following is NOT a direct benefit of using VR and AR in medical healthcare?
   a) Improved surgical precision
   b) Enhanced remote consultations
   c) Increased healthcare costs
   d) Better diagnostic accuracy

4. What ethical principle is primarily concerned with ensuring that patients fully understand and agree to the medical procedures facilitated by VR and AR technologies?
   a) Practitioner beneficence
   b) Public justice
   c) Practitioner nonmaleficence
   d) Patient autonomy

Clinical Vignette:

5. A patient is scheduled for a complex surgery and uses an AR application to visualize the procedure and its potential outcomes. This application helps the patient make an informed decision about their surgery. Which bioethical principle is primarily being addressed in this scenario?
   a) Practitioner beneficence
   b) Practitioner nonmaleficence
   c) Patient autonomy
   d) Public justice

Basic Science Vignette:

6. A medical school uses Virtual Reality (VR) to simulate complex surgeries for students. This allows them to practice procedures in a risk-free environment, improving their skills before performing real surgeries. What primary benefit of VR in medical education is highlighted in this scenario?
   a) Simplified surgical procedures
   b) Reduced need for real surgeries
   c) Enhanced surgical skills through realistic simulations
   d) Lowered educational costs

Philosophy Vignette:

7. A hospital implements Augmented Reality (AR) to provide surgeons with real-time data overlays during operations, improving precision and outcomes. However, there are concerns about potential technical failures leading to errors. Which principle of bioethics is most directly challenged by this concern?
   a) Autonomy (informed consent)
   b) Beneficence (do good)
   c) Nonmaleficence (do no harm)
   d) Justice (fair distribution of benefits and burdens)

Correct Answers:

1. True
2. False
3. c) Increased healthcare costs
4. d) Patient autonomy
5. c) Patient autonomy
6. c) Enhanced surgical skills through realistic simulations
7. c) Nonmaleficence (do no harm)

BEYOND THE CHAPTER
Virtual & Augmented Reality

- *Virtual Reality* by Samuel Greengard
- *Augmented Reality: Principles and Practice* by Dieter Schmalstieg and Tobias Hollerer
- *Learning Virtual Reality: Developing Immersive Experiences and Applications for Desktop, Web, and Mobile*by Tony Parisi

***

Chapter 32: Richard Dawkins (1941-) — The Evolutionary Biologist: Gene-Centered Evolution

Dawkins's gene-centered view of evolution, coupled with his advocacy for rationalism and skepticism, has profoundly influenced biology and public understanding of science.

Abstract: Renowned British evolutionary biologist and author Richard Dawkins has significantly impacted the scientific and literary world, particularly with his gene-centered view of evolution in "*The Selfish Gene*" (1976). A fervent advocate for atheism and scientific skepticism, he critically analyzes religious perspectives in "*The God Delusion*" (2006). This discourse examines Dawkins' rationalist foundations and empiricist orientation, his pivotal role in shaping the scientific method, and his indirect contributions to medicine and bioethics. Dawkins promotes a society that relies on reason, logic, and empirical evidence, advocating the scientific method as a systematic approach to understanding the natural world. His influence extends to medicine, fostering evidence-based practices and critical thinking aligned with modern medical principles. Furthermore, Dawkins shapes bioethical discourse, advocating for dignity, rights, and well-being in healthcare through rational discourse and evidence-based decision-making. Through his extensive work, Dawkins has fostered a global shift toward scientific skepticism and rationalism, leaving a transformative legacy in the quest for truth and enlightenment.

**

Introduction: Richard Dawkins, a renowned British evolutionary biologist has carved a significant niche in the scientific and literary realms. Born on March 26, 1941, in Nairobi, Kenya, Dawkins' family relocated to England when he was eight. He blossomed academically at Balliol College, Oxford, where he pursued zoology, later earning a doctorate under the guidance of the eminent ethologist Nikolaas Tinbergen. Dawkins catapulted to fame with his groundbreaking book "*The Selfish Gene*" (1976), where he introduced the gene-centered view of evolution and coined the term "*meme*" to describe cultural information transmission. Apart from his scientific endeavors, he is a staunch advocate for atheism and scientific skepticism, often challenging religious perspectives through works like "*The God Delusion*" (2006). As we delve deeper, we will explore Dawkins' rationalist foundations, empiricist orientation, and role in shaping the scientific method. Furthermore, we will assess his contributions to medicine and his influence on the four principles of bioethics: patient autonomy (informed consent), practitioner beneficence (do good), practitioner nonmaleficence (do no harm), and public justice (be fair).

Rationalism: Richard Dawkins firmly anchors his scientific and philosophical viewpoints in rationalist foundations, advocating for using reason and logic as primary tools in acquiring knowledge. His seminal works underscore the belief that certain cognitive faculties and instinctual behaviors are evolutionarily ingrained mechanisms facilitating understanding and knowledge acquisition. Dawkins posits that these innate faculties, encompassing basic reasoning skills, pattern recognition, and problem-solving abilities, serve as "*a priori*" principles guiding human cognition and behavior. Through a lens of evolutionary biology, he illustrates how these inherent traits have been sculpted over millennia to aid survival and adaptation. Furthermore, Dawkins employs deductive reasoning in his arguments, meticulously building upon definitional truths, axioms, or postulates to craft coherent and logically sound narratives. This rationalist approach vividly manifests in his exploration of complex biological phenomena and his critique of religious doctrines, where he emphasizes empirical evidence and logical consistency as the bedrock of true understanding and enlightenment. Through this rationalist lens, Dawkins seeks to foster a society grounded in science and reason, where beliefs are formed based on evidence and logical deduction rather than tradition or faith.

Empiricism: Richard Dawkins embodies an empiricist orientation in his scholarly pursuits, firmly advocating for acquiring knowledge through sensory experiences and environmental interactions. This approach is deeply rooted in the scientific method, where observations and experiments form the basis of understanding complex phenomena. Through his extensive work in evolutionary biology, Dawkins underscores the importance of empirical evidence in substantiating theories and hypotheses. He applies inductive reasoning, a cornerstone of empiricism, where specific observations are meticulously analyzed to formulate general conclusions about the natural world. In his literary works, Dawkins often illustrates the intricate web of life through detailed observations and analyses, thereby fostering a deeper understanding of biological processes and evolutionary dynamics. This empiricist stance is also evident in his critique of religious doctrines, where he urges individuals to rely on tangible evidence and rational inquiry rather than succumbing to faith-based beliefs. Through his empiricist lens, Dawkins endeavors to cultivate a society where knowledge is grounded in observable realities, fostering a culture of critical thinking and scientific inquiry that seeks to unravel the mysteries of the universe through evidence-based exploration.

The Scientific Method: Richard Dawkins, a luminary in evolutionary biology, has played a pivotal role in shaping and promoting the scientific method as a rigorous and systematic approach to inquiry. His works often exemplify the meticulous application of this method, starting with observing biological phenomena and formulating pertinent questions regarding their underlying mechanisms. Dawkins is known for crafting testable hypotheses, offering tentative explanations that can be verified and falsified through empirical research. His commitment to controlled experiments and data analysis is evident in his deep dives into evolutionary dynamics, where he scrutinizes data to support or refute existing hypotheses. Furthermore, Dawkins emphasizes the necessity of replicability in scientific research, advocating for consistent results under similar conditions as a benchmark for credibility. His role extends to fostering a robust peer-review process, where research undergoes scrutiny by experts in the field to evaluate its validity, methodology, and significance. Through his writings and public engagements, Dawkins has championed the scientific method as a beacon of rational inquiry, encouraging a culture of critical thinking and evidence-based exploration in both the scientific community and the broader society.

Medicine: While Richard Dawkins is primarily recognized for his contributions to evolutionary biology and his advocacy for atheism and scientific skepticism, his influence permeates into the realm of medicine as well, albeit indirectly. Through his rigorous promotion of evidence-based practices and critical thinking, Dawkins has fostered a climate encouraging medical professionals to adhere to healthcare standards of care grounded in peer-reviewed, empirical research. His emphasis on rationalism and empiricism aligns with the evolving medical healthcare standards of care, predicated on current best practices informed by meticulous research, clinical expertise, and patient needs. Moreover, Dawkins' gene-centered view of evolution offers a nuanced understanding of the genetic underpinnings of various diseases, potentially guiding research and interventions in medical genetics and personalized medicine. Although not directly involved in the medical field, Dawkins' advocacy for science and reason echoes in the corridors of modern medicine, encouraging a healthcare approach that is both compassionate and firmly rooted in evidence, thereby ensuring the delivery of care that meets the highest standards of competence and skill.

Ethics: Richard Dawkins, through his extensive body of work and public discourse, has indirectly influenced the discourse on the four principles of bioethics: patient autonomy (informed consent), practitioner beneficence (do good), practitioner nonmaleficence (do no harm), and public justice (be fair). His staunch advocacy for rationalism and empiricism fosters a climate where informed consent, a cornerstone of autonomy, is revered, encouraging individuals to make healthcare decisions based on clear, evidence-based information. In beneficence, Dawkins' evolutionary perspective underscores the intrinsic value of fostering well-being and positive outcomes in healthcare settings, grounded in a deep understanding of biological processes. His emphasis on scientific skepticism and critical thinking aligns with the principle of nonmaleficence, urging healthcare professionals to avoid harm by meticulously evaluating interventions based on empirical evidence. Furthermore, Dawkins' commitment to rational discourse and evidence-based decision-making echoes the principle of justice, promoting fairness and equity in healthcare by encouraging policies and practices grounded in factual, unbiased research. Through his influence, Dawkins indirectly nurtures a bioethical landscape deeply committed to upholding individuals' dignity, rights, and well-being in healthcare settings, guided by the beacon of science and reason.

Conclusion: Richard Dawkins, a towering figure in science and literature, has left an indelible mark on various facets of human knowledge and understanding. Dawkins has journeyed through an illustrious career with groundbreaking contributions to evolutionary biology and a relentless pursuit of rational and empirical inquiry. His pioneering works, such as "*The Selfish Gene*," have reshaped our understanding of evolutionary dynamics and introduced novel concepts like the "*meme*," offering a fresh perspective on cultural information transmission. A staunch advocate for scientific skepticism, Dawkins has been a vocal critic of religious doctrines, urging society to embrace evidence-based beliefs and critical thinking. His influence extends beyond the confines of biology, permeating the fields of medicine and bioethics, where his advocacy for rationalism and empiricism has fostered a climate of evidence-based practices and ethical considerations grounded in science and reason. Through his extensive work and public engagements, Dawkins has championed a holistic approach to knowledge acquisition, where rationalism, empiricism, and the scientific method converge to illuminate the intricacies of the natural world and human existence. His enduring legacy is a testament to the transformative power of science and reason in fostering a society grounded in truth, inquiry, and enlightenment.

Richard Dawkins' Legacy: Pioneering the gene-centered view of evolution and fostering a global shift toward scientific skepticism and rationalism through his literary and academic endeavors.

**

REVIEW QUESTIONS

True/False Questions:

1. Richard Dawkins introduced the gene-centered view of evolution in his book "*The God Delusion*."
   True or False?
2. Dawkins' advocacy for rationalism and empiricism indirectly influences modern bioethics, promoting principles such as patient autonomy and public justice.
   True or False?

Multiple-Choice Questions:

3. Which of the following books by Richard Dawkins focuses on his critique of religious beliefs?
   a) "*The Selfish Gene*"
   b) "*Climbing Mount Improbable*"
   c) "*The God Delusion*"
   d) "*The Extended Phenotype*"

4. What key concept did Richard Dawkins introduce to describe the transmission of cultural information?
   a) Gene flow
   b) Meme
   c) Natural selection
   d) Genetic drift

Clinical Vignette:

5. A medical ethics class is discussing the influence of prominent scientists on bioethics. The professor mentions Richard Dawkins and his indirect impact on bioethical principles. Which principle is most closely associated with ensuring patients are well-informed before making healthcare decisions?
   a) Practitioner beneficence
   b) Practitioner nonmaleficence
   c) Patient autonomy
   d) Public justice

Basic Science Vignette:

6. A researcher is studying the gene-centered view of evolution in a population of fruit flies. According to Richard Dawkins' theory, what is the primary unit of natural selection in this scenario?
   a) The individual fruit flies
   b) The population of fruit flies as a whole
   c) The specific genes within the fruit flies
   d) The environment in which the fruit flies live

Philosophy Vignette:

7. During a public debate on scientific skepticism, a participant references Richard Dawkins' stance on the role of evidence in shaping beliefs. According to Dawkins, what should be the primary basis for forming and holding beliefs?
   a) Tradition and cultural norms
   b) Religious texts and faith
   c) Empirical evidence and rational inquiry
   d) Intuition and personal feelings

Correct Answers:

1. False
2. True
3. c) "*The God Delusion*"
4. b) Meme
5. c) Patient autonomy
6. c) The specific genes within the fruit flies
7. c) Empirical evidence and rational inquiry

BEYOND THE CHAPTER
Richard Dawkins (1941-)

- *The Selfish Gene* by Richard Dawkins
- *The Extended Phenotype* by Richard Dawkins
- *The Blind Watchmaker* by Richard Dawkins
- *The God Delusion* by Richard Dawkins

***

Chapter 33: Climate Change — The Anthropocene Challenge: Ethical Imperatives

Climate change demands ethical imperatives, urging interdisciplinary approaches to safeguard our planet.

Abstract: Climate change, epitomizing the Anthropocene era where human activities wield a geologic impact, poses an intricate dilemma deeply rooted in industrial actions and has global repercussions. This entry explores the profound influence of climate change across various facets: rationalist cognitive faculties and ethical deliberations, the pivotal role of empiricism in shaping responsive policies, the evolution and adaptations within the scientific method to address this multi-dimensional issue, and the ensuing transformations in medical healthcare standards and practices. Additionally, it underscores the necessity of broadening our ethical compass to navigate the intricacies of climate-induced disparities and environmental stewardship. We can address climate change's colossal challenges by fostering interdisciplinary dialogues and adaptive strategies, encouraging intellectual and ethical societal evolution. This analysis serves as a testament to climate change's role as a formidable catalyst urging a holistic reassessment of our interactions with and impacts on the Earth's systems and communities.

**

Introduction: Climate change is one of the most pressing and complex issues of our time, a multifaceted problem with roots in human activity and consequences that reach every corner of the globe. A manifestation of the Anthropocene, where human actions have become a geological force, climate change challenges our understanding of Earth systems and how we engage with ethics, science, and governance. At its core, climate change is driven by an increase in greenhouse gases, particularly carbon dioxide and methane, largely due to the burning of fossil fuels and deforestation. These changes contribute to rising global temperatures, sea-level rise, extreme weather events, and disruptions to ecological balance. The ramifications are vast and disparate, unequally impacting natural ecosystems, human health, economies, and social systems. As such, confronting climate change requires technological solutions or policy interventions and careful evaluation of our scientific methods, ethical frameworks, and notions of justice. This complexity necessitates a multi-disciplinary approach that combines insights from empirical observations, rational deliberations, and ethical principles, touching upon themes of patient autonomy (informed consent), practitioner beneficence (do good), practitioner nonmaleficence (do no harm), and public justice (be fair) as we strive for solutions.

Rationalism: In grappling with the colossal issue of climate change, rationalist inclinations rooted in biological evolution provide an intriguing lens through which we can evaluate our innate and learned faculties for acquiring knowledge and making decisions. Our evolved cognitive mechanisms like pattern recognition, problem-solving, and deductive reasoning serve as "*a priori*" principles guiding our understanding of complex climate data and modeling outcomes. This evolutionary toolkit has empowered us to conceptualize and confront the abstract, temporally distant threats that climate change poses and handicap us with cognitive biases such as short-termism and in-group favoritism.

While these faculties have historically conferred survival advantages, in the context of climate change, they may skew our ethical and policy decisions toward immediate, parochial interests over long-term, collective well-being. Here, "*emergence*" becomes relevant: the collective human ability to confront climate change is a complex interplay of these basic faculties, heightened by our advanced ethical reasoning and self-awareness. Scientists, educators, and policymakers are responsible for understanding and negotiating the limitations of our rationalistic faculties. They must create frameworks and policies that mitigate these limitations, fostering environments that balance instinctual behaviors with learned ethical principles like patient autonomy (informed consent), practitioner beneficence (do good), practitioner nonmaleficence (do no harm), and public justice (be fair). Through this, we hope to construct more equitable, effective solutions to the multi-layered challenges that climate change presents.

Empiricism: Climate change is a topic where empiricism is pivotal, providing a mechanism for acquiring and refining knowledge primarily through sensory experience and environmental interactions. The vast array of climate data — from satellite images of melting polar ice caps to meticulous recordings of sea-level rise and carbon dioxide concentrations — relies on inductive reasoning. Specific observations accumulate over time to support conclusions about the Earth's shifting climate patterns and human impact. This empiricist orientation is essential for forming ethical and policy guidelines because it grounds abstract principles in tangible evidence. Cognitive biases and sociocultural factors evolved as heuristic shortcuts or shaped by environmental conditions often cloud our interpretation of these data, making it crucial for scientists and policymakers to remain vigilant in scrutinizing the data and their interpretive frameworks. The adaptive value of our empiricist faculties lies in their flexibility, enabling us to continually adapt our understanding and responses to the ever-evolving challenges posed by climate change. Regarding "*emergence*," the collective endeavor to mitigate climate change transcends individual empirical faculties and observations, generating a multi-layered, complex interaction of scientific, social, and ethical dimensions. Phenomena like public awareness, international cooperation, and even climate activism can be seen as emergent properties that arise from the complex interplay of individual sensory experiences, data interpretation, and cultural narratives. These emergent phenomena underscore the necessity for empirically grounded public policy that is both adaptive and responsive to the intricacies of a changing climate.

The Scientific Method: The challenge of understanding and mitigating climate change has significantly influenced the evolution of the scientific method, pushing it toward greater interdisciplinary collaboration, increased data complexity, and a focus on long-term, systemic phenomena. Traditional scientific disciplines, such as physics and chemistry, now converge with ecology, economics, and social sciences to holistically address the multi-faceted problem. The nature of climate change also demands the formulation of increasingly complex hypotheses that can accommodate a wide array of variables, from greenhouse gas emissions to human behavioral patterns. Given the issue's long-term implications and global scale, controlled experiments often give way to advanced computer modeling and simulations, necessitating rigorous validation methods to ensure credibility. The concept of replicability has also expanded, as climate studies' sheer scope and complexity often require multi-institutional efforts and cross-validation to achieve robust findings. Peer review has never been more critical, as the policy implications of climate research are both urgent and far-reaching, necessitating extreme diligence in evaluating methodology and data interpretation. Thus, climate change has not only served as a compelling subject for scientific inquiry but has also catalyzed adaptations and refinements in the scientific method itself, ensuring it remains fit for tackling the grand challenges of the Anthropocene.

Medicine: Climate change has profoundly influenced the field of medicine, necessitating reevaluating and adapting medical healthcare standards of care in multiple dimensions. The increasing prevalence of vector-borne diseases like malaria and dengue, aggravated by changing climatic conditions, has stimulated research into new preventive measures and treatments, thereby modifying existing standards. Similarly, the escalating incidence of heat-related illnesses and respiratory disorders due to worsening air quality has compelled medical healthcare communities to update their guidelines, incorporating climate-resilient strategies and interventions. Furthermore, climate change has catalyzed interdisciplinary approaches that merge epidemiology, environmental science, and public health, enriching evidence-based practices. Regarding legal contexts, the evolving landscape has started considering whether healthcare providers are meeting new responsibilities in preparing for and responding to climate-related health issues. Telemedicine and mobile clinics are increasingly seen as viable adaptations to address healthcare disparities exacerbated by climate change, especially in vulnerable communities. The medical healthcare community, guided by peer-reviewed, evidence-based practices, is thus not only treating the symptoms of climate change but is also actively involved in reshaping healthcare delivery models and standards of care. These evolving standards reflect a broadened understanding of medical healthcare that transcends traditional boundaries, acknowledging the interconnectedness of human health, social systems, and a rapidly changing environment.

Ethics: Climate change exerts a transformative influence on the bioethical landscape, compelling a reexamination and recalibration of core principles such as patient autonomy (informed consent), practitioner beneficence (do good), practitioner nonmaleficence (do no harm), and public justice (be fair). The principle of autonomy, traditionally expressed through informed consent, is complicated by climate change's diffuse and often intangible impacts. How do we obtain global "*informed consent*" for actions that have far-reaching, sometimes irreversible, consequences for ecosystems and human communities? Beneficence, the obligation to do good, extends beyond immediate healthcare settings to encompass a responsibility toward environmental stewardship, given the clear link between a healthy planet and human well-being. Nonmaleficence, the commitment to do no harm, likewise broadens its scope to consider the unintentional but devastating harm that human activities can inflict on the environment, affecting current and future generations. Finally, the principle of justice confronts perhaps the most profound challenge. Climate change disproportionately impacts marginalized communities, exacerbating existing inequalities. This inequity demands rethinking what justice entails in distributing resources, healthcare, and opportunities within and across nations. Therefore, climate change necessitates an expanded ethical framework that accounts for the interconnectedness of human and environmental health, urging a collective moral responsibility that transcends traditional bioethical considerations.

Conclusion: Climate change is a pivotal issue that profoundly influences our rationalist and empiricist frameworks, scientific methodology, medical practices, and ethical norms. As a daunting and ubiquitous challenge, it is a crucible for testing our innate and learned faculties for understanding and acting upon complex, interrelated phenomena. A rationalist lens pushes us to confront the limitations of our "*a priori*" cognitive faculties, urging the need for self-aware ethical reasoning that transcends immediate, parochial concerns. In the empiricist domain, climate change forces us to recognize the value of observational data while remaining cognizant of the biases that might influence our interpretations. The subject has catalyzed innovations within the scientific method, driving interdisciplinary collaborations and the evolution of more complex, systemic inquiry models. In medicine, the impacts of climate change have compelled a rethinking of healthcare standards, integrating environmental considerations, and spawning new avenues of interdisciplinary research. Ethically, it challenges us to rethink and expand upon traditional bioethical principles, considering the broader, interconnected realities of human and environmental health. Climate change demands a synergistic, multi-disciplinary approach informed by rationalist and empiricist perspectives, grounded in rigorous scientific methodology, and ever-mindful of its broad-ranging ethical implications. It beckons us to evolve, not just technologically or politically, but intellectually and ethically, pushing us toward a more holistic understanding of our place within the larger web of existence.

Climate Change's Legacy: Climate change stands as an existential reminder of the Anthropocene era, fundamentally reorienting our philosophical and scientific paradigms to grapple with the interconnected complexities of human activity, ecological balance, and ethical responsibility, thereby acting as a catalyst for a comprehensive reevaluation of our place in, and impact on, the world.

**

REVIEW QUESTIONS

True/False Questions:

1. Climate change is primarily driven by natural processes and has minimal influence from human activities.
   True or False?

2. The ethical principle of justice in the context of climate change involves ensuring that the impacts and benefits of climate policies are equitably distributed across all communities.
   True or False?

Multiple-Choice Questions:

3. Which of the following is NOT a primary driver of climate change?
   a) Burning of fossil fuels
   b) Deforestation
   c) Natural solar cycles
   d) Methane emissions

4. How has climate change influenced the scientific method?
   a) Reduced the need for interdisciplinary collaboration
   b) Increased reliance on long-term, systemic studies
   c) Simplified hypothesis testing
   d) Decreased the importance of peer review

Clinical Vignette:

5. A healthcare provider in a coastal city is noticing an increase in heat-related illnesses and respiratory conditions due to worsening air quality. According to updated healthcare standards in the context of climate change, what should the provider prioritize?
   a) Continuing with traditional treatment methods without considering environmental factors
   b) Implementing climate-resilient strategies and educating patients about preventive measures
   c) Ignoring the role of climate change and focusing solely on immediate medical treatments
   d) Referring all patients to specialists without addressing the root environmental causes

Basic Science Vignette:

6. A research team is developing an AI system to model climate change impacts on various ecosystems. After integrating extensive environmental data, the system starts predicting new patterns of species migration and habitat loss. However, it also generates a significant number of false positive results. What is the best course of action to address the false positives?
   a) Ignore the false positives as the system's overall performance has improved
   b) Review and refine the algorithm using a more comprehensive dataset
   c) Remove the new migration patterns from the system to prevent false positives
   d) Disable the AI system until the false positives can be fully understood

Philosophy Vignette:

7. An AI system in climate research uses Big Data to suggest ethical guidelines for global environmental policies. It proposes an optimal framework that maximizes environmental sustainability but overlooks the socio-economic rights of certain communities. What is the best philosophical approach to address this issue?
   a) Prioritize the AI's ability to maximize environmental sustainability
   b) Reprogram the AI to give more weight to socio-economic rights and environmental sustainability
   c) Develop an oversight committee to review the AI's recommendations before implementation
   d) Disable the AI system until it can be programmed to always respect socio-economic rights

Correct Answers:

1. False
2. True
3. c) Natural solar cycles
4. b) Increased reliance on long-term, systemic studies
5. b) Implementing climate-resilient strategies and educating patients about preventive measures
6. b) Review and refine the algorithm using a more comprehensive dataset
7. b) Reprogram the AI to give more weight to socio-economic rights and environmental sustainability

BEYOND THE CHAPTER
Climate Change

- *This Changes Everything: Capitalism vs. The Climate* by Naomi Klein
- *The Uninhabitable Earth: Life After Warming* by David Wallace-Wells
- *The Sixth Extinction: An Unnatural History* by Elizabeth Kolbert

***

Philosophy of Science

Medical Research

WEEK SEVEN: STEWARDSHIP AND SYSTEMS

WEEK SEVEN

Class Hour 1

- Chapter 34: Environmentalism — The Stewardship Ethos: Planetary Preservation
- Chapter 35: Global Health Initiatives — The Equity Imperative: Health for All
- Chapter 36: Green Revolution — The Agricultural Paradigm: Balancing Innovation

Class Hour 2

- Chapter 37: Complexity Theory — The Holistic Framework: Understanding Systems
- Chapter 38: Epigenetics — The Dynamic Influence: Nature and Nurture
- Chapter 39: Microbiome — The Symbiotic Network: Health and Disease

EXAM II: 50Q - online

Introduction

In Week Seven of our "*Philosophy of Science*" course, we delve into the interconnected themes of stewardship and systems, exploring how scientific and ethical paradigms intersect to shape our understanding of the world. This section encapsulates the evolution of environmental consciousness, global health imperatives, agricultural innovation, complexity theory, genetic regulation, and microbiome research, offering a holistic view of how these diverse fields converge to inform our approaches to science and ethics.

As we begin, the ethos of environmentalism sets the stage, highlighting the profound shift from Romantic and transcendentalist philosophies to a robust, scientific methodology dedicated to planetary preservation. This stewardship ethos is not just about conserving nature but integrates ethical considerations, influencing sectors like medicine and bioethics to adopt a more environmentally conscious approach. From this foundation of sustainability, we transition to global health initiatives, which advocate for equitable healthcare access worldwide. These initiatives embody a blend of rationalist and empirical methodologies, promoting innovations in medical research and public health policies, driven by ethical principles such as beneficence and justice.

The narrative then weaves into the Green Revolution, a transformative period in agriculture that balanced scientific innovation with ethical challenges. By introducing high-yield crop varieties and advanced agricultural practices, this movement addressed food insecurity but also raised bioethical questions about environmental impact and social equity. The complexities and unintended consequences of these advancements underscore the need for a nuanced approach to scientific progress.

Furthering our exploration, Complexity Theory offers a holistic framework to understand intricate systems in nature, economics, and social structures. This interdisciplinary approach diverges from reductionism, emphasizing the importance of nonlinear interactions and emergent phenomena. It revitalizes the scientific method with computational models and network analyses, reshaping our understanding of dynamic systems and informing more adaptive strategies in fields like medicine and ethics.

Next, we delve into the dynamic interplay of nature and nurture through the lens of epigenetics. This field reveals how environmental factors influence gene expression without altering the DNA sequence, bridging the gap between genetic predispositions and environmental influences. The insights gained from epigenetic research pave the way for personalized medicine, although they also prompt ethical considerations regarding patient autonomy and justice.

Finally, the study of the human microbiome illuminates the symbiotic relationships between our bodies and the microorganisms within us. This research transforms our understanding of health and disease, showcasing how advanced genomics and computational techniques uncover the intricate connectivity of life. The ethical implications of microbiome research, such as informed consent and nonmaleficence, highlight the need for evolving ethical frameworks to keep pace with scientific discoveries.

In this multifaceted exploration, we see how stewardship and systems thinking intertwine across various scientific domains, driving a deeper appreciation for the ethical dimensions of scientific inquiry and the intricate balance required to advance human knowledge and well-being.

***

Philosophy of Science

Medical Research

Chapter 34: Environmentalism — The Stewardship Ethos: Planetary Preservation

Environmentalism champions stewardship and sustainability, integrating scientific and ethical approaches to planetary health.

Abstract: Environmentalism, a multifaceted movement deeply rooted in rationalist and empiricist approaches, has sculpted a transformative trajectory in numerous domains, steering a more encompassing view of the interplay between humanity and the natural world. Grounded initially in Romanticism and transcendentalist philosophies, it has matured to encapsulate a comprehensive scientific methodology, assimilating empirical observations and rational assessments of ecological data. This profound evolution has reverberated through sectors like medicine and bioethics, redefining medical healthcare standards of care and expanding the boundaries of ethical considerations to embody a heightened sensitivity toward environmental health. Environmentalism has fortified the scientific method by fostering a symbiotic relationship between humans and the environment, propelling multidisciplinary collaborations, and setting new benchmarks for methodological rigor. The profound influence of environmentalism underscores a pivotal shift in societal development, marking a legacy of stewardship for Earth's ecosystems and fostering a unified responsibility toward conserving our planet's biodiversity and well-being.

**

Introduction: Environmentalism is a multifaceted social, political, and ethical movement that aims to address the profound challenges associated with the human impact on the natural world. Rooted in a concern for ecological sustainability, environmentalists advocate for policies and actions that minimize pollution, conserve natural resources, and protect vulnerable ecosystems. While its origins can be traced back to Romanticism and transcendentalist philosophies, environmentalism has evolved to incorporate a scientific approach grounded in empirical observations and rational assessments of ecological data. This shift reflects the intersection of rationalist and empiricist thought patterns in shaping the scientific methods that underpin environmental studies, including the study of climate change, biodiversity, and pollution. Furthermore, environmentalism substantially influences various domains, such as medicine and bioethics, with increasing recognition of the interdependence between environmental health and human well-being. Through this lens, bioethics principles such as patient autonomy (informed consent), practitioner beneficence (do good), practitioner nonmaleficence (do no harm), and public justice (be fair) are extended to include a responsibility toward the environment, urging a reevaluation of ethical obligations beyond human-centric concerns.

Rationalism: Environmentalism exhibits rationalist inclinations in utilizing certain innate cognitive faculties for understanding and solving environmental issues. For instance, the principles of deductive reasoning are applied in examining the cause-and-effect relationships between human activities and environmental degradation. By starting with axiomatic truths — such as the laws of thermodynamics or the principles of ecology — environmental scientists and activists can deduce the long-term consequences of resource exploitation, pollution, or biodiversity loss. This reliance on rational faculties allows for quick, reliable assessments that can be used to formulate public policies or ethical guidelines aimed at conserving ecosystems and mitigating climate change. The concept of emergence is also applicable in this context. Just as rational faculties coalesce into complex cognitive abilities, individual actions and policies in environmentalism often sum up to create holistic approaches to ecological sustainability, such as circular economies or sustainable development models. Moreover, the adaptive value of these rationalistic faculties becomes evident when they enable societies to adapt to environmental challenges quickly, enhancing communal survival and fitness. On an ethical plane, awareness of the limitations of these cognitive faculties is essential; for example, while rationalism can provide us with tools for managing resources more effectively, it must be balanced with ethical considerations about the fair distribution of these resources. This reflects the role of professionals, ethicists, and policymakers in examining and interpreting these faculties within an ethical framework, thereby fostering a more nuanced and responsible form of environmentalism.

Empiricism: Environmentalism's empiricist orientation is evident in its heavy reliance on data gathered through sensory experiences and environmental observations. This empirical approach employs inductive reasoning to formulate general conclusions based on specific instances, such as collecting climate data, to make broad statements about global warming trends. These empirical observations guide policy decisions where evidence-based approaches precede regulations like emission standards or wildlife protection acts. Cognitive biases and external factors such as culture and socialization also play a part in affecting the interpretation of data and the consequent policies that are formulated. Recognizing this, environmental professionals often emphasize the need for continuous monitoring and data collection to ensure that public policies are adaptive and responsive to new empirical evidence. This adaptability reflects the inherent value of empirical faculties in dealing with unforeseen environmental challenges, thereby contributing to both human and ecological survival and fitness. Emergence comes into play when these data points and observations are synthesized to understand complex systems like climate patterns or ecosystem balances. These emergent properties, often more intricate than the sum of individual observations, necessitate a multidisciplinary empirical approach. Language about biodiversity, the culture around sustainability, and social norms about conservation could be considered emergent phenomena that arise from individual experiences but collectively shape a society's relationship with its environment. Consequently, empirical evidence informs and evolves with the ethical frameworks and public policies that guide environmental action.

The Scientific Method: Environmentalism has played a significant role in shaping and refining the scientific method by necessitating a multidisciplinary, often large-scale approach to inquiry. Observing complex phenomena like climate change or ecosystem dynamics requires the integration of various scientific disciplines such as biology, chemistry, physics, and social sciences, each contributing unique observational methods and hypotheses. This interdisciplinary nature has also pushed for advancements in data collection techniques, ranging from satellite imagery to citizen science initiatives, to provide the comprehensive datasets needed for robust conclusions. Because of the global scale and long timelines associated with many environmental issues, environmental science has contributed to the development of innovative experimental designs that account for variables that are often uncontrolled, thereby adding a layer of complexity to traditional controlled experiments. Environmental studies frequently rely on longitudinal data, meta-analyses, and complex statistical models to test hypotheses. Additionally, the high stakes associated with environmental issues — such as public health and global sustainability — have heightened the importance of rigorous peer review and replicability. There is an acute awareness that findings must not only withstand scrutiny from experts in the same field but also be communicable to policymakers and the public, further refining the scientific method to ensure the robustness and accessibility of research. In this way, environmentalism has influenced not just the subject matter of scientific inquiry but also the methodologies, collaborative practices, and ethical considerations that comprise the scientific method.

Medicine: Environmentalism has profoundly influenced the field of medicine, expanding our understanding of health to include the impacts of environmental factors. Ecological health and environmental toxicology have gained prominence, examining how pollutants, climate change, and habitat destruction can directly or indirectly affect human health. For example, research on air quality has led to evidence-based practices for managing respiratory conditions like asthma. Understanding the link between pesticide exposure and certain cancers has informed guidelines for agricultural practices. This has implications for healthcare standards of care, as medical professionals increasingly consider environmental exposure history when diagnosing and treating patients. Furthermore, environmentalism has fueled a push for "*green medicine*," which seeks to minimize the environmental impact of healthcare itself. Hospitals and healthcare providers are increasingly adopting sustainable practices, from waste management to energy use, affecting the standards against which medical facilities are evaluated. In some cases, environmental sustainability has been integrated into the metrics that determine the adequacy of delivered care, thereby influencing legal contexts. As a result, the healthcare community recognizes that a reasonably competent and skilled healthcare professional should also be aware of the environmental determinants of health and the ecological footprint of healthcare practices. These evolutions underscore how environmentalism has enriched the medical healthcare field's scientific and ethical dimensions and informed best practices sensitive to the interconnectedness of human and environmental well-being.

Ethics: Environmentalism has significantly influenced the realm of bioethics, expanding the traditional focus from human-centered concerns to a more ecologically inclusive perspective. For instance, the principle of autonomy, usually concerned with informed consent in a clinical setting, has been broadened to incorporate the rights of communities to be informed about environmental hazards that could affect their health. This aligns with environmental justice movements that advocate for community participation in environmental decision-making processes. Beneficence, or the obligation to do good, has evolved to include responsibilities for sustainable practices that benefit individual patients and broader ecosystems, thus encouraging holistic well-being. Similarly, nonmaleficence, the principle of doing no harm, has been expanded to account for the environmental impact of medical practices. This has led to reevaluating medical waste disposal methods, energy usage in healthcare facilities, and even the ecological consequences of pharmaceuticals. The principle of justice, traditionally focused on equitable access to healthcare, now also addresses the fair distribution of environmental goods and risks. For example, it advocates against practices that disproportionately expose marginalized communities to environmental hazards. In this manner, environmentalism has significantly enriched bioethical considerations, driving a shift toward more inclusive, ecologically responsible practices. These adaptations reflect a growing recognition that ethical principles cannot be fully realized in isolation from their environmental context; rather, they are deeply interconnected, demanding a more integrated approach to human and ecological well-being.

Conclusion: Environmentalism's sweeping impact across multiple domains speaks to its transformative power as a social, ethical, and scientific movement. Embodying rationalist and empiricist approaches has enhanced our scientific understanding of intricate ecological issues through deductive reasoning and empirical data. This has fortified the scientific method, setting new benchmarks for multidisciplinary collaboration and methodological rigor. Beyond science, environmentalism has also permeated the medical field, pushing it toward a more ecologically sensitive paradigm that considers the environmental determinants of health and the ecological impact of healthcare practices. In doing so, it has shifted standards of care, melding clinical effectiveness with environmental stewardship. Moreover, it has redefined the boundaries of bioethics, extending traditional principles like patient autonomy (informed consent), practitioner beneficence (do good), practitioner nonmaleficence (do no harm), and public justice (be fair) to encompass environmental dimensions. These principles are no longer confined to human interactions but are increasingly considered integral to the more extensive ecological systems within which humans exist. Environmentalism has become an indispensable framework for understanding and responding to the complex challenges that define our relationship with the natural world by instigating shifts in scientific inquiry, healthcare practices, and ethical considerations. Through its rationalist and empiricist inclinations, impact on scientific methodology, contributions to medicine, and influence on bioethics, environmentalism is a compelling testament to the interconnectivity of human existence and the environment, urging a more holistic approach to societal development and well-being.

Environmentalism's Legacy: Advocates for the planet's preservation, fostering a collective responsibility toward Earth's ecosystems and biodiversity.

**

REVIEW QUESTIONS

True/False Questions:

1. Environmentalism is primarily focused on preserving natural resources and minimizing pollution through a combination of empirical observations and rational assessments.
   True or False?

2. The principle of nonmaleficence in environmentalism includes ensuring that medical practices do not harm the environment.
   True or False?

Multiple-Choice Questions:

3. Which of the following is NOT typically associated with the goals of environmentalism?
   a) Minimizing pollution
   b) Conserving natural resources
   c) Promoting economic growth at any cost
   d) Protecting vulnerable ecosystems

4. How has environmentalism influenced the scientific method?
   a) By reducing the need for interdisciplinary collaboration
   b) By introducing advancements in data collection techniques and statistical models
   c) By simplifying experimental designs
   d) By discouraging long-term data collection

Clinical Vignette:

5. A community living near an industrial area is experiencing high rates of respiratory illnesses. According to the principles of environmentalism, what action should healthcare providers and policymakers prioritize?
   a) Focusing solely on treating the symptoms without addressing the environmental cause
   b) Relocating the affected community without addressing the industrial pollution
   c) Implementing stricter regulations on industrial emissions and providing comprehensive care for the affected residents
   d) Ignoring the issue and waiting for more conclusive evidence

Basic Science Vignette:

6. A research team uses environmental data to study the effects of pollution on local ecosystems. After collecting extensive data, the system starts identifying new patterns of species decline and habitat degradation. However, it also generates a significant number of false positive results. What is the best course of action to address the false positives?
   a) Ignore the false positives as the system's overall performance has improved
   b) Review and refine the data collection and analysis methods using a more comprehensive dataset
   c) Remove the new patterns from the study to prevent false positives
   d) Disable the data collection system until the false positives can be fully understood

Philosophy Vignette:

7. An AI system in environmental ethics uses data to suggest guidelines for sustainable development. It proposes an optimal framework that maximizes environmental sustainability but overlooks community participation in decision-making. What is the best philosophical approach to address this issue?
   a) Prioritize the AI's ability to maximize environmental sustainability
   b) Reprogram the AI to give more weight to community participation and environmental sustainability
   c) Develop an oversight committee to review the AI's recommendations before implementation
   d) Disable the AI system until it can be programmed to always respect community participation

Correct Answers:

1. True
2. True
3. c) Promoting economic growth at any cost
4. b) By introducing advancements in data collection techniques and statistical models
5. c) Implementing stricter regulations on industrial emissions and providing comprehensive care for the affected residents
6. b) Review and refine the data collection and analysis methods using a more comprehensive dataset
7. b) Reprogram the AI to give more weight to community participation and environmental sustainability

BEYOND THE CHAPTER

Environmentalism

- *Silent Spring* by Rachel Carson
- *This Changes Everything: Capitalism vs. The Climate* by Naomi Klein
- *The Overstory* by Richard Powers

***

Philosophy of Science
© 2024 Jeffrey W. Bulger
Medical Research

Chapter 35: Global Health Initiatives — The Equity Imperative: Health for All

Global Health Initiatives advocate for equitable healthcare, blending scientific rigor and ethical principles.

Abstract: Global Health Initiatives (GHIs) embody international strategies and collaborations to enhance global health and well-being. These initiatives intersect rationalist philosophies, empirical methodologies, scientific rigor, medical advancements, and ethical principles. Anchored in rationalistic approaches, they foster cognitive advancements, collective intelligence, and ethical consciousness to address global health issues. They also emphasize data-driven insights and adaptive strategies, contributing to culturally sensitive, ethically grounded healthcare models. GHIs have influenced the scientific method, facilitating innovations in medical research and rigorous peer review, catalyzing universal healthcare solutions. These initiatives have propelled medical advancements, standardizing care and fostering research that transcends borders, improving global healthcare quality. Ethically, GHIs adhere to principles of patient autonomy (informed consent), practitioner beneficence (do good), practitioner nonmaleficence (do no harm), and public justice (be fair), setting bioethical standards. GHIs significantly influence healthcare practices, public policies, and ethical norms worldwide, underscoring the necessity for equitable health access and highlighting global health disparities.

**

Introduction: Global Health Initiatives encompass a broad range of international programs, policies, and strategies designed to improve health and well-being across nations. Rooted in a commitment to public health, these initiatives aim to tackle systemic challenges that often transcend borders, such as infectious diseases, malnutrition, maternal and child health, and access to healthcare services. Collaborative by nature, these efforts involve a complex web of stakeholders, including governmental organizations, non-governmental organizations — NGOs, academic institutions, and private sector partners, all working together to solve deeply ingrained health disparities. Whether eradicating polio, combating tuberculosis, or improving mental health services, Global Health Initiatives are a testament to what can be achieved when the global community unites behind a common goal. Understanding these initiatives' impact and ethical implications is critical for their long-term success and scalability amid ever-changing geopolitical landscapes and rising healthcare needs.

Rationalism: Global Health Initiatives often exemplify rationalist inclinations by deploying cognitive faculties like problem-solving, pattern recognition, and deductive reasoning to address health challenges on a global scale. Through a rationalist lens, these initiatives draw from a set of "*a priori*" principles — such as the inherent value of human life, the obligation to alleviate suffering, and the potential for scientific innovation to solve health problems — that guide the formulation of policies, strategies, and interventions. For example, the application of deductive reasoning is evident in how vaccination campaigns are planned and executed. Public health experts use foundational knowledge on herd immunity and infectious disease spread to logically deduce the minimum percentage of a population that needs to be vaccinated to achieve collective protection. Furthermore, these initiatives' complex, multi-stakeholder nature often leads to the emergence of collective intelligence and ethical awareness that is more advanced than what individual entities might achieve independently. Public policy, in turn, reflects these rationalistic tendencies by setting regulations that align with these fundamental principles, from ensuring vaccine equity to ethical considerations like informed consent. These initiatives enhance societal fitness and survival by acknowledging the adaptive value of rational faculties in quickly addressing global health challenges. However, it is crucial for stakeholders involved to remain aware of the limitations and ethical ramifications of their cognitive faculties, ensuring that principles such as patient autonomy (informed consent), practitioner beneficence (do good), practitioner nonmaleficence (do no harm), and public justice (be fair) are not compromised in the pursuit of global health goals.

Empiricism: Global Health Initiatives are deeply grounded in an empiricist orientation, relying heavily on sensory experiences and environmental interactions through data collection, field research, and real-world trials to inform policies and interventions. Inductive reasoning is central to these efforts; specific observations about health metrics, such as disease prevalence or vaccination rates, lead to more general conclusions and strategies for addressing healthcare issues. For instance, the success of anti-malarial campaigns in one region may serve as empirical evidence to adapt and implement similar strategies elsewhere. The role of cognitive biases and external factors like culture and socialization is also acknowledged; these variables can significantly impact the efficacy of health interventions and must be considered when interpreting data. From a public policy perspective, the empiricist approach mandates the creation of adaptive, evidence-based regulations that can be fine-tuned in response to new findings or societal changes. The adaptive value of this empiricist orientation is its agility in addressing unforeseen challenges and rapidly evolving health crises, thereby enhancing societal survival and fitness. Just as the concept of "*emergence*" in rationalism highlights the collective intelligence greater than the sum of individual faculties, the emergent properties in empiricism can be seen in phenomena like the development of culturally sensitive healthcare models or the complex dynamics of pandemic response, which can't be easily predicted by studying individual sensory experiences or variables in isolation. Therefore, empiricism plays a critical role in shaping the ethical and practical dimensions of Global Health Initiatives.

The Scientific Method: Global Health Initiatives have played a seminal role in shaping and refining the scientific method, particularly in the context of medical and public health research. These initiatives often begin by observing a health disparity or an outbreak, leading to hypothesis formulation about the most effective interventions. For instance, randomized controlled trials — considered the gold standard in medical research — are frequently used to test interventions like new vaccines, treatment modalities, or public health campaigns. Data are meticulously collected, quantitatively and qualitatively, and analyzed to reach conclusions that either support or refute the initial hypothesis. Moreover, the multi-stakeholder nature of these initiatives allows for rigorous peer review, as findings are scrutinized by researchers and policymakers, NGOs, and sometimes even the communities involved. Importantly, the scale and diversity of Global Health Initiatives provide unique opportunities to replicate studies across different populations and settings, further fortifying the scientific method by ensuring that results are consistent and universally applicable. Additionally, these initiatives often lead to advancements in research methodologies tailored to the unique challenges posed by global health crises, such as ethical considerations in resource-poor settings or the development of rapid diagnostic tools for emerging diseases. Global Health Initiatives serve as both a crucible and a catalyst for the scientific method, driving its evolution to meet global complex, multi-dimensional healthcare challenges.

Medicine: Global Health Initiatives have significantly contributed to the advancement of medicine in clinical practice and the establishment of healthcare standards of care. Often operating at the intersection of multiple disciplines, these initiatives have fostered groundbreaking research in infectious diseases, maternal and child health, and chronic conditions, shaping evidence-based practices widely recognized in the medical community. For instance, the development and deployment of multidrug-resistant tuberculosis (MDR-TB) treatments or the global distribution of vaccines for diseases like polio and HPV were accelerated through these initiatives. Moreover, these programs often serve as large-scale, real-world laboratories for testing new interventions, providing invaluable data that further refines standards of care. In doing so, they help ensure that healthcare delivery aligns with current best practices informed by cutting-edge research, clinical expertise, and a nuanced understanding of patient needs across diverse populations. Importantly, these initiatives' international scope and collaborative nature make them highly influential in harmonizing medical practices across borders and elevating healthcare quality globally. This is crucial for raising the standard of care in resource-poor settings and provides a benchmark against which legal frameworks can assess the adequacy of medical care, thus having far-reaching implications for both the global healthcare system and individual patient outcomes.

Ethics: Global Health Initiatives wield significant influence in shaping and reinforcing the four cornerstone principles of bioethics: patient autonomy (informed consent), practitioner beneficence (do good), practitioner nonmaleficence (do no harm), and public justice (be fair). Autonomy is often promoted through initiatives prioritizing informed consent, particularly in research studies or vaccination campaigns, ensuring that participants know the risks and benefits of a given intervention. Beneficence is manifested through the very mission of these initiatives: to improve health outcomes and quality of life, guided by evidence-based practices. The principle of nonmaleficence is meticulously safeguarded through rigorous pre-implementation assessments and post-implementation evaluations, ensuring that interventions, even when well-intentioned, do not inadvertently cause harm. Ethical review boards, often multi-disciplinary in composition, act as gatekeepers to maintain beneficence and nonmaleficence. Lastly, justice, the principle advocating for fairness and equitable treatment, is deeply ingrained in these initiatives. Whether it is equitable access to vaccines, addressing social determinants of health, or prioritizing vulnerable populations, the ethical underpinning of justice is often at the forefront of policy considerations. However, it's crucial to note that these principles can sometimes become tense, necessitating complex ethical deliberations. By actively engaging with these ethical principles, Global Health Initiatives contribute to the well-being of populations and serve as leading models in applying and evolving bioethical standards in healthcare.

Conclusion: Global Health Initiatives is a place where rationalism, empiricism, the scientific method, medicine, and ethics converge to create a transformative impact on a global scale. Guided by rationalist inclinations, these initiatives lay the foundation of logical reasoning and principles to frame strategies that address urgent health needs. Their empiricist orientation ensures that evidence and data serve as the bedrock upon which interventions are constantly fine-tuned based on real-world outcomes. The scientific rigor they bring advances medical practices and shapes and refines the scientific method, making it more responsive to complex health challenges. Moreover, they are instrumental in evolving standards of care in medicine, improving healthcare quality, and setting international benchmarks. Ethically, these initiatives work within a framework that values patient autonomy (informed consent), practitioner beneficence (do good), practitioner nonmaleficence (do no harm), and public justice (be fair), thereby pushing forward the ethical standards by which all healthcare endeavors should be measured. Global Health Initiatives exemplify an interdisciplinary, evidence-based, and ethically guided approach to solving some of the world's most daunting health challenges. Their work serves as a testament to what can be achieved when different facets of human knowledge and ethics collaborate, and they continue to set the stage for the ongoing evolution of healthcare practices, public policy, and ethical standards on a global level.

Global Health Initiatives' Legacy: Focuses on equity in health access and outcomes, spotlighting interconnectedness and disparities in global health.

**

REVIEW QUESTIONS

True/False Questions:

1. Global Health Initiatives are solely focused on eradicating infectious diseases.
   True or False?

2. Empiricism in Global Health Initiatives emphasizes the importance of data-driven decisions and real-world observations.
   True or False?

Multiple-Choice Questions:

3. Which ethical principle is challenged by ensuring equitable access to healthcare interventions across different populations?
   a) Beneficence
   b) Autonomy
   c) Justice
   d) Nonmaleficence

4. What is one significant contribution of Global Health Initiatives to the scientific method?
   a) Eliminating the need for peer review
   b) Introducing new experimental designs tailored to global health issues
   c) Reducing the importance of replicability in research
   d) Simplifying data collection processes

Clinical Vignette:

5. During a global vaccination campaign, a healthcare provider must ensure that all participants understand the risks and benefits of the vaccine. Which ethical principles are primarily at play in this scenario?
   a) Autonomy and beneficence
   b) Nonmaleficence and justice
   c) Beneficence and nonmaleficence
   d) Autonomy and justice

Basic Science Vignette:

6. Researchers are investigating the effectiveness of a new intervention to reduce the spread of malaria in a high-risk region. They decide to implement the intervention in multiple randomly selected communities and compare the outcomes with communities that did not receive the intervention. What is the primary study design they are using?
   a) Cross-sectional study
   b) Cohort study
   c) Randomized controlled trial
   d) Case-control study

Philosophy Vignette:

7. A health policy strategist is designing a global vaccination program. They begin by determining fundamental principles and use these to develop a logical framework for achieving herd immunity. What philosophical approach are they primarily employing in their planning?
   a) Empiricism
   b) Inductive reasoning
   c) Rationalism
   d) Pragmatism

Correct Answers:

1. False
2. True
3. c) Justice
4. b) Introducing new experimental designs tailored to global health issues
5. a) Autonomy and beneficence
6. c) Randomized controlled trial
7. c) Rationalism

BEYOND THE CHAPTER
Global Health Initiatives

- *Global Health: An Introduction to Current and Future Trends* by Kevin McCracken and David R Phillips
- *Introduction to Global Health* by Kathryn H. Jacobsen
- *Global Health 101* by Richard Skolnik

***

Philosophy of Science

Medical Research

Chapter 36: Green Revolution — The Agricultural Paradigm: Balancing Innovation

The Green Revolution revolutionized agriculture, balancing scientific innovation with ethical challenges.

Abstract: The "*Green Revolution*," spearheaded by Norman Borlaug and others in the mid-20th century, revolutionized agriculture in developing nations by incorporating scientific innovation to enhance crop yields and fight food insecurity. This movement integrated high-yielding crop varieties, synthetic fertilizers and pesticides, and advanced irrigation methodologies, significantly increasing global food supplies and mitigating hunger. It showcased a blend of rationalism, grounded in deductive reasoning and "*a priori*" principles, and empiricism, emphasizing observation and evidence-based practices. These philosophies crystallized in a refined scientific method that underpinned breakthroughs in agriculture and medicine, influencing public health and nutritional science. Despite its achievements, the Green Revolution raised bioethical dilemmas, embodying tensions between beneficence and nonmaleficence and raising questions about justice and autonomy, due to its environmental impact and marginalization of smallholders. This initiative serves as a multifaceted lesson in the interplay between scientific innovation, ethics, and societal implications, urging an ethically informed approach to addressing global challenges.

**

Introduction: The "*Green Revolution*" refers to the significant agricultural transformation primarily in developing nations between the 1940s and 1960s. Pioneered by agronomist Norman Borlaug and supported by various organizations, the movement aimed to combat food insecurity and improve crop yields through scientific innovation. Combining high-yielding crop varieties, synthetic fertilizers, pesticides, and improved irrigation techniques dramatically increased agricultural productivity. Initially centered on staples such as wheat and rice, the Green Revolution's methods were adopted worldwide to include a variety of crops, substantially raising global food supplies and reducing hunger. While it has been hailed as a triumph of scientific ingenuity that saved millions from starvation, it has also drawn criticisms for its environmental impact, its tendency to favor larger farms over smallholders, and its ethical considerations around biotechnology and corporate influence. In the forthcoming sections, we will delve into the rationalist and empiricist perspectives that guided this movement, its role in shaping the scientific method, its contributions to medicine, and its influence on the ethical principles of bioethics: patient autonomy (informed consent), practitioner beneficence (do good), practitioner nonmaleficence (do no harm), and public justice (be fair).

Rationalism: The Green Revolution embodies a rationalist inclination by applying a framework of deductive reasoning, axiomatic principles, and problem-solving to address the issue of food scarcity. Rooted in the notion that scientific innovation could be a universally applicable solution, pioneers like Norman Borlaug utilized innate cognitive faculties of pattern recognition to identify the traits that would lead to higher-yielding, disease-resistant crop varieties. This relied heavily on the "*a priori*" principles of genetics and plant biology, which were assumed to hold across different agricultural contexts. The approach was underpinned by the epigenetic assumption that the interaction of genes with the environment — in this case, optimized through fertilizers, pesticides, and irrigation — would improve crop yields. On a broader scale, the movement's reliance on technological solutions reflects a collective form of rationalistic emergence. Societal institutions, policymakers, and scientific communities joined forces, leveraging their collective cognitive abilities to address a complex problem more effectively than any individual or isolated effort could. However, the ethical implications of these rationalistic tendencies are significant. In prioritizing high yields, the Green Revolution often sidestepped environmental sustainability, and its focus on larger-scale farming systems has been criticized for marginalizing smaller, subsistence farmers. This highlights the need for ethical frameworks to evaluate and guide the deployment of rationalistic faculties in such transformative endeavors. Professionals and policy-makers are responsible for recognizing the limitations of a purely rationalist approach, ensuring that it aligns with broader ethical considerations and social justice. Thus, while the Green Revolution serves as an example of rationalism's adaptive value in swiftly addressing food scarcity, it also underscores the need for a more nuanced and ethically-aware application of these innate faculties.

Empiricism: The Green Revolution was profoundly empiricist in its orientation, emphasizing the importance of observation, experimentation, and evidence-based practice. Its breakthroughs in high-yield crops were rooted in empirical research, including field trials, data collection, and statistical analysis, all designed to translate sensory experiences and environmental interactions into generalizable knowledge. Inductive reasoning played a crucial role as specific observations about crop yields, disease resistance, and responsiveness to fertilizers and pesticides led to broader conclusions about optimizing agriculture on a global scale. External factors, such as cultural practices in farming and societal expectations for food security, were also considered, showing how empiricism could be sensitive to social and environmental contexts. Regarding ethics and public policy, the Green Revolution's empiricist tendencies underscore the necessity for guidelines and regulations grounded in empirical evidence that are adaptive and responsive to societal and environmental changes. This aligns with empiricism's adaptability to new and unforeseen challenges, enhancing both food security and, by extension, human survival. The concept of "*emergence*" is also manifest in the Green Revolution, as its collective impact — both positive and negative — on social systems, economies, and the environment cannot be entirely predicted or explained by its components, such as specific technologies or practices. These emergent properties may include changes in rural-urban migration, shifts in labor markets, and even alterations in cultural traditions around food and farming. Therefore, while the Green Revolution showcased the value of an empiricist approach in solving pressing food security issues, it also highlighted the need for a comprehensive understanding considering such a large-scale initiative's complex, emergent impacts.

The Scientific Method: The Green Revolution made substantial contributions to the refinement and application of the scientific method, particularly in the fields of agronomy, genetics, and ecology. The challenge of global food insecurity necessitated a systematic approach to inquiry, starting with observation and questioning. Agronomists observed the limitations of traditional farming methods and crop varieties, leading to hypotheses about improving yields, resilience, and nutrient content. Controlled experiments — often in field trials — were rigorously designed to test these hypotheses. Data collection methodologies were advanced to capture complex variables such as soil quality, water availability, and crop yield. Statistical tools were employed to analyze these data meticulously, leading to conclusions that either supported or refuted the initial hypotheses. Notably, the international scope of the Green Revolution demanded that findings be replicable across different geographical and cultural contexts. This criterion of replicability pushed researchers to develop standardized testing procedures that could be adopted globally, effectively universalizing certain aspects of the scientific method in agricultural research. Finally, given the enormous social, economic, and environmental stakes, the peer review process was particularly rigorous, involving experts in agricultural science and related fields like ecology, economics, and ethics. This multidisciplinary scrutiny made the scientific method in the context of the Green Revolution more robust, influencing how it is understood and practiced in tackling complex global challenges.

Medicine: While the Green Revolution is most commonly associated with advancements in agriculture, its ripple effects have had consequential contributions to the field of medicine, particularly in public health and nutritional science. The dramatic increase in food production helped mitigate malnutrition and associated health conditions, easing the burden on healthcare systems. This transformation was facilitated through an evidence-based approach akin to medical standards of care, and it involved close monitoring, data collection, and iterative improvements. With the alleviation of food scarcity and improved nutritional quality, conditions like stunted growth, vitamin deficiencies, and other malnutrition-related diseases saw a marked decrease. The methodologies used in crop research — like controlled trials and peer-reviewed studies — have applications in medical research protocols, thereby contributing to evolving healthcare standards. Moreover, as the Green Revolution propelled the need for understanding the environmental impact of pesticides and fertilizers, it indirectly influenced toxicological research, contributing to our understanding of how chemicals interact with biological systems — a vital aspect of pharmacology. While the primary focus was agriculture, the rigorous scientific and ethical frameworks developed provided a template that could translate into medical research and healthcare delivery practices. Thus, the Green Revolution contributed to immediate improvements in public health through enhanced food security and influenced the methodologies and ethical considerations central to medical healthcare standards of care.

Ethics: The Green Revolution's impact on bioethical principles presents a nuanced landscape. On the side of beneficence, the initiative indisputably aimed to "*do good*" by significantly enhancing food security and reducing starvation, thereby improving the quality of life for millions around the globe. However, regarding nonmaleficence or "*do no harm*," the narrative becomes more complex. The use of synthetic fertilizers and pesticides, though increasing yield, has had long-term environmental consequences, including soil degradation and water pollution. These ecological impacts can be considered harmful to the environment and communities reliant on these ecosystems. Justice or fairness is another ethical dimension impacted by the Green Revolution. While the movement aimed to alleviate global hunger, the benefits have yet to be evenly distributed. Smallholder farmers often needed help to afford the new technologies, leading to increased economic disparity and marginalization. This raises questions about the equitable distribution of benefits and risks, thus challenging the principle of justice. As for autonomy and informed consent, the Green Revolution did not traditionally operate within a framework that prioritized these ethical principles. Farmers, especially in developing countries, were often introduced to new technologies and practices without fully understanding the beneficial and harmful long-term implications. While the Green Revolution made significant strides in promoting the principle of beneficence, its legacy presents ethical challenges regarding nonmaleficence, justice, and autonomy that offer crucial lessons for any large-scale scientific endeavor.

Conclusion: The Green Revolution is a monumental chapter in scientific, ethical, and social history annals. Driven by a blend of rationalism and empiricism, it was a groundbreaking endeavor that dramatically increased food production and alleviated hunger for millions. Through the lens of the scientific method, it offered a rigorously tested, replicable model for enhancing agricultural output, thereby influencing other disciplines, including medicine and public health. However, this triumph of human ingenuity also brought forth a range of ethical dilemmas and unintended consequences.

While it fulfilled the bioethical principle of beneficence by combating food scarcity, it posed challenges to principles of nonmaleficence, justice, and autonomy. The environmental degradation and social inequities that followed its implementation illuminate the limitations and ethical responsibilities intrinsic to large-scale scientific interventions. As such, the Green Revolution serves as a potent case study for the transformative power of science and technology while cautioning us about the need for a balanced, ethically grounded approach to addressing global challenges. It embodies the complexities that arise when scientific innovation intersects with ethical considerations, social justice, and environmental sustainability, offering enduring lessons for future endeavors to solve humanity's most pressing issues.

Green Revolution's Legacy: A remarkable stride in combating global hunger through scientific innovation, yet shadowed by environmental and socio-economic repercussions, urging a more balanced, ethically grounded approach to future global interventions.

**

REVIEW QUESTIONS

True/False Questions:

1. The Green Revolution primarily focused on increasing agricultural productivity through the development of high-yielding crop varieties, synthetic fertilizers, and advanced irrigation techniques.
   True or False?

2. The Green Revolution had no significant impact on environmental sustainability and smallholder farmers.
   True or False?

Multiple-Choice Questions:

3. Which of the following principles of bioethics was most directly challenged by the environmental impact of the Green Revolution?
   a) Autonomy
   b) Beneficence
   c) Nonmaleficence
   d) Justice

4. What methodology did the Green Revolution heavily rely on to achieve its advancements in agriculture?
   a) Randomized controlled trials
   b) Peer-reviewed theoretical models
   c) Empirical field trials and statistical analysis
   d) Meta-analysis of existing literature

Clinical Vignette:

5. During the Green Revolution, a farmer in a developing country adopts new high-yield crop varieties and synthetic fertilizers. Over time, the farmer notices soil degradation and a decrease in water quality. Which ethical principles are at play in this scenario, considering the long-term environmental impact?
 a) Beneficence and autonomy
 b) Nonmaleficence and justice
 c) Autonomy and nonmaleficence
 d) Nonmaleficence and beneficence

Basic Science Vignette:

6. A research team employs high-yield crop varieties and synthetic fertilizers to boost agricultural productivity as part of a Green Revolution initiative. After initial success, they observe significant soil degradation and water pollution. What is the best course of action to address these environmental issues?
 a) Ignore the environmental issues as the overall crop yield has improved
 b) Review and refine the agricultural practices using a more sustainable approach
 c) Remove the high-yield crop varieties from further use to prevent further degradation
 d) Disable the use of synthetic fertilizers until the environmental impact can be fully understood

Philosophy Vignette:

7. An AI system in agricultural ethics uses data to suggest guidelines for implementing Green Revolution technologies. It proposes an optimal framework that maximizes crop yields but overlooks the long-term environmental sustainability and impact on smallholder farmers. What is the best philosophical approach to address this issue?
 a) Prioritize the AI's ability to maximize crop yields
 b) Reprogram the AI to give more weight to environmental sustainability and the well-being of smallholder farmers
 c) Develop an oversight committee to review the AI's recommendations before implementation
 d) Disable the AI system until it can be programmed to always respect environmental sustainability and the well-being of smallholder farmers

Correct Answers:

1. True
2. False
3. c) Nonmaleficence
4. c) Empirical field trials and statistical analysis
5. c) Autonomy and nonmaleficence
6. b) Review and refine the agricultural practices using a more sustainable approach
7. b) Reprogram the AI to give more weight to environmental sustainability and the well-being of smallholder farmers

BEYOND THE CHAPTER
Green Revolution

- *The Doubly Green Revolution: Food for All in the Twenty-First Century* by Gordon Conway
- *The Green Revolution Revisited: Critique and Alternatives* edited by Bernhard Glaeser
- *Bread and Roses: The Radical Tradition of the Green Revolution* by Angus Wright

***

Chapter 37: Complexity Theory — The Holistic Framework: Understanding Systems

Complexity Theory offers a holistic understanding of systems, transforming scientific inquiry and ethical frameworks.

Abstract: Complexity Theory, an interdisciplinary framework, revolutionizes understanding of the intricacies and emergent phenomena in complex systems across biology, economics, and social sciences. Deviating from traditional reductionist approaches, it fosters a holistic view, exploring nonlinear and unpredictable patterns from simple interactions. This analysis navigates its rationalist foundations, using formal logic to decipher principles, and its empiricist orientation, employing data-driven methodologies and real-world case studies. Furthermore, it delineates how Complexity Theory rejuvenates the scientific method, introducing techniques like computational models and network analyses to study nonlinear and adaptive systems, reshaping paradigms. In medicine, it promotes dynamic, patient-centered strategies grounded in empirical evidence. Moreover, the theory adds depth to bioethics, urging a reevaluation of principles like autonomy and justice through a lens that considers the intricate interplay of factors. This interdisciplinary narrative underscores Complexity Theory's transformative role in fostering nuanced, robust, and adaptable frameworks in science, medicine, and ethics.

**

Introduction: Complexity Theory is a multidisciplinary framework that aims to understand how simple components can interact to form complex systems and patterns, often producing far from intuitive outcomes. Originating from mathematics, computer science, and physics, it has transcended its initial domains to offer invaluable insights into diverse fields such as biology, economics, and social sciences. This theory focuses on studying emergent phenomena, where the whole is greater than the sum of its parts, as in the case of ant colonies, neural networks, or even the intricacies of human societies. It grapples with questions about how order can arise from disorder, how complexity evolves, and how local interactions can lead to global patterns. The ideas promulgated by Complexity Theory challenge traditional reductionist approaches that seek to understand systems by breaking them down into their simplest components. Instead, it advocates for a holistic view, emphasizing complex systems' nonlinear, adaptive, and often unpredictable nature. As we delve deeper into this subject, we will explore its rationalist foundations, empiricist orientation, role in shaping the scientific method, contributions to medicine, and profound influence on the four ethical principles of bioethics: patient autonomy (informed consent), practitioner beneficence (do good), practitioner nonmaleficence (do no harm), and public justice (be fair).

Rationalism: The rationalist foundations of Complexity Theory lie in its reliance on mathematical modeling, formal logic, and axiomatic systems to elucidate the underlying principles of complex systems. At its core, Complexity Theory leans heavily on a set of axioms or postulates to explain phenomena like emergence, adaptability, and self-organization. These principles are "*a priori*" foundations for understanding how simple parts can interact to generate unpredictably complex wholes. Furthermore, Complexity Theory presupposes certain "*hardwired*" faculties, such as pattern recognition and problem-solving, are essential for interpreting and engaging with complexity. Through deductive reasoning, the theory enables us to draw far-reaching conclusions from simple initial conditions, exploring how rules at a micro-level can evolve to form complex macro-level systems. This deductive process allows Complexity Theory to generalize findings, offering insights that can be universally applied to diverse disciplines, from the natural sciences to economics and social sciences. Complexity Theory upholds the rationalist conviction that a structured, rule-based approach can yield a deep understanding of even the most intricately complicated phenomena by establishing a rational framework that systematically studies inherently complex, chaotic, and often unpredictable systems.

Empiricism: The empiricist orientation of Complexity Theory is evident in its embrace of observation, experimentation, and data-driven methodologies to understand complex systems. While its rationalist foundations provide conceptual scaffolding, empiricism adds the flesh and bones through real-world data and case studies. This is particularly noticeable in the extensive use of computer simulations and models incorporating empirical data to replicate complex systems, offering insights into their behavior and properties. For instance, agent-based models in economics or cellular automata in biology utilize real-world parameters to explore how local interactions can yield complex, global patterns. In fields like ecology and medicine, empirical studies collect vast amounts of data to examine the emergent properties that arise from complex interactions among individual components. Inductive reasoning plays a crucial role here, as specific observations of particular instances are generalized to offer broad conclusions about system behaviors. This melding of empiricism with its rationalist underpinnings allows Complexity Theory to maintain a dynamic, adaptive approach to understanding how simple interactions can lead to complex and often unpredictable outcomes. By grounding its abstract principles in empirical reality, Complexity Theory ensures that its insights are theoretically robust and practically applicable.

The Scientific Method: Complexity Theory has transformed the scientific method by introducing novel approaches to studying inherently non-linear, adaptive, and complex systems. Traditional scientific methods often hinge on reductionism, isolating variables for controlled experimentation to produce replicable results. While this approach is effective for simple systems, it becomes inadequate when tackling complex systems where the whole is greater than the sum of its parts. Complexity Theory brings computational models, network analyses, and multi-agent simulations as alternative tools for scientific inquiry. These methods permit researchers to observe emergent phenomena in a controlled setting, making it possible to formulate and test hypotheses about system behaviors that are not readily amenable to traditional experimental designs. Consequently, the theory broadens the scientific method to include simulations and modeling as acceptable forms of empirical verification, adding a layer of methodological pluralism. This shift has encouraged a more interdisciplinary approach to scientific inquiry, breaking down silos between different scientific domains. While it adheres to the fundamentals of hypothesis testing, data collection, and peer review, Complexity Theory challenges and expands our understanding of what constitutes valid scientific methodology, thereby enriching the overall toolkit of scientific inquiry.

Medicine: Complexity Theory has significantly influenced the field of medicine by offering new perspectives on healthcare systems, patient care, and even the understanding of physiological processes. Recognizing that healthcare institutions are complex adaptive systems, the theory has helped to improve healthcare delivery by focusing on system dynamics, such as the interdependencies among various departments or the ripple effects of policy changes. This systemic view enables a more nuanced approach to standardizing healthcare practices, emphasizing adaptability and responsiveness over rigid protocols. Moreover, Complexity Theory has impacted our understanding of disease pathology and treatment. For example, the complexities of cellular interactions in cancer have been modeled using principles from this theory, offering insights that could lead to more effective treatments. It also provides a framework for understanding the multi-faceted nature of patient care, where social, psychological, and biological factors interact in complex ways to impact health outcomes. By applying Complexity Theory to the existing standards of care, medical professionals can develop more holistic, patient-centered strategies that are adaptive to individual needs while still being grounded in peer-reviewed, evidence-based practices. In this way, Complexity Theory enriches medical healthcare standards by making them more dynamic and better attuned to the intricacies of both human biology and healthcare systems.

Ethics: Complexity Theory has profound implications for bioethics, adding nuance and depth to the principles of patient autonomy (informed consent), practitioner beneficence (do good), practitioner nonmaleficence (do no harm), and public justice (be fair). Bioethics has often relied on a reductionist approach, simplifying ethical issues into discrete, manageable parts for easier resolution. However, Complexity Theory posits that ethical scenarios in healthcare are often intricate systems with interconnected variables that influence outcomes. Take, for instance, the principle of autonomy, usually realized through informed consent. Complexity Theory underscores that patients' decision-making processes are affected by many factors — emotional, cognitive, and social — that interact in complex ways. Thus, ensuring true informed consent requires a more holistic approach that accounts for these variables. Beneficence and nonmaleficence are enriched by a complexity perspective, emphasizing that "*doing good*" and "*doing no harm*" are not always straightforward objectives but may involve complex trade-offs within a web of physiological and psychosocial factors. The principle of justice, concerning fairness in the distribution of healthcare resources, gains additional complexity when considering social determinants, healthcare access, and systemic factors contributing to health disparities. By applying Complexity Theory to bioethics, we can develop more robust ethical frameworks sensitive to the complexities of human behavior, social systems, and medical practice, thereby fostering more ethical and effective healthcare solutions.

Conclusion: Complexity Theory has reinvigorated our understanding across multiple disciplines, reshaping traditional paradigms and offering a sophisticated lens through which to view the world. At its core, it synthesizes rationalist and empiricist perspectives, employing mathematical models as foundational structures and using empirical data to refine our understanding of complex systems. Its implications on the scientific method are transformative, bringing computational simulations and network analyses into the fold of acceptable methodologies for scientific inquiry. In medicine, it challenges the rigidity of existing standards of care, promoting a more dynamic, responsive, and adaptive approach to healthcare delivery and disease management. Moreover, it enriches the ethical discourse in bioethics, adding complexity and nuance to our understanding of principles like patient autonomy (informed consent), practitioner beneficence (do good), practitioner nonmaleficence (do no harm), and public justice (be fair). Through its interdisciplinary nature, Complexity Theory bridges gaps between different domains of knowledge, from natural sciences to social sciences, contributing to more robust, nuanced, and adaptable frameworks. Whether it is understanding the intricate dynamics of a cellular system or the moral complexities of healthcare, Complexity Theory has proven itself as an indispensable tool, profoundly influencing our methodologies, practices, and ethical considerations.

Complexity Theory's Legacy: As a paradigm-shifting framework, Complexity Theory is poised to leave a lasting legacy by fundamentally transforming how we approach, understand, and interact with complex systems across diverse disciplines, thereby enriching scientific methodology and ethical reasoning.

**

REVIEW QUESTIONS

True/False Questions:

1. Complexity Theory primarily focuses on understanding systems by breaking them down into their simplest components.
   True or False?

2. Complexity Theory suggests that emergent phenomena arise from simple interactions among system components.
   True or False?

Multiple-Choice Questions:

3. Which of the following is NOT a method commonly associated with Complexity Theory?
   a) Computational models
   b) Network analyses
   c) Reductionist experiments
   d) Multi-agent simulations

4. How has Complexity Theory influenced the scientific method?
   a) By simplifying the study of complex systems
   b) By promoting the use of advanced computational models and simulations
   c) By reducing the need for interdisciplinary collaboration
   d) By discouraging the study of nonlinear systems

Clinical Vignette:

5. A medical team is treating a patient with a complex, multi-system illness. According to Complexity Theory, what approach should the team prioritize?
   a) Focusing solely on the most immediate symptoms
   b) Considering the interactions between different physiological and psychological factors
   c) Using a one-size-fits-all treatment protocol
   d) Ignoring social and environmental factors affecting the patient's health

Basic Science Vignette:

6. Researchers are studying the spread of an infectious disease within a large urban population. They implement a simulation model that incorporates real-world data about human movement and interactions. Their goal is to predict how the disease will spread and identify effective intervention strategies. What methodology are the researchers primarily using?
   a) Cross-sectional analysis
   b) Randomized controlled trial
   c) Computational modeling
   d) Case-control study

Philosophy Vignette:

7. A philosopher is analyzing the decision-making process of healthcare providers during a complex medical emergency. They emphasize the importance of understanding the interaction between multiple variables, such as patient condition, available resources, and team dynamics, to arrive at ethical decisions. What philosophical approach is the philosopher most likely employing?
   a) Utilitarianism
   b) Empiricism
   c) Rationalism
   d) Complexity Theory

Correct Answers:

1. False
2. True
3. c) Reductionist experiments
4. b) By promoting the use of advanced computational models and simulations
5. b) Considering the interactions between different physiological and psychological factors
6. c) Computational modeling
7. d) Complexity Theory

BEYOND THE CHAPTER
Complexity Theory

- *Complexity: A Guided Tour* by Melanie Mitchell
- *Complexity: The Emerging Science at the Edge of Order and Chaos* by M. Mitchell Waldrop
- *At Home in the Universe: The Search for the Laws of Self-Organization and Complexity* by Stuart Kauffman

***

Philosophy of Science

Medical Research

Chapter 38: Epigenetics — The Dynamic Influence: Nature and Nurture

Epigenetics bridges nature and nurture, transforming our understanding of genetic regulation and ethical implications.

Abstract: In the fast-paced expansion of biological sciences, epigenetics is transforming our understanding of the relationship between genes and environmental influences. This field explores heritable changes in gene expression without altering the DNA sequence, revealing a complex layer of genetic regulation dictating cells' identities and functions. Rationalism promotes a methodical approach to deciphering molecular intricacies, while empiricism emphasizes observation and experiential evidence. Integrating epigenetics into the scientific method requires innovative methodologies and stringent analytical processes to investigate gene-environment interplay. Epigenetics offers promising avenues for diagnosing and treating diseases through understanding epigenetic modifications, heralding targeted and personalized medicine. However, this shift brings ethical dilemmas, urging a reevaluation of patient autonomy (informed consent), practitioner beneficence (do good), practitioner nonmaleficence (do no harm), and public justice (be fair). As we enter a new scientific epoch, epigenetics promises groundbreaking discoveries and fosters thoughtful discourse on the accompanying ethical ramifications.

**

Introduction: Epigenetics, a fascinating genetic frontier, delves into the study of heritable changes in gene expression without alterations to the underlying DNA sequence. This complex layer of regulation enables cells with the same genetic code, like a neuron and a liver cell, to express distinct profiles of genes, leading to their unique identities and functions. Epigenetic modifications, including DNA methylation and histone modifications, act as intricate molecular switches that turn genes on or off, influencing various biological processes and outcomes. Environmental factors, lifestyle, and experiences can induce epigenetic changes, offering a dynamic interface between genes and the environment. As our understanding of epigenetics deepens, it reveals profound implications for understanding development, disease, and inheritance and reconceptualizing the very fabric of heredity and evolution. The landscape of genetics is no longer solely determined by the sequence of bases within our DNA but also by the intricate epigenetic codes that overlay it.

Rationalism: The exploration of epigenetics can be seen through a rationalist lens, as it emphasizes the importance of reason and systematic inquiry in understanding the natural world, particularly the intricate interplay between genes and the environment. Rationalism posits that certain truths about the universe can be deduced through pure reason, independent of sensory experience. In epigenetics, this means looking beyond the immediate and observable (the phenotype or trait manifestation) to deduce the deeper, hidden regulatory mechanisms. Instead of accepting genetic determinism — that our DNA sequence solely dictates our biological fate — epigenetics introduces a nuanced understanding, suggesting that external factors can influence gene expression. This perspective encourages a deeper, more logical analysis of the factors contributing to our genetic makeup and their interrelations. Rational inquiry into epigenetics challenges the deterministic outlook, emphasizing that understanding the broader picture requires observing the effects (phenotypes) and deciphering the underlying logic and mechanisms. Thus, epigenetics, in its exploration of the complex codes overlaying our DNA, embodies a rationalist inclination, promoting a systematic, reasoned approach to understanding the intricate dynamics of life at the molecular level.

Empiricism: Epigenetics embodies the principles of empiricism, underscoring the significance of observation and experiential evidence in deciphering biological phenomena. Empiricism contends that knowledge arises primarily from sensory experience and experimental data rather than pure reason alone. This approach is particularly evident in epigenetics, as scientists rely heavily on empirical data to uncover how external factors like nutrition, stress, and toxins can influence gene expression without changing the DNA sequence itself. Through meticulously designed experiments, researchers have observed and documented cases where environmental stimuli induce epigenetic modifications, such as DNA methylation or histone alterations. For instance, studies on identical twins with the same genetic code have shown that differences in their environments can lead to divergent epigenetic patterns and health outcomes as they age. Furthermore, breakthroughs like the discovery of the Agouti mice, where maternal diet influenced offspring's coat color and health due to epigenetic modifications, underscore the empiricist orientation. In essence, epigenetics thrives on empirical evidence, with each experiment building upon the wealth of knowledge that emphasizes the dynamic relationship between our environment, epigenome, and genome.

The Scientific Method: Epigenetics, as an emerging and rapidly evolving discipline, has had a nuanced influence on the application and understanding of the scientific method. Traditionally, the scientific method involves systematic observation, formulation of hypotheses, experimentation, and conclusion. Epigenetics, with its intricate interplay of genes and the environment, introduces complexities that demand innovative experimental designs and data interpretation methods. For instance, the realization that external factors can induce heritable changes without altering the DNA sequence challenges previous deterministic perspectives on heredity. As a result, researchers have had to employ longitudinal studies, sophisticated statistical methods, and novel techniques to discern the often subtle epigenetic changes. Furthermore, the potential reversibility of some epigenetic modifications necessitates rigorous experimental controls and replication to validate findings. In essence, epigenetics has pressed the scientific community to refine the methodological rigor and adapt the traditional scientific method to accommodate the multifaceted nature of gene-environment interactions. By doing so, epigenetics enriches our understanding of biology and underscores the necessity for adaptability and precision in the ever-evolving landscape of scientific inquiry.

Medicine: Epigenetics has ushered in a transformative medical era, reshaping our understanding of disease onset, progression, and treatment. Recognizing that gene expression can be influenced without altering the underlying DNA sequence has broadened the scope of therapeutic interventions. Epigenetic modifications, like DNA methylation and histone modifications, have been linked to numerous conditions, from cancers to neurodegenerative disorders. For instance, epigenetic dysregulation is a hallmark of many cancers, leading to the silencing of tumor suppressor genes or activating oncogenes. This understanding has paved the way for epigenetic therapies that aim to reverse these abnormal modifications. Drugs like azacitidine, used in myelodysplastic syndromes, target DNA methylation processes. Additionally, understanding the epigenetic basis of diseases provides diagnostic potential; specific epigenetic patterns can serve as biomarkers for early disease detection or prognosis. Moreover, epigenetics has illuminated how lifestyle choices, from diet to stress management, can impact health at the molecular level, underscoring the importance of preventative medicine. As research unravels the epigenome's intricacies, its profound influence on medicine promises more targeted, personalized therapeutic approaches and a deeper comprehension of disease etiology.

Ethics: Epigenetics, with its revelations about the dynamic interplay between genes and environment, presents unique ethical challenges and considerations, particularly when framed within the principles of patient autonomy (informed consent), practitioner beneficence (do good), practitioner nonmaleficence (do no harm), and public justice (be fair). Autonomy, which emphasizes an individual's right to make informed decisions about their health, is impacted as epigenetic findings might compel individuals to make lifestyle choices for their benefit and future generations, given the potential transgenerational effects. Beneficence, the duty to do good, becomes complex when considering interventions like epigenetic editing. While these technologies may promise disease prevention or treatment, the long-term effects and potential unintended consequences remain unknown. Nonmaleficence, the principle of "*do no harm*," is challenged as we grapple with the implications of altering the epigenome intentionally through medical interventions and unintentionally through environmental exposures. Lastly, justice advocates for fairness in health care and raises questions about access to potentially life-altering epigenetic treatments. As epigenetic therapies become available, ensuring equitable distribution and addressing potential disparities in access becomes paramount. In essence, the evolving field of epigenetics demands a recalibration and deep reflection on established ethical principles, ensuring they remain robust in the face of rapid scientific advancement.

Conclusion: In the multifaceted realm of biology, epigenetics has emerged as a revolutionary field, reshaping our understanding of genetics, human development, disease, and evolution. Highlighting the heritable changes in gene expression without alterations to DNA sequences, it underscores the dynamic relationship between genes and their environment, revealing a layer of complexity previously unappreciated. Epigenetics intersects with philosophical stances like rationalism and empiricism, urging a deeper and broader analysis while emphasizing the importance of tangible, experimental evidence. As it intertwines with the scientific method, it propels a more rigorous and adaptive approach to research. In medicine, epigenetics holds transformative potential, opening new avenues for diagnosis, treatment, and prevention. Yet, with its profound insights come ethical challenges, especially as we navigate the implications of potentially modifying the epigenome and consider the broader societal ramifications. Epigenetics is not just about the molecular intricacies of life but also a testament to the endless quest for knowledge, its applications, and the ethical responsibilities accompanying profound scientific discoveries.

Epigenetics' Legacy: By revealing that environment and experience can shape genetic expression without altering DNA sequences, epigenetics fundamentally challenges deterministic perspectives on heredity and evolution, underscoring the dynamic interplay of nature and nurture in shaping life's trajectory.

**

REVIEW QUESTIONS

True/False Questions:

1. Epigenetics studies heritable changes in gene expression without altering the underlying DNA sequence.
   True or False?

2. Epigenetic modifications, such as DNA methylation and histone modifications, are influenced solely by genetic factors and not by environmental factors.
   True or False?

Multiple-Choice Questions:

3. Which of the following best describes an epigenetic change?
   a) A mutation in the DNA sequence
   b) A reversible modification of DNA or histone proteins that affects gene expression
   c) The complete removal of a gene from the genome
   d) The insertion of a foreign gene into the genome

4. How has the field of epigenetics influenced the scientific method?
   a) By eliminating the need for empirical evidence
   b) By introducing innovative experimental designs to study gene-environment interactions
   c) By focusing solely on genetic determinism
   d) By reducing the importance of peer review

Clinical Vignette:

5. A patient is diagnosed with a neurodegenerative disorder linked to epigenetic changes. Considering the principles of epigenetics, what approach might a healthcare provider take to treat or manage this condition?
   a) Focus only on surgical interventions
   b) Utilize drugs that target and modify epigenetic markers such as DNA methylation
   c) Recommend ignoring environmental factors such as diet and lifestyle
   d) Advise against any medical interventions due to the complexity of epigenetic mechanisms

Basic Science Vignette:

6. A team of scientists is studying the effects of childhood nutrition on long-term health by examining changes in DNA methylation and histone modifications in a cohort of participants over several decades. They find that early dietary patterns significantly influence gene expression related to metabolism and immune function. What is the primary focus of this study?
   a) Genetic sequencing
   b) Epigenetic modifications
   c) Protein-coding genes
   d) Chromosomal abnormalities

Philosophy Vignette:

7. An ethicist is analyzing the implications of using epigenetic information in public health policies. They emphasize that ethical principles such as patient autonomy, beneficence, nonmaleficence, and justice must guide the use of epigenetic data to ensure fair and responsible health interventions. What philosophical approach does this scenario most closely align with?
   a) Utilitarianism
   b) Empiricism
   c) Rationalism
   d) Bioethics

Correct Answers:

1. True
2. False
3. b) A reversible modification of DNA or histone proteins that affects gene expression
4. b) By introducing innovative experimental designs to study gene-environment interactions
5. b) Utilize drugs that target and modify epigenetic markers such as DNA methylation
6. b) Epigenetic modifications
7. d) Bioethics

BEYOND THE CHAPTER

Epigenetics

- *The Epigenetics Revolution: How Modern Biology is Rewriting Our Understanding of Genetics, Disease, and Inheritance* by Nessa Carey
- *Epigenetics: How Environment Shapes Our Genes* by Richard C. Francis
- *The Developing Genome: An Introduction to Behavioral Epigenetics* by David S. Moore

***

Philosophy of Science — Medical Research

Chapter 39: Microbiome — The Symbiotic Network: Health and Disease

The microbiome's symbiotic network transforms health insights, emphasizing holistic and ethical perspectives.

Abstract: In recent years, the study of the human microbiome, an intricate network of microorganisms within our bodies, has gained substantial attention for its influence on health and disease. This research reveals the complex, symbiotic relationships shaping our physiological functions and mental well-being. Utilizing advanced genomics and computational techniques, scientists explore the microbiome's complexities and implications. Microbiome research has revolutionized the scientific method, fostering innovation in biology, medicine, and computational sciences, and encouraging cross-disciplinary collaborations. This field has significantly influenced medical paradigms, pioneering groundbreaking therapies and personalized interventions. However, rapid advancement presents ethical dilemmas centered on patient autonomy (informed consent), practitioner beneficence (do good), practitioner nonmaleficence (do no harm), and public justice (be fair), necessitating the adaptation of ethical frameworks. The microbiome stands as a testament to the intricate interplay between humans and microbes, poised to reshape our understanding of biology, medicine, and the intrinsic connectivity of life, fostering holistic, symbiotic perspectives on health and existence.

**

Introduction: The human microbiome, the intricate ecosystem of microorganisms inhabiting our body, has increasingly come under the scientific spotlight for its profound influence on health, disease, and behavior. Comprised of trillions of bacteria, viruses, fungi, and other microscopic entities, this complex network coexists with human cells in a delicate symbiosis that impacts everything from digestion and immune function to mental health. Researchers are delving into the microbiome's mysteries using cutting-edge genomics and computational techniques, revealing an astonishing diversity that rivals the complexity of the rainforest or coral reef ecosystems. Furthermore, the microbiome has been implicated in many conditions — from obesity and diabetes to autoimmune disorders and mental illnesses. As we continue to unravel its intricate workings, the microbiome is not only reshaping our understanding of biology and medicine but also challenging ethical frameworks around patient autonomy (informed consent), practitioner beneficence (do good), practitioner nonmaleficence (do no harm), and public justice (be fair) in applying scientific knowledge.

Rationalism: In the microbiome study, rationalist perspectives focus on innate biological mechanisms that underlie the complex relationships between microbes and their hosts. Just as rationalism posits that humans have natural faculties like reasoning and pattern recognition, microbiome researchers often begin with "*a priori*" principles about how host-microbe interactions have evolved to benefit both entities. These foundational principles guide researchers in exploring symbiotic relationships, nutrient absorption efficiencies, and immune system coordination, all of which have been optimized over evolutionary time for mutual benefit. Deductive reasoning, based on evolutionary biology postulates or molecular biology axioms, leads to hypotheses that can be empirically tested, adding layers of complexity to our understanding. Epigenetic factors and environmental interactions are also considered, offering additional dimensions to foundational principles. On a personal level, individuals engaged in microbiome research must remain conscious of the limitations of their hypotheses and the ethical implications, especially when manipulating microbial communities for potential health benefits. At the organizational and policy levels, there's a responsibility to ensure that the adaptive value of rationalistic principles in microbiome research is ethically and thoughtfully considered, particularly as discoveries could significantly impact public health. The concept of "*emergence*" is highly relevant here; the collective behavior of a microbiome's trillions of individual microbes interacting with host cells may give rise to complex physiological states or health outcomes that are more intricate than the sum of individual interactions.

Empiricism: In the context of microbiome research, an empiricist orientation heavily emphasizes observational data and experimental outcomes to understand this intricate ecosystem. Researchers use high-throughput technologies like next-generation sequencing and metabolomics to gather exhaustive data from microbial communities within diverse environments — from the human gut to soil samples. Inductive reasoning is paramount, as specific observations, such as the presence or absence of certain bacterial strains, lead to broader conclusions about health implications or environmental impact. Cognitive biases can also come into play, affecting the interpretation of data or the formulation of hypotheses. For instance, initial observations that associate specific gut microbes with obesity may lead to an undue focus on these microbes at the expense of others. On an individual level, an empiricist orientation necessitates awareness of how personal experiences or biases may affect the interpretation of data and, consequently, ethical decision-making. From a professional standpoint, empiricism underscores the need for rigorous, evidence-based guidelines in scientific research and healthcare applications related to the microbiome. Public policy must also be grounded in empirical evidence, adapting to discoveries that influence our understanding of microbial interactions, their impact on human health, and environmental sustainability. The concept of "*emergence*" is particularly salient here, as the collective behaviors and potential benefits or harms of a microbiome cannot be easily predicted by analyzing individual microbial species or their isolated interactions; instead, they often emerge from the intricate interplay of multiple factors, akin to the development of complex ecosystems or human cultures.

The Scientific Method: The study of the microbiome has had a significant impact on the refinement and application of the scientific method, particularly in the fields of biology, medicine, and computational sciences. The inherent complexity of the microbiome, involving trillions of microbes interacting in intricate networks, necessitates innovative observational and experimental techniques. Traditional methods, often designed for studying isolated organisms or simple systems, have given way to more sophisticated, multi-disciplinary approaches incorporating genomics, bioinformatics, and systems biology. This has pushed the boundaries of the scientific method, prompting researchers to develop more advanced data collection and analysis tools and encouraging collaborations between disparate scientific disciplines. Formulating testable hypotheses in this arena often involves complex, multi-variable models that can be verified and falsified through controlled experiments. Comparative studies across diverse populations have highlighted the critical importance of replication, fostering rigorous data collection and analysis standards. Moreover, the interdisciplinary nature of microbiome research has necessitated more robust forms of peer review involving experts from various scientific backgrounds, enhancing the overall rigor and credibility of findings. Thus, microbiome research advances our understanding of a crucial biological system and influences how the scientific method evolves to tackle increasingly complex questions, thereby setting new standards for scientific inquiry.

Medicine: The contributions of microbiome research to medicine are manifold and transformative, fundamentally altering our understanding of health and disease. Previously considered secondary players, microbes are now recognized as pivotal in maintaining physiological balance, aiding digestion, modulating the immune system, and even affecting mental health. Consequently, microbiome research has led to groundbreaking therapeutic avenues such as fecal microbiota transplants for treating C. difficile infections, probiotic regimes for gastrointestinal disorders, and targeted microbial interventions for metabolic and autoimmune diseases. These innovations are being integrated into standards of care, underpinned by rigorous, peer-reviewed research that meets the benchmarks for evidence-based practices. The advent of personalized medicine has also seen the microbiome playing a pivotal role; for instance, understanding an individual's gut microbiota composition can lead to more tailored treatments for diseases like irritable bowel syndrome or even certain types of cancer. This integration reflects the evolving nature of medical best practices and has significant legal implications; healthcare providers must now consider microbial interactions when diagnosing and treating a wide array of conditions, adding a new dimension to the expected level and type of care within the medical community. Thus, microbiome research has become a cornerstone of modern medicine, influencing clinical guidelines and therapeutic strategies while elevating the complexity and efficacy of healthcare delivery.

Ethics: The burgeoning field of microbiome research adds complex layers to the bioethical landscape, particularly impacting principles like patient autonomy (informed consent), practitioner beneficence (do good), practitioner nonmaleficence (do no harm), and public justice (be fair). For instance, the issue of informed consent takes on added complexity when considering that interventions like fecal transplants not only affect the individual patient but also have the potential to alter their microbial communities in ways that still need to be fully understood. Given our still-evolving knowledge base, this raises questions about how comprehensive informed consent can be. Beneficence, or the commitment to do good, is also affected. While manipulating the microbiome may provide therapeutic benefits for conditions like Crohn's disease or obesity, the long-term effects of such interventions on overall health are still under study, prompting caution. The principle of nonmaleficence — do no harm — similarly mandates rigorous testing and cautious application of microbiome-related treatments, as unintended consequences could include the development of antibiotic-resistant strains or adverse immune responses. Finally, justice comes into play when considering who has access to advanced microbiome diagnostics and treatments, which are often expensive and not universally available. This raises questions about healthcare equality and how to equitably distribute this cutting-edge science's benefits — and potential risks. As microbiome research continues to unfold, it will be critical for ethical frameworks to adapt and address these multidimensional challenges.

Conclusion: The study of the human microbiome has had a transformative impact across multiple domains, including rationalism, empiricism, the scientific method, medicine, and ethics. Its unique influence on health and disease is a quintessential example of how complex biological systems can challenge and refine existing frameworks for understanding and exploring the world. Rationalistic approaches in microbiome research consider evolved, innate mechanisms that promote host-microbe symbiosis, providing a foundational lens through which to explore this intricate ecosystem. Empirically, the field leans heavily on observational data and experimental outcomes, utilizing cutting-edge technologies to generate insights that fundamentally alter healthcare practices and policies. The microbiome's complexity also has a ripple effect on the scientific method, demanding a multidisciplinary approach that leverages advancements in genomics, bioinformatics, and systems biology. In medicine, microbiome research has catalyzed innovative therapies and diagnostic techniques, becoming integral to evolving standards of care and personalized medicine strategies. Ethically, the field presents complex challenges that necessitate re-examining and adapting principles like informed consent, beneficence, nonmaleficence, and justice. As we continue to unravel the mysteries of the microbiome, its broad-ranging impact underscores the need for a dynamic, multifaceted approach to scientific inquiry, healthcare, and ethical considerations.

The Microbiome's Legacy: The human microbiome is projected to leave a lasting legacy as a paradigm-shifter in our understanding of health, disease, and the self, fundamentally altering the dualistic perspective of "*human vs. microbe*" to a more holistic, interconnected view that emphasizes the symbiotic relationships integral to our existence.

**

REVIEW QUESTIONS

True/False Questions:

1. The human microbiome is an intricate network of microorganisms that have a significant impact on our physiological functions and mental well-being.
   True or False?
2. Microbiome research primarily focuses on the interactions between human cells and viruses only.
   True or False?

Multiple-Choice Questions:

3. Which of the following principles is NOT a primary ethical concern in microbiome research?
   a) Patient autonomy
   b) Practitioner beneficence
   c) Nonmaleficence
   d) Economic efficiency
4. What is one of the main contributions of microbiome research to personalized medicine?
   a) Standardized treatments for all patients
   b) Generalized dietary guidelines
   c) Tailored treatments based on individual gut microbiota composition
   d) Uniform diagnostic procedures

Clinical Vignette:

5. A patient with recurrent Clostridioides difficile infections is considering a fecal microbiota transplant (FMT) as a treatment option. The physician explains the potential benefits and risks involved. What ethical principle is primarily being addressed in this scenario?
   a) Reducing healthcare costs
   b) Ensuring patient autonomy through informed consent
   c) Promoting practitioner beneficence by prioritizing patient care
   d) Avoiding harm by preventing adverse reactions

Basic Science Vignette:

6. Researchers discover that altering the microbiome in mice can reduce symptoms of a specific disease. They consider translating this approach to humans but are unsure of potential long-term effects. What should be their next step?
   a) Proceed with human trials immediately to test efficacy in humans
   b) Conduct further animal studies to understand long-term effects before human trials
   c) Abandon the research due to potential unknown risks
   d) Rely on existing data from other microbiome studies to justify proceeding with human trials

Philosophy Vignette:

7. A hospital implements a new treatment that alters patients' microbiomes to improve mental health outcomes. However, there is limited understanding of the long-term impacts on patients' overall health. What is the best ethical approach to address this uncertainty?
   a) Prioritize the immediate mental health benefits and proceed with the treatment
   b) Ensure that patients are fully informed of the potential risks and obtain their consent before treatment
   c) Develop an oversight committee to continuously monitor the treatment's effects and make adjustments as necessary
   d) Halt the treatment until comprehensive long-term studies are conducted

Correct Answers:

1. True
2. False
3. d) Economic efficiency
4. c) Tailored treatments based on individual gut microbiota composition
5. b) Ensuring patient autonomy through informed consent
6. b) Conduct further animal studies to understand long-term effects before human trials
7. b) Ensure that patients are fully informed of the potential risks and obtain their consent before treatment

BEYOND THE CHAPTER
Microbiome

- *The Human Microbiome: An Introduction to the Next Frontier in Medicine* by Julian R. Marchesi and Julian Parkhill
- *The Microbiome Diet: The Scientifically Proven Way to Restore Your Gut Health and Achieve Permanent Weight Loss* by Raphael Kellman
- *I Contain Multitudes: The Microbes Within Us and a Grander View of Life* by Ed Yong

***

Philosophy of Science

Medical Research

WEEK EIGHT: COGNITIVE AND ETHICS

WEEK EIGHT

Class Hour 1

- Chapter 40: Neuroscience — The Cognitive Frontier: Understanding the Mind
- Chapter 41: Heraclitus' Ship of Theseus Paradox — The Identity Puzzle: Continuity and Change
- Chapter 42: Stem Cell Research — The Regenerative Potential: Ethical Frontiers

Class Hour 2

- Chapter 43: Systems Thinking — The Integrative Framework: Complex Problem-Solving
- Chapter 44: Imre Lakatos (1922-1976) — The Research Methodologist: Research Programs
- Chapter 45: Paul Feyerabend (1924-1994) — The Methodological Anarchist: Epistemological Pluralism

---

Introduction

Welcome to Week Eight of "*Philosophy of Science*," where we journey into the intertwining realms of cognitive science and ethics. This week, we delve into the heart of neuroscience, grapple with the paradox of identity, explore the promise and pitfalls of stem cell research, embrace the holistic approach of systems thinking, and ponder the philosophies of two great minds—Imre Lakatos and Paul Feyerabend. Each chapter offers a unique lens through which we can view the profound questions of human consciousness, identity, and the ethical dimensions of scientific advancement.

Our exploration begins at the cognitive frontier with neuroscience, an interdisciplinary field that bridges biology, psychology, and computer science to unravel the mysteries of the mind. Advanced techniques like neuroimaging reveal the intricate workings of the brain, offering new perspectives on neurological disorders and their treatments. This scientific endeavor is not just about understanding the mind but also about navigating the ethical implications that arise as we unlock these secrets. The principles of bioethics—autonomy, beneficence, nonmaleficence, and justice—are central to this discussion, as we balance the potential for groundbreaking medical interventions with the moral responsibilities that accompany them.

Next, we confront the Ship of Theseus paradox, a thought experiment from Heraclitus that challenges our notions of identity and continuity. This ancient puzzle asks whether an object that has had all its components replaced remains fundamentally the same object. In the context of modern medicine, this paradox finds relevance in organ transplantation and regenerative medicine, where the essence of the self is questioned amidst transformative changes. Here, the ethical principles we discussed in neuroscience come into play again, prompting us to consider what it means to be the same person despite physical alterations.

As we transition to stem cell research, the theme of regeneration and ethical complexity continues. Stem cells, with their remarkable ability to develop into various cell types, hold the promise of curing debilitating diseases. However, the use of embryonic stem cells raises significant ethical dilemmas about the beginnings of life and the limits of scientific intervention. The dialogue around stem cell research is a microcosm of the larger ethical landscape, requiring a careful balancing act between the potential benefits and the moral considerations of such advancements.

Integrating these ideas, we turn to systems thinking, an approach that emphasizes the interconnectedness of components within a larger whole. In medicine, this perspective encourages holistic treatment strategies that consider the myriad interactions affecting health outcomes. Systems thinking not only enhances our problem-solving capabilities but also enriches our ethical frameworks, ensuring that healthcare practices are both effective and just.

The philosophical insights of Imre Lakatos offer a bridge between rationalist and empiricist traditions through his methodology of research programs. Lakatos' framework, which includes a core set of fundamental assumptions surrounded by adaptable hypotheses, mirrors the evolving nature of scientific inquiry and ethical discourse. This dynamic approach underscores the importance of flexibility and adaptation in the pursuit of knowledge, reinforcing the need for continuous reevaluation of ethical standards in light of new discoveries.

Finally, we embrace the radical views of Paul Feyerabend, who championed epistemological pluralism and methodological anarchy. Feyerabend's critique of rigid scientific methodologies advocates for a more inclusive and diverse approach to knowledge, one that values multiple perspectives and acknowledges the chaotic nature of scientific progress. His philosophy invites us to consider the broader implications of scientific practices, advocating for a more personalized and equitable approach to medicine and ethics.

In weaving together the themes of neuroscience, identity, regenerative medicine, systems thinking, and the philosophies of Lakatos and Feyerabend, we embark on a journey that challenges us to rethink the nature of knowledge, the self, and the ethical dimensions of scientific advancement. This week promises to deepen our understanding of the complex interplay between science and ethics, guiding us toward a more nuanced and holistic view of the world.

***

Chapter 40: Neuroscience — The Cognitive Frontier: Understanding the Mind

Neuroscience explores the cognitive frontier, enhancing our understanding of the mind and ethical care.

Abstract: Neuroscience, an interdisciplinary field, stands at the forefront of unveiling the intricacies of the nervous system, melding principles of biology, psychology, computer science, and other disciplines to understand brain functions and behaviors. Grounded in rationalist and empiricist doctrines, it employs a meticulous scientific method to explore neural underpinnings driving cognition, behavior, and societal interactions. Using advanced techniques like neuroimaging and electrophysiology, it offers groundbreaking insights into neurological and psychiatric disorders, revolutionizing medical care and developing nuanced therapeutic strategies. Neuroscience profoundly influences bioethics, reshaping discussions on patient autonomy (informed consent), practitioner beneficence (do good), practitioner nonmaleficence (do no harm), and public justice (be fair) amid rapid neurotechnology advancements. Its journey, characterized by methodological rigor and pursuit of empirical truths, fosters an enriched comprehension of human consciousness and identity, with transformative implications across law, education, and healthcare. Consequently, neuroscience's legacy lies in its capacity to redefine philosophical perspectives and influence diverse applications, affirming its position as a pivotal pillar in contemporary scientific exploration.

**

Introduction: Neuroscience is the interdisciplinary field of study that delves into the complex mechanisms of the nervous system, aiming to understand how neurons and intricate neural circuits result in behavior, perception, memory, and cognition, among other functions — originating from a blend of disciplines such as biology, psychology, chemistry, physics, and computer science, neuroscience endeavors to elucidate both the microscopic workings of individual neurons and the macroscopic activities of neural networks in organisms. As a field, it not only lays the groundwork for comprehending the physiological basis of mental health conditions, neurological disorders, and other aspects of human health but also raises essential questions about human identity, free will, and consciousness. Spanning levels of analysis from molecular interactions to large-scale brain networks, neuroscience serves as an indispensable pillar for various applications, including but not limited to medicine, robotics, artificial intelligence, and ethics. Its expansive scope, methodological rigor, and profound implications for human well-being make it one of the most captivating and consequential fields of scientific inquiry today.

Rationalism: Neuroscience inherently subscribes to a rationalist foundation, acknowledging that certain cognitive faculties and instinctual behaviors are evolutionarily hardwired for adaptive advantages. This premise resonates in studying neural substrates for basic reasoning, pattern recognition, and problem-solving abilities. Neuroscientific inquiries into innate faculties often employ deductive reasoning, starting with axioms or postulates about neural structures and functions and then observing or manipulating these elements to glean new insights. Epigenetic factors and learned behaviors are also increasingly recognized for their roles in modulating innate neural pathways, adding a layer of complexity to our understanding of cognition and behavior. The concept of "*emergence*" is particularly relevant; neuroscience posits that intricate neural circuits give rise to phenomena like consciousness, complex social interactions, and advanced ethical reasoning that are far greater than the sum of individual neural activities. This acknowledgment serves as a foundational perspective for neuroscience's diverse applications. For example, the field informs the design of educational curricula that consider innate learning styles and cognitive strengths or weaknesses. Furthermore, by dissecting the neural basis of decision-making and ethical reasoning, neuroscience can inform public policy and bioethical considerations, ensuring that regulations are attuned to the limitations and potential risks associated with our evolved cognitive faculties. In essence, neuroscience's rationalist underpinnings guide our understanding of neural mechanisms and our ethical, educational, and policy-related discussions surrounding human capabilities and limitations.

Empiricism: Neuroscience is deeply rooted in an empiricist orientation, emphasizing the importance of sensory experiences and environmental interactions in shaping neural pathways, cognitive processes, and states of consciousness. Researchers in the field often employ inductive reasoning, beginning with specific observations — through neuroimaging, electrophysiological recordings, or behavioral studies — and advancing to general conclusions about brain function and its impact on behavior and cognition. Cognitive biases, often arising as evolved heuristic "*shortcuts*," are scrutinized to understand how they might influence perception and decision-making. External factors such as culture and socialization are critical variables affecting cognitive development and function. This empiricist stance has significant implications across various domains. Individually, it serves as a reminder to remain vigilant of how one's sensory experiences and environmental conditions shape one's views and ethical decisions. Professionally, it underscores the paramount importance of empirical evidence in shaping ethical guidelines and clinical best practices. From a public policy standpoint, an empiricist orientation advocates for regulations that are not only grounded in scientific evidence but are also adaptable to evolving societal norms and technological innovations. The concept of "*emergence*" is pivotal, as empiricism in neuroscience shows us that phenomena like language development, cultural mores, or complex social behaviors cannot be fully understood by examining individual sensory experiences alone. They emerge from intricate interactions between sensory experiences, cognitive faculties, and the external environment, exemplifying the profound complexity of the nervous system.

The Scientific Method: Neuroscience has played a seminal role in adopting and refining the scientific method. With its interdisciplinary nature, drawing from biology, psychology, computer science, and even philosophy, neuroscience requires a particularly rigorous approach to inquiry. Observations of complex phenomena such as cognition, memory, or consciousness lead to highly specific and testable hypotheses, often requiring innovative experimental designs to isolate variables effectively. Advanced techniques like neuroimaging, electrophysiology, and optogenetics are tools for conducting meticulous, controlled experiments, providing rich, multidimensional data. The analysis of these data frequently involves sophisticated statistical and computational methods to draw credible conclusions. Importantly, the field has been a forerunner in emphasizing the necessity of replicability, leading to the development of open science initiatives that share raw data and experimental protocols. Given the profound ethical and medical implications of neuroscience research, the peer-review process is especially stringent, scrutinizing the validity and methodology of a study, its ethical considerations, and societal impact. As a result, neuroscience has contributed to shaping a more robust, transparent, and interdisciplinary adaptation of the scientific method, setting a high standard for empirical rigor and ethical responsibility that influences scientific inquiry.

Medicine: Neuroscience has revolutionized medicine by providing invaluable insights into the neurological and psychological underpinnings of health and disease, thereby contributing to the establishment and continual refinement of standards of care. Groundbreaking discoveries in neuropharmacology have led to the development of medications for disorders such as depression, schizophrenia, and Parkinson's. Advanced neuroimaging techniques like MRI and PET scans have enhanced diagnostic accuracy and facilitated the early detection of conditions such as Alzheimer's, improving prognostic outcomes. Neuroscience has also propelled the field of neurosurgery, offering more precise mapping of brain areas for surgical interventions. Moreover, the application of neuroscience in rehabilitative medicine, through the understanding of neural plasticity, has helped to develop evidence-based practices that significantly improve recovery from strokes or spinal cord injuries. Cognitive neuroscience findings are shaping interventions for learning disabilities and contributing to the personalization of mental health treatments. These advances are entrenched in the medical healthcare standards of care, driven by peer-reviewed, evidence-based practices. As a result, neuroscience has elevated clinical expertise and helped meet complex and diverse patient needs. Its contributions are critical in legal contexts, where neuroscience-informed standards of care often serve as benchmarks to evaluate the adequacy and quality of healthcare delivered.

Ethics: Neuroscience exerts a profound influence on the four principles of bioethics: patient autonomy (informed consent), practitioner beneficence (do good), practitioner nonmaleficence (do no harm), and public justice (be fair). Autonomy, particularly in the form of informed consent, becomes complex when neuroscience delves into understanding and potentially altering cognitive processes, emotions, and even consciousness. The field plays a pivotal role in shaping how we define and assess mental competence, thereby influencing protocols for obtaining informed consent for medical procedures or participation in research. Beneficence is enhanced through neuroscience's contributions to medicine, such as developing therapies for neurodegenerative diseases or mental health conditions. However, the principle of nonmaleficence — doing no harm — raises challenging questions, particularly with advancements like neural implants, cognitive-enhancing drugs, or even potential manipulations of memory and emotion. Ethical safeguards must be implemented to ensure these technologies do not harm individuals or societies. Finally, the principle of justice is deeply affected by neuroscience, especially considering equitable access to advanced neurological treatments and how neuropsychological data might be used or misused in legal, educational, and employment contexts. Neuroscience's intersection with bioethics demands constant re-evaluation and adaptation of ethical guidelines, ensuring that the pursuit of knowledge and medical advancements adheres to principles that safeguard human dignity, well-being, and fairness.

Conclusion: Neuroscience is a monumental interdisciplinary endeavor that has significantly impacted various dimensions of human knowledge and well-being. Its rationalist foundations help us navigate the intricate neural circuits that underlie cognition, offering a window into understanding evolved cognitive faculties. Simultaneously, its empiricist orientation deepens our understanding of how external factors like sensory experiences shape the brain, enabling a more comprehensive view of human behavior and cognition. By adhering to rigorous methodologies in line with the scientific method, neuroscience advances our understanding of the brain and mind, sets high standards for empirical rigor, and contributes to medical sciences in transformative ways. These contributions to medicine have redefined standards of care, improving the diagnosis and treatment of neurological and psychiatric disorders, thereby influencing legal paradigms. Moreover, neuroscience places an indelible mark on bioethics, continually raising new questions about patient autonomy (informed consent), practitioner beneficence (do good), practitioner nonmaleficence (do no harm), and public justice (be fair), demanding our ethical frameworks evolve with scientific understanding. In its quest to understand the most intricate of biological systems, neuroscience serves as a cornerstone for a plethora of applications ranging from healthcare to law, education, and ethics, substantiating its role as one of the most vital and consequential scientific fields of our time.

Neuroscience's Legacy: By unraveling the intricate neural mechanisms that underlie cognition, emotion, and behavior, neuroscience fundamentally alters our understanding of what it means to be human, challenging and refining long-standing philosophical notions of consciousness, free will, and identity in ways that have lasting implications for ethics, law, medicine, and technology.

**

REVIEW QUESTIONS

True/False Questions:

1. Neuroscience integrates principles from various disciplines such as biology, psychology, and computer science to understand brain functions and behaviors.
   True or False?

2. The study of neuroscience does not influence ethical considerations in medical practice.
   True or False?

Multiple-Choice Questions:

3. Which of the following is NOT a technique commonly used in neuroscience research?
   a) Neuroimaging
   b) Electrophysiology
   c) Next-generation sequencing
   d) Optogenetics

4. What is one of the main contributions of neuroscience to medicine?
   a) Developing general dietary guidelines
   b) Enhancing the accuracy of diagnostic techniques like MRI and PET scans
   c) Standardizing non-specific treatment methods
   d) Replacing traditional surgery with alternative therapies

Clinical Vignette:

5. A patient is undergoing an experimental treatment involving neural implants to manage severe epilepsy. The physician explains that while the treatment may offer significant benefits, there are also risks related to potential long-term changes in cognitive function. Which ethical principle is primarily being addressed here?
   a) Reducing healthcare costs
   b) Ensuring patient autonomy through informed consent
   c) Promoting practitioner beneficence by prioritizing patient care
   d) Avoiding harm by considering nonmaleficence

Basic Science Vignette:

6. A neuroscientist conducts an experiment where they observe changes in brain activity using fMRI while participants perform memory tasks. Which scientific methodological approach does this experiment exemplify?
   a) Deductive reasoning
   b) Inductive reasoning
   c) Behavioral observation
   d) Statistical analysis

Philosophy Vignette:

7. A neuroethicist is debating the ethical implications of using deep brain stimulation to treat severe depression. They are primarily concerned with which principle of bioethics?
   a) Autonomy (informed consent)
   b) Beneficence (do good)
   c) Nonmaleficence (do no harm)
   d) Justice (fair distribution of benefits and burdens)

Correct Answers:

1. True
2. False
3. c) Next-generation sequencing
4. b) Enhancing the accuracy of diagnostic techniques like MRI and PET scans
5. b) Ensuring patient autonomy through informed consent
6. b) Inductive reasoning
7. c) Nonmaleficence (do no harm)

BEYOND THE CHAPTER
Neuroscience

- *The Tell-Tale Brain: A Neuroscientist's Quest for What Makes Us Human* by V.S. Ramachandran
- *The Man Who Mistook His Wife for a Hat: And Other Clinical Tales* by Oliver Sacks
- *Neuroscience: Exploring the Brain* by Mark F. Bear, Barry W. Connors, and Michael A. Paradiso

***

Chapter 41: **Heraclitus' Ship of Theseus Paradox** — The Identity Puzzle: Continuity and Change

The Ship of Theseus paradox challenges our understanding of identity, continuity, and ethical implications in medicine.

Abstract: The Ship of Theseus paradox, attributed to the ancient philosopher Heraclitus, is a thought experiment exploring identity's nature in the face of change. In the paradox, the ship owned by the mythical king Theseus is gradually replaced part by part, posing whether the fully refurbished ship remains the same entity or transforms into something fundamentally different. This paradox has intrigued philosophers for centuries and significantly influenced discussions in various fields, including rationalism, empiricism, medicine, and ethics. The implications of this paradox are particularly profound in medical contexts such as organ transplantation and regenerative medicine, where questions of identity and continuity are paramount. Furthermore, the paradox challenges the ethical principles of bioethics, including autonomy, beneficence, nonmaleficence, and justice, by posing complex questions about the nature of the self in the face of transformative change. Overall, the Ship of Theseus paradox continues to serve as a powerful tool for navigating the intricate web of philosophical, scientific, and ethical questions that arise when considering the nature of identity in complex, dynamic systems.

**

Introduction: The Ship of Theseus paradox, often attributed to the ancient Greek philosopher Heraclitus, is a thought experiment that raises profound questions about the nature of identity, continuity, and change. It is based on a scenario in which the ship owned by the mythical king Theseus is gradually replaced, piece by piece, with new materials. The paradox asks whether the fully refurbished ship remains the same as the original or if it becomes a fundamentally different entity. This philosophical dilemma has not only engaged the minds of thinkers for centuries, but it has also permeated various fields, influencing discussions around rationalism and empiricism and shaping the scientific method. Its implications extend to medicine, where questions about identity and change are paramount in areas like organ transplantation and regenerative medicine. Furthermore, the paradox has profound relevance to the ethical principles of bioethics, including patient autonomy (informed consent), practitioner beneficence (do good), practitioner nonmaleficence (do no harm), and public justice (be fair), challenging and enriching our understanding of these foundational concepts. The Ship of Theseus paradox is a robust framework for examining the intricate web of philosophical, scientific, and ethical questions that arise when considering identity and change in complex, dynamic systems.

Rationalism: The Ship of Theseus paradox, attributed to Heraclitus, serves as a compelling exploration within the framework of rationalism, emphasizing foundational principles and deductive reasoning. At its core, the paradox questions the persistence of identity amidst gradual change: if every part of Theseus' ship is replaced over time, does it remain the same ship? This inquiry demands a deductive analysis rooted in *a priori* truths and definitional terms — in this case, the conceptual identity of the ship, independent of its physical components. Rationalism, which prioritizes reason and logical coherence derived from axioms, finds application here as philosophers analyze whether the ship's identity can be upheld through these deductive methods despite material alterations. Moreover, the paradox extends beyond metaphysical contemplation to raise ethical considerations. As parts of the ship are replaced, ethical dilemmas emerge concerning authenticity, continuity, and the moral implications of transformation. This parallels contemporary ethical issues in fields like bioethics or technology ethics, where gradual changes or emergent properties can lead to unforeseen moral challenges. The notion of emergence, central to both philosophical discourse and scientific inquiry, further complicates the debate: emergent properties in complex systems, including ethical systems, can challenge our understanding and raise new ethical questions. The Ship of Theseus paradox illustrates how top-down deductive reasoning, rooted in rational "*a priori*" and definitional truths, can be applied to analyze the persistence of identity and ethical implications in dynamic systems. By contemplating the ship's transformation, philosophers explore the boundaries of rationalist thought while grappling with timeless questions about identity, change, and the ethical dimensions of continuity in evolving contexts.

Empiricism: The Ship of Theseus paradox, as articulated by Heraclitus, presents an intriguing perspective within the framework of empiricism, which emphasizes bottom-up inductive reasoning starting from observations to derive general principles. The paradox challenges us to consider whether the ship maintains its identity as its parts are gradually replaced over time. This inquiry necessitates an empirical approach where observations of the ship's changing physical components lead to inductive conclusions about its identity: can we infer a general principle about identity from the observed changes? Empiricism, grounded in sensory experience and empirical evidence, offers a methodology to analyze how the ship's identity might evolve or persist based on observable changes, contrasting with the deductive reasoning of rationalism. Ethical considerations also emerge from the Ship of Theseus paradox within an empirical framework. As the ship undergoes transformations, ethical dilemmas arise concerning authenticity, integrity, and the moral implications of gradual change. These ethical reflections mirror contemporary debates in applied ethics and scientific research ethics, where empirical observations of evolving situations prompt ethical assessments of continuity and integrity. The concept of emergence further complicates this analysis, as emergent properties in complex systems can lead to unforeseen ethical challenges that demand empirical observation and ethical deliberation. The Ship of Theseus paradox illustrates how bottom-up inductive reasoning, rooted in empirical observations and the synthesis of data, can be applied to explore the persistence of identity and ethical implications in evolving systems. By examining the ship's gradual transformation, philosophers and ethicists engage in empirical investigation to understand the nature of identity and address ethical dilemmas that arise from observable changes in complex contexts.

The Scientific Method: The Ship of Theseus paradox significantly informs and parallels the iterative, questioning nature of the scientific method. Just as the paradox asks us to observe the gradual transformation of the ship and question the essence of its identity, the scientific method begins with observing a phenomenon and formulating a question seeking to explain its causes or implications. The paradox then prompts us to hypothesize about the continuity of the ship's identity, much like a scientist formulates a testable hypothesis to explain the observed phenomenon. The method of replacing the ship's components can be likened to conducting controlled experiments to test the validity of a hypothesis. The meticulous collection and analysis of data in these experiments echo the careful consideration of each replacement in the ship's structure, ultimately leading to a conclusion that either supports or refutes the original hypothesis or, in the case of the paradox, the maintained identity of the ship. The necessity for replication in scientific experiments mirrors the repeatability of the ship's transformation, ensuring that the outcome is not an anomaly. Lastly, the peer review process in scientific research is akin to the broader philosophical scrutiny and debate surrounding the paradox, validating the findings or interpretations and integrating them into the collective understanding of the field. In this way, the Ship of Theseus paradox is a metaphorical embodiment of the scientific method, encapsulating the essential steps of observation, hypothesis formulation, experimentation, analysis, and peer review defining this systematic inquiry approach.

Medicine: the Ship of Theseus paradox has profound implications for medicine, particularly in evolving medical healthcare standards of care and innovative medical interventions such as organ transplantation and regenerative medicine. Just as the paradox questions the ship's identity as its components are replaced, medical practitioners grapple with similar queries about human identity and continuity in the face of transformative medical procedures. The medical healthcare standards of care in medicine, akin to the meticulous replacement of the ship's parts, require a rigorously evidence-based and peer-reviewed approach to ensure that the provided care aligns with the most current and effective practices. Thc evolution of these standards reflects the dynamic nature of medical knowledge and patient needs, much like the changing structure of the ship reflects the necessity for renewal and improvement. As new medical technologies and treatments emerge, the paradox challenges us to consider how these interventions might alter our understanding of human identity and continuity and how standards of care must adapt to accommodate these changes while ensuring the highest quality of patient care. In this way, the Ship of Theseus paradox serves as a valuable metaphor for the ongoing evolution and adaptation required in medicine to uphold the highest standards of care in the face of changing knowledge, technology, and patient needs.

Ethics: The Ship of Theseus paradox significantly informs and challenges the foundational principles of bioethics, including patient autonomy (informed consent), practitioner beneficence (do good), practitioner nonmaleficence (do no harm), and public justice (be fair). In terms of autonomy, the paradox compels us to consider how changes to the self, akin to the ship's transformation, might affect a patient's ability and right to make informed decisions about their healthcare. This is especially relevant in progressive illnesses or transformative medical interventions, where the "*self*" before and after might have different perspectives and desires. The principle of beneficence, or the obligation to do good, is echoed in the ship's refurbishment, symbolizing the medical profession's duty to improve the patient's health and well-being. Similarly, nonmaleficence, or the commitment to do no harm, is reflected in carefully replacing the ship's components, symbolizing the meticulous approach required to avoid harm in medical interventions. Lastly, the paradox challenges the principle of justice, or the imperative to be fair, as it raises questions about identity and continuity that might impact equitable access to healthcare resources and treatments. For example, should the "*new*" ship or "*new*" self be afforded the same rights and access as the original? In this way, the Ship of Theseus paradox serves as a rich metaphorical framework for exploring and challenging our understanding of these bioethical principles.

Conclusion: Heraclitus' Ship of Theseus paradox has had a profound and lasting impact on various domains, from philosophy and science to medicine and ethics. By challenging our understanding of identity and continuity, the paradox has paved the way for critically examining some of the most foundational principles in these fields. In medicine, the paradox finds resonance in the evolving medical healthcare standards of care. New medical interventions challenge our understanding of human identity and continuity, necessitating an adaptation of healthcare standards to ensure the highest quality of patient care. Additionally, the paradox has significant implications for the ethical principles of bioethics, including patient autonomy (informed consent), practitioner beneficence (do good), practitioner nonmaleficence (do no harm), and public justice (be fair), offering a rich metaphorical framework for exploring these complex and interconnected concepts. Heraclitus' Ship of Theseus paradox is invaluable for navigating the intricate web of philosophical, scientific, and ethical questions that arise when considering identity and change in complex, dynamic systems.

Herclitus' Ship of Theseus paradox's Legacy: The paradox has fundamentally shaped philosophical and scientific discussions on identity, continuity, and change, with profound implications in rationalism, empiricism, the scientific method, medicine, and ethics, challenging and enriching our understanding of these foundational concepts across various fields.

**

REVIEW QUESTIONS

True/False Questions:

1. The Ship of Theseus paradox explores whether an object that has had all its components replaced remains fundamentally the same object.
   True or False?

2. The Ship of Theseus paradox has no relevance to modern medical practices such as organ transplantation and regenerative medicine.
   True or False?

Multiple-Choice Questions:

3. Which field is directly influenced by the Ship of Theseus paradox?
   a) Philosophy
   b) Medicine
   c) Ethics
   d) All of the above

4. What is a significant ethical concern related to the Ship of Theseus paradox in the context of medicine?
   a) Determining the nutritional value of different diets
   b) Ensuring the continuity of a patient's identity in organ transplantation
   c) Standardizing medical procedures across different countries
   d) Implementing uniform healthcare policies

Clinical Vignette:

5. A patient undergoes a series of organ transplants over several years, gradually replacing many of their original organs. They express concerns about their identity and continuity. Which philosophical concept from the chapter best addresses these concerns?
   a) Cartesian Dualism
   b) The Principle of Sufficient Reason
   c) The Ship of Theseus Paradox
   d) Utilitarian Ethics

Basic Science Vignette:

6. A research team is studying a specific protein involved in cell regeneration. They replace parts of the protein sequence with synthetic peptides, one segment at a time. After several cycles of replacement, they analyze the protein's structure and function to determine if it remains the same as the original or has fundamentally changed. This scenario is analogous to which philosophical concept discussed in the chapter?
   a) Quantum Entanglement
   b) The Principle of Sufficient Reason
   c) The Ship of Theseus Paradox
   d) Occam's Razor

Philosophy Vignette:

7. A psychologist is exploring the identity of a patient who has undergone extensive cognitive-behavioral therapy, resulting in significant changes in personality traits and behaviors over time. The psychologist questions whether the patient remains the same individual they were before therapy. This scenario is best illustrated by which concept discussed in the chapter?
   a) Dualism
   b) The Allegory of the Cave
   c) The Ship of Theseus Paradox
   d) The Veil of Ignorance

Correct Answers:

1. True
2. False
3. d) All of the above
4. b) Ensuring the continuity of a patient's identity in organ transplantation
5. c) The Ship of Theseus Paradox
6. c) The Ship of Theseus Paradox
7. c) The Ship of Theseus Paradox

BEYOND THE CHAPTER

Heraclitus' Ship of Theseus Paradox

- *Heraclitus, the Identity of Opposites, and the Flicker of Time* Journal of Philosophical Research, 31, 29-45, by Grupp, Jeffrey. (2006).
- *Personal Identity* by Harold Noonan
- *A Survey of Metaphysics* by E. J. Lowe
- *How Things Persist* by Katherine Hawley
- *Personal Identity and Self-Consciousness* by Brian Garrett

***

Philosophy of Science

Medical Research

Chapter 42: Stem Cell Research — The Regenerative Potential: Ethical Frontiers

Stem cell research unlocks regenerative potential, navigating ethical frontiers and medical advancements.

Abstract: Stem cell research represents both hope and ethical dilemmas in contemporary biomedicine. This dynamic field, grounded in rationalism and empiricism, continually refines the scientific method, paving the way for revolutionary medical interventions capable of treating diseases and injuries like Parkinson's disease and spinal cord afflictions. Leveraging the regenerative prowess of undifferentiated cells, stem cell research offers vast therapeutic possibilities. However, it is mired in ethical complexities, particularly regarding embryonic stem cell usage and the boundaries of life and morality. As the field evolves, it fosters multidisciplinary dialogue integrating bioethical principles of patient autonomy (informed consent), practitioner beneficence (do good), practitioner nonmaleficence (do no harm), and public justice (be fair), urging continuous re-evaluation of scientific innovation limits. This balance between medical advancements and ethical considerations epitomizes stem cell research, cementing its legacy as a source of transformative healthcare solutions and nuanced ethical discourse.

**

Introduction: Stem cell research, a biomedical science frontier, encapsulates the promise of groundbreaking medical advancements and a quagmire of ethical complexities. These undifferentiated cells, capable of transforming into specialized cells that comprise various tissues and organs, hold enormous potential for regenerative medicine, including treatments for conditions as diverse as Parkinson's disease, spinal cord injuries, and diabetes. Yet, this field also grapples with deeply ingrained ethical and moral questions, often pivoting on the source of these stem cells, such as embryonic stem cells, which can stir debates about the sanctity of life. As science pushes the boundaries of what is medically possible, society must confront the issue of what is ethically permissible. Therefore, to fully appreciate the scope and implications of stem cell research, it is essential to examine its rationalist foundations and empiricist orientation, its role in refining the scientific method, its invaluable contributions to medicine, and its complex interaction with the ethical principles of bioethics: patient autonomy (informed consent), practitioner beneficence (do good), practitioner nonmaleficence (do no harm), and public justice (be fair).

Rationalism: Stem cell research fundamentally relies on the rationalist foundations of human cognition, employing "*a priori*" principles such as deductive reasoning, pattern recognition, and problem-solving abilities in its exploration and application. Researchers leverage their innate cognitive faculties to hypothesize how stem cells could be directed to become specific types of cells and tissues based on their understanding of cellular biology and genetics. This framework often involves setting axiomatic principles, such as the defining characteristics of stem cells, and then logically extrapolating potential applications or outcomes. Complex interactions among genes, signaling pathways, and environmental cues are deciphered using innate faculties of pattern recognition, leading to increasingly sophisticated models of stem cell behavior and, ultimately, their medical applications. While rationalist faculties have accelerated scientific discovery, they also have ethical responsibilities. Scientists and ethicists must scrutinize the implications of stem cell research within a broader moral landscape, asking questions that transcend empirical observation — like the ethical concerns surrounding embryonic stem cells. Public policy, too, needs to consider both the potential for groundbreaking medical advancements and the ethical boundaries that rationalism alone may not fully encompass. In the end, stem cell research exemplifies the concept of "*emergence*," where the collective cognitive faculties of interdisciplinary teams yield insights and ethical guidelines more complex and nuanced than any individual could conceive. This emergent phenomenon guides the rapid advancements in the field and its conscientious, ethical oversight, facilitating a balanced approach to one of modern medicine's most promising yet contentious areas.

Empiricism: Stem cell research is deeply grounded in an empiricist orientation, relying heavily on sensory experiences and environmental interactions to shape scientific and ethical frameworks. Through extensive experimentation and observation, scientists employ inductive reasoning to derive general conclusions about the potential and limitations of stem cells. For example, meticulous observation of how stem cells differentiate into specialized cells under specific conditions provides invaluable empirical data, guiding future research and application. Empirical methodologies have challenges, such as cognitive biases that might affect the interpretation of data or ethical considerations. Moreover, social and cultural factors significantly impact how research outcomes are evaluated ethically, often leading to varied international policies on what is permissible in stem cell research. Consequently, it becomes crucial for researchers and policy-makers to be aware of how their own experiences and societal contexts influence their views and decisions. Empirically grounded guidelines and regulations are essential to ensuring that the scientific community proceeds in an innovative and ethically sound manner. In this complex tapestry, "*emergence*" is particularly relevant. The collective knowledge generated through empirical observations makes phenomena far more complicated than any one experiment or set of observations could predict — be it advanced therapeutic applications or nuanced ethical frameworks. These emergent properties, such as the development of regenerative medicine or shifting cultural attitudes toward bioethics, reveal the intricate relationship between empirical evidence and stem cell research's broader, often unpredictable, consequences.

The Scientific Method: Stem cell research has played a significant role in utilizing and refining the scientific method, acting as a crucible where the fundamentals of empirical investigation meet cutting-edge biomedical innovation. Beginning with observing stem cells' unique ability to differentiate into various cell types, researchers formulate specific, testable hypotheses, often exploring the cellular and molecular mechanisms governing these processes. Controlled experiments, often utilizing advanced techniques like CRISPR gene editing or three-dimensional cell cultures, are designed to verify or falsify these hypotheses. The meticulous data collection and analysis — sometimes involving large-scale genomic studies or high-throughput screening — are integral to these endeavors. The field also places a high premium on replicability; given the potential medical applications and ethical implications, it is crucial that other researchers can consistently reproduce findings under the same conditions. Peer review is the final gatekeeper, with experts scrutinizing the research's validity, methodology, and ethical considerations. Through this rigorous process, stem cell research advances its field. It contributes to the evolution of the scientific method itself, setting new standards for data collection, ethical oversight, and experimental design. As a result, it has become a paragon for how complex biological research can be conducted in a manner that is both scientifically rigorous and ethically responsible.

Medicine: Stem cell research has profoundly impacted the field of medicine, heralding a new era of therapeutic possibilities and shaping standards of care in various specialties. This research has fueled advancements in regenerative medicine, enabling the replacement or repair of damaged tissues and organs. Conditions once deemed incurable or untreatable, such as certain types of spinal cord injuries, Parkinson's disease, and degenerative eye conditions, are now within the purview of potential stem cell-based therapies. In oncology, stem cells are used in bone marrow transplants, offering life-saving treatments for patients with leukemia and other blood disorders. These treatments are gradually being integrated into medical healthcare standards of care, and as more peer-reviewed, evidence-based practices emerge, they will further shape these benchmarks. By offering new modalities of treatment, stem cell research enriches clinical expertise, influencing diagnostic and therapeutic decision-making processes. Patient needs are also increasingly met as stem cell therapies open avenues for personalized medicine, targeting the specificities of individual cases. Consequently, stem cell research pushes the boundaries of what is medically achievable and continually refines standard, competent, and ethical care in modern medical healthcare. Its contributions are thus immeasurable regarding immediate medical healthcare applications and setting the stage for future breakthroughs.

Ethics: Stem cell research intersects intriguingly and complexly with the four cardinal bioethical principles of patient autonomy (informed consent), practitioner beneficence (do good), practitioner nonmaleficence (do no harm), and public justice (be fair). The principle of autonomy underscores the critical importance of informed consent, especially in studies involving human subjects or the donation of biological material like embryos. Researchers must provide comprehensive information, ensuring participants understand their involvement's scope, risks, and implications. Beneficence manifests in the field's enormous potential for alleviating suffering and curing debilitating diseases, fulfilling the ethical obligation to "*do good*." However, the principle of nonmaleficence, or "*do no harm*," raises serious concerns, particularly when embryonic stem cells are involved. Ethical questions emerge about the potential harm done in harvesting these cells, often sparking debates about when life is considered to begin. The principle of justice, emphasizing fairness and equity, also comes into play as the fruits of stem cell research must be accessible to all, regardless of socio-economic status, age, or nationality. There's also the justice-related issue of whose cells are being used for research and whether the benefits and burdens of such research are being equitably distributed. Thus, stem cell research is a lens through which the nuances and potential conflicts among these bioethical principles can be critically examined, requiring an ongoing, dynamic dialogue between science, ethics, and society.

Conclusion: Stem cell research is pivotal in biomedical science, embodying unprecedented promise and intricate ethical dilemmas. The field represents an amalgamation of rationalist and empiricist approaches, driving advancements in scientific methodology and medical practice. By offering transformative prospects for treating various debilitating conditions, from neurodegenerative diseases to spinal cord injuries, it sets new paradigms in healthcare standards of care. Yet, these scientific and medical achievements are inextricably bound with ethical considerations that intersect with the principles of patient autonomy (informed consent), practitioner beneficence (do good), practitioner nonmaleficence (do no harm), and public justice (be fair). These ethical questions range from informed consent to the moral dimensions of using embryonic cells and from equitable access to treatments to the societal consensus on what constitutes harm or benefit. As the field continues to evolve, it compels an ever-ongoing conversation between scientists, ethicists, and policymakers, pushing us to continually reassess what we can achieve and aim for. Stem cell research, therefore, serves as both a beacon of medical potential and a crucible for ethical discourse, demanding a nuanced, multi-disciplinary dialogue as we navigate its complex terrain.

Stem Cell Research's Legacy: Opens therapeutic avenues for degenerative diseases and injuries by harnessing the potential of cellular regeneration.

**

REVIEW QUESTIONS

True/False Questions:

1. Stem cell research relies on the regenerative ability of undifferentiated cells to treat various diseases and injuries, such as Parkinson's disease and spinal cord afflictions.
   True or False?

2. Ethical considerations in stem cell research are limited to ensuring the efficacy of medical treatments and do not involve broader questions about life and morality.
   True or False?

Multiple-Choice Questions:

3. Which of the following is NOT a significant ethical issue in stem cell research?
   a) Informed consent
   b) Beneficence
   c) Data storage methods
   d) Nonmaleficence

4. In what way does stem cell research impact the medical healthcare standards of care?
   a) By standardizing treatment protocols for all conditions
   b) By introducing new treatments that continually refine and set new standards
   c) By reducing the need for peer-reviewed research
   d) By minimizing the role of personalized medicine

Clinical Vignette:

5. A patient is considering participating in a clinical trial involving embryonic stem cells for treating a degenerative disease. The research team must ensure the patient understands the potential benefits and risks, including ethical considerations. Which ethical principle is primarily being addressed in this scenario?
   a) Reducing healthcare costs
   b) Ensuring patient autonomy through informed consent
   c) Promoting practitioner beneficence by prioritizing patient care
   d) Avoiding harm by considering nonmaleficence

Basic Science Vignette:

6. A patient with a spinal cord injury is treated with stem cell therapy. Over time, the patient shows significant improvement in motor functions due to the regeneration of damaged neurons. What fundamental property of stem cells is utilized in this treatment?
   a) Their ability to cause inflammation
   b) Their capacity for unlimited division
   c) Their ability to differentiate into specialized cells
   d) Their potential to prevent infections

Philosophy Vignette:

7. A research team uses embryonic stem cells to develop treatments for Parkinson's disease. This approach raises ethical concerns about the moral status of embryos. Which principle of bioethics is most directly challenged by using embryonic stem cells in research?
   a) Nonmaleficence (do no harm)
   b) Beneficence (do good)
   c) Autonomy (informed consent)
   d) Justice (fair distribution of benefits and burdens)

Correct Answers:

1. True
2. False
3. c) Data storage methods
4. b) By introducing new treatments that continually refine and set new standards
5. b) Ensuring patient autonomy through informed consent
6. c) Their ability to differentiate into specialized cells
7. a) Nonmaleficence (do no harm)

BEYOND THE CHAPTER
Stem Cell Research

- *The Science of Stem Cells* by Jonathan M. W. Slack
- *Stem Cells: A Very Short Introduction* by Jonathan Slack
- *Stem Cells: An Insider's Guide* by Paul Knoepfler

***

Chapter 43: Systems Thinking — The Integrative Framework: Complex Problem-Solving

Systems Thinking offers an integrative framework for complex problem-solving, enhancing scientific and ethical practices.

Abstract: Systems Thinking, a multidisciplinary approach grounded in rationalist and empiricist traditions, facilitates a nuanced understanding of complex relationships and patterns within interconnected systems, transcending conventional reductionist paradigms. It employs deductive reasoning and logical analysis to decipher universal principles while emphasizing data-driven insights and real-world validations. This approach has significantly influenced the scientific method, encouraging a perspective encompassing both components and their interactions, fostering a deeper understanding of complex phenomena. In the medical field, Systems Thinking promotes an integrative approach, encouraging practitioners to consider multifaceted interactions influencing health outcomes and develop holistic treatment strategies. It also influences bioethical considerations, offering a robust framework for informed consent, optimizing treatment efficacy and safety, and fostering equitable healthcare resource distribution. Systems Thinking promises to leave an indelible mark as a transformative tool, reshaping scientific inquiry, ethical discourse, and problem-solving across various sectors.

**

Introduction: Systems Thinking is an intellectual approach that seeks to understand the complex relationships, structures, and patterns within interconnected systems rather than dissecting them into isolated components. Rooted in various disciplines such as engineering, biology, and social sciences, this holistic methodology posits that the individual elements of a system derive their function and significance from their role in the broader network. By examining how these elements interact and influence one another, Systems Thinking aims to offer insights into the system's emergent properties as a whole, enabling better problem-solving, decision-making, and innovation. The approach is versatile, finding applications in areas ranging from business management to environmental conservation and public policy. In scientific inquiry and medicine, Systems Thinking challenges traditional reductionist paradigms by insisting that understanding a system's constituent parts requires comprehending their roles within the larger framework. As we delve deeper, we will explore its rationalist foundations, empiricist orientation, role in shaping the scientific method, contributions to medicine, and influence on the four ethical principles of bioethics: patient autonomy (informed consent), practitioner beneficence (do good), practitioner nonmaleficence (do no harm), and public justice (be fair).

Rationalism: Systems Thinking, in its quest to understand the interconnectedness and complexity of systems, leans on rationalist underpinnings. The idea that inherent principles or axioms are at work in systems — from biological ecosystems to intricate human organizations — mirrors the rationalist notion that foundational truths can be discerned through reason and logical analysis. Systems Thinking often deploys deductive reasoning, working from general principles to predict specific outcomes within a system. For instance, one might start with the axiom that "*everything is interconnected*" and then deduce how changes in one part of a system might affect the whole. The rationalist foundation is particularly visible in how Systems Thinking employs cognitive faculties such as pattern recognition and problem-solving. The ability to identify recurring structures and dynamics within various systems suggests an innate, perhaps even evolutionarily hardwired, aptitude for Systems Thinking. By relying on these "*a priori*" faculties, Systems Thinking aims to derive universal principles that can be applied across diverse domains. The systematic and structural approach encourages identifying underlying rules and applying logical rigor, standing as a testament to its rationalist roots.

Empiricism: While Systems Thinking has a rationalist framework, it also integrates an empiricist orientation, recognizing that our understanding of complex systems is deeply informed by sensory experience and empirical observation. Systems Thinking often relies on data collection, case studies, and experimental methods to validate or challenge the theoretical models in real-world applications. Here, inductive reasoning takes center stage; specific observations about how individual elements within a system behave or interact can lead to broader generalizations or principles about the system. For example, the observation that removing a top predator in an ecosystem leads to an overabundance of certain species can be generalized to understand trophic cascades in various ecosystems. Empirical methods are particularly vital when Systems Thinking is applied to fields like medicine, environmental science, or social policy, where controlled experiments and longitudinal studies provide essential insights into system behaviors over time. The empiricist inclination in Systems Thinking serves as a grounding mechanism, ensuring that the theoretical frameworks and axioms are intellectually robust and empirically validated. This empirical focus helps to evolve Systems Thinking from an abstract conceptual tool into a practical methodology for addressing complex and dynamic real-world challenges.

The Scientific Method: Systems Thinking has substantially impacted the evolution and practice of the scientific method by introducing a more holistic perspective to inquiry and research. Traditionally, the scientific method has been characterized by reductionism, where phenomena are broken down into constituent parts for easier analysis. While this approach has been immensely successful in advancing human knowledge, it often overlooks the intricate relationships and emergent properties that arise when the components interact as a whole system. Systems Thinking brings to the fore the necessity of understanding these relationships, interactions, and feedback loops that give rise to complex behaviors. It encourages scientists to formulate hypotheses about individual elements and the connections between them, thereby enriching the scope and depth of research questions. By incorporating Systems Thinking, controlled experiments can be designed to capture systemic interactions, and data analysis can be executed in a manner that considers the complexity of these interactions. This opens avenues for more comprehensive conclusions that reflect the multi-dimensional nature of the phenomena under study. While maintaining the scientific rigor, replicability, and peer review essential to the scientific method, Systems Thinking elevates the process by encouraging a focus on isolated components and the system as an integrated whole.

Medicine: Systems Thinking has significantly enriched the field of medicine by advocating for a more integrative approach to healthcare, challenging the reductionist tendency to treat symptoms in isolation from one another. This perspective is particularly salient in understanding and treating complex medical conditions like diabetes, heart disease, and even mental health disorders, where multiple factors — including genetics, lifestyle, and environmental influences — interact in complex ways to produce a given health outcome. By applying Systems Thinking, medical professionals can better identify the interconnected factors contributing to a patient's condition and develop more effective, personalized treatment plans. This is also instrumental in shaping healthcare standards of care, pushing them to evolve beyond protocols for treating isolated symptoms to guidelines that take a systemic view of patient health. For example, in a hospital setting, Systems Thinking can aid in understanding how different departments are interlinked, from diagnosis to treatment to billing, ensuring a more coordinated and efficient patient experience. Standards of care thus influenced by Systems Thinking become not just evidence-based but also systemically coherent, aligning with a holistic understanding of patient needs, clinical expertise, and the latest research. This shift is essential for the evolution of medical best practices and has legal implications in defining adequate, competent care.

Ethics: Systems Thinking profoundly influences the bioethical principles that guide medical practice and healthcare policy. Starting with autonomy, Systems Thinking facilitates a more nuanced understanding of informed consent. By comprehending how various medical interventions interact within a patient's health system, physicians can provide more comprehensive advice, enabling patients to make well-informed decisions. Regarding beneficence and nonmaleficence, Systems Thinking encourages medical professionals to consider the systemic consequences of their actions. This means weighing the immediate benefits and risks of treatment for a specific condition and understanding how this treatment might interact with other variables in a patient's health profile, thus optimizing the "*do good*" and "*do no harm*" directives. Lastly, justice, or fairness in medical contexts, is significantly enriched by Systems Thinking. In public health, for instance, it offers tools to analyze how healthcare resources are allocated and can help identify systemic inequities. This provides a more holistic view of justice that considers not just individual interactions but the larger system, making it possible to address issues like healthcare disparities more effectively. Systems Thinking enhances the ethical depth and practical efficacy of medical decision-making and policy by influencing these four bioethical principles.

Conclusion: Systems Thinking has emerged as a transformative framework, influencing multiple domains, from scientific inquiry to medicine and ethics. It reconciles the rationalist quest for universal principles with an empiricist grounding, offering a more rounded and comprehensive approach to understanding the world. Its emphasis on interconnectedness and relationships has shifted the paradigm in the scientific method, moving away from strict reductionism to embrace the complexity inherent in natural and manufactured or artificial systems. This shift has been particularly impactful in medicine, where it has redefined standards of care and ushered in a more integrated, holistic approach to diagnosis and treatment. Systems Thinking extends its reach into the ethical realm, recontextualizing key principles like patient autonomy (informed consent), practitioner beneficence (do good), practitioner nonmaleficence (do no harm), and public justice (be fair) within a systemic framework. This has given new dimensions to ethical considerations, making them more responsive to the complex realities of healthcare and social systems. Through its multi-faceted impact, Systems Thinking has proven to be more than just an academic concept; it has become a vital tool for grappling with the increasingly complex challenges of the modern world. It enriches our cognitive toolset, enabling us to navigate complexity with greater efficacy, ethical integrity, and a deeper understanding of the intricate tapestry of relationships that define our world.

Systems Thinking's Legacy: Bridging the gap between reductionism and holism, Systems Thinking is poised to leave a lasting legacy as an integrative framework that enhances scientific inquiry, enriches ethical considerations, and revolutionizes complex problem-solving across diverse fields.

**

REVIEW QUESTIONS

True/False Questions:

1. Systems Thinking emphasizes understanding the interactions and relationships within interconnected systems rather than focusing solely on individual components.
   True or False?

2. Systems Thinking exclusively relies on rationalist principles and does not incorporate empirical data or observations.
   True or False?

Multiple-Choice Questions:

3. Which of the following fields is NOT significantly influenced by Systems Thinking?
   a) Business management
   b) Environmental science
   c) Classical mechanics
   d) Public policy

4. How does Systems Thinking impact the standards of care in medicine?
   a) By focusing only on treating isolated symptoms
   b) By promoting a holistic view that considers multiple factors influencing health
   c) By reducing the need for patient-specific treatments
   d) By standardizing all medical treatments across different conditions

Clinical Vignette:

5. A hospital aims to improve patient care by analyzing the interconnectedness of various departments, from diagnosis to treatment to billing, to ensure a coordinated and efficient patient experience. Which principle of Systems Thinking is being applied here?
   a) Focusing solely on individual components
   b) Emphasizing the reductionist approach
   c) Understanding the system as an integrated whole
   d) Ignoring the interactions between departments

Basic Science Vignette:

6. A hospital uses Systems Thinking to address high patient readmission rates. By analyzing the interconnected factors such as patient follow-up care, medication adherence, and social support, they develop a comprehensive intervention plan. What is the primary advantage of using Systems Thinking in this context?
   a) Isolating the main cause of readmissions
   b) Reducing the cost of patient care
   c) Understanding and addressing multiple factors contributing to readmissions
   d) Simplifying the treatment process

Philosophy Vignette:

7. In developing public health policies, a team uses Systems Thinking to ensure fair distribution of healthcare resources. They consider socioeconomic factors, geographic accessibility, and healthcare infrastructure. Which principle of bioethics is most directly supported by this approach?
   a) Autonomy (informed consent)
   b) Beneficence (do good)
   c) Nonmaleficence (do no harm)
   d) Justice (fair distribution of benefits and burdens)

Correct Answers:

1. True
2. False
3. c) Classical mechanics
4. b) By promoting a holistic view that considers multiple factors influencing health
5. c) Understanding the system as an integrated whole
6. c) Understanding and addressing multiple factors contributing to readmissions
7. d) Justice (fair distribution of benefits and burdens)

BEYOND THE CHAPTER
Systems Thinking

- *Thinking in Systems: A Primer* by Donella H. Meadows
- *The Systems Thinker: Essential Thinking Skills for Solving Problems, Managing Chaos, and Creating Lasting Solutions in a Complex World* by Albert Rutherford
- *An Introduction to Systems Thinking* by Barry Richmond

***

Chapter 44: Imre Lakatos (1922-1976) — The Research Methodologist: Research Programs

Lakatos's methodology of research programs bridged rationalist and empiricist traditions, offering a dynamic framework for scientific progress.

Abstract: In the 20th century, Imre Lakatos distinguished himself in the philosophy of science by ingeniously blending rationalist and empiricist traditions into the innovative framework of "*research programs*." These programs consist of a resilient "*hard core*" of fundamental assumptions, supported by a "*protective belt*" of adjustable auxiliary hypotheses, offering a dynamic and nuanced understanding of scientific progression. Lakatos' conceptual infrastructure has reshaped modern perspectives on the scientific method and influenced various sectors, including medicine and bioethics.The structure of Lakatos' theory, which combines stable core principles with adaptable empirical observations, enriches ethical discourse in medicine. It impacts bioethical principles such as patient autonomy (informed consent), practitioner beneficence (do good), nonmaleficence (do no harm), and public justice (be fair). His approach also informs public policy and legal frameworks, solidifying Lakatos' legacy as a pivotal figure whose synthesis of rationalism and empiricism provides a robust blueprint for navigating the intricate terrains of science, ethics, and beyond.

**

Introduction: Imre Lakatos, a Hungarian-born philosopher of science and mathematics, emerged as a prominent figure in the 20th-century philosophy of science landscape. He is best known for his theory of "*research programs*," which sought to reconcile the rationalist and empiricist traditions in the philosophy of science. Lakatos' work responded to the limitations he saw in Karl Popper's falsificationism and Thomas Kuhn's paradigm shifts. By introducing the concept of a "*hard core*" of unchangeable assumptions and a "*protective belt*" of auxiliary hypotheses, Lakatos offered a more nuanced understanding of scientific progress that accommodated both rationalist foundations and empirical observations. His ideas have had a far-reaching impact, shaping modern conceptions of the scientific method and influencing various other disciplines, including medicine and bioethics. In the subsequent discussion, we will delve into Lakatos' rationalist foundations, empiricist orientation, role in shaping the scientific method, contributions to medicine, and his influence on the four ethical principles of bioethics: patient autonomy (informed consent), practitioner beneficence (do good), practitioner nonmaleficence (do no harm), and public justice (be fair).

Rationalism: Imre Lakatos' philosophy is deeply rooted in rationalist foundations, particularly in his emphasis on the role of "*hard core*" axiomatic principles that guide scientific research programs. In Lakatos' view, this hard core consists of a set of "*a priori*" assumptions that are not easily subjected to empirical falsification, much like the innate cognitive faculties posited by rationalism. These foundational principles are the bedrock upon which a "*protective belt*" of auxiliary hypotheses and empirical observations are built. Lakatos applies deductive reasoning to elaborate on these core principles, allowing for the development and refinement of scientific theories. His approach acknowledges the complexity added by epigenetic factors and learned behaviors, which can modify or extend the hard core over time, akin to how rational faculties evolve and adapt. Lakatos' work also has ethical implications, especially for scientists and educators who must navigate the tension between unchangeable core principles and the mutable protective belt of hypotheses. This dynamic mirrors our ethical responsibility to understand our rational faculties' limitations and adaptive advantages. Lakatos' ideas can inform public policy in a broader societal context by emphasizing regulations that respect foundational principles and allow for empirical adjustments. His philosophy also aligns with the concept of "*emergence*," suggesting that a scientific community's collective capabilities, like an organism's collaborative cognitive capabilities, are more than the sum of individual contributions. The hard core and protective belt interact intricately, like basic cognitive faculties combine to produce complex behaviors and ethical reasoning.

Empiricism: Imre Lakatos' philosophy of science incorporates a strong empiricist orientation, particularly evident in his concept of the "*protective belt*" of auxiliary hypotheses surrounding a research program's "*hard core*." Unlike the hard core, which is more resistant to change, the protective belt is subject to empirical testing and modification. Lakatos employs inductive reasoning, allowing specific observations to inform and potentially alter the auxiliary hypotheses. This approach acknowledges the role of cognitive biases, socialization, and culture, which can influence how empirical data are interpreted and integrated into the research program. Lakatos' empiricist stance has significant implications for ethics and public policy. It underscores the importance of empirical evidence in shaping ethical guidelines and decision-making, both individually and professionally. For instance, empirical data can inform ethical principles like beneficence and nonmaleficence in medicine. Regarding public policy, Lakatos' philosophy supports the need for empirically grounded regulations that are adaptive to societal changes and technological advances, much like the protective belt adapts to new empirical findings. His work also resonates with the concept of "*emergence*" in empiricism, suggesting that the collective properties and behaviors of a research program — or even a scientific community — are more complex than the sum of individual empirical observations. This complexity can give rise to emergent phenomena, such as the development of new scientific paradigms or ethical frameworks, which cannot be easily predicted by studying individual components alone.

The Scientific Method: Imre Lakatos made significant contributions to the philosophy of science that have had a lasting impact on our understanding of the scientific method. His theory of "*research programs*" offers a nuanced framework that goes beyond the traditional steps of observation, hypothesis formulation, and testing. Lakatos introduces the idea of a "*hard core*" of foundational principles that are somewhat resistant to falsification, around which a "*protective belt*" of auxiliary hypotheses is built. These hypotheses are subjected to empirical testing, thereby integrating rationalist and empiricist elements into the scientific method. His approach allows for a more dynamic and iterative process, where the protective belt can be modified based on new empirical data without necessarily discarding the entire research program. Lakatos' philosophy also addresses the importance of replicability and peer review. The protective belt serves to refine hypotheses and theories, making them more robust and replicable over time. His work emphasizes the collective endeavor of science, where research programs gain acceptance not just through individual experiments but through a sustained, community-wide effort to test and refine auxiliary hypotheses. This aligns with the rigorous peer review process, as the validity and methodology of a research program are continually scrutinized and adapted in light of new evidence. In this way, Lakatos' ideas have enriched and complicated our understanding of the scientific method, making it more reflective of scientific inquiry's actual practices and complexities.

Medicine: While Imre Lakatos was not directly involved in medicine, his philosophy of science has implications for how medical research and medical healthcare standards of care are approached.

Lakatos' concept of "*research programs*," with its "*hard core*" and "*protective belt,*" can be applied to the development and refinement of medical practices and guidelines. The "*hard core*" could be likened to the foundational principles of medical ethics and basic physiological knowledge, which are less subject to change. The "*protective belt*," on the other hand, consists of evolving methods, treatments, and diagnostic criteria that are continually updated based on new empirical evidence. Lakatos' philosophy underscores the importance of evidence-based practices subject to rigorous empirical testing and peer review in medical healthcare standards of care. Just as scientific research programs are refined over time, medical healthcare standards of care are also dynamic, adapting to new research findings, clinical expertise, and patient needs. This aligns with the legal framework that evaluates the adequacy of delivered care based on current best practices within the medical community. By emphasizing foundational principles and the need for empirical adaptability, Lakatos' philosophy offers a robust framework for understanding how medical healthcare standards of care can be both stable and responsive to new information, ensuring that healthcare delivery remains at the forefront of scientific and ethical advancements.

Ethics: Imre Lakatos' philosophy of science, though not directly aimed at bioethics, offers a conceptual framework that can be applied to the ethical principles governing medical practice and research. His idea of a "*hard core*" of foundational beliefs and a "*protective belt*" of adaptable hypotheses can be seen as analogous to the four pillars of bioethics: patient autonomy (informed consent), practitioner beneficence (do good), practitioner nonmaleficence (do no harm), and public justice (be fair). The "*hard core*" could be equated with these ethical principles, which are the foundational guidelines that are less susceptible to change. Meanwhile, the "*protective belt*" could represent the specific policies, procedures, and practices operationalizing these principles in various medical contexts. For instance, the principle of autonomy, which emphasizes informed consent, can be viewed as a core ethical tenet adapted to different medical situations through empirical evidence and societal norms. Beneficence and nonmaleficence, which advocate for doing good and avoiding harm, are similarly foundational but must be continually refined based on new medical knowledge and technologies. The principle of justice, which calls for fairness in medical treatment, can also be adapted to address evolving societal views on what constitutes equitable care. In this way, Lakatos' philosophy provides a dynamic framework for bioethics, allowing for both stability in core principles and flexibility in their application, thereby enriching the ethical discourse in medicine.

Conclusion: Imre Lakatos' philosophy of science has had a profound and far-reaching impact on various disciplines, including the philosophy of science, medicine, and bioethics. His theory of "*research programs*," with its innovative "*hard core*" and "*protective belt*" framework, has offered a nuanced understanding of scientific progress that harmonizes rationalist and empiricist traditions. This framework has enriched our knowledge of the scientific method and provided valuable insights into the dynamic nature of medical healthcare standards of care and ethical principles. Lakatos has contributed to a more robust and flexible approach to scientific inquiry and ethical decision-making by emphasizing the importance of foundational axioms and empirical adaptability. His work underscores the need for a balanced interplay between unchanging principles and evolving empirical evidence. This concept has implications for public policy, legal frameworks, and ethical guidelines in medicine and beyond. In essence, Lakatos' philosophy is a conceptual bridge that allows for stability in core principles and flexibility in their application, enriching scientific, medical, and ethical discourse. His ideas continue to resonate, offering a comprehensive and adaptable framework that addresses the complexities and challenges of modern science and ethics.

Imre Lakatos' Legacy: His theory of "*research programs*," which reconciles rationalist and empiricist approaches, has profoundly influenced modern conceptions of the scientific method, as well as ethical and medical frameworks.

**

REVIEW QUESTIONS

True/False Questions:

1. Imre Lakatos is best known for his theory of "*research programs*" in the philosophy of science.
   True or False?

2. Lakatos' concept of the "*protective belt*" includes foundational principles that are resistant to change.
   True or False?

Multiple-Choice Questions:

3. According to Lakatos, what does the "*hard core*" of a research program consist of?
   a) Empirical data that can be easily falsified
   b) Fundamental assumptions resistant to empirical falsification
   c) Hypotheses that are frequently tested and changed
   d) The latest technological advancements

4. How does Lakatos' philosophy of science contribute to the medical healthcare standards of care?
   a) By providing specific medical treatments
   b) By emphasizing evidence-based practices subject to empirical testing
   c) By eliminating the need for clinical trials
   d) By focusing solely on theoretical principles without practical application

Clinical Vignette:

5. A new experimental drug is being developed, and the medical team needs to decide whether to proceed with clinical trials. Applying Lakatos' philosophy, how should they approach this decision?
   a) Proceed with the trials without further testing
   b) Rely solely on the theoretical principles behind the drug
   c) Incorporate a "*protective belt*" of auxiliary hypotheses and subject them to empirical testing
   d) Avoid any empirical testing and base the decision on prior successful drugs

Basic Science Vignette:

6. A scientist is developing a new research program to study cancer treatment efficacy. To align with Imre Lakatos' framework, what approach should the scientist prioritize?
   a) Develop a set of unchangeable foundational principles and avoid testing them
   b) Establish a "*hard core*" of foundational principles and use a "*protective belt*" of auxiliary hypotheses that are tested and modified based on empirical evidence
   c) Rely solely on theoretical models without empirical testing
   d) Focus exclusively on empirical observations without a guiding theoretical framework

Philosophy Vignette:

7. A bioethics committee is formulating guidelines for integrating new medical technologies into clinical practice. Inspired by Imre Lakatos' principles, what should be the main focus to ensure ethical and effective integration?
   a) Implement the technologies without modifying existing ethical guidelines
   b) Ensure a core set of ethical principles while adapting specific practices based on empirical evidence and outcomes
   c) Use theoretical ethical principles without regard for empirical outcomes
   d) Focus exclusively on the empirical success of the technologies without considering ethical implications

Correct Answers:

1. True
2. False
3. b) Fundamental assumptions resistant to empirical falsification
4. b) By emphasizing evidence-based practices subject to empirical testing
5. c) Incorporate a "*protective belt*" of auxiliary hypotheses and subject them to empirical testing
6. b) Establish a "*hard core*" of foundational principles and use a "*protective belt*" of auxiliary hypotheses that are tested and modified based on empirical evidence
7. b) Ensure a core set of ethical principles while adapting specific practices based on empirical evidence and outcomes

BEYOND THE CHAPTER
Imre Lakatos (1922-1996)

- *For and Against Method: Including Lakatos's Lectures on Scientific Method and the Lakatos-Feyerabend Correspondence* by Imre Lakatos and Paul Feyerabend
- *The Methodology of Scientific Research Programmes* by Imre Lakatos
- *Mathematics, Science and Epistemology* by Imre Lakatos

***

Philosophy of Science

Medical Research

Chapter 45: Paul Feyerabend (1924-1994) — The Methodological Anarchist: Epistemological Pluralism

Feyerabend's advocacy for methodological pluralism challenged rigid scientific methodologies, promoting a more inclusive and dynamic approach to knowledge.

Abstract: In this discourse, we explore the intricate philosophical landscape delineated by Paul Feyerabend, a notable critic of the orthodox scientific methodology of the 20th century. We examine Feyerabend's nuanced stance on rationalism and empiricism, highlighting his advocacy for methodological pluralism and a broadened scope of rationality while valuing empirical investigation without elevating it to a doctrinal status. Feyerabend's contributions to the philosophy of science sparked a paradigm shift by challenging the monolithic view of scientific methodology. His call for epistemological anarchism emphasized the diverse, and sometimes chaotic, paths leading to scientific innovation, urging a move away from strict adherence to a uniform method. This treatise also examines the implications of Feyerabend's philosophy in medicine, advocating for a balanced approach that combines evidence-based practices with clinical intuition and patient-specific considerations. Additionally, we consider the influence of his philosophy on bioethics, endorsing a more personalized and equitable approach. Conclusively, we honor Feyerabend's enduring legacy of intellectual openness, which reshaped our understanding of scientific inquiry and its multifaceted nature.

**

Introduction: Paul Feyerabend stands as one of the most controversial philosophers of science of the 20th century, challenging conventional wisdom on the nature and method of scientific inquiry. Born in 1924 in Vienna, his intellectual journey led him to critique the prevailing scientific methodologies, urging a more pluralistic and decentralized approach to scientific practice. Feyerabend posited that science doesn't progress through a uniform method but rather through a diverse mix of methodologies and ideas; in this sense, he challenged the established narratives of rationalism and empiricism. He famously argued against any overarching, methodological dogma in science, captured in his provocative declaration that "*anything goes*." Such radical perspectives made Feyerabend a figure of both admiration and criticism. As we delve deeper into his thoughts, we shall assess his leanings toward rationalism or empiricism, his contributions and critique of the scientific method, implications for the field of medicine, and the influence he might have exerted on the foundational principles of bioethics.

Rationalism: Paul Feyerabend's relationship with rationalism is complex and multifaceted. While he vehemently critiqued the notion of a single, unified scientific method and argued for a pluralistic approach to science, he was not entirely dismissive of rationalist tenets. His critique primarily revolved around the hegemony of a particular rationalism style in the scientific discourse. Feyerabend contended that rigid adherence to rationalistic methods could stifle scientific progress and innovation. Instead, he called for an "*epistemological anarchism*," where multiple methods, including those traditionally considered "*irrational*," could coexist and be employed in the scientific endeavor. While this perspective may seem directly opposed to classical rationalism, Feyerabend wasn't entirely rejecting reason and logic; he advocated for a broader conception of what rationality could entail. He believed in the power of human reason but thought it was not limited to one strict methodological form. In this way, Feyerabend's views might be better characterized as challenging the orthodoxy of a stringent rationalistic approach in science rather than a wholesale rejection of rationalism itself.

Empiricism: Paul Feyerabend's views on empiricism, like many aspects of his philosophy, took a nuanced stance. While he was not an empiricist in the traditional sense, Feyerabend deeply respected observational data and the richness of experiences that empirical investigations bring. However, he was critical of the rigidity that sometimes accompanied empirical approaches, particularly when empirical methods were considered the sole arbiter of truth in scientific investigations. For Feyerabend, empiricism was one of many tools in the scientist's toolkit, valuable but not exclusive. He argued that the history of science was replete with instances where departures from strict empiricism led to significant breakthroughs. Moreover, he believed that scientific theories often shape observations, hinting at a complex interplay between theory and empirical data. This interdependence suggests that empirical observations are not always neutral but can be influenced by existing theoretical frameworks. While Feyerabend acknowledged the indispensable role of empirical investigation in science, he resisted its elevation to an unquestionable doctrine. He called for a more flexible, open approach that recognizes the intricacies of the relationship between theory and observation.

The Scientific Method: Paul Feyerabend's contributions to the philosophy of science brought forth a radical rethinking of the scientific method. His work significantly departed from the conventional reverence for a standardized, objective approach to scientific inquiry. Feyerabend famously argued that no universal method underlies all scientific progress, challenging the deeply held notion of a monolithic scientific methodology. His proposition, summarized by the statement "*anything goes*," was a call for epistemological anarchism, asserting that science advances not by strict adherence to a single method but often through a chaotic blend of approaches, intuition, and even breaches of established norms. Historical examples illustrate his point. For instance, the serendipitous discovery of penicillin by Alexander Fleming might have been missed had he been strictly adherent to his initial study design. Similarly, accepting the idea of tectonic plate movement required data accumulation from diverse fields before its groundbreaking acceptance. The rediscovery of Mendelian genetics through varied paths by different scientists demonstrated that innovations can spring from various, even coincidental, methodologies rather than a strict, singular approach. While Feyerabend's perspective drew criticism, his intention was not to discredit science but to emphasize its richness, diversity, and context-dependent nature. He highlighted such deviations and serendipitous events as integral to the history of scientific progress. In effect, Feyerabend urged the scientific community to reassess the rigidity of its methodologies and to embrace a more pluralistic view, recognizing that different contexts and problems might require diverse approaches and that rigid adherence to a single method could potentially limit the vast potential of scientific innovation.

Medicine: While Paul Feyerabend wasn't a medical doctor and didn't directly contribute to medical practices per se, his philosophy had implications for the field of medicine, particularly in how evidence-based practices are conceptualized and adopted. In line with his critiques of the rigidity in scientific methodology, Feyerabend would likely question the strict adherence to established medical standards without room for deviation, innovation, or context-specific tailoring. He might have seen the strictly evidence-based approach as potentially limiting, arguing for the importance of clinical intuition, individual patient needs, and the socio-cultural context in medical decision-making. In a Feyerabendian view, while evidence-based standards of care are crucial for ensuring quality and consistency, there should also be a recognition that medicine, like other scientific endeavors, is multifaceted and not always reducible to a single best approach. Hence, his philosophy underscores the importance of balancing rigorous evidence-based standards with the flexibility to address unique patient circumstances and needs. In contemporary medicine, these insights can be particularly valuable in personalized medicine, where individualized care and patient-specific interventions are paramount.

Ethics: Paul Feyerabend's philosophy, emphasizing pluralism, context-sensitivity, and skepticism toward overarching doctrines, has implications for bioethics, even though he did not explicitly engage with these principles in depth. Regarding autonomy and informed consent, his advocacy for diverse perspectives and skepticism toward any authoritative narrative aligns with the idea that patients should be informed about various treatment options and be actively involved in decision-making rather than merely following a doctor's directives. Beneficence and nonmaleficence, principles rooted in promoting well-being and avoiding harm, resonate with Feyerabend's warnings against the dangers of rigid dogmas that can inadvertently harm progress or limit potential beneficial interventions. His ethos also supports a broader understanding of "*doing good*" and "*avoiding harm*," recognizing that these concepts can be context-dependent and pluralistic. Finally, regarding justice, Feyerabend's distrust of uniform methods and his call for methodological diversity can be seen as a call for fairness and equity, ensuring that varied contexts and individuals are considered rather than a one-size-fits-all approach. While Feyerabend did not directly address bioethical principles, his philosophy champions a more individualistic, patient-centered, and context-aware approach that aligns with these core ethical values.

Conclusion: Paul Feyerabend's indelible mark on the philosophy of science emerged as a fierce critique against stringent methodological frameworks and an affirmation of the diverse, rich tapestry that underpins scientific progress. Arguing against dogmatic adherence to singular scientific methods, he championed a more pluralistic, context-sensitive approach, emphasizing the importance of accommodating various epistemological standpoints. While this made him a contentious figure, Feyerabend's voice was crucial in initiating debates on the nature of scientific inquiry, driving home the point that science, in its essence, is a human endeavor subject to varied influences, strategies, and methodologies. Beyond pure science, his ideas found resonance in medicine and bioethics, underlining the importance of adaptability, individual-centric approaches, and the constant reevaluation of established norms. While he may not have offered a neat package of doctrines, Feyerabend's legacy lies in his challenge to orthodoxy and his enduring call for an open, reflective, and inclusive approach to understanding the world.

Paul Feyerabend's Legacy: Critiqued strict scientific methodology with his "*anything goes*" philosophy, emphasizing scientific discovery's diverse and sometimes anarchistic paths.

**

REVIEW QUESTIONS

True/False Questions:

1. Paul Feyerabend advocated for a single, unified scientific method as the only way to achieve scientific progress.
   True or False?

2. Feyerabend believed that the history of science shows a diverse and sometimes chaotic mix of methodologies leading to scientific innovation.
   True or False?

Multiple-Choice Questions:

3. What is the core idea behind Feyerabend's philosophy of science?
   a) There is one best scientific method that must be strictly followed.
   b) Scientific progress happens only through empirical observation.
   c) Scientific advancements can come from a variety of methods and approaches.
   d) Rationalism is the only valid approach to scientific inquiry.

4. How does Feyerabend's philosophy impact medical practices, particularly in evidence-based medicine?
   a) It suggests that evidence-based practices should be the only consideration in medical decision-making.
   b) It emphasizes the importance of strict adherence to medical protocols without deviation.
   c) It highlights the need for balancing evidence-based standards with clinical intuition and patient-specific considerations.
   d) It advocates for the elimination of evidence-based medicine in favor of traditional practices.

Clinical Vignette:

5. A physician is treating a patient with a rare condition and considers using an unconventional treatment that lacks extensive empirical support but has shown promise in anecdotal reports. According to Feyerabend's philosophy, what should the physician consider in making this decision?
   a) Disregard the unconventional treatment entirely and stick to standard protocols.
   b) Consider the unconventional treatment if it aligns with the patient's preferences and shows potential benefit, even if it lacks extensive empirical support.
   c) Wait for extensive clinical trials before considering any new treatment options.
   d) Use the unconventional treatment without informing the patient of the lack of empirical support.

Basic Science Vignette:

6. A researcher is exploring the effects of a new drug on bacterial growth. Instead of adhering strictly to traditional methodologies, they decide to combine various experimental approaches, including observational studies, intuition-based hypotheses, and unconventional techniques. This approach reflects which philosophical stance?
   a) Strict empiricism
   b) Rationalism
   c) Methodological pluralism
   d) Positivism

Philosophy Vignette:

7. In a debate about scientific progress, one scientist argues that breakthroughs often occur not through a strict, uniform methodology but through a mix of different approaches, including those considered non-traditional. This perspective aligns with the views of which philosopher?
   a) Karl Popper
   b) Thomas Kuhn
   c) Paul Feyerabend
   d) Immanuel Kant

Correct Answers:

1. False
2. True
3. c) Scientific advancements can come from a variety of methods and approaches
4. c) It highlights the need for balancing evidence-based standards with clinical intuition and patient-specific considerations
5. b) Consider the unconventional treatment if it aligns with the patient's preferences and shows potential benefit, even if it lacks extensive empirical support
6. c) Methodological pluralism
7. c) Paul Feyerabend

BEYOND THE CHAPTER
Paul Feyerabend (1924-1994)

- *Against Method* by Paul Feyerabend
- *Science in a Free Society* by Paul Feyerabend
- *Farewell to Reason* by Paul Feyerabend

***

Philosophy of Science

Medical Research

WEEK NINE: PHILOSOPHY AND FOUNDATIONS

WEEK NINE

Class Hour 1

- Chapter 46: Daniel Dennett (1942-) — The Cognitive Philosopher: Interdisciplinary Dialogue
- Chapter 47: Consciousness — The Emergent Marvel: Beyond the Brain
- Chapter 48: Free Will and Determinism — The Dynamic Interplay: Freedom in Complexity

Class Hour 2

- Chapter 49: Medicine — The Dynamic Fusion: Science, Technology, Ethics
- Chapter 50: Hippocratic Oath — The Ethical Bedrock: Medical Integrity
- Chapter 51: Nuremberg Code — The Ethical Framework: Human Dignity

---

Introduction

In week nine of "*Philosophy of Science*," we embark on a journey through the profound interplay between philosophy and the foundational elements of scientific inquiry and medical ethics. This week, we will delve into the contributions of key thinkers and pivotal documents that have shaped our understanding of consciousness, free will, medical practice, and bioethics. Our exploration begins with the cognitive philosopher Daniel Dennett and traverses the emergent marvel of consciousness, the dynamic interplay of free will and determinism, the evolving nature of medicine, the ethical bedrock of the Hippocratic Oath, and the human dignity safeguarded by the Nuremberg Code.

Daniel Dennett's interdisciplinary dialogue sets the stage, blending rationalism and empiricism to offer a rich analysis of human consciousness. His approach lays the groundwork for examining consciousness as an emergent property, extending beyond the brain and incorporating a network of neural interactions. This perspective revolutionizes our understanding of the mind and medicine, emphasizing the intricate symphony of neural networks and their influence on our cognitive faculties and health.

As we delve deeper, the dynamic interplay between free will and determinism unfolds through the lens of complexity theory and systems thinking. Free will emerges from deterministic roots, highlighting the intricate balance between genetic and environmental factors. This nuanced view reshapes our grasp of autonomy and ethics, influencing discussions and practices across rationalism, empiricism, and the scientific method.

The narrative seamlessly transitions into the dynamic fusion of science, technology, and ethics in medicine. Here, we witness the continual evolution of healthcare, driven by structured, analytical approaches and evidence-based practices. Ethical principles such as patient autonomy, practitioner beneficence, nonmaleficence, and public justice anchor this evolution, ensuring that medical advancements are both scientifically sound and morally grounded.

The timeless Hippocratic Oath emerges as the ethical bedrock of modern medicine, emphasizing integrity and patient-centered care. This oath, rooted in ancient Greece, continues to influence contemporary medical practice, integrating rationalist principles with empirical healthcare experiences. It fosters a patient-centered approach, guiding clinical expertise with moral integrity and shaping modern bioethical principles.

Finally, the Nuremberg Code stands as a testament to humanity's commitment to ethical integrity in scientific research and medical practice. Instituted in the aftermath of World War II, the Code safeguards human dignity through its emphasis on informed consent and ethical integrity. Its principles continue to shape contemporary bioethical standards, preventing past atrocities and fostering a human-centric approach in the confluence of science, ethics, and human rights.

Together, these chapters weave a compelling narrative that underscores the profound connections between philosophy, science, and ethics. They illustrate how interdisciplinary dialogue, emergent properties, complexity theory, and foundational ethical principles collectively advance our understanding of human existence and enhance the integrity of scientific and medical practices.

***

Philosophy of Science

Medical Research

Chapter 46: Daniel Dennett (1942-) — The Cognitive Philosopher: Interdisciplinary Dialogue

Dennett's exploration of human consciousness, integrating rationalism and empiricism, has fostered interdisciplinary dialogue and enriched our understanding of cognitive science.

Abstract: Daniel Dennett, a preeminent American philosopher and cognitive scientist, has considerably influenced the intellectual spheres of the late 20th and early 21st centuries. Fusing rationalism's analytical rigor with empiricism's grounded methodologies, Dennett offers a multifaceted exploration of human consciousness and cognitive evolution, emphasizing a physicalist stance that roots the mind and consciousness in material processes. His fervent advocacy for integrating the scientific method into philosophical inquiries has fostered collaborative and evidence-based dialogues, extending his influence to medicine and bioethics. Dennett's interdisciplinary approach has not only catalyzed critical discourse on science and philosophy intersectionality but also potentially informed medical treatment approaches and shaped bioethical principles. This comprehensive analysis illuminates Dennett's remarkable legacy, spotlighting his pivotal role in pioneering a deeper, nuanced understanding of the intricacies of human existence through the convergence of philosophy and science.

**

Introduction: Daniel Dennett, a distinguished American philosopher, writer, and cognitive scientist, has carved a significant niche in the intellectual landscape of the 20th and 21st centuries. Born on March 28, 1942, Dennett dedicated his career to exploring complex topics such as the philosophy of mind, evolutionary psychology, and the philosophy of science. A staunch materialist, he contends that consciousness and the mind are grounded in physical processes, a perspective that has propelled extensive discourse in both philosophical and scientific circles. His prolific career, marked by seminal works such as "*Consciousness Explained*" and "*Darwin's Dangerous Idea*," has fostered a deeper understanding of human cognition and catalyzed critical discussions on the intersectionality of science and philosophy. In the subsequent sections, we will delve deeper into Dennett's rationalist foundations, empiricist orientation, and profound influence on scientific methods, medicine, and bioethics.

Rationalism: In philosophy, Daniel Dennett stands as a beacon of rationalist thought, a perspective deeply embedded in his theories and works. Rationalism, which emphasizes the role of innate cognitive faculties as primary sources of knowledge, finds a resonant echo in Dennett's explorations of the mind and consciousness. Through his lens, elements like basic reasoning skills and pattern recognition are not mere constructs but evolutionarily hardwired mechanisms that facilitate understanding and knowledge acquisition. Dennett posits that consciousness might be an emergent property of these innate faculties, a complex tapestry woven from the threads of "*a priori*" principles that govern our cognition and perception. His approach often employs deductive reasoning, drawing from axioms and definitional truths to build robust arguments. For instance, in his seminal work "*Darwin's Dangerous Idea,*" he applies a rationalist lens to dissect the implications of Darwinian evolution, portraying it as an algorithmic process, a notion grounded in logical and mathematical foundations. Through his rationalist foundations, Dennett seeks to unravel the complexities of the human mind, offering a perspective that marries the intricacies of philosophy with the empirical rigor of science.

Empiricism: Daniel Dennett, a prominent figure in philosophy and cognitive science, ardently embraces an empiricist orientation in his scholarly pursuits. This approach, grounded in the belief that knowledge is primarily acquired through sensory experiences and interactions with the environment, is a cornerstone of his theories on consciousness and the human mind. Dennett employs inductive reasoning as a fundamental methodology, extrapolating nuanced observations about specific phenomena to formulate broader, general conclusions. In his dissection of consciousness, he leans heavily on empirical data, analyzing how sensory inputs craft our perception of reality and influence our cognitive processes. This empiricist stance allows Dennett to explore the intricacies of the mind with scientific rigor, fostering a rich dialogue between philosophy and the empirical sciences. Through this lens, he seeks to unravel the complex tapestry of human cognition, emphasizing the pivotal role of sensory experiences in shaping our understanding of the world and ourselves. This approach not only underscores the adaptive value of empirical faculties but also paves the way for a deeper, more nuanced exploration of the multifaceted phenomenon of consciousness.

The Scientific Method: Daniel Dennett has played a pivotal role in shaping and endorsing the scientific method within philosophy and cognitive science. His approach to understanding complex phenomena, particularly consciousness, embodies the systematic and empirical inquiry that characterizes the scientific method. Dennett begins by meticulously observing a phenomenon, often drawing from interdisciplinary sources to formulate cogent questions about its underlying causes or explanations. This leads to the development of testable hypotheses, which are then scrutinized through rigorous analysis and experimentation. His commitment to empirical validation is evident in his insistence on verifiable and falsifiable claims, fostering a dialogue grounded in evidence and rationality. Furthermore, Dennett advocates for the replicability of experiments, a cornerstone of credible scientific research, ensuring that findings can withstand the scrutiny of repeated testing under consistent conditions. His work often undergoes rigorous peer review, a process that attests to the validity and significance of his contributions. Through integrating the scientific method into philosophical inquiry, Dennett has facilitated a more nuanced and grounded exploration of complex topics like consciousness, thereby bridging the gap between philosophy and science and fostering a collaborative and evidence-based approach to understanding the intricacies of the human mind.

Medicine: While Daniel Dennett is primarily known for his contributions to philosophy and cognitive science, his work also intersects with medicine, particularly neuroscience and psychology. Dennett's exploration of consciousness and the human mind has implications for understanding mental health and neurological conditions. His emphasis on empirical research and evidence-based practices aligns with the standards of care in the medical community, which rely on peer-reviewed, scientifically grounded methodologies to establish best practices. By advocating for a rigorous, scientific approach to understanding the complexities of the human mind, Dennett indirectly contributes to the evolving standards of medical healthcare. His theories, which often delve into the intricacies of cognitive processes and consciousness, can offer insights into the neurological underpinnings of various mental health conditions, potentially informing treatment approaches and patient care. Furthermore, his emphasis on integrating research, clinical expertise, and patient needs into the formulation of medical healthcare standards of care echoes the principles that govern modern medicine, fostering a collaborative and evidence-based approach to healthcare that seeks to enhance the quality and efficacy of medical interventions.

Ethics: Daniel Dennett's philosophical discourse, particularly in cognitive science and consciousness, has profound implications for the bioethical principles that govern healthcare and medical research. Firstly, his emphasis on autonomy resonates deeply with the principle of informed consent, advocating for a nuanced understanding of human cognition to facilitate individuals' capacity to make informed decisions about their health. Secondly, his approach to beneficence, grounded in empirical research and rational discourse, encourages the pursuit of interventions and policies that are fundamentally aimed at promoting well-being and positive outcomes. In the context of nonmaleficence, Dennett's analytical scrutiny of human consciousness can be seen as a call to avoid harm by fostering a deeper understanding of the complexities of the human mind, thereby informing more compassionate and effective healthcare strategies. Lastly, his commitment to justice, or fairness, is reflected in his interdisciplinary approach to philosophy and science, which seeks to bridge gaps and foster inclusivity in the discourse surrounding human cognition and consciousness. Through his work, Dennett encourages a bioethical framework grounded in empirical evidence and is deeply committed to promoting the well-being and dignity of individuals, thereby influencing the ethical landscape of healthcare and medicine.

Conclusion: In the pantheon of contemporary thinkers, Daniel Dennett emerges as a luminary whose contributions span philosophy, cognitive science, and beyond. Dennett has dedicated his life to dissecting complex phenomena such as consciousness, evolutionary psychology, and the intricacies of the human mind, often bridging the gap between science and philosophy. His rationalist foundations have paved the way for a deeper understanding of innate cognitive faculties, fostering a dialogue that intertwines philosophical intricacies with empirical rigor. Simultaneously, his empiricist orientation emphasizes the pivotal role of sensory experiences in shaping our perception of reality, fostering a rich and nuanced exploration of human cognition. Dennett's endorsement and integration of the scientific method into philosophical inquiries have facilitated a collaborative and evidence-based approach to understanding the complexities of the human mind. Moreover, his insights into neuroscience and psychology have indirectly contributed to the evolving standards of medical healthcare, potentially informing treatment approaches and patient care. Furthermore, his philosophical discourse has significantly influenced the bioethical principles governing healthcare and medical research, promoting a framework grounded in empirical evidence and committed to fostering the well-being and dignity of individuals. Through his prolific career, Dennett has not only catalyzed critical discussions on the intersectionality of science and philosophy but has also emerged as a beacon of intellectual inquiry, fostering a deeper understanding of the multifaceted tapestry of human existence.

Daniel Dennett's Legacy: Pioneering a profound interdisciplinary dialogue that bridges philosophy and science, particularly in consciousness and cognitive evolution, fostering a deeper understanding of the human mind through a blend of rationalist and empiricist approaches.

**

REVIEW QUESTIONS

True/False Questions:

1. Daniel Dennett's philosophical perspective is grounded in a physicalist stance, emphasizing that the mind and consciousness are rooted in material processes.
   True or False?

2. Dennett's work primarily dismisses the scientific method in favor of purely rationalist and philosophical inquiries.
True or False?

Multiple-Choice Questions:

3. Which book by Daniel Dennett explores the implications of Darwinian evolution through a rationalist lens?
a) "*Consciousness Explained*"
b) "*Darwin's Dangerous Idea*"
c) "*Brainstorms*"
d) "*Elbow Room*"

4. What methodological approach does Daniel Dennett often use to derive broader conclusions from specific observations?
a) Deductive reasoning
b) Abductive reasoning
c) Inductive reasoning
d) Analogical reasoning

Clinical Vignette:

5. A patient with a complex neurological disorder is undergoing evaluation. The medical team decides to integrate a more nuanced understanding of consciousness and cognitive processes into the treatment plan. Which philosopher's work, emphasizing empirical evidence and rational analysis of consciousness, might influence this approach?
a) Immanuel Kant
b) Daniel Dennett
c) John Locke
d) René Descartes

Basic Science Vignette:

6. A neurologist is studying a patient with a rare brain disorder that affects consciousness. Drawing on Daniel Dennett's theories, what should the neurologist focus on to understand the patient's condition better?
a) The patient's subjective reports and personal experiences
b) The religious and cultural beliefs about consciousness
c) The underlying physical processes and neural mechanisms in the brain
d) The philosophical debates about the nature of the soul

Philosophy Vignette:

7. During a seminar on the philosophy of mind, a student asks how Daniel Dennett's views on consciousness challenge traditional Cartesian dualism. What is the most accurate response based on Dennett's perspective?
a) Dennett argues that consciousness is a non-physical entity separate from the brain
b) Dennett supports Cartesian dualism, believing in the separation of mind and body
c) Dennett asserts that consciousness arises from physical processes in the brain and rejects the idea of a separate, non-physical mind
d) Dennett believes consciousness is solely a philosophical concept without any basis in physical reality

Correct Answers:

1. True
2. False
3. b) "*Darwin's Dangerous Idea*"
4. c) Inductive reasoning
5. b) Daniel Dennett
6. c) The underlying physical processes and neural mechanisms in the brain
7. c) Dennett asserts that consciousness arises from physical processes in the brain and rejects the idea of a separate, non-physical mind

BEYOND THE CHAPTER
Daniel Dennett (1942-)

- *Consciousness Explained* by Daniel Dennett
- *Sweet Dreams: Philosophical Obstacles to a Science of Consciousness* by Daniel Dennett
- *From Bacteria to Bach and Back: The Evolution of Minds* by Daniel Dennett

***

Philosophy of Science — Medical Research

Chapter 47: **Consciousness** — The Emergent Marvel: Beyond the Brain

Consciousness is an intricate symphony of neural networks, extending beyond the brain to revolutionize our understanding of mind and medicine.

Abstract: The examination of consciousness as an emergent property is driving a paradigm shift in neuroscience, philosophy, and medicine. This inquiry, viewed comprehensively, argues that consciousness is a multifaceted phenomenon born from the intricate interplay of neural networks, including the brain, spinal cord, and peripheral and enteric nervous systems. This perspective invites a nuanced approach to rationalism, expanding intrinsic cognitive faculties into an interconnected model of rationality influenced by complex neural circuits. It enhances empiricism by emphasizing environmental interactions in shaping consciousness. This understanding propels interdisciplinary methodologies, engaging advanced technologies to unravel consciousness's nuances. In medicine, it fosters a holistic approach, integrating mental and physical health and revolutionizing diagnostic and therapeutic strategies. It enriches bioethical discussions, integrating patient autonomy, practitioner beneficence, nonmaleficence, and public justice, fostering a more adaptive approach to patient care. Ultimately, this perspective fosters a comprehensive, dynamic understanding of cognition and ethics, driving evidence-based advancements across fields.

**

Introduction: The concept of human consciousness has long intrigued scholars, researchers, and thinkers across disciplines, but its intricacies remain enigmatic. Traditionally ascribed as a function of the brain, emerging perspectives argue that consciousness is not the sole purview of this singular organ but rather an emergent property arising from a complex symphony of interconnected neural systems. These include the brain, spinal cord, peripheral nerves, and the often-overlooked enteric nervous system that governs the gut. As we grapple with this expanded view, the implications reverberate through multiple domains, provoking us to reevaluate longstanding frameworks. This inquiry will delve into how this nuanced understanding of consciousness as an emergent property reframes our rationalist and empiricist philosophies, reshapes our grasp of the scientific method, impacts medical practice, and necessitates a more comprehensive ethical framework in bioethics, encompassing patient autonomy (informed consent), practitioner beneficence (do good), practitioner nonmaleficence (do no harm), and public justice (be fair).

Rationalism: Understanding consciousness as an emergent property offers a new lens through which to explore the rationalist perspective within the framework of biological evolution. According to rationalism, certain cognitive faculties such as reasoning, pattern recognition, and problem-solving abilities are not just developed through experience; they are evolutionarily hardwired into the neural architecture as "*a priori*" principles. When these innate faculties are viewed as components in a vast network that includes the brain, spinal cord, peripheral nerves, and the enteric system, the idea of rationality expands to incorporate a more dynamic, interconnected model. This emergent rationality suggests that even basic cognitive faculties, which have evolved as adaptive mechanisms to enhance survival and reproductive success, become part of a collective greater than the sum of its parts. For example, consider the phenomenon of the "*gut feeling*." The enteric nervous system, often referred to as the "*second brain*," consists of a complex network of neurons in the gut. This system can influence emotional states and decision-making processes, demonstrating that rational thought is not solely a cerebral function but an emergent property of multiple interacting neural networks. Such a view necessitates integrated ethical and scientific scrutiny. On an individual level, it challenges us to be aware of the limitations and adaptive biases of our evolved rational faculties. At the level of public policy and professional ethics, a comprehensive understanding of emergent rationality would call for regulations and educational paradigms that acknowledge the intricate choreography of these faculties, each contributing differentially to the complex tapestry of human rationality and consciousness.

Empiricism: The empiricist orientation toward consciousness posits that organisms primarily acquire knowledge through sensory experiences and environmental interactions. This perspective aligns with inductive reasoning, where individual observations pave the way for more general conclusions. In this framework, consciousness is not simply an intrinsic quality but emerges as a complex, adaptive response to a plethora of external stimuli. Cognitive biases, often evolutionarily developed as heuristic "*shortcuts*," add a layer of complexity to the empirical approach. These biases interact dynamically with external factors like socialization and culture, making the nature of emergent consciousness even more multifaceted. Empiricism underscores the crucial role that lived experiences play in shaping not just individual cognition but also collective behaviors and ethical norms. For instance, the development of language skills in children demonstrates how sensory input and social interactions contribute to cognitive development, illustrating the emergent nature of consciousness. On a professional level, empiricism prioritizes the collection and analysis of data to form ethical guidelines, implying that our ethical frameworks should be flexible and responsive to new information. Similarly, public policy ought to be grounded in empirical evidence and designed to be adaptive to societal shifts and technological advancements.

The Scientific Method: The exploration of consciousness plays a pivotal role in shaping and refining the scientific method, especially in fields like neuroscience, psychology, and even philosophy. Observations of complex cognitive behaviors or states of awareness in organisms naturally incite queries about their origins, mechanisms, and implications. These queries often result in testable hypotheses about the nature of consciousness and its emergent characteristics. For example, neuroscientists may hypothesize the relationship between neural networks and states of awareness. At the same time, psychologists may explore how social and environmental factors contribute to the emergence of collective behaviors. The intricacy of the subject demands rigorous controlled experiments, often involving interdisciplinary approaches that utilize advanced technologies like fMRI scans, artificial intelligence algorithms, or ethological studies. Data from such experiments are meticulously scrutinized to deduce whether emergent properties like consciousness can be linked to specific biological, psychological, or environmental factors. Given the complex interplay of variables, these experiments' replicability is vital to establish credible findings. Peer review further strengthens the endeavor by ensuring that the research methodology is sound and that the conclusions drawn significantly contribute to our understanding of emergent properties like consciousness.

Medicine: Grasping consciousness, rationality, and sense perceptions as emergent properties from the body's intricate neural networks reshapes the practice and philosophy of medicine. Historically segmented into "*mental*" and "*physical*" health, these boundaries blur when we appreciate that all cognitive and emotional experiences originate from the physical interactions within our neural systems. The gut-brain axis, for instance, highlights how disorders like irritable bowel syndrome, once strictly seen as gastrointestinal, can influence and be influenced by mood and cognition. This intertwined understanding necessitates a holistic approach to treatment, addressing the broader network rather than isolated symptoms. Recognizing the emergent nature of consciousness and its associated cognitive and emotional states in psychiatric care provides a framework for understanding complex disorders like schizophrenia or bipolar disorder. Treatment protocols, thus, become more comprehensive, aiming not only at isolated symptoms but also at the interaction between various cognitive faculties. This integrated approach is a product of evidence-based medicine influenced by the study of emergent properties in consciousness. Ethically, recognizing the complexity of consciousness as an emergent property enriches the ethical dimensions of medical decision-making, from end-of-life choices to evaluations of patient capacity and obtaining informed consent. It adds layers of sophistication to legal evaluations concerning medical malpractice or adequacy of care, given that the standards for "*reasonable care*" become deeply interwoven with advanced scientific understanding of consciousness as an emergent property.

Ethics: The concept of consciousness has significant ramifications for bioethics, particularly the principles of patient autonomy (informed consent), practitioner beneficence (do good), practitioner nonmaleficence (do no harm), and public justice (be fair). In the realm of autonomy and informed consent, understanding consciousness as an emergent property compels medical healthcare providers to consider the complex interplay of cognitive faculties when explaining medical procedures or research participation to patients. This adds depth to the ethical requirement for truly "*informed*" consent, acknowledging that cognitive capacities and conscious awareness are complex and multifaceted. The principle of beneficence gains nuance when medical healthcare providers grasp the interconnected nature of cognitive faculties, allowing for more targeted interventions to enhance overall well-being rather than addressing isolated symptoms. This understanding likewise enriches nonmaleficence; recognizing the complexity of consciousness helps prevent potential harms associated with medical interventions, such as unintentional cognitive or emotional side effects of treatment. Lastly, the principle of justice — fair distribution of healthcare resources — is influenced by understanding the emergent nature of consciousness. Decisions about resource allocation can consider not only physical suffering but also complex mental and emotional states.

Conclusion: The concept of consciousness has multidisciplinary implications that fundamentally transform our approaches to rationalism, empiricism, the scientific method, medicine, and ethics. By seeing consciousness not as a monolithic entity but as an emergent, intricate result of multiple interacting systems and faculties, we add complexity to our human cognition and behavior models. In the rationalist tradition, this view challenges us to think beyond individual, innate faculties to consider how they combine and evolve into more complex forms of awareness and reasoning. For empiricism, it highlights the dynamic interplay between environmental interactions and cognitive development, recognizing that our understanding of consciousness must be continuously revised as new data emerges. The concept similarly informs the scientific method, necessitating interdisciplinary approaches to explore the complexities of human consciousness. In medicine, it has paved the way for more nuanced standards of care, particularly in neurology and psychiatry, as well as in bioethical considerations. It enriches ethical discourse by providing a more integrated framework for the bioethical principles of patient autonomy (informed consent), practitioner beneficence (do good), practitioner nonmaleficence (do no harm), and public justice (be fair). Consciousness as an emergent property is a cornerstone for a more interconnected, dynamic understanding of human cognition, ethical considerations, and healthcare practices, emphasizing the need for adaptive, evidence-based approaches across these domains.

Consciousness — Emergent Properties' Legacy: The concept of consciousness as an emergent property disrupts traditional localized models, compelling a revolutionary rethinking across neuroscience, ethics, philosophy, and medicine, necessitating a more holistic, adaptive approach to understanding human cognition and its applications.

**

REVIEW QUESTIONS

True/False Questions:

1. The chapter suggests that consciousness is solely a function of the brain and not influenced by other neural systems.
   True or False?

2. Understanding consciousness as an emergent property enriches the ethical principle of beneficence by allowing for more targeted interventions.
   True or False?

Multiple-Choice Questions:

3. According to the chapter, which system is referred to as the "*second brain*" due to its influence on emotional states and decision-making processes?
   a) Central nervous system
   b) Peripheral nervous system
   c) Enteric nervous system
   d) Autonomic nervous system

4. The chapter emphasizes that in the scientific method, the study of consciousness requires:
   a) Reductionist methodologies
   b) Isolated experiments
   c) Interdisciplinary approaches
   d) Traditional psychological theories

Clinical Vignette:

5. A 60-year-old patient presents with symptoms of both gastrointestinal discomfort and anxiety. The medical team considers the patient's gut health and mental state. According to the chapter, which approach should the medical team adopt for a comprehensive treatment plan?
   a) Focus solely on gastrointestinal treatment
   b) Address only the patient's anxiety
   c) Integrate treatment for both gastrointestinal and mental health issues
   d) Prescribe a generic medication for overall health

Basic Science Vignette:

6. Dr. Smith is studying the impact of gut microbiota on mental health. She notices that patients with irritable bowel syndrome (IBS) often exhibit symptoms of anxiety and depression. Considering the enteric nervous system's role, which approach should she take to best understand the relationship between gut health and mental health?
   a) Focus exclusively on psychological interventions for anxiety and depression
   b) Treat IBS with gastrointestinal medications only
   c) Adopt a holistic approach that includes both gastrointestinal and psychological treat ments
   d) Ignore the gut-brain connection and treat symptoms independently

Philosophy Vignette:

7. Emily is writing a paper on rationalism and wants to argue that rationality extends beyond traditional cognitive faculties. She decides to use the concept of "*gut feeling*" as an example. How should she best articulate her argument?
   a) Rationality is solely a function of the brain, with no input from other neural systems
   b) The enteric nervous system's influence on decision-making shows that rationality is an emergent property of interconnected neural networks
   c) Emotional states are entirely separate from rational decision-making processes
   d) Cognitive faculties are fixed and do not adapt or interact with other bodily systems

Correct Answers:

1. False
2. True
3. c) Enteric nervous system
4. c) Interdisciplinary approaches
5. c) Integrate treatment for both gastrointestinal and mental health issues
6. c) Adopt a holistic approach that includes both gastrointestinal and psychological treatments
7. b) The enteric nervous system's influence on decision-making shows that rationality is an emergent property of interconnected neural networks

BEYOND THE CHAPTER
Consciousness

- *The Conscious Mind: In Search of a Fundamental Theory* by David J. Chalmers
- *Consciousness Explained* by Daniel Dennett
- *The Astonishing Hypothesis: The Scientific Search for the Soul* by Francis Crick

***

Chapter 48: Free Will and Determinism — The Dynamic Interplay: Freedom in Complexity

Free Will emerges from deterministic roots, blending complexity theory and systems thinking to reshape our grasp of autonomy and ethics.

Abstract: The centuries-old philosophical discourse on Free Will and Determinism evolves significantly through complexity theory and systems thinking, proposing a nuanced reconciliation between the opposing viewpoints. This perspective posits Free Will as an emergent property, rooted in deterministic elements like genetics and environmental factors, but not entirely reducible to them, unfolding from the complexities of human neural networks, cultural backgrounds, and individual experiences. This write-up explores the implications of this perspective across rationalism, empiricism, the scientific method, medicine, and ethics. Each section delves into the nuanced layers of Free Will and Determinism, illustrating how recognizing Free Will as an emergent property reshapes discussions and practices in these fields. By embracing this multidimensional approach, we pave the way for more inclusive, interdisciplinary dialogues that impact ethical considerations, scientific methodologies, and healthcare practices, fostering a broader understanding of the legacy of the Free Will versus Determinism debate.

**

Introduction: The philosophical debate between Free Will and Determinism has been a focal point of intellectual discourse for centuries, and its ramifications span many disciplines, including ethics, science, and medicine. At its core, this debate centers on the nature of human agency and the extent to which our choices are freely made or determined by external factors such as genetics or the environment. Free Will argues for the capacity of rational agents to make choices independent of deterministic forces. In contrast, Determinism contends that all effects, including human decisions, result from prior causes, as no uncaused effects exist. A pivotal concept in this discussion is that Free Will could be an "*emergent property*," a concept rooted in complexity theory and systems thinking, much like how "*wetness*" emerges from the interaction of individual water molecules but isn't a property of any single molecule. This phenomenon arises from complex systems like the brain but is not reducible to its components, such as specific neurochemical reactions, genetic predispositions, and environmental factors. Suppose Free Will is indeed an emergent property. In that case, it suggests a reconciliation between Free Will and a Deterministic universe by positing that while various factors may constrain our actions, the capacity for choice emerges from the complexities of our neural networks, cultural imprints, and experiential learnings. In the sections that follow, we will delve into how the tension between Free Will and Determinism manifests itself in rationalism, empiricism, the scientific method, medicine, and ethics and how a nuanced understanding of Free Will as an emergent property has the potential to reframe discussions and practices across these disciplines. This perspective reorients our understanding of human agency and has significant implications for ethical principles and scientific methodology.

Rationalism: In the context of rationalism, the debate between Free Will and Determinism gains nuanced layers. Rationalist foundations suggest that human cognitive faculties like reasoning, pattern recognition, and problem-solving are innate and hardwired into our neural architecture as evolutionarily adaptive mechanisms. These faculties operate under deterministic principles; they result from genetic encoding and biochemical reactions that have evolved to optimize survival and reproductive success. Yet, the intricate interaction of these deterministic faculties gives rise to what might be considered Free Will. From this viewpoint, Free Will could be seen as an emergent property, similar to how a "*flock*" emerges from individual birds following simple innate rules. The flock is not reducible to any particular bird yet arises as a collective behavior. For instance, while individual neurons operate under strict biochemical laws, the collective interactions among billions of neurons in the brain produce the emergent phenomena of consciousness and decision-making that feel autonomous. The example of a "*gut feeling*" further illustrates this point. The enteric nervous system, or "*second brain*," consists of a complex network of neurons in the gut that can influence emotional states and decision-making processes. This demonstrates that rational thought is not solely a cerebral function but an emergent property of multiple interacting neural networks.

Empiricism: From an empiricist standpoint, the question of Free Will versus Determinism takes on an orientation deeply rooted in sensory experiences and environmental interactions. Empiricism posits that knowledge is primarily gained through the senses and honed through inductive reasoning, where specific observations shape broader conclusions. Cognitive biases, socialization, and cultural factors further complicate this landscape, introducing variables that shape behavior and cognition in non-deterministic ways. Free Will might surface as an emergent property within this intricate mesh of experiences and influences. Drawing from complexity theory, while deterministic elements like genetic predispositions may set the stage, unpredictability arises from the non-linear interactions and feedback loops among varied sensory experiences and these innate tendencies. For example, think of how individual cars on a highway operate under deterministic rules but lead to the emergent property of "*traffic flow*," which can be quite unpredictable. Each car's movement is influenced by specific rules and the behavior of surrounding cars, yet the resulting traffic pattern is not easily reducible to these individual elements. Similarly, each decision may appear as a personal act of Free Will but is a point within a more complex web of causality that includes deterministic and unpredictable factors. As individuals navigate these complicated networks, their choices — the manifestation of what we might label as Free Will — arise from this emergent complexity. This view aligns closely with the empiricist emphasis on adaptability and responsiveness to environmental challenges, allowing for a form of agency rooted in deterministic principles and responsive to many unpredictable variables. Therefore, from an empirical perspective, Free Will can be viewed as an emergent property of complex systems like social dynamics and neural networks, a phenomenon that is constructed on deterministic foundations but evolves into something far more intricate and unpredictable when these elements interact in real-world contexts.

The Scientific Method: The debate between Free Will and Determinism has profound implications for the scientific method, particularly in how scientists study complex systems and human behavior. If Free Will is considered an emergent property — a concept supported by complexity theory — arising from lower-level deterministic processes such as neurochemical reactions, genetic predispositions, and environmental factors, the scientific method must adopt a systems-based methodology to account for the multilayered complexity inherent in such phenomena. While traditional scientific approaches rely on isolating variables for controlled testing, an emergent property like Free Will demands an interdisciplinary, systems-based methodology drawing from chaos theory or complexity theory. Consider, for instance, the challenges meteorologists face in predicting weather patterns. While individual components like temperature and humidity operate under deterministic laws, their interactions create a complex, emergent system where small initial differences can yield vastly different outcomes, a feature known as "*sensitive dependence on initial conditions*" in chaos theory. Meteorology, therefore, relies on a multitude of scientific disciplines and computational models to predict a phenomenon that, although based on deterministic processes, is inherently complex and difficult to predict. These theories explore how deterministic systems can yield unpredictable, complex outcomes, making them particularly applicable to the study of Free Will. Given the intricacy and multifactorial nature of emergent properties, formulating a testable hypothesis that can be fully verified or falsified becomes daunting. Even if an experiment is replicable, the high number of influencing variables — many of which may not yet be understood — can make drawing firm conclusions challenging. Peer review, too, becomes complex; experts from multiple disciplines may need to scrutinize a single study to assess its validity adequately. The challenge extends to the concept of replicability, as small initial differences in complex systems can lead to significantly different outcomes. While the scientific method's core principles remain unchanged, the debate pushes it toward a more nuanced, interdisciplinary approach that accounts for the unpredictable nature of emergent properties, like Free Will, even within systems operating under deterministic laws.

Medicine: The implications of the Free Will and Determinism debate extend into medicine, complicating ethical considerations and shaping medical healthcare standards of care. If Free Will is understood as an emergent property arising from complex interactions between neurochemical reactions, genetic predispositions, and environmental factors, medical practice must evolve to accommodate this nuanced perspective. The concept aligns well with the burgeoning field of personalized medicine, which advocates for treatments tailored to an individual's unique genetic makeup and life circumstances. A deterministic view would suggest that medical outcomes are largely predictable based on initial conditions like genetic predispositions and environmental factors, which might inadvertently encourage a one-size-fits-all approach to healthcare. In contrast, recognizing Free Will as an emergent property necessitates a more holistic and adaptable medical healthcare standard of care. Take, for example, the human immune response. While individual components like white blood cells and antibodies operate under deterministic principles, their collective behavior results in a highly adaptive and personalized immune response. This complexity reflects an emergent property of the system, where the "*choice*" — in this case, how to respond to a pathogen — is not fixed but arises from numerous influencing factors like genetic makeup, previous exposures, and even current stress levels. Just as the immune system's response is an emergent property, a notion understood well within systems thinking, so too is the concept of Free Will, requiring a medical healthcare approach that is informed by the principles of complexity theory and is similarly nuanced. This perspective encourages an understanding of patient care that considers the intricate interplay of factors influencing an individual's health, advocating for more personalized and flexible treatment plans that account for the emergent nature of human decision-making and behavior.

Ethics: The debate between Free Will and Determinism profoundly influences bioethical principles, offering a complex tapestry upon which ethical considerations are drawn. When viewed as an emergent property, Free Will enriches our understanding of these principles by demanding that they be as dynamic and multifaceted as the systems from which Free Will arises. For instance, consider the concept of "*mental well-being*," a complex outcome shaped by genetics, life experiences, social interactions, and many other factors. While each element operates under deterministic principles, their collective interaction gives rise to an emergent property — mental well-being — that cannot be easily predicted or reduced to its components. The principle of autonomy, which underpins the practice of informed consent, becomes far more nuanced when we recognize that a patient's mental well-being (and, by extension, their decision-making capacity) is emergent rather than fixed. Informed by complexity theory, this emergent complexity demands a more dynamic and systems-oriented approach to beneficence and nonmaleficence. What does "*doing good*" mean in the context of a patient whose mental well-being is shaped by a myriad of interacting factors? It pushes healthcare providers to consider a more holistic, individualized approach to care. Similarly, the principle of justice must consider the complex systems influencing each individual's well-being, necessitating medical healthcare systems that are not merely equitable on the surface but deeply attuned to the emergent complexities of human life.

Conclusion: The complex interplay between Free Will and Determinism is not an "*either-or*" scenario but rather a "*both-and*" situation that, informed by complexity theory and systems thinking, requires a multifaceted approach across different disciplines. Free Will as an emergent property, a cornerstone in complexity theory and systems thinking, invites us to look beyond the simplistic binaries and embrace a more layered, intricate understanding of human agency. It encourages us to reconsider how we approach ethics, conduct scientific research, and practice medicine by adopting a systems-oriented perspective that acknowledges the multifaceted nature of human experience. By embracing Free Will as an emergent property deeply rooted in complexity theory, we bridge the dichotomy between Free Will and Determinism, opening the door to more integrative and comprehensive approaches across various fields. This nuanced view allows for a more complete understanding of human agency and significantly impacts ethical considerations, scientific methodologies, and healthcare practices. Free Will as an emergent property invites us to delve into the complexities of human experience more deeply, acknowledging the intricate Deterministic systems that serve as the scaffolding for our emergent Free Will choices.

Free Will and Determinism's Legacy: The Free Will and Determinism debate catalyzes interdisciplinary dialogue, pushing philosophy, science, medicine, and ethics toward a nuanced understanding of human agency as a complex interplay between deterministic and emergent properties.

**

REVIEW QUESTIONS

True/False Questions:

1. The chapter argues that Free Will can be seen as an emergent property, much like how "*wetness*" emerges from the interaction of water molecules.
   True or False?

2. According to the chapter, Determinism contends that human decisions result from prior causes and no uncaused effects exist.
   True or False?

Multiple-Choice Questions:

3. According to the chapter, how does complexity theory contribute to the understanding of Free Will?
   a) It refutes the existence of Free Will entirely
   b) It shows that Free Will is predetermined by genetics
   c) It posits Free Will as an emergent property from complex deterministic systems
   d) It suggests Free Will does not exist

4. According to the chapter, what methodology is suggested to study complex systems like Free Will?
   a) Isolated variable testing
   b) Reductionist approach
   c) Systems-based methodology
   d) Purely empirical methods

Clinical Vignette:

5. A 45-year-old man with a genetic predisposition to hypertension is undergoing treatment. Despite the deterministic influences of his genetics and lifestyle, he actively participates in a personalized care plan that includes medication, diet changes, and exercise. Over time, his blood pressure improves significantly as he adheres to the plan. This scenario exemplifies how Free Will emerges from deterministic systems. Which concept does this approach illustrate?
   a) Genetic determinism
   b) Free Will as an emergent property
   c) Standardized treatment protocol
   d) Environmental determinism

Basic Science Vignette:

6. A neuroscientist is conducting a study on how different brain regions interact to produce decision-making processes. She finds that no single region is responsible for decisions, but rather, decisions emerge from complex interactions across various networks. Which theory best explains her findings?
   a) Genetic determinism
   b) Complexity theory
   c) Classical conditioning
   d) Behavioral determinism

Philosophy Vignette:

7. In her philosophy class, Maria is exploring the concept of Free Will as an emergent property. She wants to argue that Free Will can coexist with deterministic elements such as genetics and environment. How should she best present her argument?
   a) Free Will is entirely independent of genetic and environmental influences
   b) Deterministic elements like genetics and environment completely negate the existence of Free Will
   c) Free Will emerges from the complex interactions between deterministic elements, like genetics and environment, making it more than just the sum of these parts
   d) Free Will is solely determined by individual choices without any influence from external factors

Correct Answers:

1. True
2. True
3. c) It posits Free Will as an emergent property from complex deterministic systems
4. c) Systems-based methodology
5. b) Free Will as an emergent property
6. c) Adopt a holistic approach that considers both genetic predispositions and environmental factors
7. c) Free Will emerges from the complex interactions between deterministic elements, like genetics and environment, making it more than just the sum of these parts

BEYOND THE CHAPTER
Free Will and Determinism

- *Freedom Evolves* by Daniel Dennett
- *Elbow Room: The Varieties of Free Will Worth Wanting* by Daniel Dennett
- *Free Will* by Sam Harris

***

Philosophy of Science

Medical Research

Chapter 49: **Medicine** — The Dynamic Fusion: Science, Technology, Ethics

Medicine synthesizes science, technology, and ethics, continually evolving to enhance human health and ethical standards.

Abstract: Medicine is a dynamic system at the crossroads of science, technology, and humanity's endeavor to enhance well-being and counteract illness. Rooted in rationalism and empiricism, it employs structured, analytical approaches and evidence-based practices, fostering healthcare innovations and interdisciplinary interventions. These approaches, refined through rigorous experimentation, data analysis, and peer-reviewed insights, enhance scientific discourse and policy formulations. Medicine adapts to evolving technologies and knowledge, developing healthcare standards that reflect contemporary best practices and legal benchmarks. It is anchored by ethical principles such as patient autonomy (informed consent), practitioner beneficence (do good), practitioner nonmaleficence (do no harm), and public justice (be fair), shaping patient-practitioner interactions and societal perspectives on healthcare equity. These principles influence legal frameworks and public policies, promoting an ethically responsible healthcare landscape. Medicine is a continually evolving, self-reflective entity committed to human health and ethical constructs, significantly advancing science and society and leaving an indelible legacy on human well-being.

**

Introduction: Medicine stands at the intersection of science, technology, and human well-being, serving as a testament to humanity's quest to combat illness and improve quality of life. It is a multidisciplinary field that draws from a broad spectrum of scientific disciplines, including biology, chemistry, physics, and psychology, among others, to understand the complex mechanisms that govern health and disease. Evolving through millennia, from the rudimentary practices of ancient civilizations to today's advanced clinical interventions, medicine has been guided by an ever-expanding body of empirical evidence. Rooted in rigorous observation, experimentation, and analysis, the medical field has significantly shaped and been shaped by the scientific method. It has given rise to various branches — from primary care to specialized areas like oncology, neurology, and cardiology — all aimed at enhancing healthcare delivery. Moreover, as medicine continually pushes the boundaries of what is medically possible, it constantly grapples with ethical dimensions. The biomedical principles of patient autonomy (informed consent), practitioner beneficence (do good), practitioner nonmaleficence (do no harm), and public justice (be fair) serve as moral compasses, guiding healthcare professionals in their interactions with patients and influencing policy decisions at the societal level.

Rationalism: The medical field's rationalist foundations are evident in its structured, analytical approaches to understanding and treating complex biological systems. Rooted in rationalism, medicine posits that human physiology and pathology can be understood through certain axiomatic principles derived from basic sciences like biology and chemistry. Medical professionals are trained to employ deductive reasoning, starting with broader theories or principles to arrive at specific diagnostic or treatment plans. For example, the principle that antibiotics treat bacterial infections is an axiomatic truth in medicine, established through years of empirical evidence but rationalized through biochemical mechanisms of action. In clinical practice, a physician employs "*a priori*" reasoning — pattern recognition, problem-solving skills, and basic medical axioms — to diagnose a patient's symptoms and prescribe appropriate treatments. However, medicine also recognizes the limitations and complexities added by epigenetic factors, environmental influences, and individual variations. This multi-layered complexity demands a nuanced application of rationalist principles, often requiring interdisciplinary input from genetics, psychology, and social sciences. Ethically, healthcare providers must be acutely aware of the limitations of their rational faculties, particularly when dealing with matters as variable and individualized as human health. Regulatory bodies also play a role in creating guidelines that ensure the limitations of rationalistic thinking are acknowledged and that medical practices are updated based on new evidence. The adaptive value of these rational faculties in medicine lies in their capacity for quick, reliable responses to the recurring challenges of diagnosing and treating illness, thereby enhancing healthcare outcomes. Ultimately, the rationalistic foundations of medicine interact with empiricist methods and ethical considerations to form a more comprehensive, emergent system for understanding and improving human health.

Empiricism: Medicine's empiricist orientation is manifested in its deep commitment to evidence-based practice, where conclusions are drawn from direct observations, experiments, and data analysis. The role of empiricism is particularly pronounced in clinical trials and epidemiological studies, where inductive reasoning takes center stage. From specific observations — be it a drug's efficacy in a small group or the prevalence of a disease in a certain demographic — general conclusions are made to inform broader medical practice. However, this empirical methodology is not without its limitations. Cognitive biases, shaped by evolution as heuristic shortcuts, can sometimes skew the interpretation of empirical data. Social, cultural, and environmental factors also introduce variability that can affect the generalizability of medical findings. From an ethical standpoint, empiricism helps shape guidelines based on real-world outcomes, ensuring that the most effective and safest treatments are recommended. On a public policy level, empiricism informs the creation of healthcare regulations, such as drug approvals or vaccination programs, designed based on extensive empirical evidence to be both adaptive and responsive. The adaptive value of this empirical orientation is in its ability to quickly integrate new information and adapt to unforeseen challenges, thereby enhancing healthcare outcomes. As a complex emergent system, medicine combines the insights gained from individual sensory experiences, such as patient feedback and clinical observations, to form a comprehensive, nuanced understanding of health and disease far greater than the sum of its empirical parts. This emergent property extends to phenomena like the rapid global response to pandemics, the development of interdisciplinary treatments, and the establishment of complex healthcare systems.

The Scientific Method: Medicine has been both a beneficiary and a shaper of the scientific method, underscoring the importance of rigorous, empirical investigation in healthcare practices. From its early days, medicine moved from anecdotal observations and traditional remedies to a systematic understanding of disease. Physicians and researchers began to employ the scientific method to investigate everything from the physiology of the human body to the efficacy of treatments. This led to formulating testable hypotheses about the causes of diseases, conducting controlled experiments to verify these hypotheses, and collecting data to analyze the results. As a result, randomized controlled trials (RCTs) have become the gold standard in medical research, offering robust evidence for or against the effectiveness of medical interventions. These RCTs embody the scientific method in their design, ensuring that hypotheses are testable and replicable under controlled conditions. Moreover, medicine's complex and high-stakes environment has pushed for innovations in the scientific method, including more stringent requirements for statistical analysis, ethical considerations, and long-term follow-up studies. The peer-review process in medicine is particularly rigorous, as the stakes are often a matter of life and death. The insights generated through this meticulous application of the scientific method have been pivotal in shaping guidelines and policies, setting standards within the medical community, and influencing the broader scientific discourse. Thus, medicine has adapted and contributed to refining the scientific method, making it more robust, ethical, and reliable.

Medicine: The contributions of medicine to its field are profound, driving a continuous cycle of innovation, improvement, and refinement in healthcare delivery. Central to this is medical healthcare "*standards of care*," a dynamic set of guidelines that embody the medical community's collective wisdom and current best practices. These standards are derived from a rigorous combination of empirical research, clinical trials, and expert consensus, all of which serve as the foundation for evidence-based medicine. By setting these guidelines, medicine creates a framework that healthcare professionals must follow, ensuring minimum competence and care across diverse medical settings. The standards also adapt to new information and technologies, accommodating advancements such as telemedicine, personalized medicine, and artificial intelligence in diagnostics and treatment plans. Importantly, these standards serve a legal function, acting as benchmarks against which the adequacy of care can be assessed in cases of medical malpractice or negligence. As a result, they not only guide clinical decision-making but also have broader societal implications, shaping healthcare policy and legal frameworks. Medicine's contributions are a complex, self-regulatory ecosystem that balances scientific rigor, ethical considerations, and societal needs to deliver the highest possible quality of care. This constant self-scrutiny and adaptation make medicine not just a field of practice but also a discipline committed to its ongoing improvement and the well-being of the communities it serves.

Ethics: Medicine wields a profound influence on the bioethical principles that underpin healthcare interactions and policy decisions. Autonomy, exemplified by the practice of informed consent, has evolved due to medicine's growing emphasis on patient-centered care, ensuring that patients have comprehensive information and agency in making healthcare decisions. Beneficence is another cornerstone, guiding medical professionals to pursue interventions that optimize patient welfare. This principle is operationalized through evidence-based practices that refine treatment modalities to maximize benefits while minimizing risks. Nonmaleficence, encapsulated by the maxim "*do no harm*," has led to stringent clinical guidelines and rigorous protocols designed to reduce errors, adverse effects, and unintended consequences. This principle often dovetails with beneficence, as medical professionals weigh the risks and benefits of interventions to provide the most appropriate care. Justice, the commitment to fairness and equitable treatment, has increased prominence with the advent of public health ethics and healthcare policy. Medicine has been pivotal in highlighting health disparities and advocating equitable access to quality healthcare across socio-economic strata. Legal frameworks and guidelines often draw on these bioethical principles, giving them societal and policy-level importance beyond the patient-practitioner interaction. Through these principles, medicine contributes to a holistic ethical framework that not only shapes healthcare providers' conduct but also informs regulatory standards, influences public policy, and impacts societal views on healthcare equity and justice. Thus, medicine serves as both a practitioner and a custodian of these ethical principles, constantly grappling with their nuances in an ever-changing healthcare landscape.

Conclusion: Medicine represents a dynamic interplay between science, ethics, and societal needs, undergirded by empirical observation and rational thought. Serving as both a subject and an object of scientific scrutiny, medicine has been pivotal in shaping and being shaped by the scientific method. It has set rigorous standards for data collection, experimental design, and peer review, thus influencing broader scientific practices while refining its methodologies. At the heart of medical practice are evidence-based standards of care, which serve as evolving benchmarks for healthcare delivery. These standards reflect the current state of medical knowledge and hold professionals accountable in legal settings. Ethically, medicine has been instrumental in advancing the principles of patient autonomy (informed consent), practitioner beneficence (do good), practitioner nonmaleficence (do no harm), and public justice (be fair), integrating them into clinical practice and public policy. These principles offer a moral framework beyond the individual patient-provider relationship to influence broader societal perspectives on healthcare justice and equity. Medicine is not just a field concerned with treating illness but a complex system that continually interrogates its practices. It adapts to new information and technologies, confronts ethical dilemmas, and engages with societal issues while striving to improve the human condition.

Medicine's Legacy: Medicine is a dynamic and intricate system that merges science, technology, and ethics to continually enhance human well-being, grounded in rationalism and empiricism, and guided by steadfast ethical principles.

**

REVIEW QUESTIONS

True/False Questions:

1. Medicine relies heavily on both rationalism and empiricism to develop and refine its practices and approaches.
   True or False?

2. The ethical principles guiding medical practice include patient autonomy, beneficence, nonmaleficence, and public justice.
   True or False?

Multiple-Choice Questions:

3. The empiricist orientation in medicine emphasizes:
   a) Deductive reasoning and innate cognitive faculties
   b) Evidence-based practices derived from direct observations and experiments
   c) Theoretical models and philosophical postulates
   d) Traditional and anecdotal medical practices

4. Which of the following best describes the role of "*standards of care*" in medicine?
   a) Guidelines based on anecdotal evidence and historical practices
   b) Recommendations that are seldom updated and rarely based on new research
   c) Dynamic guidelines derived from rigorous research, clinical trials, and expert consensus
   d) Protocols primarily focused on minimizing costs rather than patient outcomes

Clinical Vignette:

5. Dr. Smith is treating a patient with a rare condition. He decides to follow the standard protocol for this condition, which has been developed through randomized controlled trials and peer-reviewed studies. This protocol ensures the treatment is evidence-based and aligns with best practices. What principle of the scientific method is Dr. Smith primarily adhering to?
   a) Hypothesis formulation
   b) Deductive logic
   c) Replicability and peer review
   d) Theoretical modeling

Basic Science Vignette:

6. Dr. Lee is developing a new diagnostic tool for detecting early-stage cancer. She follows the Scientific Method to ensure the tool's accuracy and reliability. Which steps should Dr. Lee take to adhere to this method?
   a) Rely solely on anecdotal evidence and patient testimonials
   b) Formulate a hypothesis, conduct controlled experiments, collect and analyze data, and subject the tool to peer review and replication studies
   c) Skip the data collection phase and proceed directly to publishing the results
   d) Base her conclusions on initial observations without further testing

Philosophy Vignette:

7. In a philosophy seminar, Mark argues that medicine's adherence to bioethical principles is crucial for ethical healthcare delivery. How should he best support his argument using these principles?
   a) Bioethical principles are not relevant to modern medical practice
   b) Medicine upholds patient autonomy, beneficence, nonmaleficence, and justice by ensuring informed consent, pursuing beneficial treatments, avoiding harm, and promoting fair access to healthcare
   c) Ethical considerations in medicine only apply to experimental research, not everyday clinical practice
   d) Bioethical principles are secondary to achieving quick and efficient healthcare outcomes

Correct Answers:

1. True
2. True
3. b) Evidence-based practices derived from direct observations and experiments
4. c) Dynamic guidelines derived from rigorous research, clinical trials, and expert consensus
5. c) Replicability and peer review

. b) Formulate a hypothesis, conduct controlled experiments, collect and analyze data, and subject the tool to peer review and replication studies

7. b) Medicine upholds patient autonomy, beneficence, nonmaleficence, and justice by ensuring informed consent, pursuing beneficial treatments, avoiding harm, and promoting fair access to healthcare

BEYOND THE CHAPTER

Medicine

- *The Emperor of All Maladies: A Biography of Cancer* by Siddhartha Mukherjee
- *The Laws of Medicine: Field Notes from an Uncertain Science* by Siddhartha Mukherjee
- *Medical Apartheid: The Dark History of Medical Experimentation on Black Americans from Colonial Times to the Present* by Harriet A. Washington

***

Philosophy of Science™ — Medical Research

Chapter 50: Hippocratic Oath — The Ethical Bedrock: Medical Integrity

The Hippocratic Oath, a timeless ethical guide, anchors modern medicine in principles of integrity and patient-centered care.

Abstract: The Hippocratic Oath, a historical and moral cornerstone in the medical profession, has remained influential from ancient Greece to contemporary practice. This analysis explores the Oath's foundational role, integrating rationalist principles with empiricist orientations based on tangible healthcare experiences. The Oath has shaped scientific inquiry and research protocols, emphasizing a rigorous and ethical approach to medical advancements. It has profoundly impacted medicine's ethical and professional landscapes, setting the groundwork for modern healthcare standards that blend clinical expertise and moral integrity. Additionally, it has developed modern bioethical principles, including patient autonomy (informed consent), practitioner beneficence (do good), practitioner nonmaleficence (do no harm), and public justice (be fair), fostering a patient-centered approach to healthcare. This write-up underscores the enduring legacy of the Hippocratic Oath as a beacon of ethical guidance, highlighting its adaptability and continuous relevance in the evolving medical landscape.

**

Introduction: The Hippocratic Oath, often considered one of the foundational documents of Western medicine, dates back to ancient Greece and is traditionally attributed to the physician Hippocrates, who lived around the 5th century BCE. This solemn vow sets forth a code of ethics and moral guidelines for physicians, serving as a rite of passage for those entering the medical profession. Its essence focuses on the commitment to patient welfare, confidentiality, and the imperative to "*do no harm*," linking medicine to an ethical framework beyond mere technical proficiency. While the specific language and focus may differ across various renditions, the Oath has been broadly adopted and adapted over millennia, speaking to its enduring relevance. As medical science has evolved, the Hippocratic Oath has been scrutinized and evaluated against developing standards of rationalism, empiricism, the scientific method, and bioethics. The principles embedded within it — from autonomy and beneficence to nonmaleficence and justice — continue to be pivotal in shaping medical practice, ethics, and governance, even as medical professionals grapple with modern complexities like technological innovation, patient rights, and global health challenges.

Rationalism: The Hippocratic Oath's rationalist foundations lie in its attempt to systematize medical ethics and practice through axiomatic principles and guidelines intended to be universally true and applicable. The Oath presupposes that physicians have a moral and professional obligation to act in certain prescribed ways by their specialized training and innate faculties for reason, pattern recognition, and problem-solving. This aligns with rationalist thought, which posits that innate cognitive faculties can guide us toward moral or empirical truths. For example, the Oath's injunction to "*do no harm*" is a fundamental axiom from which other ethical imperatives may be deductively reasoned. If harm is to be avoided, then it follows that physicians should prioritize patient well-being, uphold confidentiality, and exercise prudence in their treatments. The Oath appeals to a universal human faculty for ethical reasoning, presupposing its self-evidently beneficial principles can be apprehended through rational contemplation. In this way, the Hippocratic Oath employs a rationalist framework to establish a foundational ethic for medical practice, which assumes that physicians, guided by reason and moral intuition, will recognize the intrinsic value of its principles.

Empiricism: While the Hippocratic Oath is primarily an ethical document, it possesses an empiricist orientation in that its tenets are shaped by the experiences and observations drawn from medical practice. The principles it sets forth are not abstract, detached concepts but guidelines deeply rooted in the concrete interactions between healthcare providers and patients. For instance, the injunction to "*do no harm*" is a moral axiom and a practical guideline that emerges from centuries of observation about the consequences of unethical or incompetent medical practice. In an empiricist vein, the Oath can be seen as a codification of "*best practices*," developed through the accumulated experience of what works to promote patient well-being and what does not. Inductive reasoning is evident in how the Oath's principles generalize from specific medical cases to broader ethical guidelines. The emphasis on patient confidentiality, beneficence, and nonmaleficence is borne out of particular experiences where violations of these principles led to tangible harm. Thus, the Oath recognizes that ethical and effective medical practice is not solely an intellectual exercise but is deeply intertwined with empirical realities, continually refined through interaction with the real world.

The Scientific Method: The Hippocratic Oath may not directly engage with the scientific method. However, its ethos has profoundly influenced the systematic approach to medical inquiry that we associate with modern science. The Oath's emphasis on beneficence and nonmaleficence translates into a broader ethical framework that underpins scientific research involving human subjects. It demands a rigorous, responsible approach to experimentation and data collection, implicitly encouraging physicians to seek evidence-based treatments that "*do no harm*." While the Oath predates the formalization of the scientific method, its insistence on ethical integrity and patient welfare has been integrated into the protocols of scientific inquiry. For instance, principles from the Oath have informed the ethical guidelines for clinical trials, where testable hypotheses are formed and investigated through controlled, replicable experiments. These guidelines also extend to the peer review process, ensuring that medical research's methodology, validity, and ethical considerations are meticulously scrutinized. The Oath, thus, serves as a moral compass that complements the scientific method, ensuring that the pursuit of knowledge doesn't compromise the well-being of individuals and communities, which is fundamental for the credibility and social contract of the scientific enterprise.

Medicine: The Hippocratic Oath has played an indispensable role in shaping medicine's ethical and professional landscape, providing a foundational framework that has deeply influenced healthcare standards of care. By establishing key ethical principles such as "*do no harm*," patient confidentiality, and the prioritization of patient welfare, the Oath has laid down the moral groundwork that medical professionals are expected to uphold. These principles have been institutionalized in various guidelines, protocols, and laws governing medical practice. Notably, the concept of standards of care, which outlines the expected level of competence and ethical conduct, draws heavily from the Hippocratic Oath's moral imperatives. Modern healthcare standards of care are determined based on evidence-based practices and are continually evolving to reflect new scientific insights and societal needs. However, the essence of these standards remains rooted in the ethical principles articulated in the Hippocratic Oath. For example, the principle of nonmaleficence directly correlates with the requirement for healthcare providers to refrain from actions that could harm patients, a criterion often examined in legal contexts to assess the adequacy of delivered care. Thus, the Oath serves as both a moral and practical compass, steering the practice of medicine toward a human-centered approach that balances clinical expertise with ethical integrity.

Ethics: The Hippocratic Oath has been instrumental in shaping the four cornerstone principles of bioethics: patient autonomy (informed consent), practitioner beneficence (do good), practitioner nonmaleficence (do no harm), and public justice (be fair). The Oath's most direct influence is perhaps on the principle of nonmaleficence, encapsulated in the phrase "*do no harm*," which serves as a guiding maxim for physicians and has been embedded deeply into medical ethics and practice. Similarly, the principle of beneficence, or the ethical mandate to "*do good*," can be traced back to the Oath's commitment to patient welfare and the active promotion of healing. Although not explicitly mentioned in the Oath, the principle of autonomy has evolved as a natural extension of its ethical framework. The Oath's emphasis on the sacredness of the doctor-patient relationship and the confidentiality it demands implies respect for the patient, which contemporary bioethics has elaborated into the principle of informed consent. Finally, while the Oath does not explicitly address justice issues, its underlying ethos encourages fairness and equity in medical treatment. Physicians swear to apply their skills to benefit the sick, acknowledging an ethical obligation beyond individual patients to encompass broader societal concerns about justice and fairness in healthcare. In this way, the Hippocratic Oath has provided the moral scaffolding upon which modern bioethical principles stand, continuing to guide and inform ethical debates in contemporary medicine.

Conclusion: The Hippocratic Oath stands as a monumental touchstone in the history of medicine, its influence seeping into various aspects of medical practice, ethics, and even the scientific method. Originating in ancient Greece and attributed to the physician Hippocrates, the Oath is a rite of passage for medical professionals and a moral charter guiding patient interactions. Its principles, deeply rooted in rationalism and empiricism, have acted as axioms and empirical guides, providing a sense of ethical and practical coherence in a field that continually evolves with technological and scientific advancements. The Oath's ethical imperatives have been pivotal in shaping healthcare standards of care, infusing them with a moral gravitas that goes beyond technical competence to encompass empathy, integrity, and human dignity. Moreover, its principles have been essential in formulating the bioethical pillars of patient autonomy (informed consent), practitioner beneficence (do good), practitioner nonmaleficence (do no harm), and public justice (be fair). These have informed everything from clinical trials to healthcare policies, making the Oath an ancient document and a living framework that continues to guide and challenge the medical community. As modern medicine grapples with complex ethical dilemmas brought on by advancements like gene editing, artificial intelligence, and global healthcare disparities, the Hippocratic Oath remains a foundational text, its principles reverberating through the corridors of hospitals, research labs, and ethical debates, reminding all involved of the profound human dimensions of medical practice.

Hippocratic Oath's Legacy: The Hippocratic Oath serves as the foundational ethical code for medical practice, bridging ancient wisdom with modern bioethics, and continues to guide the moral, scientific, and practical imperatives that govern healthcare professionals and research.

**

REVIEW QUESTIONS

True/False Questions:

1. The Hippocratic Oath emphasizes the principle of "*do no harm*" as a fundamental ethical guideline for physicians.
   True or False?

2. The Hippocratic Oath does not address patient confidentiality in its principles.
   True or False?

Multiple-Choice Questions:

3. Which principle of bioethics directly originates from the Hippocratic Oath's emphasis on avoiding harm to patients?
   a) Autonomy
   b) Justice
   c) Nonmaleficence
   d) Beneficence

4. The Hippocratic Oath has most directly influenced which aspect of modern medical practice?
   a) The development of new surgical techniques
   b) The establishment of healthcare standards of care
   c) The integration of artificial intelligence in diagnostics
   d) The advancement of telemedicine technologies

Clinical Vignette:

5. Dr. Smith is preparing to perform a surgical procedure and ensures that the patient fully understands the risks, benefits, and alternatives beforc obtaining consent. Which bioethical principle is Dr. Smith primarily upholding?
   a) Beneficence
   b) Nonmaleficence
   c) Autonomy
   d) Justice

Basic Science Vignette:

6. Dr. Johnson is developing a new cancer treatment protocol. To uphold the Hippocratic Oath, which action should he prioritize to adhere to the principle of nonmaleficence?
   a) Ensuring all treatments are backed by extensive clinical research
   b) Offering the treatment to a wide range of patients
   c) Designing the protocol to minimize potential side effects
   d) Providing emotional support to patients throughout the treatment

Philosophy Vignette:

7. In developing a new research protocol, Dr. Adams wants to ensure it upholds the principles of the Hippocratic Oath. Which practice most strongly embodies the principle of beneficence?
   a) Ensuring all participants receive a thorough debriefing after the study
   b) Designing the study to provide maximum benefit to participants
   c) Securing funding from unbiased sources
   d) Publishing results in open-access journals to benefit the broader community

Correct Answers:

1. True
2. False
3. c) Nonmaleficence
4. b) The establishment of healthcare standards of care
5. c) Autonomy
6. c) Designing the protocol to minimize potential side effects
7. b) Designing the study to provide maximum benefit to participants

BEYOND THE CHAPTER
Hippocratic Oath

- *The Hippocratic Oath and the Ethics of Medicine* by Steven H. Miles
- *Hippocratic Writings* edited by G.E.R. Lloyd
- *The Birth of the Clinic: An Archaeology of Medical Perception* by Michel Foucault

***

Philosophy of Science

Medical Research

Chapter 51: Nuremberg Code — The Ethical Framework: Human Dignity

The Nuremberg Code safeguards human dignity in research, emphasizing informed consent and ethical integrity.

Abstract: The Nuremberg Code, instituted in 1947 in response to the horrific medical experiments of World War II, delineated ten principles to safeguard the dignity, safety, and consent of human research subjects. This research investigates the Nuremberg Code's significant implications on the scientific method, medicine, and ethics, highlighting its rationalist foundations and empiricist approach. This study explores how the Code has revolutionized the ethical parameters of scientific research and medical practices, fostering a human-centric approach grounded in patient autonomy (informed consent), practitioner beneficence (do good), practitioner nonmaleficence (do no harm), and public justice (be fair). We also examine the Code's enduring legacy as a dynamic document that prevents past atrocities and shapes contemporary bioethical principles and policies. The Nuremberg Code epitomizes humanity's collective commitment to upholding ethical integrity in scientific and medical advancements, marking a milestone in the confluence of science, ethics, and human rights.

**

Introduction: The Nuremberg Code, established in 1947 during the Nuremberg Trials, serves as a seminal framework for ethical medical and scientific research involving human subjects. Born from the grim revelations of medical experiments conducted on prisoners during World War II — most notoriously in Nazi concentration camps — the Code laid down ten principles to ensure research participants' dignity, safety, and consent. This monumental ethical guideline was a watershed moment in the intersection of science, ethics, and human rights. It fundamentally shifted the way research projects had to consider the well-being of their subjects, instigating a global dialogue that moved ethics to the forefront of medical and scientific research. Over the years, the Nuremberg Code has profoundly influenced international regulations, helped define best practices in diverse fields from psychology to epidemiology, and laid the groundwork for the ethical principles of bioethics: patient autonomy (informed consent), practitioner beneficence (do good), practitioner nonmaleficence (do no harm), and public justice (be fair). As such, it serves as an enduring legacy of our collective commitment to prevent the repetition of past atrocities, guiding the moral compass of scientific endeavors.

Rationalism: The Nuremberg Code embodies a rationalist foundation by appealing to innate moral and ethical reasoning faculties generally acknowledged as universal across human cultures. While it emerged from a specific historical context marked by atrocities committed under the guise of scientific research, the principles it espouses draw upon "*a priori*" notions of human dignity, autonomy, and ethical conduct. These principles are not derived merely from empirical observations but rather from axiomatic truths about human worth and moral responsibility. For instance, the Code's emphasis on voluntary consent rests on the fundamental idea that individuals possess an innate right to control their bodies and destinies — a rationalist notion central to ethical theories going back to Enlightenment thinkers like Kant. Similarly, the Code's tenets against causing harm and its advocacy for beneficence echo natural law theories, grounded in the rational understanding that ethical conduct is intrinsic to human well-being. These axioms serve as the base from which further ethical guidelines and medical protocols are deduced, using deductive reasoning to elaborate on the ethical implications for various research contexts. Therefore, the Nuremberg Code serves as a rationalist document, setting moral norms based on logical principles seen as inherently and universally true.

Empiricism: Although the Nuremberg Code has rationalist foundations, it also displays a strong empiricist orientation. The Code was conceived in response to actual, observable abuses committed during medical experiments in World War II, particularly those conducted by Nazi physicians. These grim empirical realities forced the international community to address the ethical gaps in the existing medical and scientific research framework. This historical context provided specific observations that led to the general principles outlined in the Code. Furthermore, ongoing empirical assessments have sustained the Code's effectiveness and continued relevance. Over the years, it has been revised and fine-tuned based on real-world observations and experiences in medical research. Its principles have been tested in varied contexts, ranging from clinical trials for new drugs to social psychology experiments, and they have often been refined in light of further data. Ethical review boards, which operate under guidelines heavily influenced by the Nuremberg Code, rely on case studies and empirical data to make informed decisions. Therefore, while the Code may be rooted in "*a priori*" ethical principles, its enduring impact and practical applicability are deeply entwined with an empiricist approach to ethics and human well-being.

The Scientific Method: The Nuremberg Code has played an instrumental role in shaping the ethical dimensions of the scientific method, adding a layer of human-centered concern to a process primarily focused on empirical inquiry. While the scientific method emphasizes observation, hypothesis formulation, controlled experimentation, and data analysis, the Nuremberg Code introduces ethical principles that must be integrated when human subjects are involved at each stage. For example, before formulating a hypothesis, researchers must consider the ethical implications, such as potential harm or benefit to participants. This pre-emptive ethical consideration has become a standard part of research design, often requiring approval from ethics review boards that operate under guidelines influenced by the Nuremberg Code. The Code also stresses the importance of informed consent, ensuring that participants know what the experiment entails, thereby adding an ethical dimension to the observation and data collection stages. The focus on ethics has, in turn, influenced the peer-review process; research publications now often include a section detailing ethical considerations and approvals, and studies that do not meet ethical standards are generally not published. Thus, the Nuremberg Code has broadened the scope of the scientific method to include not just empirical rigor but also moral responsibility, fundamentally altering how scientific research is designed, conducted, and validated.

Medicine: The Nuremberg Code has made indelible contributions to medicine, particularly in defining and elevating the medical healthcare standards of care during medical research involving human subjects. Before its establishment, there were limited ethical guidelines governing such research, often leading to abuses and questionable practices. The Code set a new international precedent by emphasizing principles like informed consent, risk minimization, and the necessity of scientifically sound research. These principles have been integrated into peer-reviewed, evidence-based practices, becoming a cornerstone of modern medical ethics. The Code has been instrumental in shaping Institutional Review Boards (IRBs) and ethics committees, which evaluate the design and conduct of medical studies. These boards have become integral to the research landscape, ensuring that studies meet scientific and ethical criteria before initiating. This has led to a more humane and patient-centered approach in clinical trials and epidemiological studies. Moreover, in legal contexts, the Code's principles often serve as a yardstick against which the adequacy of medical care can be judged, particularly in cases that involve research on human subjects. By influencing practice and policy, the Nuremberg Code has thus been pivotal in harmonizing medical ethics with scientific ambition, ensuring that the quest for knowledge does not compromise human dignity and well-being.

Ethics: The Nuremberg Code has had a profound and lasting impact on the four fundamental principles of bioethics: patient autonomy (informed consent), practitioner beneficence (do good), practitioner nonmaleficence (do no harm), and public justice (be fair). The principle of autonomy is most directly addressed by the Code's emphasis on informed consent, requiring that subjects be fully aware of the risks and benefits of participating in research and that they participate voluntarily. This underpins the ethical and legal standards for patient engagement in modern medical practice. Beneficence, the ethical obligation to do good, is reflected in the Code's stipulation that medical experiments should aim for fruitful results for the well-being of society, ensuring that the research has a beneficent intent. Nonmaleficence, the directive to do no harm, is echoed in the Code's insistence that experiments should avoid all unnecessary physical and mental suffering. These two principles shape the risk-benefit assessments now standard in medical research ethics. Finally, the principle of justice, which demands fairness in distributing benefits and burdens, can be traced back to the Code's prohibition of experiments in which death or disabling injury is expected. This has paved the way for ethical guidelines that require the equitable selection of research subjects, ensuring that no group is unfairly burdened or excluded. Thus, the Nuremberg Code serves as a foundational text, the principles of which have been woven into contemporary bioethics.

Conclusion: The Nuremberg Code, formulated in 1947, has had a transformative impact on many fields — science, medicine, ethics, and law — setting forth a standard that places the dignity and well-being of the human subject at the center of research ethics. The Code was born from a harrowing historical context marked by gross ethical misconduct, but its principles have transcended that period to serve as universal ethical guidelines. The Code has entrenched itself as an indispensable framework for research involving human subjects, influencing international norms and legislation. It intersects with rationalist traditions by appealing to universal ethical principles while embodying empiricism by evolving based on real-world outcomes. In reshaping the ethical considerations intrinsic to the scientific method, the Nuremberg Code has compelled researchers to integrate humanistic values into empirical inquiry. Its principles have deeply influenced medicine, serving as a yardstick for ethical conduct and defining the gold standard of patient care. Most importantly, the Code has been a guiding light in elaborating bioethical principles — patient autonomy (informed consent), practitioner beneficence (do good), practitioner nonmaleficence (do no harm), and public justice (be fair) — that govern contemporary medical ethics. As a landmark document in human history, the Nuremberg Code remains an enduring legacy that helps ensure the ethical integrity of scientific and medical endeavors.

Nuremberg Code's Legacy: The Nuremberg Code stands as a foundational ethical framework that has revolutionized the conduct of human subject research, embedding the principles of human dignity and moral integrity into the scientific method and medical practice.

**

REVIEW QUESTIONS

True/False Questions:

1. The Nuremberg Code was established in response to ethical breaches in medical research during World War II.
   True or False?

2. The Nuremberg Code only applies to medical experiments conducted in clinical settings. True or False?

Multiple-Choice Questions:

3. Which principle of bioethics is most directly emphasized by the Nuremberg Code's requirement for voluntary consent?
   a) Beneficence
   b) Nonmaleficence
   c) Autonomy
   d) Justice

4. What was one of the primary reasons for the establishment of the Nuremberg Code?
   a) To regulate pharmaceutical companies
   b) To address the ethical breaches of medical experiments during World War II
   c) To improve the efficacy of medical treatments
   d) To promote international scientific collaboration

Clinical Vignette:

5. Dr. Lee is conducting a clinical trial and ensures that all participants fully understand the study's purpose, risks, and benefits before they agree to participate. Which principle from the Nuremberg Code is Dr. Lee primarily upholding?
   a) Beneficence
   b) Nonmaleficence
   c) Autonomy
   d) Justice

Basic Science Vignette:

6. A clinical researcher is designing a new study involving human participants. To comply with the principles of the Nuremberg Code, he ensures that participants are fully informed about the nature of the study and voluntarily agree to participate. What principle of the Nuremberg Code is primarily demonstrated by his actions?
   a) Risk minimization
   b) Informed consent
   c) Scientific validity
   d) Beneficence

Philosophy Vignette:

7. A medical researcher is designing a study on a new surgical technique. To comply with the Nuremberg Code's ethical guidelines, which principle should they prioritize when considering the selection of research subjects?
   a) Selecting subjects based on their availability and convenience
   b) Ensuring a fair distribution of risks and benefits in the selection process
   c) Choosing subjects who are likely to benefit the most from the new technique
   d) Recruiting subjects only from specific demographic groups

Correct Answers:

1. True
2. False
3. c) Autonomy
4. b) To address the ethical breaches of medical experiments during World War II
5. c) Autonomy
6. b) Informed consent
7. b) Ensuring a fair distribution of risks and benefits in the selection process

BEYOND THE CHAPTER
Nuremberg Code (1946)

- *The Nuremberg Code: Human Rights in Human Experimentation* by George J. Annas and Michael A. Grodin
- *Doctors from Hell: The Horrific Account of Nazi Experiments on Humans* by Vivien Spitz
- *The Nazi Doctors and the Nuremberg Code: Human Rights in Human Experimentation* edited by George J. Annas and Michael A. Grodin

***

Philosophy of Science

Medical Research

WEEK TEN: Standards in Practice

WEEK TEN

Class Hour 1

- Chapter 52: Declaration of Geneva — The Universal Guide: Medical Ethics
- Chapter 53: Declaration of Helsinki — The Ethical Compass: Biomedical Research
- Chapter 54: 45CFR46 — The *Common Rule*: Ethical Research

Class Hour 2

- Chapter 55: *Belmont Report* — The Ethical Blueprint: Human Research
- Chapter 56: IACUC — The Guardian of Ethics: Animal Research
- Chapter 57: 3Rs — Thc Ethical Triad: Humane Research

EXAM III: 50Q - online

Introduction

In the tenth week of "*Philosophy of Science*," we turn our attention to the critical standards that govern ethical practice in medicine and scientific research. This week's focus is on the documents and principles that set the benchmarks for responsible and humane conduct in these fields. Our journey through these standards begins with the Declaration of Geneva and traverses the ethical compasses provided by the Declaration of Helsinki, 45CFR46, the *Belmont Report*, the Institutional Animal Care and Use Committee (IACUC), and the 3Rs guiding principles.

The Declaration of Geneva serves as a universal guide to medical ethics, emphasizing respect and responsibility within the medical profession. Established after World War II, it underscores the sanctity of life and human dignity, fostering a standard of care and respect that binds physicians globally. This declaration's dynamic approach to ethics ensures that healthcare providers uphold the principles of patient autonomy, practitioner beneficence, nonmaleficence, and public justice, thereby fostering empathy and responsibility in medical practice.

Building on this foundation, the Declaration of Helsinki, created in 1964, provides an ethical compass for biomedical research involving human subjects. This declaration balances rational principles with empirical adaptability, influencing the scientific method and ensuring that research is conducted with informed consent, ethical review, and a thorough risk-benefit analysis. It safeguards the dignity, safety, and well-being of research participants, guiding medical science and human research with a respect for human life.

The regulatory framework of 45CFR46, known as the *Common Rule*, further ensures ethical integrity in human research. Instituted in 1974, it blends rationalist and empiricist foundations to protect human dignity in scientific studies. This framework shapes methodologically sound and ethically responsible research, emphasizing the four central bioethical principles and aligning scientific pursuits with ethical norms. The *Common Rule's* enduring legacy lies in its ability to harmonize the pursuit of knowledge with the respect for human dignity.

The *Belmont Report*, instituted in 1979, serves as an ethical blueprint for human research. Catalyzed by historical ethical transgressions, it outlines principles of respect for persons, beneficence, and justice. These principles have evolved into the four pillars of bioethics, reshaping the scientific method and emphasizing patient autonomy and ethical scrutiny. The *Belmont Report* turns abstract principles into concrete guidelines, fostering a humane and ethical approach to biomedical research and healthcare.

In parallel, the Institutional Animal Care and Use Committee (IACUC) ensures humane treatment in animal research. Emerging in response to heightened concerns for animal welfare, the IACUC integrates scientific rigor with ethical compassion. Guided by the 3Rs—Replacement, Reduction, and Refinement—the committee ensures that research complies with the highest ethical and scientific standards. This holistic approach to animal research combines empirical scrutiny with philosophical reflection, shaping the future of scientific progress with moral responsibility.

The 3Rs themselves—Replacement, Reduction, and Refinement—serve as an ethical triad guiding humane research. These principles encourage minimizing harm and optimizing benefits through alternative methods to animal testing, limiting animal usage, and enhancing animal welfare. Rooted in rationalist and empiricist traditions, the 3Rs harmonize empirical evidence with moral axioms, reshaping research to foster innovation, compassion, and ethical integrity. Their impact on bioethics principles underscores the alignment of scientific advancement with moral accountability and humanitarian stewardship.

Together, these chapters weave a narrative that underscores the profound connections between ethical standards, scientific research, and medical practice. They illustrate how foundational documents and guiding principles ensure that scientific and medical advancements are conducted with respect for human and animal dignity, fostering a culture of ethical responsibility and compassion in the pursuit of knowledge.

***

Chapter 52: Declaration of Geneva — The Universal Guide: Medical Ethics

The Declaration of Geneva sets global ethical standards, fostering respect and responsibility in the medical profession.

Abstract: The Declaration of Geneva, established by the World Medical Association in 1948, outlines the ethical obligations guiding physicians globally. Created after World War II to deter unethical practices like nonconsensual human experimentation, it sets universal mandates based on human dignity and the sanctity of life, fostering a standard of care and respect for all patients. Adapting with lessons from the medical community, it reflects a dynamic approach to ethics. While not directly shaping the scientific method, it influences ethical frameworks in medical research, safeguarding human rights and dignity. The Declaration advances medicine by fostering responsibility and empathy among healthcare providers, setting benchmarks for ethical conduct and quality care. It has influenced bioethical principles such as patient autonomy (informed consent), practitioner beneficence (do good), practitioner nonmaleficence (do no harm), and public justice (be fair). As a globally recognized guide, the Declaration unifies healthcare practitioners around shared principles, fostering a commitment to human dignity and well-being in the medical profession.

**

Introduction: The Declaration of Geneva, adopted by the World Medical Association (WMA) in September 1948, serves as a contemporary restatement of the Hippocratic Oath, outlining the ethical obligations and principles that guide physicians worldwide. Created in the aftermath of World War II, the declaration was designed to address and prevent unethical medical practices, including human experimentation without consent. Over the years, it has undergone several revisions to remain up-to-date with evolving medical, ethical, and social landscapes. The Declaration sets a global standard for medical ethics, acting as a moral compass for healthcare providers in their clinical practice, research, and relationships with patients. By examining the Declaration's rationalist foundations, its empiricist orientation, its role in shaping the scientific method, its contributions to the advancement of medicine, and its influence on the four foundational principles of bioethics — patient autonomy (informed consent), practitioner beneficence (do good), practitioner nonmaleficence (do no harm), and public justice (be fair) — we can gain a comprehensive understanding of its enduring impact on medical ethics and practice.

Rationalism: The Declaration of Geneva, emphasizing universally applicable ethical principles, has strong rationalist underpinnings. Rationalism posits that certain fundamental truths or principles can be discerned through reason, independent of sensory experience. In the case of medical ethics, the Declaration embodies these rationalist ideals by setting forth axiomatic principles that are considered self-evident and universally binding on physicians, regardless of cultural or social context. These principles are deduced from basic postulates about human dignity, the sanctity of life, and the duties and responsibilities of those entrusted with the healthcare of others. For example, commitments such as "*I will maintain the utmost respect for human life*" or "*I will not permit considerations of age, disease or disability, creed, ethnic origin, gender, nationality, political affiliation, race, sexual orientation, social standing, or any other factor to intervene between my duty and my patient*" arise from "*a* *priori*" notions about the inherent worth and equality of all human beings. Such statements serve as universal ethical mandates guiding physicians in their decision-making, irrespective of the particulars of any situation or empirical circumstances. In this way, the Declaration of Geneva can be seen as applying rationalist philosophy to medical ethics.

Empiricism: While the Declaration of Geneva is largely grounded in rationalist principles, it also incorporates an empiricist orientation, reflecting the experiential lessons of medical history. Empiricism focuses on knowledge primarily derived from sensory experience, observational data, and environmental interaction. The formulation and subsequent revisions of the Declaration were informed by real-world events and practical experience, most notably the ethical violations and atrocities committed during World War II. These experiences served as specific instances from which general principles were inductively derived. In essence, the empirical evidence of what could go wrong without ethical guidelines prompted the creation of universal standards. Furthermore, the Declaration's amendments over the years represent an ongoing responsiveness to empirical data and societal shifts. For example, changes in our understanding of patient autonomy, informed by empirical observations in healthcare settings, have led to revisions in the Declaration to reflect this more nuanced understanding. Similarly, the rise of medical technologies and treatment modalities, each with its own empirically observed risks and benefits, has influenced the evolving text. Therefore, although the Declaration of Geneva may be rooted in "*a priori*" ethical principles, its orientation and revisions reflect an empirical approach considering the lessons learned from real-world medical practice and ethical dilemmas.

The Scientific Method: The Declaration of Geneva doesn't directly shape the scientific method, but it does have an indirect influence on the ethical framework within which scientific research, particularly medical research, is conducted. The scientific method relies on empirical observation and testing to arrive at verifiable conclusions. However, the ethical parameters for these activities, especially those involving human subjects, must be clearly defined to prevent abuse or misconduct. The Declaration of Geneva, focusing on principles like the sanctity of human life, autonomy, and beneficence, provides a foundational ethical guideline that researchers can refer to when designing and conducting experiments. This ethical overlay ensures that pursuing knowledge through the scientific method does not override basic human rights and dignity. By doing so, the Declaration contributes to the credibility and acceptability of scientific findings. Ethical lapses in medical research can harm participants and discredit the scientific endeavor, reducing public trust and impeding progress. Thus, while the Declaration of Geneva may not directly address the scientific method's techniques, its ethical principles serve as crucial guardrails that help maintain the integrity and societal trust in medical research, indirectly shaping how the scientific method is applied in the context of healthcare.

Medicine: The Declaration of Geneva has had a profound and enduring impact on medicine, serving as a cornerstone for the ethical standards and norms that guide physicians and healthcare professionals globally. Although not a legal document, the Declaration provides a universal ethical framework that supplements and often informs local and national healthcare policies and standards of care. By articulating principles such as respect for human life, patient confidentiality, and the primacy of patient welfare, the Declaration has helped shape the expected level and type of care that medical professionals should provide. Its principles are harmonious with the evidence-based, peer-reviewed standards of care that evolve in response to advances in medical science and practice. For healthcare providers, the Declaration serves as a moral and ethical guide, complementing the technical and scientific standards of care. In legal contexts, while it may not be directly enforceable, the Declaration's principles often echo the standards against which the adequacy of care is judged. Therefore, the Declaration of Geneva sets ethical benchmarks and indirectly influences the quality of medical care by promoting a culture of responsibility, empathy, and moral integrity among healthcare professionals. In this way, it contributes to improving healthcare delivery and patient outcomes.

Ethics: The Declaration of Geneva has been instrumental in shaping the four foundational principles of bioethics: patient autonomy (informed consent), practitioner beneficence (do good), practitioner nonmaleficence (do no harm), and public justice (be fair). In terms of autonomy, the Declaration underscores the necessity of respecting patients as individuals with their values and rights, thereby paving the way for the concept of informed consent. This principle insists that patients have the right to know the details of their medical treatment, including risks and alternatives, and make decisions accordingly. Beneficence and nonmaleficence — doing good and doing no harm — are deeply embedded in the Declaration's commitment to the well-being and dignity of the patient. Phrases like "*the health and well-being of my patient will be my first consideration*" echo these ethical imperatives as a reminder that the primary responsibility of healthcare providers is to act in the patient's best interests. The principle of justice, which emphasizes fairness and equitable treatment, is embodied in the Declaration's insistence that physicians must not discriminate based on age, gender, ethnicity, or social standing. By laying out these principles, the Declaration of Geneva serves as an ethical roadmap for medical professionals, influencing medical practice and broader debates and policies that shape healthcare ethics globally. It acts as a universal ethical guideline that harmonizes diverse perspectives and practices, setting a global standard for ethical decision-making in healthcare.

Conclusion: The Declaration of Geneva has had a far-reaching impact on the medical profession, setting the ethical and moral framework within which physicians operate. Established by the World Medical Association in 1948, this seminal document has continually evolved to reflect changing societal norms and advancements in medical science. By emphasizing ethical principles rooted in rationalist philosophy and informed by empirical evidence, the Declaration serves as an invaluable guide to medical practitioners. It contributes to the scientific method by establishing ethical guidelines that uphold human dignity and protect against abuse in medical research. It complements and enhances medical standards of care, integrating moral considerations into healthcare delivery. Moreover, the Declaration has been instrumental in shaping bioethical principles, providing a comprehensive ethical paradigm encompassing patient autonomy (informed consent), practitioner beneficence (do good), practitioner nonmaleficence (do no harm), and public justice (be fair). These guidelines guide clinicians' everyday interactions with patients and inform broader healthcare policy and legal frameworks. Thus, the Declaration of Geneva is a cornerstone in medical ethics, unifying a global community of healthcare providers around shared principles that transcend cultural and national boundaries. It has become a global constitution for physicians, setting standards for ethical behavior and ensuring that the physician's primary commitment is always to the patient's well-being.

The Declaration of Geneva's Legacy: The Declaration serves as a universal ethical compass for physicians, harmonizing diverse medical practices around globally recognized principles of patient care and human dignity, thereby setting enduring moral and ethical standards in medical science and healthcare.

**

REVIEW QUESTIONS

True/False Questions:

1. The Declaration of Geneva was established by the World Medical Association in 1948 as a response to unethical medical practices during World War II.
   True or False?

2. The Declaration of Geneva is legally enforceable in all countries where it is adopted. True or False?

Multiple-Choice Questions:

3. A hospital ethics committee is revising its guidelines to align with the Declaration of Geneva. To uphold the principle of public justice, which policy should they prioritize?
a) Ensuring all medical staff receive regular ethics training
b) Providing equal access to medical treatments for all patients regardless of their background
c) Increasing the hospital's research funding for innovative treatments
d) Implementing the latest medical technologies across all departments

4. What is one of the primary roles of the Declaration of Geneva in the medical profession?
a) To provide legal guidelines for medical malpractice
b) To offer a universal ethical framework for physicians
c) To standardize medical treatments globally
d) To regulate pharmaceutical companies

Clinical Vignette:

5. Dr. Smith, a physician, ensures that his treatment plans are always in the best interest of his patients, avoiding any potential harm while promoting their well-being. Which principles from the Declaration of Geneva is Dr. Smith upholding?
a) Autonomy and Justice
b) Beneficence and Nonmaleficence
c) Autonomy and Nonmaleficence
d) Beneficence and Justice

Basic Science Vignette:

6. Dr. Taylor is about to start a clinical trial for a new medication and wants to ensure the study complies with the Declaration of Geneva's ethical guidelines. Which of the following actions best reflects the principle of patient autonomy?
a) Collecting detailed health histories of all participants
b) Ensuring participants fully understand the study and give informed consent
c) Using state-of-the-art technology for accurate data collection
d) Recruiting a diverse group of participants for the trial

Philosophy Vignette:

7. A philosopher is analyzing the rationalist foundations of the Declaration of Geneva, which states that physicians must not allow any considerations to interfere with their duty to patients. Which principle of the Declaration best aligns with this rationalist perspective?
a) Patient autonomy
b) Nonmaleficence
c) Public justice
d) Beneficence

Correct Answers:

1. True
2. False
3. b) Providing equal access to medical treatments for all patients regardless of their background
4. b) To offer a universal ethical framework for physicians
5. b) Beneficence and Nonmaleficence
6. b) Ensuring participants fully understand the study and give informed consent
7. c) Public justice

BEYOND THE CHAPTER
Declaration of Geneva (1948)

- *The World Medical Association, The Declaration of Geneva, and the Physician's Oath* by John P. Howe III
- *The Declaration of Geneva: A Modern-Day Hippocratic Oath?* by R. R. Bhugra
- *The Declaration of Geneva: A Revision and Reaffirmation* by John P. Howe III

***

Chapter 53: Declaration of Helsinki — The Ethical Compass: Biomedical Research

The Declaration of Helsinki guides ethical biomedical research, balancing rational principles and empirical adaptability.

Abstract: The Declaration of Helsinki, created in 1964 by the World Medical Association, is a cornerstone in establishing ethical guidelines for biomedical research involving human subjects. Grounded in rationalist principles of universal moral norms and innate human faculties for ethical decision-making, it also adapts to new knowledge and changing circumstances. By balancing rationalism and empiricism, the Declaration has influenced the scientific method, setting ethical prerequisites for hypothesis formation, experiment design, and data collection, thereby enhancing research credibility and humanitarian impact. It has profoundly impacted global healthcare practices, emphasizing informed consent, ethical review, and risk-benefit analysis. As a guide in bioethics, it embodies principles such as patient autonomy, practitioner beneficence, practitioner nonmaleficence, and public justice, safeguarding research participants' dignity, safety, and well-being. The Declaration of Helsinki remains an adaptive, foundational ethical compass, guiding medical science and human research while fostering respect for human life.

**

Introduction: The Declaration of Helsinki, adopted in 1964 by the World Medical Association, is a foundational ethical guideline for medical researchers and healthcare professionals involved in clinical trials and biomedical research with human subjects. Created in response to ethical lapses in medical research, such as the notorious Tuskegee Syphilis Study and the revelations of the Nuremberg Trials, the document aims to safeguard research participants' dignity, safety, and well-being. Emphasizing principles like informed consent, scientific integrity, and the necessity for ethical review boards, the Declaration has undergone multiple revisions to adapt to the evolving landscapes of science and medical ethics. Its global influence is pervasive, shaping research protocols, legislative frameworks, and medical healthcare standards of care. Through this document, the medical community seeks to harmonize the often conflicting demands of advancing scientific knowledge and maintaining inviolable respect for human life.

Rationalism: The rationalist underpinnings of the Declaration of Helsinki can be traced back to its foundational attempt to systematize ethical considerations in medical research through principles that are deemed universal. The Declaration assumes that ethical behavior in medical research can be guided by certain "*a priori*" principles — axiomatic truths that don't need empirical verification but are accepted as inherently valid. For example, the "*informed consent*" principle assumes that individuals have an innate capacity for autonomy and rational decision-making, implying that they have the right and ability to make decisions about their own bodies and personal information. Similarly, the notion that research must have a sufficient scientific basis draws from the rationalist ideal that scientific knowledge can be advanced through deductive reasoning, built on solid theoretical foundations. The document's guidelines aim to be logically coherent and universally applicable, much like the axioms in a rationalist framework. By laying down these principles, the Declaration attempts to leverage innate human faculties for ethical decision-making and rational problem-solving. It presupposes that certain moral norms are universal and can be understood through reason, thereby serving as a stable framework to guide the complex, ever-evolving field of medical research.

Empiricism: While the Declaration of Helsinki is steeped in principles that adhere to rationalist ideals, it also has a significant empiricist orientation. This is most evident in its flexibility and adaptability to new data and changing circumstances. Since its inception, the Declaration has undergone multiple revisions, each informed by real-world experiences, ethical dilemmas, and advances in medical technology. The guidelines are not fixed dogma but are meant to evolve as more is learned about the implications and outcomes of medical research on human subjects. The principle of "*risk-benefit analysis*," for example, implicitly relies on empirical data — researchers must collect evidence about potential risks and benefits before commencing a study. Likewise, the mandate for ethical review boards involves continuous empirical evaluation of research protocols to ensure they adhere to ethical norms. These layers of oversight and ongoing evaluation represent inductive reasoning in action: specific observations about individual research projects contribute to general conclusions about ethical best practices. Therefore, the Declaration embodies an empiricist approach by allowing for the revision and refinement of its guidelines in response to accumulating empirical evidence, aiming to safeguard human dignity and well-being effectively.

The Scientific Method: The Declaration of Helsinki plays a pivotal role in shaping the scientific method, particularly in biomedical research involving human subjects. While it doesn't lay out the scientific method per se, it establishes a rigorous ethical framework that researchers must adhere to, thereby influencing how hypotheses are formed, experiments are designed, and data are collected and analyzed. For instance, the Declaration's insistence on prior ethical review ensures that research questions and hypotheses meet ethical standards before empirical testing. It also emphasizes the necessity of informed consent, which impacts how subjects are selected and data are collected, safeguarding against potential biases and ethical lapses. The Declaration's stipulation that research must contribute to the advancement of scientific knowledge ensures that research projects are meaningful, not trivial or redundant, thereby enhancing the quality of hypotheses and the ensuing analysis. Furthermore, its call for public dissemination of research findings aligns with the scientific method's emphasis on peer review and replicability. By setting these ethical guidelines, the Declaration of Helsinki serves as a crucial adjunct to the scientific method, reinforcing its inherent rigor and ensuring its ethical soundness, thereby fortifying scientific investigations' credibility and humanitarian impact.

Medicine: The Declaration of Helsinki has had a profound and lasting impact on medicine, particularly in establishing ethical standards for research that have trickled down to influence healthcare practices globally. Serving as a cornerstone for ethical research, it has informed medical healthcare standards of care, particularly in how patients engage in research and treatment decisions. Its principles of informed consent, ethical review, and risk-benefit analysis have been integrated into broader medical protocols, informing evidence-based practices and improving the standard of care. The Declaration serves as a moral yardstick against which medical research is measured, shaping institutional review boards, ethics committees, and research protocols. By insisting on rigorous ethical scrutiny, it also indirectly upholds the scientific integrity of medical research, ensuring that the evidence generated is scientifically robust and ethically sound. Consequently, the treatments and interventions that eventually become medical healthcare standards of care are more likely to be effective and ethical. As medical healthcare standards of care continually evolve to incorporate new research findings, the Declaration's ethical principles are a constant guide, aligning the rapid advancements in medical science with the consistent ethical considerations essential for patient well-being. This integration of ethics into the core of medical research and practice has had a sweeping, global impact, setting the stage for a more humane, ethical, and effective medical landscape.

Ethics: The Declaration of Helsinki is a seminal text in bioethics, directly influencing the core ethical principles of patient autonomy (informed consent), practitioner beneficence (do good), practitioner nonmaleficence (do no harm), and public justice (be fair). It enshrines the principle of autonomy through its strong emphasis on informed consent, insisting that all research subjects must be fully aware of the risks and benefits involved and voluntarily agree to participate. This cornerstone principle ensures respect for individual agency and personal freedom, embodying the bioethical commitment to autonomy. The principle of beneficence is echoed in the Declaration's requirement that medical research have the potential for beneficial outcomes for the research subjects themselves or the broader community, underscoring the ethical obligation to contribute positively to human welfare. Nonmaleficence, or the principle to "*do no harm*," is deeply embedded in the document's insistence on ethical review and risk-benefit analysis before undertaking any research. Compelling researchers to rigorously weigh potential harms against benefits safeguards against unnecessary suffering or exploitation. Finally, the principle of justice is reflected in its call for equitable selection of research subjects, prohibiting the targeting of vulnerable populations for risky experiments. By codifying these principles, the Declaration of Helsinki has profoundly influenced the ethical landscape of biomedical research, serving as a universal guideline for maintaining moral integrity in pursuing medical advancement.

Conclusion: The Declaration of Helsinki, first established in 1964, has reverberated throughout the global medical and research communities, establishing itself as the gold standard for ethical considerations in biomedical research involving human subjects. It straddles both rationalist and empiricist philosophies, seeking to combine inherent ethical principles with evolving empirical data. Doing so continually shapes the scientific method, adding ethical rigor to hypothesis formation, experiment design, data collection, and peer review. Importantly, the Declaration has sweepingly impacted medical practice, elevating healthcare standards through key principles like informed consent, ethical review, and risk-benefit analysis. These principles have shaped medical ethics and indirectly influenced medical research's scientific quality, driving ethical and methodological advancements. Furthermore, the Declaration has been seminal in outlining bioethical principles — patient autonomy (informed consent), practitioner beneficence (do good), practitioner nonmaleficence (do no harm), and public justice (be fair) — that serve as the moral cornerstones for any research involving human subjects. Its influence has transcended academia and clinical settings, making its way into legislative frameworks and public consciousness, thereby solidifying its role as a foundational, adaptive, and enduring ethical guide in the ever-evolving fields of medicine and human research.

The Declaration of Helsinki's Legacy: Serving as the ethical compass for biomedical research, the Declaration has fundamentally shaped the integration of ethical considerations into the scientific method and medical practice, thereby setting a global standard for the moral treatment of human subjects in research.

**

REVIEW QUESTIONS

True/False Questions:

1. The Declaration of Helsinki, established in 1964, is a foundational document for ethical guidelines in biomedical research involving human subjects.
   True or False?

2. The Declaration of Helsinki primarily emphasizes legal standards over ethical considerations in medical research.
True or False?

Multiple-Choice Questions:

3. Which principle of bioethics is most strongly emphasized by the Declaration of Helsinki through its focus on informed consent?
a) Beneficence
b) Nonmaleficence
c) Autonomy
d) Justice

4. How has the Declaration of Helsinki influenced the scientific method?
a) By establishing experimental techniques
b) By setting ethical guidelines that affect hypothesis formation and data collection
c) By standardizing statistical analysis methods
d) By regulating the funding of medical research

Clinical Vignette:

5. Dr. Lee is conducting a clinical trial and ensures that all participants are fully aware of the study's risks and benefits before obtaining their consent to participate. Which principle from the Declaration of Helsinki is Dr. Lee upholding?
a) Beneficence
b) Autonomy
c) Nonmaleficence
d) Justice

Basic Science Vignette:

6. A researcher is designing a clinical trial and follows the Declaration of Helsinki's guidelines, ensuring that participants provide informed consent, and the study is approved by an ethics committee. Which principle of the Declaration is primarily demonstrated by obtaining informed consent?
a) Confidentiality
b) Autonomy
c) Beneficence
d) Nonmaleficence

Philosophy Vignette:

7. A philosopher analyzes the Declaration of Helsinki, focusing on its rationalist foundation that requires researchers to prioritize human dignity and ethical integrity. Which principle of the Declaration best aligns with the rationalist perspective that human subjects should be treated fairly?
a) Scientific validity
b) Informed consent
c) Risk-benefit analysis
d) Public justice

Correct Answers:

1. True
2. False
3. c) Autonomy
4. b) By setting ethical guidelines that affect hypothesis formation and data collection
5. b) Autonomy
6. b) Autonomy
7. d) Public justice

BEYOND THE CHAPTER

Declaration of Helsinki (1964)

- *The Declaration of Helsinki: The Cornerstone of Research Ethics* by R. R. Bhugra
- *The Declaration of Helsinki and Public Health* by S. O. S. Agre and others
- *The Declaration of Helsinki: Past, Present and Future* by Urban Wiesing

***

Philosophy of Science

Medical Research

Chapter 54: 45CFR46 — The *Common Rule*: Ethical Research

45CFR46 ensures ethical integrity in human research, blending rational and empirical foundations to protect human dignity.

Abstract: This study examines the multi-faceted role of 45CFR46, or the "*Common Rule*," a regulatory framework instituted in 1974 to govern ethical conduct in research involving human subjects. It delineates the rationalist and empiricist foundations of the *Common Rule*, illustrating its role in shaping methodologically sound and ethically responsible research. The analysis explores its influence on the scientific method, fostering research that upholds human dignity and integrity while facilitating innovation. In medicine, the *Common Rule* ensures clinical trials adhere to stringent ethical standards, influencing best practices and healthcare standards. It operationalizes the four central bioethical principles: patient autonomy (informed consent), practitioner beneficence (do good), practitioner nonmaleficence (do no harm), and public justice (be fair), aligning scientific pursuits with ethical norms. The study highlights 45CFR46's enduring legacy as an ethical guide that blends deductive reasoning with empirical evidence, reinforcing the harmony between knowledge pursuit and human dignity.

**

Introduction: 45CFR46 is a crucial federal policy in the United States that establishes the ethical framework and guidelines for researchers engaged in human subject research. Originally put into effect in 1974 and frequently updated, this Code of Federal Regulations Title 45, Part 46 — better known as the "*Common Rule*" — aims to protect individuals who volunteer for research studies by mandating informed consent, ethical review by Institutional Review Boards (IRBs), and continuous oversight during research. The *Common Rule's* significance extends beyond regulatory compliance; it shapes the ethical and methodological foundation of research involving human subjects. In response to a history marred by unethical research practices, like the Tuskegee Syphilis Study, 45CFR46 serves as a moral and legal cornerstone that holds researchers accountable while fostering scientific innovation. As we delve deeper into its rationalist foundations, empiricist orientation, role in shaping the scientific method, contributions to medicine, and influence on the four core ethical principles of bioethics — patient autonomy (informed consent), practitioner beneficence (do good), practitioner nonmaleficence (do no harm), and public justice (be fair) — we will appreciate the comprehensive role it plays in intertwining ethics with scientific research.

Rationalism: The rationalist foundations of 45CFR46 can be observed in its structured, deductive approach to establishing ethical norms and procedures for research involving human subjects. The *Common Rule* is undergirded by principles — respect for persons, beneficence, and justice — that serve as axiomatic starting points for ethical deliberation. These principles are akin to the "*a priori*" tenets in rationalism, providing a framework for deductive reasoning in the ethical conduct of research. For instance, the necessity for informed consent is derived logically from the axiom of respect for persons, which posits that individuals have an innate right to autonomy and self-determination. Similarly, the requirement for Institutional Review Boards (IRBs) to evaluate the risks and benefits of research proposals stems from the axiomatic principle of beneficence, which urges doing good and avoiding harm. When applied through a systematic, logical framework, these principles lead to ethical guidelines universally applicable to varied research settings. In this way, 45CFR46 embodies a rationalist approach by using deductive reasoning to build from foundational ethical principles to specific guidelines and rules. This structured approach reflects an understanding that certain innate faculties of reason and moral intuition serve as hardwired mechanisms for defining what is morally acceptable in the realm of human subject research.

Empiricism: The empiricist orientation of 45CFR46 is evident in its commitment to adapting ethical guidelines based on observed outcomes, empirical evidence, and societal changes. While the *Common Rule* is rooted in foundational ethical principles, it also recognizes that ethical guidelines must evolve based on real-world data, experiences, and shifting cultural norms. Institutional Review Boards (IRBs), for instance, operate under the empiricist methodology by gathering specific data about each research proposal, evaluating the risks and benefits based on empirical evidence, and making general ethical conclusions for the proposed research. The very existence and periodic updating of 45CFR46 itself serve as an example of inductive reasoning; specific historical instances of ethical misconduct in human subject research, such as the Tuskegee Syphilis Study, led to the general conclusion that comprehensive federal guidelines were necessary to protect individual rights and welfare. The policy also responds to new technologies and methods under scrutiny, adjusting its guidelines to maintain ethical integrity amid scientific advancements. Additionally, 45CFR46 considers social variables like cultural differences and individual biases, acknowledging that external factors can influence ethical decision-making. Using empirical evidence and feedback loops, 45CFR46 embodies the empiricist approach, demonstrating the necessity for adaptable, evidence-based human-subject research regulations.

The Scientific Method: 45CFR46 plays a significant role in shaping the scientific method by establishing ethical guidelines that ensure the integrity and credibility of human subject research. While the scientific method prioritizes empirical observation, hypothesis formulation, and data-driven conclusions, 45CFR46 adds a layer of ethical scrutiny to this process. This scrutiny begins at the stage of hypothesis formulation, as researchers must first gain the approval of Institutional Review Boards (IRBs) before initiating any study involving human subjects. The IRBs assess the scientific validity of the proposed research and its ethical implications, ensuring that the study's methodology adheres to principles such as informed consent, minimization of risk, and equitable selection of subjects. This ethical review effectively becomes a precondition for the subsequent stages of experimentation, data collection, and analysis. By setting these standards, 45CFR46 ensures that the scientific method is conducted within an ethical framework, enhancing credibility and societal acceptance of research findings. Moreover, the ethical guidelines prescribed by 45CFR46 indirectly foster replicability, as researchers who follow these guidelines are more likely to produce methodologically sound studies that others can repeat. In essence, 45CFR46 serves as a cornerstone that integrates ethics into the scientific method, safeguarding human subjects' dignity and the integrity of scientific inquiry.

Medicine: 45CFR46 has profoundly impacted medicine by setting ethical standards that guide clinical research, thus influencing medical best practices and healthcare standards of care. By requiring the rigorous ethical review of human subject research, the *Common Rule* ensures that clinical trials are conducted under stringent ethical conditions, thereby enhancing the reliability of the resulting data. This is critical for medicine, as clinical trials are the foundation for evidence-based practices. Without ethical guidelines, the validity of research could be compromised, potentially leading to flawed medical protocols and jeopardizing patient safety. By embedding ethical oversight into the research process, 45CFR46 indirectly helps establish peer-reviewed, evidence-based practices that inform healthcare standards. The guidelines also promote patient autonomy through informed consent, advocating for patient-centered care within a research context. Moreover, the ethical framework ensures that research does not disproportionately burden or benefit specific populations, reflecting principles of justice and beneficence that are integral to healthcare standards of care. Therefore, 45CFR46 not only shapes medical research but also has a cascading influence on the broader healthcare landscape. Its ethical stipulations serve as a template for adequate and ethical care, forming an integral part of the legal and professional frameworks that define medical standards.

Ethics: 45CFR46 plays a pivotal role in operationalizing and reinforcing the four foundational bioethical principles of patient autonomy (informed consent), practitioner beneficence (do good), practitioner nonmaleficence (do no harm), and public justice (be fair) within research involving human subjects. Starting with autonomy, the *Common Rule* mandates informed consent, ensuring that participants have complete awareness of the risks, benefits, and objectives of a research study before willingly participating. This epitomizes the principle of autonomy by empowering individuals to make informed decisions about their involvement in research. The principle of beneficence is manifested through stringent ethical review processes conducted by Institutional Review Boards (IRBs). These boards weigh the potential benefits against the risks of a proposed study, thereby ensuring that the research aims to do good and contribute positively to society or scientific knowledge. Nonmaleficence, the principle of doing no harm, is also upheld through IRB reviews, requiring researchers to minimize participants' risks to the greatest extent possible. Lastly, justice is highlighted by guidelines that ensure a fair selection of research subjects, preventing the exploitation of vulnerable populations while ensuring that the burdens and benefits of research are equitably distributed. By embedding these principles into its guidelines, 45CFR46 serves as a concrete framework for ethical conduct in research, thereby shaping the science of human subject research and its ethics.

Conclusion: 45CFR46, colloquially known as the "*Common Rule*," has emerged as an indispensable framework in shaping the ethical landscape of human subject research, not just within the United States but globally. Its principles influence diverse realms, from the rationalist foundations that provide moral axioms to the empiricist orientation that accommodates societal changes and technological advancements. Acting as a regulatory scaffolding ensures that research involving human subjects aligns with the scientific method, offering credibility and ethical soundness to experimental findings. Its guidelines also profoundly impact medicine, contributing to the ethical rigor of clinical trials and affecting healthcare standards of care. Furthermore, the *Common Rule* operationalizes the four cornerstone principles of bioethics — patient autonomy (informed consent), practitioner beneficence (do good), practitioner nonmaleficence (do no harm), and public justice (be fair) — into actionable protocols, harmonizing ethical ideals with practical research. Thus, 45CFR46 is far more than a regulatory checklist; it is an ethical manifesto that blends deductive reasoning and empirical evidence to form a comprehensive guide for conducting scientifically rigorous and morally sound research. This intricate blend makes it a linchpin in the ongoing dialogue between ethics and science, reinforcing that pursuing knowledge should never compromise human dignity.

45CFR46's Legacy: As a pioneering ethical framework, 45CFR46 stands as a monumental legacy in integrating ethics with scientific rigor, rigorously enforcing the bioethical principles of patient autonomy (informed consent), practitioner beneficence (do good), practitioner nonmaleficence (do no harm), and public justice (be fair) to ensure that the pursuit of scientific knowledge harmoniously coexists with human dignity and ethical integrity in research involving human subjects.

**

REVIEW QUESTIONS

True/False Questions:

1. The *Common Rule*, or 45CFR46, was established in 1974 to provide ethical guidelines for research involving human subjects.
   True or False?

2. The primary purpose of 45CFR46 is to streamline the process of human subject research by reducing ethical oversight.
   True or False?

Multiple-Choice Questions:

3. Which of the following principles is most directly addressed by the *Common Rule* through its mandate for informed consent?
   a) Beneficence
   b) Nonmaleficence
   c) Autonomy
   d) Justice

4. How does 45CFR46 influence the scientific method in research involving human subjects?
   a) By establishing experimental techniques
   b) By setting ethical guidelines that must be adhered to during hypothesis formation and data collection
   c) By standardizing statistical analysis methods
   d) By regulating the funding of medical research

Clinical Vignette:

5. Dr. Smith is conducting a clinical trial to test a new medication. The Institutional Review Board (IRB) reviews his proposal and finds that while the potential benefits of the medication are promising, the study design includes a high risk of harm to participants. Which principle of bioethics is the IRB primarily concerned with in this situation?
   a) Autonomy
   b) Beneficence
   c) Nonmaleficence
   d) Justice

Basic Science Vignette:

6. A research coordinator is preparing a study under the guidelines of 45CFR46. She ensures that participants are fully informed about the study's procedures, risks, and benefits before obtaining their voluntary participation. What principle of 45CFR46 is primarily demonstrated by this action?
   a) Risk minimization
   b) Informed consent
   c) Equitable selection
   d) Scientific validity

Philosophy Vignette:

7. In a philosophical discussion, Dr. Thompson argues that the principles underpinning 45CFR46 demonstrate a rationalist approach to ethical research. She supports her argument by pointing out that the necessity for informed consent is derived logically from the respect for persons, which posits that individuals have an innate right to autonomy. How does this illustrate the rationalist foundation of 45CFR46?
   a) It shows how empirical evidence is used to adapt ethical guidelines
   b) It reflects the use of inductive reasoning based on historical ethical misconduct
   c) It demonstrates the application of *a priori* principles to derive ethical norms
   d) It highlights the consideration of social variables in ethical decision-making

Correct Answers:

1. True
2. False
3. c) Autonomy
4. b) By setting ethical guidelines that must be adhered to during hypothesis formation and data collection
5. c) Nonmaleficence
6. b) Informed consent
7. c) It demonstrates the application of *a priori* principles to derive ethical norms

BEYOND THE CHAPTER
45CFR46 (1974)

- *The Common Rule, Updated* by Holly Fernandez Lynch, I. Glenn Cohen, and Barbara E. Bierer
- *The Revised Common Rule: Implications for Schools of Social Work* by Sarah Butts and others
- *The Common Rule and Continuous Improvement in Health Care: A Learning Health System Perspective* by Mary Ann Baily and Melissa Bottrell

***

Chapter 55: Belmont Report — The Ethical Blueprint: Human Research

The Belmont Report shapes ethical research with principles of respect, beneficence, and justice, fostering humane scientific inquiry.

Abstract: The *Belmont Report*, instituted in 1979, is a pivotal bioethics document catalyzed by historical ethical transgressions in scientific studies. It outlines foundational ethical principles — respect for persons, beneficence, and justice, which evolved into the four bioethics pillars: patient autonomy (informed consent), practitioner beneficence (do good), practitioner nonmaleficence (do no harm), and public justice (be fair). By blending rationalist and empiricist approaches, the report reshaped the scientific method, particularly in medicine and psychology. It set stringent ethical norms grounded in rational principles and empirical observations, adaptable to emerging data and societal shifts. Its principles are integrated into the scientific method, fostering a humanistic and ethical approach to research. The *Belmont Report* refined medical practices by emphasizing patient autonomy and ensuring ethical scrutiny through Institutional Review Boards. It turned abstract principles into concrete guidelines, shaping the moral backbone of biomedical research and healthcare, and remains a robust ethical compass, upholding human dignity, rights, and well-being.

**

Introduction: The *Belmont Report*, published in 1979 by the *National Commission for the Protection of Human Subjects of Biomedical and Behavioral Research* in the United States, is a seminal document in bioethics and human subjects research. It was created in response to ethical breaches in scientific studies, such as the Tuskegee Syphilis Study. The report aimed to establish a comprehensive ethical framework to guide researchers in treating participants with dignity and respect. This groundbreaking text introduced the basic ethical principles governing research involving human subjects — respect for persons, beneficence, and justice. Over the years, these principles have evolved into the four foundational pillars of bioethics: patient autonomy (informed consent), practitioner beneficence (do good), practitioner nonmaleficence (do no harm), and public justice (be fair). The *Belmont Report* has been instrumental in shaping the scientific method, particularly in medicine and psychology, through its rationalist foundations and empiricist orientation. Its guidelines and recommendations have become the gold standard for ethical conduct in research, influencing legislation, academic protocols, and the broader discourse on ethics in scientific investigation.

Rationalism: The rationalist foundations of the *Belmont Report* can be traced to its emphasis on applying "*a priori*" principles and deductive reasoning to shape ethical guidelines for human subjects research. The report's ethical framework rests on universal, axiomatic principles such as respect for persons, beneficence, and justice — principles that, in a rationalist sense, could be viewed as evolved cognitive faculties geared toward fostering ethical behavior. By stipulating that all human subjects have intrinsic value and deserve to be treated with dignity and respect, the report appeals to a form of moral reasoning that may be evolutionarily hardwired into our cognitive apparatus to improve societal cohesion and individual well-being. Moreover, the *Belmont Report* holds researchers to rigorous ethical scrutiny, effectively recognizing the potential limitations and ethical implications of their cognitive faculties. In doing so, the report serves as a regulatory guide, aiming to foster an environment that understands and mitigates the downsides of evolved rational faculties, especially in research ethics. This rationalist approach to ethics complements the empirical methods commonly employed in scientific research, offering a balanced framework that respects the necessity for empirical inquiry and the inherent dignity of research participants.

Empiricism: The *Belmont Report* embodies an empiricist orientation by leveraging inductive reasoning and empirical evidence to establish ethical guidelines for human subjects research. Rooted in the lessons learned from egregious ethical lapses, such as the Tuskegee Syphilis Study, the report represents a comprehensive synthesis of specific instances of ethical misconduct, leading to general principles to prevent future maleficience. This empirical foundation calls for ongoing scrutiny, modification, and validation through sensory experiences, namely observations of how research affects human subjects in real-world contexts. This approach acknowledges that cognitive biases, shaped by evolutionary pressures, societal norms, and individual experiences, can skew ethical decision-making. The *Belmont Report* emphasizes the critical role of data and lived experiences in shaping ethics by setting forth guidelines informed by empirical observations of human vulnerability and institutional failures. It implicitly argues for not only philosophically sound regulations but also empirically verifiable, adaptive to new findings, and responsive to societal shifts. From an empiricist perspective, the report underscores the importance of a dynamic, evidence-based ethical framework that accommodates the complex, emergent nature of human behavior and social systems. For example, the report's influence on Institutional Review Boards (IRBs) has led to empirically grounded, continually evolving procedures considering human psychology's complexities, cultural variables, and technological advancements. Thus, the *Belmont Report* is a compelling example of how empiricism enriches ethical deliberation, offering ethically and empirically robust guidelines.

The Scientific Method: The *Belmont Report* has been instrumental in refining the scientific method, particularly by embedding ethical considerations into the fabric of empirical inquiry. While the scientific method has always been grounded in observation, hypothesis formation, experimentation, and data analysis, the *Belmont Report* adds an ethical layer now considered indispensable in research involving human subjects. By introducing ethical principles like respect for persons, beneficence, and justice, the report essentially mandates that these considerations be woven into the initial planning stages of research, integrated into the formulation of hypotheses, and included in the data collection and analysis methods. For instance, the Belmont principles have been operationalized into ethical review processes, such as Institutional Review Boards (IRBs), that scrutinize research proposals for ethical soundness before they proceed, thereby becoming a critical "*gatekeeping*" mechanism in the scientific method. Additionally, the report influences the replication phase; ethical considerations are not merely a one-off process but should be consistently applied across all iterations of research. In the realm of peer review, the report's guidelines serve as a benchmark for assessing the ethical quality of research, adding another dimension to the criteria by which scientific work is evaluated. Therefore, the *Belmont Report* has not just supplemented the scientific method with ethical considerations but has fundamentally reshaped it into a more comprehensive, humane approach to research.

Medicine: The *Belmont Report* has profoundly impacted the field of medicine, particularly in defining and elevating the ethical standards of care for human subjects in medical research. It is the ethical cornerstone that complements medical healthcare standards of care, dictating how research should be conducted to respect patient autonomy, ensure practitioner beneficence and nonmaleficence, and strive for public justice. In medical healthcare trials, these principles influence everything from informed consent processes to risk-benefit analyses, aligning research practices with ethical imperatives. Over time, the *Belmont Report* has influenced academic research and the broader medical healthcare landscape, as its principles have been integrated into clinical guidelines and institutional protocols. This has contributed to evolving scientifically rigorous and ethically sound medical healthcare standards of care. By shaping the formation and function of Institutional Review Boards (IRBs), the report ensures that medical research proposals undergo stringent ethical scrutiny, effectively shaping the methodology and execution of medical healthcare studies. These ethical considerations have fed back into clinical practices, informing standards of care that are more holistically centered on the patient's dignity, autonomy, and well-being. Consequently, the *Belmont Report* has helped align medical science with societal values, reinforcing that ethical considerations are not peripheral but integral to good medical healthcare.

Ethics: The *Belmont Report* has been a seminal work in the establishment and propagation of the four foundational principles of bioethics: patient autonomy (informed consent), practitioner beneficence (do good), practitioner nonmaleficence (do no harm), and public justice (be fair). While the *Belmont Report* originally articulated three principles — respect for persons, beneficence, and justice — these have evolved to include nonmaleficence and have profoundly influenced ethical conduct in biomedical research and healthcare. Autonomy, closely related to the Belmont principle of "*respect for persons*," laid the groundwork for informed consent, revolutionizing the relationship between medical healthcare providers and patients by empowering the latter to make educated decisions about their care. Beneficence, the ethical obligation to "*do good*," has been operationalized in the risk-benefit analyses now standard in clinical trials, aiming to maximize positive outcomes while minimizing harm. Nonmaleficence, or "*do no harm,*" although not explicitly named as a separate category, was also in the original *Belmont Report* but under the category of beneficence. Part of the reason for this might be that the principle of nonmaleficence is often thought of as inaction, whereas autonomy, beneficence, and justice are actionable principles. It serves as a cautionary guideline, especially in research contexts, ensuring that the quest for knowledge does not override the well-being of participants. Justice, focusing on fairness and equitable treatment, has influenced criteria for subject selection in research, ensuring that benefits and burdens are distributed across various societal sectors. Thus, the *Belmont Report* has been instrumental in shaping the ethical landscape of biomedical research and healthcare, turning abstract principles into actionable guidelines that have become the moral backbone of these fields.

Conclusion: The *Belmont Report* is a cornerstone document that has significantly shaped the ethical dimensions of biomedical and behavioral research involving human subjects. Born from a need to address ethical breaches in scientific studies, the report has successfully interwoven foundational principles of rationalism and empiricism to form a moral scaffolding that has influenced the scientific method, medicine, and bioethical principles. It has championed rationalist ethics by laying down axiomatic principles such as respect for persons, beneficence, and justice, emphasizing "*a priori*" principles that could be universally applied. Its empiricist leanings are manifest in its call for ongoing scrutiny and adaptation of these guidelines based on real-world experiences and empirical data, thus balancing philosophical rigor with empirical evidence. In doing so, it has profoundly impacted the medical field by setting new ethical standards of care that align with evolving best practices and patient needs. The report has also revolutionized bioethics, transforming its three core principles into the pillars of patient autonomy (informed consent), practitioner beneficence (do good), practitioner nonmaleficence (do no harm), and public justice (be fair). These have since become the ethical backbone of biomedical research and healthcare. As a result, the *Belmont Report* serves as a multidimensional ethical compass that guides the conduct of scientists and healthcare providers and shapes policy, research protocols, and public discourse on the ethical underpinnings of research and medical healthcare practice.

The *Belmont Report's* Legacy: The *Belmont Report* is esteemed for its groundbreaking integration of rationalist and empiricist principles into an ethical principlistic framework based on patient autonomy (informed consent), practitioner beneficence (do good), practitioner nonmaleficence (do no harm), and public justice (be fair), setting a gold standard for the ethical treatment of human subjects in research and healthcare.

**

REVIEW QUESTIONS

True/False Questions:

1. The Belmont Report introduced the basic ethical principles governing research involving human subjects — respect for persons, beneficence, and justice.
   True or False?

2. The Belmont Report emphasizes the importance of applying "*a priori*" principles and deductive reasoning to shape ethical guidelines for human subjects research.
   True or False?

Multiple-Choice Questions:

3. What historical ethical breach significantly influenced the creation of the Belmont Report?
   a) The Stanford Prison Experiment
   b) The Milgram Obedience Study
   c) The Tuskegee Syphilis Study
   d) The Little Albert Experiment

4. How does the *Belmont Report* influence the scientific method in research involving human subjects?
   a) By standardizing statistical analysis methods
   b) By setting ethical guidelines that must be considered during hypothesis formation and data collection
   c) By regulating the funding of medical research
   d) By establishing experimental techniques

Clinical Vignette:

5. A clinical trial is conducted to test a new medication for hypertension. Before the study begins, the Institutional Review Board (IRB) reviews the research proposal to ensure that it includes proper informed consent procedures and methods to minimize risks to participants. Halfway through the study, the IRB requires the research team to modify the protocol to address unexpected side effects that have emerged. What principle from the Belmont Report is primarily being upheld by the IRB's intervention?
   a) Beneficence
   b) Justice
   c) Autonomy
   d) Nonmaleficence

Basic Science Vignette:

6. A clinical researcher is planning a study and follows the *Belmont Report's* guidelines. She ensures that all participants understand the risks and benefits of the study and voluntarily agree to participate. Which principle of the *Belmont Report* is primarily demonstrated by this action?
a) Risk minimization
b) Informed consent
c) Equitable selection
d) Scientific validity

Philosophy Vignette:

7. A philosopher is analyzing the ethical implications of the Belmont Report's principle of justice. They focus on a case where a clinical trial was criticized for disproportionately recruiting participants from a low-income community without ensuring that the benefits of the research would also be available to this community. Which principle from the Belmont Report does this scenario primarily illustrate?
a) Autonomy
b) Beneficence
c) Nonmaleficence
d) Justice

Correct Answers:

1. True
2. True
3. c) The Tuskegee Syphilis Study
4. b) By setting ethical guidelines that must be considered during hypothesis formation and data collection
5. d) Nonmaleficence
6. b) Informed consent
7. d) Justice

BEYOND THE CHAPTER
The *Belmont Report* (1979)

- *The Belmont Report: Ethical Principles and Guidelines for the Protection of Human Subjects of Research* by The *National Commission for the Protection of Human Subjects of Biomedical and Behavioral Research*
- *Ethical Considerations in Research* by Florence Nightingale and others
- *The Belmont Report: A Paradigm for Ethical Behaviors and Decision Making in Nursing* by Nola A. Schmidt and Janet M. Brown

***

Chapter 56: IACUC — The Guardian of Ethics: Animal Research

The IACUC ensures humane animal research, integrating scientific rigor with ethical compassion.

Abstract: Amidst rising awareness of animal rights and heightened concern for animal welfare in the late 20th century, the Institutional Animal Care and Use Committee (IACUC) emerged as a pivotal entity, integrating science, ethics, and legislation to ensure humane treatment of research animals. Formed in response to mandates like the amended Animal Welfare Act of 1985, the IACUC epitomizes a holistic approach to animal research, combining empirical scrutiny with philosophical reflection. Comprising diverse experts, it ensures research complies with the highest ethical and scientific standards, guided by the 3Rs: Replacement, Reduction, and Refinement. This piece explores the committee's roots in rationalist and empiricist traditions, its role in shaping the scientific method and medicine, and its influence on bioethics principles: patient autonomy (informed consent), practitioner beneficence (do good), practitioner nonmaleficence (do no harm), and public justice (be fair). The IACUC's enduring legacy is a future where scientific progress aligns with moral responsibility and compassion.

**

Introduction: In the wake of mounting public concern over animal welfare and societal shifts during the latter half of the 20th century, the necessity for a structured approach to the ethical treatment of animals in research became increasingly apparent. Stemming from a burgeoning awareness of animal rights and a reaction to documented instances of mistreatment and unethical practices in research laboratories, a compelling call emerged to institute mechanisms to safeguard animal welfare comprehensively. This period of awakening culminated in the formation of the Institutional Animal Care and Use Committee (IACUC), a pivotal entity tasked with safeguarding the welfare and ethical treatment of animals used in research and experimentation. Established as a consequence of evolving societal values and stringent legislative mandates — such as the Animal Welfare Act of 1966, which was later amended to specifically include provisions for creating IACUCs in 1985 — the committee operates as a watchdog, meticulously scrutinizing and overseeing animal research protocols. Functioning as the convergence of science, ethics, and legislation, the IACUC ensures that all research endeavors involving animal subjects adhere to the highest ethical standards and scientific rigors, guided by the central philosophy of the 3Rs — Replacement, Reduction, and Refinement. Comprising a diverse panel of experts, including veterinarians, scientists, and community members, the committee champions a holistic approach to animal research, merging empirical scrutiny with philosophical contemplation. As we delve deeper into the topic, we shall dissect the IACUC's roots in rationalist and empiricist traditions, its undeniable role in shaping the scientific method, its monumental contributions to the field of medicine, and its profound influence on the articulation and adherence to the four principles of bioethics: patient autonomy (informed consent), practitioner beneficence (do good), practitioner nonmaleficence (do no harm), and public justice (be fair).

Rationalism: The underpinning rationalist foundations of the Institutional Animal Care and Use Committee (IACUC) can be seen as a response to the complex interplay between scientific exploration and ethical responsibility. In a rationalist context, establishing IACUC can be perceived as a testament to human beings' innate faculty to develop structures and systems grounded in justice, compassion, and responsibility principles. Through deductive reasoning, one can argue that the moral obligation to prevent unnecessary suffering and ensure welfare is a logical extension of a foundational, axiomatically established ethical premise. These "*a priori*" principles guide the intricate processes of assessing, monitoring, and overseeing animal research activities, fostering a culture of responsibility and cognizance. In essence, the formation and functioning of the IACUC symbolize humanity's intrinsic capability to recognize patterns of ethical concerns and respond with structured, well-reasoned solutions, thereby ensuring the delicate balance between scientific advancement and moral obligation is maintained through rational, systematic oversight. This approach reflects a deeper, inherent understanding and respect for life and its intrinsic value, showcasing the human capacity for nuanced problem-solving and moral reasoning.

Empiricism: Within empiricism, the Institutional Animal Care and Use Committee (IACUC) manifests as an evolving entity that continually refines its protocols and guidelines based on sensory experiences, environmental interactions, and tangible evidence gleaned from the practical aspects of animal research. Its empiricist orientation is evident in its systematic approach to overseeing animal care and use in research environments, heavily relying on accumulating data from ongoing studies and systematic reviews. Operating on principles of inductive reasoning, IACUC extrapolates insights from specific cases to inform broader policies and regulations, fostering an environment where knowledge is consistently acquired and refined through empirical observations. The Committee applies rigorous evaluation methods, facilitating a continuous improvement cycle driven by real-world experiences and data, thereby nurturing a responsive and adaptable framework. This cycle not only encompasses the physical well-being of the animal subjects but also extends to the analysis of environmental factors, fostering a comprehensive perspective that ensures the integrity of both the scientific process and the ethical treatment of animal subjects, cementing its role as a beacon of empirical vigilance in the ever-evolving landscape of scientific research.

The Scientific Method: In shaping the scientific method, the Institutional Animal Care and Use Committee (IACUC) is an indispensable pillar ensuring that animal research adheres to a stringent, ethical, and scientifically sound framework. For instance, the development of high-throughput screening methods for potential drug compounds has been significantly influenced by the guidance of IACUC. By enforcing the 3Rs principle, the committee has fostered the adaptation and incorporation of *in vitro* methods, greatly reducing the reliance on animal models in the preliminary stages of drug discovery, thus enhancing the speed and efficiency of these processes without compromising ethical considerations. Enforcing rigorous standards for the ethical treatment and well-being of animal subjects helps foster a research environment where observations and experiments are grounded in respect for life and empirical accuracy. IACUC's involvement ensures that the hypotheses formulated in animal research are scientifically robust and ethically conceived, promoting a holistic approach to scientific inquiry. Moreover, by overseeing the proper design and execution of experiments, the committee helps maintain the credibility and replicability of findings, ensuring consistent results and preventing potential biases from improper animal care or use. Furthermore, IACUC's role facilitates rigorous peer reviews and fosters a culture of accountability and precision within the research community. This iterative process of scrutiny and adaptation continually refines the scientific method, promoting advancements in knowledge that are both ethically responsible and scientifically valid, thus safeguarding scientific research's integrity and progressive trajectory in a multifaceted and accountable manner. IACUC operationalizes the widely recognized principles of the 3Rs

— Replacement, Reduction, and Refinement — as a guiding philosophy in its oversight of animal research.

Replacement: IACUC actively encourages the research community to explore and adopt alternatives to animal use wherever possible. This could involve utilizing *in vitro* methods, computer simulations, or other innovative techniques that can replace the need for animal subjects, thereby aligning with the growing focus on humane science.

Reduction: IACUC plays a critical role in minimizing the number of animals used in research. By advocating for precise experimental design and the implementation of statistical methods, the committee ensures that the minimum number of animals is used to obtain reliable data, consequently reducing the overall animal usage in research projects.

Refinement: The committee continually seeks ways to refine experimental procedures to minimize distress and improve the well-being of animal subjects. This entails endorsing practices that enhance animal housing conditions, implementing pain management strategies, and promoting the overall welfare of the animals involved in research, thus minimizing suffering and improving the quality of life for these animals.

Medicine: The Institutional Animal Care and Use Committee (IACUC) has been pivotal in steering the trajectory of modern medicine toward a path of ethical solidity and scientific vigor. One monumental instance of IACUC's influence can be seen in oncology. Over the past few decades, animal research overseen by IACUC has led to the development of targeted therapies for various types of cancer. Through carefully monitored and humane conditions prescribed by IACUC, researchers could understand the complexities of different cancers at the molecular level, paving the way for the development of innovative treatments that have significantly improved survival rates and quality of life for cancer patients. Ensuring that animal research is conducted under such carefully monitored and humane conditions has facilitated countless other medical breakthroughs rooted in evidence-based practices. Through its vigilant oversight, IACUC assures that the medical community can rely on the credibility and ethical soundness of research findings, thereby influencing the formulation of scientifically grounded and ethically responsible standards of care. Its insistence on ethical treatment and the rigorous analysis of experimental protocols means that the data procured from these studies can stand up to the highest scrutiny, paving the way for medical advancements that are both innovative and ethically commendable. Consequently, the role of IACUC transcends mere regulatory oversight; it becomes a cornerstone in the cultivation of medical healthcare standards that are in sync with contemporary best practices, fostering an environment where medical progress is harmoniously aligned with the principles of compassion, dignity, and respect for all life forms. This approach has undeniably influenced the very ethos of the medical healthcare community profoundly, molding it into a sector that balances pioneering advancements with moral responsibility and empathy.

Ethics: The Institutional Animal Care and Use Committee (IACUC) profoundly influences the tapestry of bioethics, embodying and promulgating principles that resonate deeply with the cardinal tenets of patient autonomy (informed consent), practitioner beneficence (do good), practitioner nonmaleficence (do no harm), and public justice (be fair). While the principle of autonomy, which usually manifests as informed consent, may seem less applicable to animal subjects, IACUC ensures representation for these beings, acknowledging their intrinsic value and safeguarding their welfare in the research environment. Through the lens of beneficence, IACUC obliges researchers to maximize benefits and advancements from the experiments conducted, advocating for humane and compassionate treatment at every stage. The nonmaleficence principle finds staunch support in IACUC's guidelines, which unequivocally mandate the minimization of harm and distress to animal subjects, striving for the highest levels of welfare. Lastly, the concept of justice is echoed in the committee's persistent efforts to ensure fairness and equity in the treatment of research animals, promoting just and equitable research. Thus, IACUC is a bulwark of bioethical integrity, fostering a research milieu where the pursuit of knowledge harmonizes with the unwavering commitment to ethical excellence and moral responsibility.

Conclusion: In the complex labyrinth of scientific research, the Institutional Animal Care and Use Committee (IACUC) emerges as a beacon of ethical stewardship and scientific integrity. By weaving the philosophies of rationalism and empiricism into its foundational structure, it fosters an environment that harmonizes instinctual cognizance with empirical analyses, thereby sculpting a more nuanced and conscientious approach to research involving animals. It bolsters the scientific method by instilling a systematic, ethically grounded framework that scrutinizes research from inception to conclusion, thereby enhancing the validity and replicability of scientific investigations. In medicine, its diligent oversight has been instrumental in ushering in advancements that resonate with scientific credibility and ethical consideration, reshaping the healthcare landscape into one that embodies compassion and respect in congruence with scientific progress. Moreover, it exemplifies the epitome of bioethical adherence, echoing the principles of patient autonomy (informed consent), practitioner beneficence (do good), practitioner nonmaleficence (do no harm), and public justice (be fair), ensuring that the sanctuary of ethics is not breached in the pursuit of knowledge. IACUC is a testament to humanity's commitment to marrying scientific advancement with moral responsibility and forging a future where progress does not come at the cost of compassion.

IACUC's Legacy: The Institutional Animal Care and Use Committee (IACUC) embodies a steadfast guardian of ethical principles in animal research, perpetuating a legacy of fostering humane scientific inquiry that intertwines with a deep-seated commitment to the welfare and dignified treatment of animal subjects, thus molding a future where ethics and progress walk hand in hand.

**

REVIEW QUESTIONS

True/False Questions:

1. The Animal Welfare Act of 1966, was amended to specifically include provisions for creating IACUCs in 1985.
   True or False?
2. The IACUC's role is limited to ensuring scientific rigor in animal research, without focusing on ethical considerations.
   True or False?

Multiple-Choice Questions:

3. Which of the following principles is NOT one of the 3Rs that guide IACUC's ethical oversight of animal research?
   a) Replacement
   b) Reduction
   c) Refinement
   d) Rehabilitation

4. How does IACUC influence the scientific method in animal research?
   a) By directly conducting experiments
   b) By setting guidelines for ethical review and humane treatment
   c) By providing funding for research projects
   d) By publishing research findings

Clinical Vignette:

5. Dr. Smith is designing an experiment to test a new drug's effects on mice. To comply with the principles of the Institutional Animal Care and Use Committee (IACUC), which of the following actions should Dr. Smith prioritize?
   a) Using the maximum number of mice to ensure robust results
   b) Replacing animal testing with computer simulations wherever possible
   c) Conducting experiments without anesthesia to avoid confounding results
   d) Minimizing the number of mice used by optimizing the experimental design

Basic Science Vignette:

6. Dr. Martinez is planning a series of experiments to test a new cancer drug on mice. According to IACUC guidelines, which approach should Dr. Martinez take?
   a) Using a large number of mice to cover all possible variables
   b) Implementing alternative methods like tissue cultures where feasible
   c) Skipping pain relief to avoid influencing the drug's effects
   d) Keeping the mice in the simplest possible environment to reduce cost

Philosophy Vignette:

7. A philosopher examines the rationalist foundations of the Institutional Animal Care and Use Committee (IACUC), emphasizing its role in ensuring ethical treatment of animal subjects in research. Which principle of IACUC best aligns with the rationalist perspective that animal subjects should be treated with inherent respect and dignity?
   a) Risk-benefit analysis
   b) Ethical review by IACUC
   c) Replacement
   d) Public justice

Correct Answers:

1. True
2. False
3. d) Rehabilitation
4. b) By setting guidelines for ethical review and humane treatment
5. b) Replacing animal testing with computer simulations wherever possible
6. b) Implementing alternative methods like tissue cultures where feasible
7. b) Ethical review by IACUC

BEYOND THE CHAPTER
Institutional Animal Care and Use Committee (IACUC)

- *Institutional Animal Care and Use Committee Guidebook* by the Office of Laboratory Animal Welfare
- *The Animal Research War* by P. Michael Conn and James V. Parker
- *The IACUC Handbook* edited by Jerald Silverman, Mark A. Suckow, and Sreekant Murthy

***

Chapter 57: 3Rs — The Ethical Triad: Humane Research

The 3Rs — Replacement, Reduction, Refinement — guide ethical animal research, balancing innovation with moral responsibility.

Abstract: In this analytical discourse, the profound influence of the 3Rs — Replacement, Reduction, and Refinement — in scientific research, particularly animal studies, is meticulously examined. Central to fostering ethical and responsible experimentation, these guiding principles encourage minimizing harm and optimizing benefits through alternative methods to animal testing, limiting animal usage, and enhancing animal welfare. Rooted in rationalist and empiricist traditions, the 3Rs serve as a moral compass, guiding scientific investigations by harmonizing empirical evidence with moral axioms. This discourse explores how these principles reshape research to foster innovation, compassion, and ethical integrity. Moreover, the 3Rs' role in shaping the four cardinal principles of bioethics — patient autonomy (informed consent), practitioner beneficence (do good), practitioner nonmaleficence (do no harm), and public justice (be fair) — is highlighted, demonstrating their impact on aligning scientific advancement with moral accountability and humanitarian stewardship.

**

Introduction: In the ever-evolving sphere of scientific research, the principle of the 3Rs — Replacement, Reduction, and Refinement — is a cornerstone to fostering ethical and responsible experimentation, particularly in animal research. Rooted deeply in the moral imperative to minimize harm and maximize benefits, the 3Rs guide researchers to substitute animal testing with alternative methods (Replacement), limit the number of animals used in studies (Reduction), and enhance animal welfare when their use is unavoidable (Refinement). This dynamic approach champions the welfare of animals and potentially fortifies the quality of scientific outputs. As we delve deeper into this critical topic, we will explore its underpinnings in rationalism and empiricism, its instrumental role in shaping the scientific method, and its significant contributions to medicine. Moreover, we will assess its profound influence on the four principles of bioethics: patient autonomy (informed consent), practitioner beneficence (do good), practitioner nonmaleficence (do no harm), and public justice (be fair), thus weaving a comprehensive narrative on its pervasive impact in contemporary science and ethics.

Rationalism: The 3Rs — Replacement, Reduction, and Refinement — echo a rationalist ethos, contending that how we approach scientific research should be premised on certain innate faculties and moral axioms that guide the advancement of knowledge. This framework can be perceived as an embodiment of evolutionarily ingrained mechanisms that steer researchers toward ethically sound and cognitively coherent methods of investigation. Through the lens of rationalism, the principles of the 3Rs serve as "*a priori*" truths that are intrinsically known and must guide our interactions with sentient beings. Deductive reasoning becomes a vital tool in this context, helping delineate ethical research boundaries by drawing from definitional truths and axioms about suffering, empathy, and the potential for non-harmful alternatives. These foundational principles, theorized to be hardwired into the human cognitive apparatus, facilitate the fostering of an environment where scientific innovation progresses hand-in-hand with moral responsibility and humane treatment of animals is not just an afterthought but a primary consideration, shaping the pathway to a more compassionate and rational future in scientific exploration.

Empiricism: The 3Rs — Replacement, Reduction, and Refinement — find substantial grounding in empiricism, where the continual adaptation and enhancement of these principles are deeply rooted in sensory experiences and environmental interactions that feed into the collective knowledge base of scientific research and ethics. This empiricist approach leverages inductive reasoning, wherein meticulous observations and data harvested from real-world interactions can lead to the formulation of generalized principles and guidelines that govern the ethicality of research protocols. Cognitive biases, developed as heuristic shortcuts, might influence this trajectory, emphasizing the necessity of a cautious and self-aware approach to ethical decision-making. As empiricism allows for assimilating various external influences, such as cultural and societal nuances, into its framework, it establishes a fluid and adaptive foundation that responds keenly to changing technological landscapes and societal expectations. In a professional realm, it underscores the pivotal role that empirical evidence plays in sculpting ethical guidelines that are both relevant and grounded in reality. Furthermore, on a public policy front, it advocates for the crafting of regulations that are receptive to societal changes and technological advancements, fostering an environment where ethics and empiricism collaborate to anticipate and navigate the complex, emergent properties of scientific endeavors, such as the evolution of language or the intricacies of ecological systems. This empiricist orientation thus engenders a robust, adaptable, and responsive ethos in implementing and continuously refining the 3Rs, facilitating a progressive trajectory in both scientific and ethical dimensions.

The Scientific Method: The principles of the 3Rs — Replacement, Reduction, and Refinement — play a pivotal role in shaping the scientific method, embedding within it a layer of ethical consciousness that seeks to harmonize the pursuit of knowledge with the welfare of sentient beings involved in research. The 3Rs influence the initial stages of the scientific method, encouraging researchers to ponder the necessity of animal usage and prompting them to explore alternative avenues that might offer similar or even superior data without involving animal distress. Moreover, they guide the formulation of testable hypotheses, steering them in more ethically sound and responsible directions. As controlled experiments are designed and executed, these principles mandate establishing protocols that minimize harm and distress, optimizing the animals' welfare. Additionally, they champion the meticulous collection and analysis of data to ensure the reliability and replicability of the experiments and the validity of adopting the 3Rs without compromising the quality of research outputs. Furthermore, the peer-review process becomes a platform where the adherence to these principles is scrutinized rigorously, fostering a scientific community that is as committed to ethical considerations as it is to the advancement of knowledge, thus carving a research landscape that is both progressive and compassionate.

Medicine: In medicine, the doctrine of the 3Rs — Replacement, Reduction, and Refinement — has markedly contributed to elevating the standards of care, fostering an environment where ethical treatment and innovation coalesce harmoniously. By promoting alternative research models, like *in vitro* methods and computational simulations (Replacement), they have fostered a shift toward more humane and often more precise investigative techniques. The reduction principle encourages the scientific community to minimize animal usage, thereby nurturing a culture of resourcefulness and efficiency, where each study is designed to extract maximal information while utilizing the minimum number of living subjects. Moreover, Refinement advocates for optimizing experimental procedures to lessen the potential distress and enhance the welfare of animals when their involvement is indispensable. This concerted effort resonates profoundly within the medical community, influencing healthcare professionals to adhere to peer-reviewed, evidence-based practices that reflect the current best standards of care. By intertwining research innovation with ethical responsibility, the 3Rs have played a pivotal role in steering the medical field toward practices that are not only legally sound but also morally upright, embodying a dynamic evolution of care that is cognizant of patient needs and ethical imperatives.

Ethics: The framework of the 3Rs — Replacement, Reduction, and Refinement — serves as a beacon in the complex landscape of bioethics, profoundly influencing the cardinal principles of patient autonomy (informed consent), practitioner beneficence (do good), practitioner nonmaleficence (do no harm), and public justice (be fair). Fostering a climate where the replacement of animal testing with alternative methods is prioritized resonates deeply with the tenet of nonmaleficence, mitigating harm and distress in experimental subjects. This aspect, in tandem with the commitment to reduce the number of animals used in research, aligns congruously with the principle of beneficence, urging researchers to act in a manner that maximizes positive outcomes and minimizes adverse consequences. Furthermore, the adherence to the principle of refinement reverberates with the ethos of justice, endorsing fair treatment and enhanced welfare standards for animals enlisted in scientific studies. Concurrently, these 3Rs ingrain a deep respect for autonomy within the research milieu, cultivating a culture where informed decisions and consensual participation are paramount, thus echoing the spirit of informed consent extended to human subjects. Collectively, these principles reshape the ethical landscape of scientific research, sculpting a path where moral accountability and the pursuit of knowledge walk hand in hand, ensuring that the advances in the field are underpinned by compassion, justice, and a steadfast commitment to ethical integrity.

Conclusion: In an era of rapid scientific evolution, the doctrine of the 3Rs — Replacement, Reduction, and Refinement — acts as a moral compass, guiding researchers toward ethical and humane research practices, particularly in animal studies. Embodying the dual ethos of rationalism and empiricism, the 3Rs entwine innate moral axioms with an adaptive approach grounded in real-world experiences, nurturing a landscape where scientific inquiries are ethically sound and empirically robust. By infusing the scientific method with an ethical consciousness, they encourage a deeper reflection on the necessity of animal use, fostering innovative alternatives that do not compromise the quality of research outputs. Furthermore, their substantial contributions to medicine are unequivocal, catalyzing a shift toward higher standards of care characterized by a harmonious blend of innovation and ethical responsibility. Their profound influence permeates the fundamental principles of bioethics — patient autonomy (informed consent), practitioner beneficence (do good), practitioner nonmaleficence (do no harm), and public justice (be fair) — crafting a narrative where scientific advancement marches in lockstep with moral accountability. As we stand at the intersection of knowledge and ethics, the 3Rs provide a beacon of wisdom, illuminating a path toward a future where the pursuit of scientific knowledge is both compassionate and just, heralding a new epoch of ethical responsibility and humanitarian stewardship in the scientific community.

3Rs' Legacy: In bringing together the ethical imperatives and scientific advancements, the 3Rs — Replacement, Reduction, and Refinement — herald a paradigm shift in research methodologies, epitomizing the harmonious confluence of compassion, innovation, and moral stewardship in contemporary science and ethics.

**

REVIEW QUESTIONS

True/False Questions:

1. The principle of the 3Rs — Replacement, Reduction, and Refinement — is solely based on the rationalist approach to scientific research.
   True or False?

2. The 3Rs influence the scientific method by encouraging researchers to minimize harm and explore alternative research methods.
   True or False?

Multiple-Choice Questions:

3. Which of the following principles of the 3Rs aims to substitute animal testing with alternative methods?
   a) Refinement
   b) Reduction
   c) Replacement
   d) Replication

4. How do the 3Rs contribute to the medical field?
   a) By increasing the number of animals used in experiments
   b) By promoting alternative research models and minimizing animal usage
   c) By eliminating the need for peer-reviewed practices
   d) By reducing the standards of care in medical research

Clinical Vignette:

5. A research team is designing an experiment that involves animal testing. They aim to adhere to the 3Rs principles. During the planning phase, they decide to use computer simulations instead of live animals for preliminary testing. Which principle of the 3Rs does this decision primarily reflect?
   a) Reduction
   b) Refinement
   c) Replacement
   d) Repetition

Basic Science Vignette:

6. A research team is designing an experiment involving animal subjects and follows the 3Rs principles. They decide to use cell culture techniques instead of live animals whenever possible. Which principle of the 3Rs is primarily demonstrated by this action?
   a) Reduction
   b) Refinement
   c) Replacement
   d) Risk minimization

Philosophy Vignette:

7. A philosopher examines the ethical foundation of the 3Rs, emphasizing the principle that seeks to minimize the number of animals used in research without compromising the integrity of the scientific findings. Which principle of the 3Rs best aligns with this ethical consideration?
   a) Replacement
   b) Reduction
   c) Refinement
   d) Public justice

Correct Answers:

1. False
2. True
3. c) Replacement
4. b) By promoting alternative research models and minimizing animal usage
5. c) Replacement
6. c) Replacement
7. b) Reduction

BEYOND THE CHAPTER
3Rs (Replacement, Reduction, and Refinement)

- *The Principles of Humane Experimental Technique* by W. M. S. Russell and R. L. Burch
- *The 3Rs in Research: A Modern Success Story* by Michael F. W. Festing
- *Implementing the 3Rs in Neuroscience Research: A Reasoned Approach* by Merel Ritskes-Hoitinga and others

***

Philosophy of Science

Medical Research

WEEK ELEVEN: PHILOSOPHERS AND ETHICS

WEEK ELEVEN

Class Hour 1

- Chapter 58: Thales (THAY-leez, 624-546 BCE) — The Water Philosopher: Transforming Myth to Science
- Chapter 59: Pythagoras (py-THAG-or-as, 570-495 BCE) — The Mathematical Metaphysicist: Numbers Governing Reality
- Chapter 60: Xenophanes (zeh-NOF-uh-neez, 570-475 BCE) — The Rational Critic: One God, Many Questions

Class Hour 2

- Chapter 61: Alcmaeon (Alk-MAY-on, 500-450 BCE) — The Anatomist: Brain as the Seat of Understanding
- Chapter 62: Hippocratic Corpus (460-unknown BCE) — The Medical Founders: Do No Harm
- Chapter 63: Socrates (469-399 BCE) — The Dialectical Philosopher: Unyielding Pursuit of Truth

---

Introduction

In the eleventh week of our journey through Philosophy of Science, we delve into the transformative ideas of early Greek thinkers whose contributions laid the groundwork for scientific inquiry, medicine, and ethics.

Thales of Miletus, often hailed as the Water Philosopher, marks the beginning of our exploration. Thales' bold proposition that water is the fundamental substance of the universe signaled a profound shift from mythological explanations to naturalistic ones. This transition to rational and empirical approaches set the stage for future scientific endeavors. His influence extended beyond natural philosophy, impacting medicine and ethics by encouraging empirical observation and logical deduction, laying a foundation that future intellectuals would build upon.

Next, we encounter Pythagoras, the Mathematical Metaphysicist, who established the mathematical foundations that continue to underpin modern science. Pythagoras' belief in numerical relationships governing reality, although intertwined with mystical elements, significantly influenced the scientific method. His blend of rationalism and empiricism resonated through the ages, shaping ethical frameworks and promoting a balanced lifestyle and virtues that align with bioethical principles.

Xenophanes, the Rational Critic, challenged the anthropomorphic depictions of gods prevalent in his time, proposing instead a singular, abstract deity. His critical stance fostered a culture of skepticism and evidence-based inquiry, essential components of scientific investigation. Xenophanes' rational and empirical approach not only contributed to the development of systematic investigation but also subtly shaped the emerging fields of medicine and ethics by promoting critical thinking and a quest for knowledge grounded in evidence.

Alcmaeon of Croton, the Anatomist, brought a new dimension to our understanding of human physiology by pioneering anatomical dissections and identifying the brain as the seat of cognition. His rational and empirical investigations into the human body laid the groundwork for neuroscience and systematic medical inquiry. Alcmaeon's contributions were instrumental in the early development of scientific methodologies and subtly influenced ethical considerations in medical practice, emphasizing the importance of understanding the human body through direct observation and study.

The Hippocratic Corpus, a seminal collection of medical writings, marks a significant shift in medical thought by attributing diseases to natural causes rather than supernatural forces. This corpus emphasized observation and reason, laying the foundation for evidence-based medicine. The ethical principles introduced, particularly the concept of nonmaleficence, or "do no harm," have had a lasting impact on medical practice and the scientific method, establishing a moral framework that continues to guide medical professionals today.

Finally, we turn to Socrates, the Dialectical Philosopher, whose unyielding pursuit of truth through dialogue and critical questioning profoundly shaped Western thought. Socrates' dialectical method underscored the importance of critical thinking and self-examination, influencing the scientific method and medical ethics. His emphasis on autonomy and beneficence has left an indelible mark on the ethical principles that guide scientific inquiry and medical practice.

As we explore the contributions of these early Greek philosophers, we will see how their pioneering ideas and methodologies laid the intellectual foundations for the development of science, medicine, and ethics. Their legacy is a testament to the enduring power of rational and empirical inquiry, shaping the way we understand the world and our place within it.

**

Pre-SOCRATICS (6th-4th centuries BCE)

Determining the precise birth of philosophical history resembles discerning the exact moment dawn breaks. Rather than a sudden event, philosophy emerged through a gradual evolution of thought — a shift from an anthropomorphic, myth-centric worldview to a more abstract, analytical, and reflective mode of thinking. This new form of contemplation, recognized today as philosophy, is grounded in rational deductive investigation and empirical inductive observation.

Though the mythological worldview offered a rich tapestry of narrative and symbolism, it presented significant barriers to developing scientific thought. By attributing natural phenomena to the whims of deities, myths stifled the emergence of scientific disciplines. For example, as long as thunderstorms were considered expressions of Zeus' wrath, the inception of meteorology — a scientific understanding of weather patterns — remained unattainable. Broadly speaking, while the cause-and-effect outcomes inherent to the scientific method were dismissed in favor of myths, divine intervention, or other non-scientific explanations, the progression and broadening of scientific knowledge remained out of reach.

***

Philosophy of Science

Medical Research

Chapter 58: Thales (THAY-leez, 624-546 BCE) — The Water Philosopher: Transforming Myth to Science

Thales of Miletus ushered in a new era, replacing myths with rational inquiry and empirical observation, marking the dawn of Western science.

Abstract: The journey from mythology to scientific inquiry in Western civilization begins with Thales of Miletus, an ancient Greek philosopher. Thales, hailed as the pioneer of Western science and philosophy, initiated an era where rationality and empirical observation replaced conjectural narratives about deities and myths. His assertion of water as the foundational element of existence, grounded in rationalism and empiricism, marked a seminal shift toward seeking natural explanations for the universe's complexities. Thales' proto-scientific method, emphasizing observation and logical deduction, set the stage for subsequent philosophical discourses and indirectly influenced medicine and ethics, encouraging an approach rooted in rational contemplation and sensory experiences. Although not directly contributing to ethics and medicine, his work laid the foundation for future explorations in moral philosophy and health science. Thales' enduring legacy lies in his revolutionary shift toward critical thinking and logical inquiry, igniting a beacon of reason that guided intellectual pursuits for centuries.

**

Introduction: Thales of Miletus, often recognized as the inaugural philosopher in the Western intellectual tradition, was not only a thinker of profound depth but also a man of diverse talents, residing in ancient Greece during the late 7th and early 6th centuries BCE. His influence spanned from philosophy and mathematics to astronomy and engineering. Widely considered the father of Western science and philosophy, Thales challenged prevailing mythological explanations of natural phenomena, advocating instead for reason and empirical observations. His revolutionary proposition — that water is the fundamental underlying substance of the world — set the stage for subsequent philosophers to posit their theories about the primary elements of existence. Though his specific contributions and the accuracy of various anecdotes about his life remain subjects of debate, Thales' overarching legacy is undeniable: he catalyzed a seismic shift in human thought, nudging humanity toward a path of rational inquiry and critical thinking.

Rationalism: Thales of Miletus stands as a monumental figure in the annals of philosophy, primarily for his pioneering embrace of rationalism. At a time when mythological narratives dominated explanations for natural occurrences, Thales sought a more reasoned, logical foundation for understanding the world. Eschewing tales of capricious deities, he posited that water was the essential substance from which all things emerged, an early attempt to pinpoint a singular, empirical basis for the universe's complexity. This inclination toward reason and observation over superstition was groundbreaking. His endeavors in astronomy, notably his alleged solar eclipse prediction, further underscored his commitment to empirical inquiry. Instead of viewing the cosmos through a purely mystical lens, Thales approached it as a realm governed by discoverable principles. Thales laid the groundwork for the Western rationalist tradition through his shift from mythos to *logos*, championing the belief that the universe's mysteries could be unraveled through reason and observation.

Empiricism: Thales of Miletus, while often praised for his rationalist tendencies, also displayed a distinct empiricist orientation in his approach to understanding the world. Empiricism emphasizes knowledge derived from sensory experience, and Thales showcased this by grounding his philosophical inquiries in observations of the natural world. His assertion that water was the primordial substance of all things wasn't a mere speculative idea; it was influenced by his empirical observations of the omnipresence and essentiality of water in life and nature — from the sustenance of life to the nourishment of crops. In astronomy, Thales' reputed prediction of a solar eclipse and his studies of celestial patterns showcased a reliance on systematic observation rather than mere theoretical musings. Moreover, his explorations in geometry, such as recognizing the properties of circles and triangles, were likely based on practical measurements and tangible applications. Thus, while Thales undoubtedly valued reason as a tool for understanding, he equally championed the importance of deriving knowledge directly from sensory experiences and actual observations of the world around him.

The Scientific Method: Thales of Miletus, though existing centuries before the formal codification of the scientific method, undeniably laid some of its foundational stones. Often celebrated as the first philosopher in the Western tradition, his most notable contribution to the proto-scientific method was his shift from mythological explanations to empirical and logical inquiry. By proposing that water was the fundamental substance underpinning all existence, Thales demonstrated an early iteration of hypothesis formulation based on observations — namely, water's critical role in sustaining life and shaping the environment. His approach emphasized the need to seek natural, consistent principles behind phenomena, sidestepping the whimsical explanations of deities and myths that had previously held sway. Furthermore, in astronomy, Thales' efforts to predict celestial events, like solar eclipses, showcased the nascent stages of systematic observation and prediction. While the scientific method, as we understand it, would undergo significant refinements over subsequent millennia, Thales' commitment to reason and observation over superstition marked a pivotal step in humanity's journey toward a systematic approach to understanding the natural world.

Medicine: It's important to note that while Thales of Miletus made significant strides in fields like philosophy, mathematics, and astronomy, his direct contributions to medicine, as recorded in historical texts, are scant. However, Thales' emphasis on rational inquiry and empirical observation undeniably set the stage for later advancements in various fields, including medicine. His paradigm shift from attributing natural phenomena to deities and myths to seeking natural and logical explanations was revolutionary. This ethos would be adopted and expanded upon by subsequent thinkers, leading to a more systematic and observational approach to understanding human health and the natural world. Additionally, his belief in water as the primary substance or *arche* could have indirectly influenced medical thought regarding the balance of bodily humors and the essential role of hydration for health. In essence, while Thales himself might not have made explicit contributions to medical knowledge, his overarching philosophy of seeking knowledge through reason and observation paved the way for the eventual development of medicine as a discipline rooted in empirical evidence and systematic study.

Ethics: Thales of Miletus, primarily renowned for his contributions to natural philosophy and epistemology, did not leave behind a detailed ethical system comparable to later figures such as Socrates, Plato, or Aristotle. Consequently, directly linking him to principles like patient autonomy (informed consent), practitioner beneficence (do good), practitioner nonmaleficence (do no harm), and public justice (be fair) is challenging. However, his foundational belief in the power of reason and rationality as tools for understanding the world inadvertently laid the groundwork for later philosophical inquiries into ethics and human behavior. By emphasizing the value of reason over myth and superstition, Thales sowed the seeds for an approach to life that would favor rational introspection and reflection, key components in ethical deliberation. The act of seeking to understand the world through reason rather than ascribing events to the caprices of deities encourages a sense of personal responsibility and consideration. While Thales might not have explicitly addressed these ethical principles, his championing of reason and observation as paths to knowledge fostered an environment where later Greek philosophers would delve deeply into questions of morality, justice, and the good life.

Conclusion: Thales of Miletus, a luminary of the ancient world, profoundly influenced Western thought across diverse fields. Grounding his inquiries in the empiricism of the natural world, Thales rejected mythological explanations, asserting instead the primacy of water as the universe's fundamental substance. This approach heralded the dawn of Western rationalism and empiricism, with Thales valuing deductive reasoning and sensory observation to unravel nature's mysteries. His endeavors laid the rudiments for the scientific method, emphasizing empirical and logical investigations over pre-existing myths. While Thales' imprints on medicine are indirect, his empirical ethos laid the groundwork for subsequent developments, hinting at the importance of natural explanations in understanding health. Thales' reliance on reason became a cornerstone in ethics, fostering an environment conducive to later Greek explorations into moral philosophy. Though he didn't articulate the principles of patient autonomy (informed consent), practitioner beneficence (do good), practitioner nonmaleficence (do no harm), and public justice (be fair) directly, his emphasis on reason over superstition inevitably underscored the essence of personal responsibility. In totality, Thales' legacy isn't just about specific theories but an epochal shift in humanity's approach to knowledge and understanding, positioning reason and observation at the heart of intellectual pursuits.

Thales' Legacy: Advocated for the primacy of water as the fundamental substance, marking an early shift from mythological explanations to naturalistic, philosophical inquiry about the universe's essence.

**

REVIEW QUESTIONS

True/False Questions:

1. Thales of Miletus proposed that water is the fundamental substance of all existence, marking a shift from mythological to rational explanations.
   True or False?

2. Thales' contributions to philosophy and science directly included detailed ethical systems comparable to those of later philosophers like Socrates and Aristotle.
   True or False?

Multiple-Choice Questions:

3. Which aspect of Thales' approach is primarily linked to his empirical orientation?
   a) Deductive reasoning
   b) Prediction of solar eclipses
   c) Emphasis on mythological explanations
   d) Formulation of ethical principles

4. Thales' assertion that water is the primary substance of the universe is an example of:
   a) Empiricism
   b) Rationalism
   c) Ethical deliberation
   d) Scientific experimentation

Clinical Vignette:

5. A 55-year-old male patient comes to the clinic with persistent dizziness and a general feeling of unwellness that has lasted for several months. His medical history is unremarkable, and initial physical examinations show no obvious abnormalities. The patient mentions that he has recently read about Thales of Miletus and is fascinated by the idea that water is the fundamental substance of life. He wonders if his symptoms could be related to dehydration or some imbalance related to water in his body. In light of Thales' emphasis on rational inquiry and empirical observation, how should the physician proceed with the patient's evaluation?
a) Suggest that the patient significantly increases his water intake and monitors his symptoms for any improvement.
b) Perform a detailed clinical assessment, including blood tests and electrolyte panels, to investigate potential underlying causes.
c) Reassure the patient that his symptoms are likely psychological and advise him to reduce stress and relax.
d) Recommend a series of alternative therapies such as acupuncture or herbal treatments to address the symptoms.

Basic Science Vignette:

6. Thales of Miletus proposed that water is the fundamental substance underlying all existence. Imagine a scenario where an ancient Greek naturalist is investigating the different forms of water and their roles in nature. The naturalist observes water in rivers, lakes, and as rain, and notes its importance in nourishing crops and sustaining life. Based on these observations, the naturalist wants to establish a hypothesis about water's role in the ecosystem. What is the best course of action for the naturalist to take?
a) Conclude that water is magical and its behavior cannot be explained by natural laws
b) Propose that water is an essential element and forms the basis of all living things
c) Ignore the observations and rely on mythological stories to explain water's behavior
d) Suggest that water's importance is a coincidence and does not warrant further study

Philosophy Vignette:

7. In a discussion about the origins of Western philosophy, a group of students debates Thales' contribution. One student argues that Thales' assertion that water is the fundamental substance marked the beginning of rationalist thought, while another contends that it was an early form of empiricism because it was based on observable phenomena. Which interpretation best aligns with Thales' legacy?
a) Thales' idea was purely speculative and should be seen as mythological
b) Thales' focus on water as the fundamental substance is primarily a rationalist concept
c) Thales' emphasis on observable phenomena suggests an empirical approach
d) Thales' contributions are irrelevant to the development of rationalism and empiricism

Correct Answers:

1. True
2. False
3. b) Prediction of solar eclipses
4. b) Rationalism
5. b) Perform a detailed clinical assessment, including blood tests and electrolyte panels, to investigate potential underlying causes
6. b) Propose that water is an essential element and forms the basis of all living things
7. b) Thales' focus on water as the fundamental substance is primarily a rationalist concept

BEYOND THE CHAPTER
Thales (THAY-leez, 624-546 BCE)

- *Thales of Miletus: The Beginnings of Western Science and Philosophy* by Patricia F. O'Grady
- *Ancient Philosophy: From 600 BCE to 500 CE* by Brian Duignan (Section on Thales)
- *The First Philosophers: The Presocratics and Sophists* translated by Robin Waterfield (Section on Thales)

***

Chapter 59: Pythagoras (py-THAG-or-as, 570-495 BCE) — The Mathematical Metaphysicist: Numbers Governing Reality

Pythagoras' integration of mathematics and metaphysics bridged empirical observation with rational inquiry, influencing ethics and medicine.

Abstract: Pythagoras emerges as a pivotal figure in ancient history, laying the groundwork for numerous disciplines, from mathematics to philosophy, and forging a synergy between the metaphysical and tangible. Born around 570 BCE in Samos, he propelled a revolutionary wave of thought, venerating numerical relationships as the cosmos' governing force, thus crystallizing rationalism's foundations. His approach, integrating elements of empiricism, guided exploration through keen observations and experiments, giving rise to a unique synthesis of rationalist and empiricist tendencies. This duality fostered a culture of systematic inquiry, seeking universal truths through mathematical determinism and empirical investigation. Pythagoras' holistic philosophy influenced medicine and ethics, advocating for balanced lifestyles and developing ethical frameworks resonating with principles like patient autonomy, practitioner beneficence, nonmaleficence, and public justice. Through a blend of intellectual rigor and spirituality, Pythagoras left an enduring legacy, epitomizing the quest for knowledge and well-being, inspiring modern intellectual and ethical landscapes.

**

Introduction: Pythagoras, a legendary figure from ancient Greece, is often regarded as one of history's most influential mathematicians and philosophers. Born around 570 BCE in Samos and later settling in Croton in southern Italy, he established a religious group known for its strict code of conduct and fervent belief in the transmigration of souls. While Pythagoras' writings have not survived the ravages of time, the contributions attributed to him and his followers, collectively known as Pythagoreans, have been foundational in numerous disciplines. Among his many achievements, the Pythagorean theorem, which relates the sides of a right triangle, remains a cornerstone in geometry. The theorem's elegant proof and widespread applications testify to Pythagoras' enduring legacy, showcasing a brilliant confluence of mathematical rigor and abstract thought. Through his teachings, Pythagoras carved a path for advancing mathematics and fostered a philosophical tradition that probed the intricate relationship between numbers, nature, and the cosmos.

Rationalism: Pythagoras' inclination toward rationalism manifested profoundly in his belief in the immutable nature of numbers and their role in explaining the universe. To him, numbers were not mere abstract entities or tools for calculation; they were the essence of reality, representing order, harmony, and the foundational principles of the cosmos. The Pythagoreans posited that everything, from celestial bodies' harmony to musical notes' intricacies, could be understood through numerical relationships. This deterministic view that the universe could be comprehended through reason and mathematical principles exemplifies a key tenet of rationalism. In a world where mysticism and mythological explanations held sway, Pythagoras' approach was a pioneering shift toward a more systematic and logical understanding of the universe. His assertion that knowledge could be derived from reason and that the cosmos was governed by an inherent order accessible through intellectual exploration marked him as a foundational figure in the rationalist tradition, setting the stage for later philosophers who would further champion reason as the supreme path to knowledge.

Empiricism: It's important to note that while Pythagoras is primarily celebrated for his rationalist inclinations, his empiricist contribution to science is also second to none. For instance, the Pythagoreans' keen observations in music led them to discover the relationship between the lengths of vibrating strings and the pitches of notes they produced, ultimately leading to the discovery of the harmonic series. This foundational understanding of musical intervals was derived from empirical observation and experimentation. Similarly, their interests in geometry and astronomy were grounded in studying concrete phenomena, seeking patterns and relationships from the tangible world. Yet, Pythagoras' approach was different from a modern empiricist. While he and his followers gathered data from the observable world, they often infused it with a profound sense of mysticism and mathematical determinism. So, while elements of empiricism are present in the Pythagorean tradition, it is best to understand them as interwoven with a dominant rationalist and mystical framework, reflecting a unique synthesis of observed experience and abstract thought in their pursuit of knowledge.

The Scientific Method: While the scientific method, as we understand it today, was developed long after Pythagoras, his teachings and approaches were undoubtedly precursors to this systematic way of understanding the world. With his profound emphasis on numbers and their role in deciphering the cosmos, Pythagoras promoted a structured way of thinking that sought universal truths through consistent principles. His belief in the importance of mathematical relationships as the underpinning of natural phenomena laid the groundwork for a more logical and methodical examination of the world. The Pythagoreans' investigations, such as their exploration of musical harmonies through empirical observation and their analysis of geometric properties, reflect an early form of hypothesis testing and validation. While Pythagoras' approach was interspersed with mysticism and lacked the rigorous experimentation and falsifiability criteria central to the modern scientific method, his emphasis on the consistent and ordered nature of the universe undeniably paved the way for future thinkers. Pythagoras and his followers planted early seeds for the systematic inquiry that would later mature into the scientific method by championing reason and seeking patterns and principles behind observable phenomena.

Medicine: Pythagoras' influence on medicine, though not as direct as his contributions to mathematics or philosophy, is rooted in his holistic view of well-being and the interconnectedness of mind, body, and cosmos. For Pythagoras, good health wasn't just the absence of illness but a harmonious balance between the physical, mental, and spiritual dimensions. His dietary recommendations, which leaned toward vegetarianism, were based not only on the principle of non-harm to animals, influenced by his beliefs in the transmigration of souls, but also on the idea that certain foods contributed to a clearer mind and a more harmonious spirit. Additionally, he acknowledged the therapeutic role of music, advocating for specific melodies and rhythms to evoke particular emotional states and heal ailments, an early hint at today's music therapy. While not a physician in the modern sense, Pythagoras' holistic approach to health, emphasis on dietary practices, and recognition of the therapeutic potential of music laid foundational ideas that would later influence the development of various medical and therapeutic paradigms, emphasizing the inseparable relationship between lifestyle, mental state, and physical health.

Ethics: Pythagoras' impact on ethical principles is profound, stemming from his holistic worldview that integrated philosophy, spirituality, and daily practices. While the specific terms "*autonomy*," "*beneficence*," "*nonmaleficence*," and "*justice*" as used in modern ethical discourse weren't part of Pythagoras' lexicon, his teachings encompassed their essence. His advocacy for vegetarianism, rooted in the belief in the transmigration of souls, reflects a form of nonmaleficence, highlighting the importance of not causing harm to other living beings. The Pythagorean community's emphasis on living harmoniously, avoiding excess, and promoting well-being among members hints at the principles of beneficence and justice. Pythagoras also valued individual introspection and self-knowledge, which aligns with the modern notion of autonomy, as he believed individuals should actively seek personal enlightenment and spiritual growth. However, the strict codes of conduct within the Pythagorean community might appear to limit individual autonomy from a contemporary perspective. While Pythagoras did not explicitly articulate these ethical principles in the manner we understand today, his teachings and lifestyle recommendations reflect a deep concern for morality, the welfare of others, and the pursuit of a just and harmonious society, thereby influencing subsequent ethical thought.

Conclusion: Pythagoras stands as a beacon of intellectual prowess and holistic vision in the pantheon of ancient thinkers. His contributions spanned an array of disciplines, reflecting a mind deeply attuned to the mysteries of the universe yet grounded in the tangible world. Whether probing the abstract realm of numbers, seeking understanding through observation, or devising precursors to the scientific method, Pythagoras was driven by an insatiable curiosity. His emphasis on holistic health intertwined the physical with the metaphysical, offering insights that are still resonant in today's discussions on well-being. Ethically, his teachings transcended mere doctrines, encapsulating broader ideals of non-harm, balance, and the pursuit of a harmonious existence. Though the Pythagorean community had distinct practices, the universal essence of Pythagoras' teachings, from the Pythagorean theorem to ethical vegetarianism, continues to inspire and challenge. His legacy is not just in the specifics of his contributions but in the ethos of integrated thinking, bridging reason and mysticism, observation, and introspection. Pythagoras' enduring influence is a testament to the power of a mind unbounded by convention, forever seeking the harmonies that underpin existence.

Pythagoras' Legacy: Established mathematical foundations while suggesting that numerical relationships govern reality, blending mathematics and metaphysical inquiry.

**

REVIEW QUESTIONS

True/False Questions:

1. Pythagoras believed that numbers were the essence of reality, representing order and harmony in the universe.
   True or False?

2. Pythagoras' contributions to philosophy and science were limited to theoretical concepts, with no emphasis on empirical observation.
   True or False?

Multiple-Choice Questions:

3. Pythagoras' belief that musical intervals could be understood through numerical relationships is an example of:
   a) Pure mysticism
   b) Rationalist reasoning
   c) Religious doctrine
   d) Superstitious belief

4. The Pythagorean theorem, which relates the sides of a right triangle, is an example of:
   a) Empirical observation
   b) Theoretical speculation
   c) Mathematical proof
   d) Mystical revelation

Clinical Vignette:

5. A modern doctor, inspired by Pythagorean principles, integrates music therapy into their treatment plan for patients with anxiety. This approach reflects which aspect of Pythagoras' philosophy?
   a) Pure mathematical reasoning
   b) Empirical observation
   c) Holistic health and harmony
   d) Religious ritual

Basic Science Vignette:

6. Pythagoras and his followers discovered the relationship between the lengths of vibrating strings and the pitches of notes they produced. Which principle did this discovery lead to in the field of music?
   a) The Golden Ratio
   b) The Pythagorean Scale
   c) The Harmonic Series
   d) The Fibonacci Sequence

Philosophy Vignette:

7. Pythagoras believed that numbers were not just abstract entities but the essence of reality. This belief led to the idea that numerical relationships govern all aspects of the cosmos. What philosophical tradition does this idea primarily align with?
   a) Empiricism
   b) Existentialism
   c) Rationalism
   d) Pragmatism

Correct Answers:

1. True
2. False
3. b) Rationalist reasoning
4. c) Mathematical proof
5. c) Holistic health and harmony
6. c) The Harmonic Series
7. c) Rationalism

BEYOND THE CHAPTER
Pythagoras (py-THAG-or-as, 570-495 BCE)

- *Pythagoras: His Life, Teaching, and Influence* by Christoph Riedweg
- *Pythagoras and the Pythagoreans: A Brief History* by Charles H. Kahn
- *The Pythagorean Sourcebook and Library* compiled and translated by Kenneth Sylvan Guthrie

***

Philosophy of Science

Medical Research

Chapter 60: Xenophanes (zeh-NOF-uh-neez, 570-475 BCE) — The Rational Critic: One God, Many Questions

Xenophanes challenged anthropomorphic gods and championed reason and empirical evidence, laying groundwork for systematic inquiry.

Abstract: Xenophanes, a towering figure in the ancient Greek intellectual landscape of the 6th century BCE, left a significant mark on philosophy and early scientific inquiry. His profound insights into theology and epistemology disrupted conventional paradigms, urging a reevaluation of anthropomorphic representations of gods and highlighting human knowledge and perception limitations. Xenophanes envisioned the divine as singular, unchanging, and universal, contrasting the multifaceted gods of Homeric and Hesiodic traditions. As an early rationalist, he championed reason and empirical evidence in discerning truths, nurturing the roots of the scientific method. His observations of natural phenomena, especially fossils, epitomized a shift toward evidence-based inquiry, fostering a conducive environment for systematic medicine's evolution. While not a direct commentator on ethics, his theological critiques and emphasis on reasoned understanding fostered a fertile ground for ethical inquiry. Xenophanes stands as a beacon of critical thinking, urging humanity to question, observe, and engage with the world's complexities.

**

Introduction: Xenophanes, an ancient Greek philosopher and poet from the 6th century BCE, was a pivotal figure in the early development of philosophical thought, particularly in epistemology and theology. Born in Colophon in Ionia, his works have largely been pieced together from fragments found in the writings of later philosophers and historians. Challenging the anthropomorphic representations of the gods common in Homeric and Hesiodic traditions, Xenophanes posited that the divine was singular, unchanging, and universal. He argued that humans, constrained by their limited perspectives, projected their traits onto deities, leading to a multitude of culturally specific gods. Moreover, Xenophanes famously commented on the tentative nature of human knowledge, suggesting that even if one were to stumble upon the absolute truth, one would never truly know it. Through these critiques, Xenophanes laid the groundwork for subsequent philosophers, urging them to question deeply held beliefs and recognize the limitations of human understanding.

Rationalism: Xenophanes is often regarded as an early rationalist, primarily for his departure from traditional mythological explanations in favor of reason and evidence-based understanding. Rejecting the anthropomorphic depictions of gods common in his time, he championed a unified and unchanging god, a conception derived not from tales and legends but from rational contemplation. Beyond theology, Xenophanes demonstrated his rationalist inclinations by commenting on the natural world in ways that challenged prevailing notions. For instance, his observations on fossils led him to infer the existence of a time when the Earth was covered by water, a significant departure from traditional beliefs. However, his epistemological stance captures his most poignant reflection on knowledge. Xenophanes opined that while humans can reach likely true beliefs, absolute certainty remains elusive. This acknowledgment of human fallibility and the limits of knowledge did not deter the search for truth but advocated for a continuous, rigorous, and reasoned inquiry. Through his assertions, Xenophanes signaled the importance of critical thinking and a reasoned approach to understanding the world, core tenets of rationalism that would greatly influence subsequent philosophical thought.

Empiricism: While Xenophanes is often celebrated for his rationalist leanings, elements of his philosophy also display a distinct empiricist orientation. Empiricism, the view that knowledge derives primarily from sensory experience, finds a home in Xenophanes' approach to discerning truths about the natural world. He posited that observations and the evidence of the senses are indispensable tools for understanding. This perspective is exemplified in his comments about fossils. Rather than relying on traditional myths or speculative reasoning, Xenophanes drew inferences about the Earth's history based on the fossilized remains of marine creatures found in mountainous terrains, suggesting a time when these regions were submerged under water. He showcased an empirical methodology by prioritizing observable phenomena and drawing conclusions. However, this did not make him a naive empiricist; he was acutely aware of the limits of human knowledge and believed that, while truths could be approached, absolute certainty was unattainable. Thus, Xenophanes' empiricist orientation is characterized by a commitment to sensory evidence tempered by a humble acknowledgment of the inherent limitations of human cognition.

The Scientific Method: Xenophanes, while not a scientist in the modern sense, played a pivotal role in the evolution of systematic inquiry that would eventually culminate in the scientific method. He eschewed prevailing mythological interpretations of natural phenomena, advocating for explanations grounded in observation and reason. His examination of fossils, inferring ancient sea-covered landscapes, exemplifies an early form of empirical investigation — the foundational bedrock of the scientific method. Instead of unquestioningly accepting handed-down traditions, Xenophanes promoted critical examination and reliance on tangible evidence. Furthermore, his acknowledgment of the limits of human understanding, emphasizing that even if one arrived at the truth, one might never be sure of it, indirectly foreshadowed scientific theories' tentative and revisable nature. While it would be an overstatement to claim that Xenophanes developed the scientific method, his approach undeniably laid some of its philosophical groundwork. By championing evidence-based reasoning over mythological narratives and embracing human knowledge's imperfections, Xenophanes helped steer ancient Greek thought toward a path of systematic inquiry, paving the way for future thinkers and scientists to build upon.

Medicine: Xenophanes, primarily known for his contributions to philosophy and critiques of traditional religious beliefs, did not directly contribute to the field of medicine in the way that figures like Hippocrates or Galen did. Most of his surviving fragments pertain to theology, epistemology, and critiques of his contemporaries. However, his emphasis on empirical observation and critical thinking indirectly influenced a broader intellectual movement in ancient Greece, recognizing the importance of evidence-based practices. This budding empiricism and skepticism toward traditional beliefs set the stage for more systematic approaches in various fields, including medicine. As thinkers began prioritizing observation and evidence over inherited wisdom and myth, medicine, too, gradually shifted from being intertwined with religious rituals and superstition to a discipline that sought to understand the natural causes of diseases and their remedies. While Xenophanes did not directly contribute to medical knowledge or practice, the intellectual environment he helped foster was crucial in evolving a more empirical and systematic approach to medicine in the following centuries.

Ethics: Xenophanes' most prominent contributions lie in theology and epistemology rather than direct discussions on ethical principles. However, his overarching philosophical approach, which emphasized critical thinking and a departure from uncritically accepting traditional views, has implications for moral thought. By challenging the anthropomorphic and often capricious depictions of gods in ancient Greek religion, Xenophanes indirectly critiqued the morality derived from these deities' actions, pushing for a more universal and stable conception of the divine. This can be linked to the principle of justice, suggesting a universal standard of rightness. Additionally, his emphasis on human limitations and the need for humility in our claims to knowledge touches on the principles of autonomy and nonmaleficence; by recognizing the fallibility of human judgment, one might be more inclined to respect individual choices and exercise caution in actions that might harm others. While Xenophanes did not directly address the ethical principles of patient autonomy (informed consent), practitioner beneficence (do good), practitioner nonmaleficence (do no harm), and public justice (be fair) as they are understood in contemporary bioethics, the seeds of critical ethical inquiry can be traced back to his broader challenges to conventional wisdom and his call for a more reasoned understanding of the world and divinity.

Conclusion: Xenophanes is a seminal figure in the annals of ancient Greek thought, significantly impacting various dimensions of intellectual pursuit. His unorthodox perspectives on divinity, advocating for a singular, universal god, reframed theological discourse and challenged established norms. At the heart of his philosophy lay a commitment to reason and empirical evidence, demonstrated through his rationalist interpretations of the divine and empirical evaluations of the natural world. These tendencies not only set the groundwork for the development of the scientific method but also indirectly shaped the evolution of medicine by fostering an environment that prized evidence and systematic inquiry. While not a direct contributor to ethical discourse regarding contemporary principles, his emphasis on critical thinking, humility in claims to knowledge, and his challenge to the arbitrary morality derived from anthropomorphic gods laid early foundations for ethical discussions. Through his diverse contributions, Xenophanes undeniably cultivated an intellectual ethos reverberating through various disciplines, prompting us to question, observe, and think critically.

Xenophanes' Legacy: Critiqued anthropomorphic conceptions of the divine and proposed a unified, abstract God, prompting reflections on the nature and limitations of human understanding.

**

REVIEW QUESTIONS

True/False Questions:

1. Xenophanes believed that the gods depicted in Homeric and Hesiodic traditions were accurate representations of divine beings.
   True or False?

2. Xenophanes' observations on fossils led him to infer that the Earth had once been covered by water, showcasing his commitment to empirical evidence.
   True or False?

Multiple-Choice Questions:

3. Xenophanes' concept of the divine was characterized by which of the following qualities?
   a) Anthropomorphic and culturally specific
   b) Singular, unchanging, and universal
   c) Mythical and capricious
   d) Polytheistic and varied

4. Which aspect of Xenophanes' philosophy highlights his rationalist tendencies?
   a) His creation of a comprehensive medical treatise
   b) His critique of anthropomorphic gods and emphasis on a singular, universal divine principle
   c) His detailed descriptions of the gods' actions in Greek mythology
   d) His belief in multiple, changing deities

Clinical Vignette:

5. A modern philosopher, inspired by Xenophanes, argues that medical practices should not solely rely on traditional beliefs but must be grounded in empirical evidence and critical thinking. This approach reflects which aspect of Xenophanes' philosophy?
   a) His focus on the supernatural
   b) His rationalist critique of traditional beliefs and emphasis on observation
   c) His mystical interpretations of the divine
   d) His adherence to mythological explanations

Basic Science Vignette:

6. Xenophanes made observations about fossils, leading him to infer a significant conclusion about Earth's history. What was his inference?
   a) Fossils were created by the gods as a test of faith
   b) Fossils indicated that the Earth had once been covered by water
   c) Fossils were remnants of mythical creatures that once roamed the Earth
   d) Fossils were placed by earlier civilizations as a form of artistic expression

Philosophy Vignette:

7. Xenophanes challenged the anthropomorphic representations of gods in his time. What was his primary argument against these representations?
   a) Anthropomorphic gods were too simplistic to explain natural phenomena
   b) Anthropomorphic gods encouraged immoral behavior
   c) Anthropomorphic gods were an indication of human limitations in understanding the divine
   d) Anthropomorphic gods were outdated and needed modernization

Correct Answers:

1. False
2. True
3. b) Singular, unchanging, and universal
4. b) His critique of anthropomorphic gods and emphasis on a singular, universal divine principle
5. b) His rationalist critique of traditional beliefs and emphasis on observation
6. b) Fossils indicated that the Earth had once been covered by water
7. c) Anthropomorphic gods were an indication of human limitations in understanding the divine

BEYOND THE CHAPTER
Xenophanes (zeh-NOF-uh-neez, 570-475 BCE)

- *Xenophanes of Colophon: Fragments* by J.H. Lesher
- *Xenophanes: A Textual and Philosophical Study* by James H. Lesher
- *The Philosophy of Xenophanes* by Kathleen Freeman

***

Philosophy of Science

Medical Research

Chapter 61: Alcmaeon (Alk-MAY-on, 500-450 BCE) — The Anatomist: Brain as the Seat of Understanding

Alcmaeon's empirical dissections revealed new anatomical insights, shifting cognition's locus to the brain and marking the dawn of systematic medical inquiry.

Abstract: In ancient Greece, Alcmaeon of Croton transitioned from myth-based explanations to systematic, empirical inquiry in medicine and natural philosophy. Flourishing in the 5th century BCE, he introduced rationalism and empiricism, notably through animal dissections that revealed new insights into human anatomy and physiology. He shifted the locus of cognition and sensation from the heart to the brain, laying a cornerstone in neuroscience and medical philosophy. Alcmaeon's emphasis on direct observation and empirical evidence marked the embryonic stages of the scientific method. While not directly formulating medical ethics, his evidence-based paradigm subtly sowed seeds for principles like autonomy and beneficence in medical practice. Alcmaeon stands as a visionary whose legacy marked a monumental shift in the epistemological frameworks of his time, paving the way for empirically grounded scientific and philosophical inquiries that shaped the Western intellectual tradition.

**

Introduction: Alcmaeon of Croton, a preeminent figure in the landscape of ancient Greek thought, is often recognized as one of the earliest pioneers in medicine and natural philosophy. Flourishing around the 5th century BCE, he was a contemporary of Pythagoras and operated chiefly in the Italian city of Croton. Remarkably, Alcmaeon delved into the mysteries of human anatomy, physiology, and the ethereal connection between the body and the soul. While fragments and secondary references remain our main sources of knowledge about his works, they suggest groundbreaking ideas, from dissecting animal bodies to recognize certain physiological functions to positing that the brain, rather than the heart, is the epicenter of human sensation and thought. His insights, laying the foundation for later medical and philosophical inquiry, mark Alcmaeon as a visionary, intertwining empirical observations with philosophical profundity, thus bridging the chasm between the tangible and the intangible in the ancient world.

Rationalism: Alcmaeon of Croton is emblematic of the budding rationalist movement that began to burgeon in ancient Greece around the 5th century BCE. His approach to understanding the natural world and human physiology was rooted in observation, empirical evidence, and logical deduction rather than mythological explanations or divine interventions. In contrast to the dominant paradigms of his time, which often attributed physiological processes and diseases to the whims of the gods, Alcmaeon sought naturalistic explanations. This inclination was evident in his groundbreaking anatomical dissections of animals, wherein he gleaned insights into the workings of the human body. His assertion that the brain, not the heart, was the central organ of sensation and thought was profoundly rationalist. It showcased a drive to understand the underlying mechanisms of perception and cognition through logical reasoning and direct observation. Alcmaeon's endeavors epitomize the shift from mythos to *logos* in ancient Greek thought — a move toward systematic inquiry and a belief that the universe, including the complexities of the human body and soul, could be comprehended through reasoned investigation and the application of intellect.

Empiricism: Alcmaeon of Croton is an early embodiment of empiricism, a philosophical orientation that values knowledge derived from sensory experience and observation. In an era when mythological and supernatural explanations were the norms, Alcmaeon championed the importance of direct investigation into the natural world. His decision to physically dissect animal bodies to uncover the mysteries of anatomy and physiology underscores his commitment to empirical methods. Such hands-on exploration was revolutionary, granting him insights previously obscured by speculative thought alone. Among his notable findings, the recognition that the brain was the center of perception and cognition underscores the significance of his empirical endeavors. Alcmaeon's work served as a testament to the belief that genuine understanding could be achieved by directly engaging with and studying the world rather than merely contemplating or theorizing about it. This marked a significant departure from purely speculative thought, paving the way for the scientific methodologies that would shape the future of medical and natural inquiry. Alcmaeon's empiricist leanings were a beacon, heralding the importance of observation and experiential knowledge in the quest for truth.

The Scientific Method: While the modern scientific method, with its systematic procedures and rigorous demands for reproducibility, took centuries to crystallize, early thinkers like Alcmaeon played pivotal roles in its nascent development. In the 5th century BCE, Alcmaeon's explorations can be precursors to the systematic methodologies that would later be central to scientific inquiry. His decision to dissect animal bodies represented early experimentation, a core component of the scientific method. Instead of solely relying on received wisdom or philosophical musings, Alcmaeon recognized the value of direct observation and investigation in formulating conclusions about the natural world. His determination to discern the functions of various organs, especially his insight regarding the brain as the seat of thought, showcases a foundational commitment to hypothesis testing through empirical means. While it would be anachronistic to claim Alcmaeon developed the scientific method as we understand it today, his approach undeniably sowed its early seeds. Through his pioneering endeavors, Alcmaeon underscored the importance of observation, experimentation, and empirical evidence as the bedrock for understanding nature, principles that would come to be integral to the evolution of the scientific method.

Medicine: Alcmaeon of Croton, a luminary in the annals of ancient medicine, pioneered investigative approaches that would lay crucial groundwork for subsequent medical advances. Flourishing in the 5th century BCE, Alcmaeon's notable foray into animal dissections illuminated facets of anatomy and physiology previously obscured by conjecture. These dissections allowed him to challenge prevailing views, most notably positing that the brain, not the heart, was the central organ responsible for sensation and cognition, thus laying the groundwork for neurology. Furthermore, Alcmaeon introduced the concept of equilibrium in health, suggesting that a balance of certain bodily factors, akin to the later Hippocratic idea of the four humors, was crucial for well-being. His emphasis on naturalistic explanations over supernatural or divine causations for diseases was a leap toward the diagnostic and therapeutic practices that would later evolve in classical medicine. By intertwining empirical observations with philosophical insights, Alcmaeon enhanced the understanding of the human body and catalyzed a shift in medical thought from speculative and mystical interpretations to a more systematic and observational methodology, foregrounding medicine's journey toward becoming a more evidence-based discipline.

Ethics: While Alcmaeon is predominantly celebrated for his contributions to medicine and natural philosophy, direct evidence of his explicit impact on the ethical principles of patient autonomy (informed consent), practitioner beneficence (do good), practitioner nonmaleficence (do no harm), and public justice (be fair) must be more detailed. However, the broader ethos of his approach, rooted in empirical observation and rational inquiry, undoubtedly influenced the burgeoning ethical framework of medical practice in ancient Greece. By prioritizing empirical understanding over myth and superstition, Alcmaeon implicitly promoted autonomy — empowering patients with knowledge derived from tangible evidence. His investigations into the natural causes of diseases and the importance of balance in health can be seen as early intimations of beneficence, focusing on what truly benefits the patient. Conversely, by steering away from mystical treatments that lacked empirical foundation, he inherently championed nonmaleficence, avoiding potential harm from untested interventions. While justice, as a principle ensuring equitable care, might not be directly attributable to Alcmaeon's work, his emphasis on natural explanations over divine determinism subtly nudged the medical field toward viewing illnesses more impartially, without the biases of divine punishment or favor. Thus, while not directly articulating these ethical principles, Alcmaeon's approach and philosophy laid down rudimentary foundations that would later evolve into more explicit ethical guidelines in medicine.

Conclusion: Alcmaeon of Croton's legacy is a testament to the transformative power of empirical observation and rational inquiry in ancient Greece. As a trailblazer in diverse fields — from medicine and natural philosophy to the rudiments of scientific methodology — he ushered in revolutionary paradigms that countered prevailing mystical and speculative notions. His dedication to dissecting animals and gleaning novel insights into human anatomy, notably the brain's primacy in cognition, exemplified his empirical and rationalist dispositions. These endeavors advanced medical understanding and instigated the initial formulations of what would later become the scientific method, emphasizing the importance of observation and experimentation. Moreover, while not overtly addressing medical ethics, his empirical orientation subtly laid the groundwork for principles like autonomy and beneficence that would become central to the discipline. Collectively, Alcmaeon's contributions exemplify a profound synthesis of observation and philosophy, illustrating his indelible mark on the evolution of thought in the Western world.

Alcmaeon's Legacy: Pioneered anatomical dissections and proposed that the brain is the seat of understanding, laying foundational insights for philosophical inquiries into cognition and the development of medical science.

**

Review Questions:

True/False Questions:

1. Alcmaeon of Croton is recognized for his contributions to both medicine and natural philosophy through his use of empirical observation and rational inquiry.
   True or False?

2. Alcmaeon believed that the heart, not the brain, was the central organ responsible for sensation and thought.
   True or False?

Multiple-Choice Questions:

3. Which of the following best describes Alcmaeon's approach to understanding the human body and disease?
a) Relying on mythological explanations
b) Using animal dissections and empirical observation
c) Consulting divine oracles
d) Following traditional humoral theory without question

4. Alcmaeon's emphasis on naturalistic explanations and empirical evidence is a precursor to which modern concept?
a) The scientific method
b) Astrology
c) Mythological storytelling
d) Alchemy

Clinical Vignette:

5. A modern medical student is studying the early history of neurology and learns about a pioneering figure who proposed that the brain, not the heart, is the central organ for sensation and cognition. This figure also conducted animal dissections to gather evidence. Who is this figure?
a) Hippocrates
b) Alcmaeon of Croton
c) Galen
d) Aristotle

Basic Science Vignette:

6. Alcmaeon of Croton was among the first to propose that the brain is the seat of understanding and cognition, based on his dissections and observations. Which modern scientific discovery aligns with Alcmaeon's findings about the brain?
a) The discovery of the circulatory system
b) The identification of neurons and their role in transmitting information
c) The theory of evolution by natural selection
d) The development of antibiotics

Philosophy Vignette:

7. Alcmaeon emphasized empirical observation and rational inquiry in his investigations, shifting the focus from mythological explanations to naturalistic ones. How might this approach influence contemporary philosophical discussions about the nature of knowledge?
a) It would support the idea that knowledge is purely subjective and based on individual beliefs
b) It would emphasize the importance of empirical evidence and logical reasoning in acquiring knowledge
c) It would advocate for the reliance on traditional myths and stories to understand the world
d) It would reject the use of scientific methods in philosophical inquiries

Correct Answers:

1. True
2. False
3. b) Using animal dissections and empirical observation, 4. a) The scientific method, 5. b) Alcmaeon of Croton
6. b) The identification of neurons and their role in transmitting information
7. b) It would emphasize the importance of empirical evidence and logical reasoning in acquiring knowledge

Beyond The Chapter
Alcmaeon (Alk-MAY-on, 500-450 BCE)

- *Alcmaeon of Croton* by Malcolm Wilson
- *Alcmaeon in Early Greek Philosophy: Nature, Health, Disease, and Death* by Hynek Bartos
- *Alcmaeon of Croton: On Life and Death* translated by Malcolm Wilson

***

Philosophy of Science

Medical Research

Chapter 62: Hippocratic Corpus (460-unknown BCE) — The Medical Founders: Do No Harm

The Hippocratic Corpus established rational, empirical medical practices, laying the foundation for evidence-based medicine and medical ethics.

Abstract: The Hippocratic Corpus, a pivotal collection of ancient Greek medical treatises, marks a fundamental shift in medicine from supernatural attributions to rationality, empiricism, and an early form of the scientific method. Authored between 460 BCE and an undetermined end date, the corpus represents early systematic observation, diagnosis, and ethical treatment, laying the groundwork for evidence-based medicine. Attributed to Hippocrates and his contemporaries, these writings encapsulated empirical data, treatment protocols, and a robust ethical framework, epitomized by the Hippocratic Oath. This shift from mysticism to a science-infused methodology heralded a new epoch in medical thought, establishing foundational principles that continue to influence contemporary healthcare and medical ethics. The Hippocratic Corpus stands as a testament to human ingenuity in understanding health and disease, illuminating the progressive evolution of medical philosophy and fostering a legacy that profoundly impacts medicine, ethics, and scientific inquiry.

**

Introduction: The Hippocratic Corpus, a seminal collection of ancient Greek medical texts, represents one of the earliest comprehensive attempts to categorize, understand, and methodically approach the science of medicine. Attributed to Hippocrates – often hailed as the "*father of medicine*" – and his students, this compilation is neither a single text nor the work of one author but rather an amalgamation of works from various authors spanning different periods. The texts within the corpus delve into numerous aspects of medicine, ranging from clinical observations and surgical procedures to philosophical musings on the nature of health and disease. Its underlying ethos, particularly as captured in the Hippocratic Oath, emphasizes ethical practice and the importance of patient welfare. Beyond its historical and scientific significance, the Hippocratic Corpus is an enduring testament to the foundational principles of Western medicine. It is a beacon highlighting the evolution of medical thought and practice throughout the ages.

Rationalism: The Hippocratic Corpus is a groundbreaking shift from supernatural to rational explanations for disease and medical phenomena. Before its inception, illnesses were commonly attributed to divine wrath or supernatural influences. However, the writings within the Hippocratic Corpus introduced a more empirical and observational approach to medicine, emphasizing that illnesses could be understood and treated based on natural causes. This rationalist inclination sought to investigate health conditions by observing symptoms, physical manifestations, and environmental factors rather than resorting to mystical or religious explanations. For instance, in "*On Airs, Waters, and Places*," considerations are given to the influence of environment and climate on health, underscoring an early comprehension of epidemiology. This nascent form of evidence-based medicine laid the groundwork for subsequent medical inquiry. By shifting the paradigm from viewing disease as a mystical curse to a natural occurrence that could be studied, diagnosed, and treated, the Hippocratic Corpus not only revolutionized medical thought but also provided a foundational perspective that was integral to the broader philosophical movement of rationalism that characterized the Classical era of ancient Greece.

Empiricism: The Hippocratic Corpus is emblematic of an early empiricist approach to medicine, emphasizing observation, experience, and the tangible evidence of the senses as pivotal in understanding and treating disease. Distinct from merely theorizing about the human body and health, the texts within the corpus exhibit a genuine commitment to grounding medical practice in directly observing symptoms and the progression of illnesses. Practical treatments and prognoses were derived from repeated and consistent patterns seen in patients. For instance, in "*Prognostics*," the emphasis is laid on observing the signs and symptoms to predict the course of a disease, underscoring the importance of empirical data in medical assessments. Similarly, texts such as "*Epidemics'* document specific cases, detailing symptoms and outcomes in a manner reminiscent of modern clinical case studies. The Hippocratic Corpus, with its empiricist orientation, thus shifted medicine from speculative thought to a discipline rooted in the meticulous collection of data and hands-on experience. It championed the belief that reliable medical knowledge emerges primarily from direct experience and systematic observation, laying foundational stones for the evolution of clinical medicine and research.

The Scientific Method: While not outlining the scientific method as we understand it today, the Hippocratic Corpus played an essential role in pioneering systematic approaches to observation and reasoning that would later inform its development. The corpus signaled a departure from attributing diseases to divine or supernatural forces. Instead, it championed the idea that medical phenomena could be systematically studied, explained, and addressed through naturalistic means. The texts encouraged an empirical approach to knowledge acquisition through detailed observations of patients, rigorous categorization of symptoms, and disease progression and outcomes assessment. Furthermore, the Hippocratic Corpus nudged medicine toward a more structured, analytical approach by emphasizing the necessity of repeatability and consistency in observations — fundamental tenets in today's scientific method. The underlying ethos of the corpus insisted on evidence over unfounded speculation and the importance of continuously refining knowledge based on new findings. In this way, the principles embedded within the Hippocratic writings laid the early groundwork. They provided an intellectual backdrop for the eventual formulation of the scientific method, cementing its role as a precursor to the systematic inquiry that characterizes modern science.

Medicine: The Hippocratic Corpus stands as a monumental leap in the annals of medicine, laying foundational principles and practices that still reverberate today. Transcending the prevalent belief of diseases as divinely ordained or supernaturally caused, this collection firmly rooted medical understanding in naturalistic observation and logical reasoning. The texts espoused the idea that diseases had tangible causes that could be studied and countered. One of its most enduring contributions is the Hippocratic Oath, a seminal code of ethics that emphasized physicians' moral obligations and responsibilities, many tenets of which continue to guide medical professionalism today. Beyond ethics, the corpus introduced diagnostic techniques, stressed the importance of diet and environment in health, and provided insights into surgeries and treatments, setting a precedent for evidence-based practices. It advocated for the importance of patient history and physical examination, methods central to clinical trials even in contemporary medicine. In totality, the Hippocratic Corpus transitioned medicine from a realm of mysticism and anecdote into a discipline based on empirical observation, ethical rigor, and systematic practice, setting a course for the evolution of medical thought and healthcare practices for millennia.

Ethics: The Hippocratic Corpus, particularly the famed Hippocratic Oath, has profoundly impacted the ethical bedrock of medical practice. It introduced key principles that are precursors to the central tenets of modern medical ethics. The Oath's exhortation to "*do no harm*" resonates with the principle of nonmaleficence, underscoring the physician's responsibility to avoid causing harm. Its emphasis on treating the sick to the best of one's ability echoes the concept of beneficence, encouraging the active promotion of patient well-being. While the notion of patient autonomy, as we understand it today, was not explicitly defined in the Hippocratic texts, the implicit expectation of maintaining patient confidentiality, as stated in the Oath, can be seen as an early recognition of the patient's rights and dignity. With its universal principles, the corpus also touches upon justice, emphasizing an unbiased approach to care, even if it doesn't dissect societal health disparities in the modern sense. Collectively, the ethical foundations set by the Hippocratic Corpus have been instrumental in shaping the moral compass of medicine, providing a cornerstone for developing the intricate, multi-faceted ethical codes that guide contemporary healthcare.

Conclusion: In reflecting upon the multifaceted impact of the Hippocratic Corpus, it's evident that this collection of ancient Greek medical writings significantly influenced the field of medicine and the broader trajectories of science and philosophy. The corpus revolutionized how disease and wellness were perceived by establishing a foundational shift from mysticism to a rational and empirical approach. This watershed moment in history reshaped medicine into a discipline rooted in systematic observation, setting early precedents for the scientific method. Furthermore, the profound ethical principles enshrined within, especially the Hippocratic Oath, set the tone for future medical ethics, emphasizing nonmaleficence, beneficence, justice, and patient dignity. These transformative insights have stood the test of time, with their echoes still reverberating in modern medical practice, ethics, and research. Ultimately, the Hippocratic Corpus serves as an enduring testament to humanity's pursuit of knowledge, embodying the timeless endeavor to meld observation, ethics, and rationality into a cohesive understanding of the human condition.

Hippocratic Corpus' Legacy: Introduced principles such as "*do not harm*" and championed that diseases have natural causes, shifting medical thought from supernatural explanations to a framework grounded in observation and reason.

**

REVIEW QUESTIONS

True/False Questions:

1. The Hippocratic Corpus is a collection of medical treatises that marks a fundamental shift from supernatural explanations of disease to a framework grounded in observation and reason.
   True or False?

2. The Hippocratic Oath, which originated from the Hippocratic Corpus, emphasizes the importance of patient confidentiality and ethical practice in medicine.
   True or False?

Multiple-Choice Questions:

3. Which of the following best describes the primary focus of the Hippocratic Corpus?
   a) Explaining diseases through divine intervention
   b) Establishing a systematic approach to medical practice based on observation and empirical evidence
   c) Providing a comprehensive guide to ancient Greek mythology
   d) Discussing philosophical theories unrelated to medicine

4. The principle of "*do no harm*," as emphasized in the Hippocratic Oath, is most closely associated with which modern ethical principle?
   a) Autonomy
   b) Justice
   c) Nonmaleficence
   d) Beneficence

Clinical Vignette:

5. A physician is treating a patient and decides to carefully observe the patient's symptoms, take a detailed medical history, and consider environmental factors before making a diagnosis. This approach reflects the teachings of which historical medical text?
   a) The Canon of Medicine by Avicenna
   b) The Yellow Emperor's Classic of Internal Medicine
   c) The Hippocratic Corpus
   d) The Book of Healing by Al-Razi

Basic Science Vignette:

6. The Hippocratic Corpus introduced the idea that diseases have natural causes and can be understood through systematic observation and empirical evidence. Which modern medical practice aligns most closely with this approach?
   a) Faith healing
   b) Using antibiotics to treat bacterial infections
   c) Relying on astrological signs for diagnosis
   d) Performing rituals to appease gods

Philosophy Vignette:

7. The ethical principles in the Hippocratic Corpus, particularly the Hippocratic Oath, emphasize the importance of nonmaleficence, or "*do no harm*." How might this principle influence contemporary ethical debates in medical practice?
   a) It suggests that all medical interventions should prioritize patient safety and avoid causing harm
   b) It advocates for ignoring patient consent in treatment decisions
   c) It supports the idea that doctors should always follow traditional practices without questioning them
   d) It promotes the use of untested and experimental treatments without evidence of safety

Correct Answers:

1. True
2. True
3. b) Establishing a systematic approach to medical practice based on observation and empirical evidence
4. c) Nonmaleficence
5. c) The Hippocratic Corpus
6. b) Using antibiotics to treat bacterial infections
7. a) It suggests that all medical interventions should prioritize patient safety and avoid causing harm

BEYOND THE CHAPTER
Hippocratic Corpus (460-unknown BCE)

- *Hippocratic Writings* translated by G.E.R. Lloyd
- *The Hippocratic Corpus: Content and Context* by Elizabeth M. Craik
- *The Hippocratic Oath and the Ethics of Medicine* by Steven H. Miles

***

Philosophy of Science

Medical Research

Chapter 63: Socrates (469-399 BCE) — The Dialectical Philosopher: Unyielding Pursuit of Truth

Socrates' method revolutionized intellectual discourse, emphasizing systematic questioning and critical examination, laying groundwork for the scientific method and ethics.

Abstract: Socrates, the venerated ancient Greek philosopher, laid the foundation for Western philosophy through his critical approach to knowledge, moral reasoning, and the pursuit of truth. His Socratic method revolutionized intellectual discourse by systematically questioning individuals' beliefs, fostering a profound understanding of ethics and epistemology. While not directly contributing to medicine, his work influenced medical ethics and healthcare practices, emphasizing a harmonious care for mind, body, and spirit. Aligned with rationalism, his teachings also valued real-world experiences as vital sources of knowledge. Socrates' approach to inquiry laid the groundwork for the scientific method, urging critical examination of claims based on evidence and logic. His exploration of ethical principles anticipated discussions on patient autonomy, practitioner beneficence, nonmaleficence, and public justice, deeply embedded in fields like medicine and law. Despite opposition and execution, Socrates' commitment to intellectual integrity and moral virtue endures, championing inquiry, introspection, and the relentless pursuit of knowledge and truth.

**

Introduction: Socrates, a classical Greek philosopher born in Athens around 469 BCE, is a foundational figure in Western philosophy. Though he left no writings of his own, his teachings, conversations, and dialectical method have been immortalized through the works of his students, most notably Plato. Operating in the heart of Athens during the 5th century BCE, Socrates developed a distinct method of inquiry, now known as the Socratic method, which sought to uncover truth by asking probing questions, leading interlocutors to acknowledge their ignorance. This approach revolutionized philosophical dialogue and frequently challenged Athenian norms and the beliefs of those around him. While revered for his contributions to ethics and epistemology, Socrates' outspoken nature and nonconformist attitude ultimately led to his trial and execution on charges of impiety and corrupting the youth. His life and death are poignant reminders of the enduring tension between innovative thought and societal conventions.

Rationalism: Socrates is often heralded as a cornerstone figure in the development of Western rationalism. At a time when Athens was deeply steeped in traditional myths and superstitions, Socrates posited that genuine knowledge and understanding could be achieved through reason and critical inquiry. Rejecting that truth is grounded solely in religious narratives or sensory perceptions, he emphasized the role of the intellect and reasoned discourse in discerning the nature of reality and morality. His method of *elenchus*, or the Socratic method, encapsulated this rationalist approach; by posing a series of probing questions and examining the contradictions in one's answers, he sought to stimulate critical thinking and guide individuals toward philosophical insights. For Socrates, true wisdom was not about accumulating information but recognizing ignorance and continuously seeking truth through rational deliberation. This emphasis on reason as the primary tool for understanding and navigating the world set the groundwork for subsequent philosophical movements. It positioned Socrates as a pioneer of rationalist thought in the Western intellectual tradition.

Empiricism: It's essential to clarify that Socrates, as primarily depicted through the writings of his student Plato, is more commonly associated with rationalism than empiricism. However, some of his philosophical approaches can be viewed as having empiricist inclinations. Socrates believed in the importance of direct engagement with the world and individuals around him as a means of understanding and knowledge acquisition. His philosophical method, rooted in dialectical conversations, relied heavily on examining his interlocutors' everyday beliefs and experiences. By challenging them to define concepts like justice, love, and virtue, Socrates emphasized the value of real-world examples and practical implications. This insistence on starting from commonly held beliefs or experiences as a basis for philosophical inquiry hints at an empiricist orientation. While he certainly prioritized the role of reason and introspection, Socrates acknowledged the necessity of grounding philosophical discourse in the tangible, experiential world, suggesting a harmonization of both rational and empirical avenues of understanding.

The Scientific Method: While Socrates is not directly associated with formulating the scientific method as understood in modern terms, his intellectual legacy played a foundational role in shaping the spirit of inquiry that eventually led to its development. The Socratic method, characterized by rigorous questioning and critical examination of beliefs, laid the groundwork for a systematic approach to inquiry and the quest for knowledge. Socrates prioritized reasoned argument and evidence-based thinking, urging individuals to challenge assumptions and avoid unquestioned acceptance of prevailing beliefs. This insistence on critically evaluating claims and seeking logical coherence parallels the scientific method's demand for empirical evidence and falsifiable hypotheses. Moreover, Socrates' acknowledgment of human ignorance ("*I know that I know nothing*") resonates with the scientific method's recognition of the provisional nature of knowledge, always open to revision in light of new evidence. While Socrates primarily focused on ethics and epistemology rather than natural sciences, his dialectical approach and commitment to intellectual honesty indirectly paved the way for later thinkers' empirical and systematic methods in their scientific pursuits.

Medicine: Socrates, in the strictest sense, did not directly contribute to the field of medicine in the way that he influenced philosophy. However, his philosophical approach had indirect implications for how ancient Greeks thought about well-being, both of the mind and body. He often drew analogies between the care of the soul and the care of the body, emphasizing the need for knowledge and understanding in both domains. In Plato's dialogues, Socrates engages with notable physicians of his time, and through these interactions, he underscores the importance of an ethical approach to medical practice. His insistence on introspection and self-awareness also carries implications for psychological well-being, potentially foreshadowing aspects of modern psychotherapy. Additionally, his dialectical approach, which revolves around questioning and critical inquiry, can be seen as a precursor to the diagnostic methods in medicine — where symptoms are probed and assumptions are tested. While not a physician himself, Socrates' emphasis on the interconnection of mind, body, and spirit and his quest for knowledge indirectly shaped holistic and ethical considerations in the broader realm of health and well-being.

Ethics: Socrates's relentless pursuit of virtue and the good life profoundly shaped Western ethical thought. Central to his philosophy was the belief that knowledge and virtue are intimately connected; to know the good is to do the good. While Socrates did not explicitly detail principles like patient autonomy (informed consent), practitioner beneficence (do good), practitioner nonmaleficence (do no harm), and public justice (be fair) in the manner of later ethical traditions, his dialogues laid the philosophical groundwork for them. The principle of autonomy resonates with Socratic ideals about the importance of self-awareness and introspection in making moral decisions. Beneficence, or the commitment to doing good, aligns with Socrates' lifelong quest for understanding virtue and the best life. Nonmaleficence, or the principle of "*do no harm*," can be traced back to Socratic discussions on injustice and the harm it inflicts on the doer and the receiver. Justice was a central theme in many Socratic dialogues, most notably in Plato's "*Republic*," where Socrates deliberated on the nature of justice and its role in an ideal society. Through his explorations of these concepts, Socrates laid a foundation for later philosophical discussions on ethics and their application in various fields, including medicine and law.

Conclusion: The indomitable spirit of Socrates has left an indelible mark on Western thought, spanning various domains from philosophy to medicine. Through his unwavering commitment to rational inquiry and the Socratic method, he set the stage for future generations to question, probe, and challenge established beliefs. Embodying both rationalist and empiricist orientations, Socrates emphasized the interconnectedness of knowledge and virtue, arguing that one cannot genuinely act unjustly when equipped with true knowledge. His explorations paved the way for the structured investigative approach seen in the scientific method and imparted a philosophical touchstone to medical ethics and practice. Moreover, his profound engagement with ethical principles laid the groundwork for modern discussions around patient autonomy (informed consent), practitioner beneficence (do good), practitioner nonmaleficence (do no harm), and public justice (be fair). Despite facing societal opposition and ultimately paying the price with his life, Socrates' teachings remain reminders of the power of critical thinking, intellectual humility, and the relentless pursuit of truth. In every sense, his legacy serves as a beacon for inquiry, introspection, and intellectual courage.

Socrates' Legacy: Championing the dialectical method of inquiry, Socrates underscored the importance of critical thinking and self-examination, fostering a timeless foundation for ethics, epistemology, and the relentless pursuit of truth in the face of societal conventions.

**

REVIEW QUESTIONS

True/False Questions:

1. Socrates is best known for his contributions to the field of medicine through direct advancements in medical practices.
   True or False?

2. The Socratic method involves asking systematic questions to uncover contradictions and stimulate critical thinking.
   True or False?

Multiple-Choice Questions:

3. Which principle closely aligns with Socrates' belief that true knowledge leads to virtuous actions?
   a) Nonmaleficence
   b) Autonomy
   c) Beneficence
   d) Justice

4. The philosophical groundwork laid by Socrates indirectly contributed to which of the following modern concepts?
   a) Randomized controlled trials
   b) The scientific method
   c) Quantum mechanics
   d) Genetic engineering

Clinical Vignette:

5. A medical practitioner emphasizes understanding patients' perspectives and encouraging their active participation in decision-making. Which ethical principle, rooted in Socratic thought, is the practitioner demonstrating?
   a) Nonmaleficence
   b) Autonomy
   c) Beneficence
   d) Justice

Basic Science Vignette:

6. Socrates' emphasis on critical questioning and logical reasoning laid the groundwork for the systematic approach to inquiry that characterizes the scientific method. Which of the following practices aligns with Socratic principles in scientific research?
   a) Accepting hypotheses without questioning
   b) Conducting experiments and revising theories based on empirical evidence
   c) Relying solely on traditional beliefs without empirical validation
   d) Disregarding logical consistency in scientific explanations

Philosophy Vignette:

7. Socrates believed that knowledge and virtue are intimately connected, and true understanding leads to moral action. How might this principle influence contemporary discussions on medical ethics, particularly in the context of informed consent?
   a) It suggests that patients should be fully informed to make autonomous decisions about their care
   b) It advocates for making decisions on behalf of patients without their input
   c) It supports ignoring patient preferences in favor of quicker treatment
   d) It emphasizes the use of traditional methods over patient education

Correct Answers:

1. False
2. True
3. c) Beneficence
4. b) The scientific method
5. b) Autonomy
6. b) Conducting experiments and revising theories based on empirical evidence
7. a) It suggests that patients should be fully informed to make autonomous decisions about their care

BEYOND THE CHAPTER
Socrates (469-399 BCE)

- *The Trial and Death of Socrates* by Plato, translated by G.M.A. Grube
- *Socrates: A Man for Our Times* by Paul Johnson
- *Socrates: A Very Short Introduction* by C.C.W. Taylor

***

Philosophy of Science

Medical Research

WEEK TWELVE: EMPIRICISM AND INQUIRY

WEEK TWELVE

Class Hour 1

- Chapter 64: Plato (427-347 BCE) — The Idealist: Forms and the Academy
- Chapter 65: Aristotle (384-322 BCE) — The Systematizer: Empiricism and Causality
- Chapter 66: Galen of Pergamum (GAY-len, 129-216 CE) — The Medical Synthesizer: Humoral Theory and Medical Ethics

Class Hour 2

- Chapter 67: Avicenna (Ibn Sina, 980-1037) — The Medical Synthesizer: The Canon of Medicine
- Chapter 68: Thomas Aquinas (1225-1274) — The Theological Synthesizer: Faith and Reason
- Chapter 69: Nicolaus Copernicus (1473-1543) — The Heliocentric Revolutionary: Shifting the Cosmic Paradigm

---

Introduction

As we embark on the twelfth week of our exploration into Philosophy of Science, we delve into the profound contributions of seminal thinkers whose ideas have shaped the trajectory of Western thought, scientific inquiry, and ethical frameworks.

Plato, the Idealist, stands as a towering figure in ancient philosophy. Establishing the Academy, he introduced the theory of Forms, positing that eternal, immutable ideas are the true reality, transcending the material world. Plato's rationalist philosophy, intertwined with empirical elements, laid a foundation for systematic exploration and holistic approaches to health and well-being. His emphasis on the pursuit of knowledge and the contemplation of ideal forms set a precedent for scientific and ethical inquiry, advocating for a deeper understanding of the world beyond mere appearances.

In stark contrast, Aristotle, the Systematizer, brought a more grounded approach to knowledge. Systematizing various fields, from biology to ethics, Aristotle championed empiricism and introduced the concepts of causality that have become fundamental to Western science and philosophy. His meticulous observations and logical analyses shaped medical practices and ethical principles, demonstrating the power of empirical investigation and rational thought in understanding the natural world and human behavior.

Moving forward in time, Galen of Pergamum emerges as a pivotal figure in the realm of medicine. Known as the Medical Synthesizer, Galen consolidated and expanded upon existing medical knowledge, emphasizing the humoral theory. His rationalist and empirical approach established a foundation for systematic medical inquiry, influencing medical ethics through principles like beneficence and nonmaleficence. Galen's synthesis of medical practices and ethical considerations left an indelible mark on medical practice, guiding physicians for centuries.

Centuries later, Avicenna, also known as Ibn Sina, continued this legacy of medical synthesis with his monumental work, the *Canon of Medicine*. By integrating Greek and Islamic knowledge, Avicenna revolutionized healthcare practices and philosophical epistemology. His rationalist and empirical methodology emphasized systematic inquiry and the ethical principles of autonomy and beneficence, profoundly impacting both medical ethics and the practice of medicine across cultures.

Thomas Aquinas, the Theological Synthesizer, sought to harmonize faith and reason. In his "*Summa Theologica*," Aquinas fused Aristotelian philosophy with Christian theology, profoundly influencing Catholic doctrine. His rationalist and empirical approach laid the groundwork for the scientific method and ethical discourse, fostering systematic inquiry and integrating moral principles with theological insights. Aquinas' synthesis of faith and reason provided a framework for exploring ethical questions within the context of religious belief.

Finally, Nicolaus Copernicus, the Heliocentric Revolutionary, shifted the cosmic paradigm with his proposal of the heliocentric model of the solar system. Challenging the geocentric views that had dominated for centuries, Copernicus' rationalist and empirical approach combined meticulous observation with bold theoretical insights. This paradigm shift laid key foundations for the scientific method and fostered a spirit of fearless inquiry and intellectual rigor, transforming humanity's understanding of the cosmos and paving the way for future astronomical discoveries.

Through the lens of these remarkable thinkers, we witness the evolution of scientific and ethical thought from ancient Greece to the dawn of the modern era. Their contributions not only advanced our understanding of the natural world and human health but also established enduring principles that continue to guide scientific inquiry and ethical practice today. As we reflect on their legacy, we gain a deeper appreciation for the interplay between rationalism and empiricism, the integration of diverse knowledge systems, and the relentless pursuit of truth that defines the human quest for knowledge.

***

Chapter 64: Plato (427-347 BCE) — The Idealist: Forms and the Academy

Plato's philosophy bridged metaphysics and empirical inquiry, fostering systematic inquiry and influencing medical ethics and holistic health.

Abstract: Plato, a monumental figure from the classical era, remains a cornerstone of Western philosophy, influencing various domains, including philosophy, science, and medicine. Born in Athens around 428/427 BCE, Plato's work traversed metaphysics, ethics, and rationalism. He focused on the immutable realm of forms, advocating intellectual and philosophical inquiry to uncover underlying realities beyond the physical world, as depicted in his Allegory of the Cave. While primarily aligned with rationalism, his philosophy also acknowledged sensory experiences as catalysts for awakening innate knowledge, blending empirical and rational approaches. Plato's insights contributed to the early stages of the scientific method, fostering systematic inquiry that evolved into modern scientific investigation. His philosophy extended to medicine, promoting a holistic approach to health by envisioning a symbiotic relationship between soul and body. Plato's dialogues on justice, beneficence, and autonomy laid foundational stones for modern ethics. Through the Academy and his writings, Plato's legacy continues to shape contemporary thought, affirming the quest for truth and virtue.

**

Introduction: Plato, a luminary of the classical era, is a foundational figure in Western philosophy. Born in Athens around 428/427 BCE, his intellectual pursuits led him to explore vast domains of thought, from ethics and politics to metaphysics and epistemology. As a pupil of the enigmatic Socrates and the mentor to the eminent Aristotle, Plato was a vital link in the philosophical chain of the Greco-Roman world. His dialogues, wherein Socrates often serves as the central interlocutor, employ a dialectical method to uncover truths about reality, virtue, and the nature of the good life. Perhaps most renowned for his theory of forms, which posits a realm of eternal and unchanging ideas beyond the physical world, Plato's ideas have deeply influenced subsequent philosophical, religious, and literary traditions. As both a philosopher and founder of the Academy in Athens — the Western world's first institution of higher learning — his legacy continues to resonate, shaping the contours of contemporary thought and discourse.

Rationalism: Plato's philosophy is emblematic of the rationalist tradition, emphasizing the primacy of reason as the path to genuine knowledge. Rooted in his distrust of sensory perception, Plato believed that our senses often deceive us, presenting mere shadows of the true, eternal, and unchanging forms or ideas. For him, the physical world we perceive is a transient replica of this higher realm of forms, accessible only through intellectual and philosophical inquiry. Plato's Allegory of the Cave in his work "*The Republic*" vividly illustrates this belief: prisoners confined to a cave, seeing only shadows on the wall, are unaware of the more real entities causing these shadows outside the cave. Just as the freed prisoner ascends to see the outside world and recognizes it as more genuine than the shadows, the philosopher, through the rigorous application of reason, ascends to understand the realm of the forms. In Plato's view, this ascent is a journey of the mind, transcending sensory limitations to grasp eternal truths. For Plato, true knowledge is innate and recollective, requiring an internal, rational awakening rather than empirical observation. This rationalist stance would heavily influence subsequent philosophical thought, particularly the works of thinkers like Descartes and Leibniz.

Empiricism: While Plato is primarily celebrated as a rationalist, strictly categorizing him might oversimplify his rich philosophical tapestry. Indeed, while Plato emphasized the superiority of the realm of forms and the intellect's capacity to grasp them, he did not entirely dismiss the value of sensory experience. In several of his dialogues, notably the "*Meno*" and "*Phaedo*," Plato suggests that while our souls innately possess knowledge from before our birth, sensory experiences can act as catalysts that awaken or "*recollect*" this pre-existing knowledge. Empirical experiences, in this sense, play an auxiliary role in facilitating the mind's journey to the truth. His "*Timaeus*" also presents a cosmology where the Demiurge shapes the material world using the forms as an ideal template, implying an inherent connection between the sensory world and the eternal truths. Though it's clear that Plato prioritized intellectual insights over empirical observations, he recognized the intricate dance between experience and reason in the quest for knowledge. Still, compared to his student Aristotle — who would become a paragon of empiricism — Plato's occasional nods to sensory experience remain subordinate to his overarching rationalist inclinations.

The Scientific Method: While Plato is predominantly renowned for his contributions to philosophy and metaphysics, his influence on the early stages of the scientific method cannot be understated. Plato's insistence on the need for a structured approach to inquiry, grounded in reason and rigorous logical examination, laid foundational thought for the systematic processes that would later define scientific investigation. His Academy in Athens became a crucible for intellectual discussions in philosophy, mathematics, and early natural sciences. Key figures like Eudoxus developed their theories within its walls, embedding systematic inquiry into the fabric of Hellenistic intellectual tradition. Moreover, Plato's emphasis on abstract forms and ideals indirectly stressed the importance of seeking underlying principles and patterns in the observable world, a key aspect of scientific endeavors. While he did not outline the scientific method as we understand it today, the dialectic method used in his dialogues — wherein hypotheses are proposed, examined, and refined through discourse — resembles the iterative process of hypothesis testing in modern science. Thus, while Aristotle is often more directly credited with influencing empirical study and observation, Plato's emphasis on reason, abstraction, and methodical examination set vital intellectual precedents for the evolution of systematic inquiry.

Medicine: Plato's contributions to medicine are indirect but profound, primarily rooted in his philosophical discourses about the nature of the human soul and body and their interconnectedness. Unlike his contemporaries, Hippocrates deemed the "*father of medicine*," Plato did not engage directly with medical practices or techniques. Instead, his writings in dialogues like the "*Timaeus*" present a holistic view of health, emphasizing the balance and harmony between the soul and body. For Plato, physical ailments were not just disruptions of bodily mechanisms; they were often intertwined with the state of one's soul. This perspective underscored the importance of mental well-being in overall health, a notion that resonates with contemporary understandings of psychosomatic medicine. Moreover, in "*The Republic,*" Plato discusses the ideal society's physicians, arguing that they should cure ailments rather than merely alleviate symptoms and suggesting that sometimes letting nature take its course might be the best approach. This is an early exploration into medical ethics and the philosophy of medical intervention. While Plato did not make direct advancements in medical techniques or knowledge, his philosophical musings laid a foundation for a more holistic, ethical, and soul-centered approach to health and wellness.

Ethics: Plato's philosophical endeavors deeply shaped the realm of ethics, and while he didn't explicitly coin terms like patient autonomy (informed consent), practitioner beneficence (do good), practitioner nonmaleficence (do no harm), and public justice (be fair), his writings undeniably touched upon these principles. Plato's "*Republic*" is an exhaustive exploration of justice, seeking to understand its essence and manifestation in individual souls and the broader polis. For Plato, a just person or society is one where all components (soul parts or societal classes) fulfill their appropriate roles in harmony. In the medical and ethical context, this harmonious approach can be seen as a precursor to beneficence — acting for the benefit of others. Furthermore, in various dialogues, Plato's emphasis on the importance of the soul's health and the dangers of leading an unjust life can be extrapolated to the principle of nonmaleficence — "*do no harm*." The Socratic dialogues, where individuals are encouraged to question, reflect, and understand the reasons behind beliefs and actions, resonate with the principle of autonomy, championing the individual's right to informed, self-directed decision-making. While rooted in metaphysics and virtue ethics, Plato's ethical vision provides foundational insights into these core ethical principles, which have since evolved and adapted within various professional and societal contexts.

Conclusion: Plato's intellectual breadth and depth have etched an indelible mark on the annals of human thought. Transcending the boundaries of philosophy, his ideas have touched diverse realms — from ethics and epistemology to science and medicine. Central to his philosophy is the belief in an immutable realm of forms, shaping his rationalist and, to a lesser extent, empiricist tendencies. Through his allegories and dialogues, Plato established the bedrock for systematic inquiry, underpinning the foundations of the scientific method. His holistic perspective on health, intertwining the soul and body's well-being, laid philosophical underpinnings for a more comprehensive approach to medicine. Above all, his exploration of ethical constructs — justice, beneficence, and autonomy — has sculpted moral discourses that continue to guide societal decisions today. Plato's legacy endures through the Academy and his writings, reminding us of the timeless quest for truth, goodness, and the essence of existence. His multifaceted contributions continue to challenge and inspire, offering intellectual nourishment to past, present, and future generations.

Plato's Legacy: Established the Academy and introduced the theory of Forms, asserting that true reality consists of eternal and unchanging ideas rather than the material world. This perspective profoundly influenced subsequent philosophical thought.

**

A Closer Look

Plato's Knowledge: Divided Line

Abstract: In exploring Plato's philosophical doctrine, this narrative navigates the nuanced journey from the mutable world of sensory perception to the immutable realm of Forms, representing a synthesis of Heraclitus' and Parmenides' seemingly contradictory philosophies. Through a meticulous analysis of Plato's Divided Line, the research delves into the progressive stages of knowledge: Conjecture, characterized by transient shadows of reality; Belief, which introduces space as a conduit for the physical manifestations of abstract Forms; Understanding, the realm harboring mathematics and scientific laws rooted in the physical world yet constrained by unexamined principles; and finally, the pinnacle of Reason, where dialectic discourse facilitates a holistic comprehension of the universe, anchored in the immutable and unified Form of the Good. This study presents a holistic vista into Plato's complex and revolutionary perception of reality, embodying both change and permanence, offering a comprehensive grasp of what Plato thought was the true nature of existence.

**

Plato's philosophical journey led him to a fascinating conclusion that reconciled the seemingly contradictory views of Heraclitus and Parmenides. Heraclitus had posited that reality was in constant flux, while Parmenides had argued for an unchanging, eternal oneness. At first glance, these views were at complete odds with each other. Yet, Plato, the great synthesizer, proposed a resolution that was as elegant as it was revolutionary.

What if reality was not a monolith, as the Milesians had believed, but rather dual, as the Pythagoreans had proposed? This dual reality, according to Plato, consisted of a world of sense perception, which was in constant Heraclitian flux, and a nonphysical, nonspatial, nontemporal, immutable world of Forms, which was the unchanging and unmoving Parmenidian one.

In this framework, both Heraclitus and Parmenides were correct, but they spoke of different aspects of reality. The world of sense perception, akin to Heraclitus' view, was ever-changing and could only provide us with Opinions at best. In contrast, the world of Ideas or Forms, echoing Parmenides, was unchanging and eternally one, offering us a realm of absolute Knowledge.

Thus, Plato's philosophical journey led him to a profound understanding of reality that embraced change and constancy, both the physical and the metaphysical. Our journey takes us from the tangible to the abstract, from the seen to the unseen, into the heart of a unified and complete Knowledge.

Conjecture

In the realms of Knowledge and Opinion, Plato proposed a hierarchy. At the base of this hierarchy, we find Conjecture, which Plato considered to hold the lowest degree of Opinion. This level is populated by the shadows and reflections of physical objects, the ephemeral and subjective experiences twice removed from reality.

Parmenides	Knowledge	REASON (Dialectic, Interrelations)	HIGHER FORMS (Good)
		UNDERSTANDING (Axioms, Mathematics)	FORMS (Axioms, Mathematics)
Heraclitus	Opinion	BELIEF (Perception)	THINGS (Spatial Displacement)
		CONJECTURE (Subjective)	SHADOWS (Twice removed)

To understand this, let's consider Plato's concept of Forms. For Plato, the Form of an object is its true reality. It's the perfect, unchanging essence of the thing, existing in a realm beyond our physical world. The physical objects we interact with daily are merely spatial displacements of these Forms. They are imperfect copies, participating in true reality to a much lesser extent than the Forms themselves.

Now, let's take this a step further. Imagine a tree. According to Plato, the physical tree you see in the park is a spatial displacement of the Form of the tree. The tree participates in reality, but not as fully as the Form of the tree. Now, consider a picture, painting, or a reflection of the tree in a pool of water. These images are various representations or "*shadows*" of the physical tree, which is already a lesser participant in reality. Therefore, the tree's picture, painting, or reflection is twice removed from the Form of the tree, participating in true reality even less than the physical tree.

Belief

When we see a tree, touch a stone, or smell a flower, we engage in an Opinion of Belief. We interact with the physical manifestations of the Forms, the perfect, unchanging essences in a realm beyond the physical world.

But how do these Forms, these abstract, ideal concepts, make their presence felt in our physical world? How do they cast their shadows on the cave wall of our reality to borrow an image from Plato's famous allegory? This question led Plato to postulate the existence of a medium, a conduit through which the Forms are reflected or make their appearances in our world.

Plato named this medium "*space*." It is the canvas upon which the Forms paint their reflections, the stage where the drama of our physical world unfolds. In Plato's philosophy, space is not just a void but an emptiness to be filled. Space is a necessary medium in the interplay between abstract Forms and their spatial representations.

Yet, despite space's crucial role, space remains a mystery. It is a brute factuality, existing without explanation or justification. Because space itself is not a Form or a representation of a Form, space is "*unknowable*" as all Knowledge is only of Forms, and space is not a Form. Space "*is*," and it "*must be*." It is the silent, unexplained enigma at the heart of Plato's philosophical universe. The four basic Forms — earth, air, fire, and water — are the fundamental building blocks of the physical world, each representing its corresponding Form through spatial displacement.

Imagine each element as a unique geometric shape: earth as a cube (6-faces), "*fire*" as a pyramid (5-faces), water as an icosahedron (20-faces), and air as an octahedron (8-faces). These geometric shapes are not just random choices but the spatial configurations of their respective Forms. The Form of Fire, for instance, is not represented in a material pyramid but in a spatial one.

Plato's philosophy doesn't stop at these four solids. He further reduces them to two triangles — the half-equilateral and the half-square. In this sense, Plato's philosophy is a form of extreme Pythagoreanism, where "*things*" (spacial displacements) are numbers. In its most fundamental form, sense perception reduces to spatial geometry, and geometry, in turn, reduces to arithmetic.

Understanding

Crossing the divided line between Opinion and Knowledge, we now transition from Heraclitus' world of spatial flux, where everything constantly changes, to Parmenides' realm of immutability, where unchanging concepts reign supreme.

This third level of knowledge is the realm of Understanding, the domain of mathematics and scientific laws, that focuses on the enduring truths of abstract concepts. Yet, as we navigate this realm of Understanding, we encounter certain limitations.

Firstly, this mathematical knowledge and scientific law level rests on unexamined first principles such as axioms and postulates. These foundational assumptions underpin our understanding, yet they need to be more substantiated and taken as given without further inquiry.

Secondly, despite mathematics' and scientific laws' abstract nature, this level of knowledge remains tethered to the physical world. Take geometry, for instance. While it deals with abstract shapes and spaces, its roots are firmly planted in the spatial world we perceive around us. This connection to the tangible world, while providing a bridge between the abstract and the concrete, also limits the degree of abstraction we can achieve.

Lastly, while offering deep insights into specific areas, the realm of Understanding fails to provide a unified vision of the cosmos. It is fragmentary, offering us puzzle pieces but not the complete picture. Each mathematical theorem or scientific law provides a glimpse into the universe's workings, yet they still need to merge into a comprehensive understanding of how everything works together.

Reason

Continuing our path across the divided line of Opinion and Knowledge, past the first level of Understanding, we now arrive at the pinnacle of Knowledge — a realm illuminated by the power of Reason and the method of dialectic. This level transcends the limitations of the ever-changing physical world of spatial displacement, offering a holistic vision of the universe — the Parmenidian one.

At this stage, we are no longer confined to the realm of concrete, changing, and particular objects of perception, a domain characterized by the flux of Heraclitus. Instead, we ascend into the realm of universal, abstract comprehension of unchanging concepts, echoing the immutability of Parmenides.

Here, we engage in dialectic, a form of discourse that seeks to establish true first principles. This is the realm that studies the Forms, the ultimate reality. At this level, all divisions and steps melt away, and we experience the Form of the Good, the ultimate unifying reality. This is the pinnacle of Plato's philosophical system, a state of knowledge where we grasp the unity of all things and comprehend the true nature of reality.

Plato's Knowledge: Physics

Abstract: In exploring Plato's perspective on physics, we delve into his interpretation of the physical world as a realm fundamentally grounded in Opinion rather than true Knowledge, which is reserved for the unchanging, eternal Forms. Plato's Theory of Forms conceptualizes the material world as a fluctuating space where individuals perceive sensory images, unique to the motions of their sense organs, as mere shadows or reflections of the true Forms. These Forms exist in a transcendent space — a resistant, largely unintelligible "*Receptacle of Becoming*," unaffected by the temporal changes that characterize the physical world. Consequently, as per Plato, physics is viewed as a "*likely story*," an acknowledgment of the existence of space and its manifestations, yet devoid of the comprehensive analytical grasp intrinsic to the understanding of the immutable Forms. This discourse underscores the limitations of empirical approaches in grasping the depth of Plato's largely metaphysical and idealistic philosophy, which places supreme reality in the perfect, unchangeable realm of abstract concepts or Forms.

**

In the grand theatre of Plato's philosophy, the ever-changing physical world of spatial displacement takes a backseat to the realm of the eternal and unchanging Forms. For Plato, if physics is about the Opinion of the physical world and its processes, and Knowledge is about the eternal and unchanging Forms, then physics is Opinion, not Knowledge.

Plato's concept of sensible images is tied to his theory of Forms. He agreed with Protagoras that every individual perceives the world differently due to the unique motions of their sense organs. In this ever-changing flux, what we perceive are not the objects themselves but their reflections or shadows in a medium Plato called "*space*." This "*space*" is the "*Receptacle of Becoming*," an everlasting entity that provides a situation for all things that come into being, or more appropriately, Forms represented by spatial displacement. Space is apprehended without the senses, is resistant to rational analysis, and remains largely unintelligible as space is not a Form.

Thus, for Plato, physics is only a "*likely story*," a narrative that acknowledges the brute factuality of space but cannot fully explain it.

Plato's philosophical ideas, particularly his Theory of Forms, don't lend themselves easily to the empirical and experimental approach of the scientific method as we understand it today. Plato's philosophy was largely metaphysical and idealistic, positing that the highest form of reality exists in abstract concepts or "*Forms*," which are perfect and unchanging and that the physical world we perceive is merely a shadow or imperfect copy of this ideal realm.

Plato's Scientific Method

Abstract: Plato's approach to scientific inquiry, grounded in ancient philosophy, carves out a unique and foundational position in the history of scientific thought. Central to this approach is deductive reasoning, a principle where conclusions logically follow from established premises, a method particularly influential in mathematics and theoretical physics. Additionally, Plato emphasized the significant role of rationality and dialogues, the latter serving to explore and refine complex ideas through discussion, resembling an early form of peer review seen in modern scientific approaches. Moreover, his Theory of Forms, albeit more philosophical, forms a developing conception of scientific laws, illustrating the immutable principles governing reality. Despite facing critiques, notably for its non-empirical nature and its focus on immutable Forms, Plato's scientific method significantly influenced the evolution of scientific thought, laying the groundwork that fostered the development of structured, systematic inquiry seen in modern scientific methods. This methodology is a testament to the rich, diverse intellectual traditions that have historically shaped the pursuit of knowledge and scientific discourse.

**

Deductive Reasoning: At the heart of Plato's scientific method is deductive reasoning, a process where conclusions are reached by logical inference from premises regarded as true. If the initial premises in a deductive argument are true, then the conclusion must also be true. This emphasis on logical consistency and sound argumentation is instrumental in developing numerous scientific areas, notably mathematics and theoretical physics. Plato's focus on deduction can be observed in his mathematical investigations, where he stressed the importance of axiomatic systems, which later became foundational in geometry.

The Role of Rationality: Plato underscored the importance of rationality in understanding the world. According to his philosophy, the rational mind can access and comprehend the Forms, which are reality's eternal and unchangeable truths. This deep-seated belief in rationality as the path to knowledge laid the groundwork for the rationalist tradition in philosophy. It influenced the scientific method's focus on logic and systematic inquiry.

Dialogues as Method of Inquiry: In Plato's works, dialogues often serve as the medium to explore and dissect complex ideas. These dialogues, characterized by deep discussions and examinations of various viewpoints, are an early form of peer review, where ideas are scrutinized, tested, and refined through discourse. This method aligns with the collaborative and critical spirit of scientific inquiry, where peer review has become a cornerstone in validating and advancing scientific knowledge.

The Concept of Ideal Forms: While the Theory of Forms is more philosophical than scientific, it contains elements that can be seen as antecedents to scientific laws. In Plato's philosophy, the Forms represent reality's perfect and immutable aspects, transcending the physical world. This notion bears semblance to the concept of scientific laws — unchangeable principles that govern the operations of the universe. Through this lens, Plato's idea of Forms may be perceived as a philosophical precursor to the abstraction and generalization that characterize scientific theories and laws.

Limitations and Critiques: It is crucial to note the limitations of Plato's approach from a modern scientific perspective. Plato's focus on immutable Forms and his disdain for the mutable physical world can be seen as counter to the empirical basis of the scientific method, which relies on observation and experimentation. Furthermore, his reliance on "*a priori*" reasoning, without empirical validation, is often seen as a deviation from the empirically grounded methods that characterize modern science.

Conclusion: While Plato's scientific method may not align directly with modern science's empirical and experimental foundations, it has undoubtedly contributed significantly to the evolution of scientific thought. Through deductive reasoning, the focus on rationality, the use of dialogue as a method of inquiry, and the early notions of immutable laws represented by the Forms, Plato's approach to understanding the world sowed seeds that would later flourish into the structured, systematic method of inquiry that is the scientific method. Plato's philosophy is an important reminder of the deep historical roots and diverse intellectual traditions that have shaped the pursuit of knowledge and the development of science.

Plato's Ethics

Abstract: In the intricate framework of Plato's ethical philosophy, the notion of the Good and the Theory of Forms stands central, portraying a transcendental realm of immutable entities, of which the Form of the Good is the epitome, serving as a beacon of ultimate truth and knowledge. Plato advocates for cultivating virtues — wisdom, courage, temperance, and justice — positioning them as facets of knowledge essential in guiding individuals toward morally upright living. Central to achieving this moral living is the harmonious functioning of the soul, comprising rational, spirited, and appetitive parts, orchestrated through reason and an alignment with the Good. Underpinning this is a transformative educational journey, vividly illustrated in the "*Allegory of the Cave*," aiming to nurture philosopher-kings endowed with wisdom and justice to lead society. This ethical outline is complemented by Plato's moral psychology and the Theory of Recollection, suggesting that learning is a rediscovery of innate knowledge held in the immortal soul, steering individuals toward ethical conduct. Consequently, Plato's philosophy delineates a path of personal and societal moral evolution, urging individuals to union with the Good, thereby fostering a society rooted in wisdom and justice.

**

The Good and the Forms: Central to Plato's philosophy is the Theory of Forms, which posits a world of eternal, unchanging Forms or Ideas that exist beyond the sensible world. The highest of these Forms is the Good, the source of all truth and understanding. Every soul aspires to understand this ultimate Form of the Good, as it represents the pinnacle of truth and the ultimate object of knowledge. Understanding the Good forms the basis of ethical behavior, allowing one to align oneself with the highest form of good.

Virtue Ethics and the Role of Virtue: In Plato's ethical framework, virtue is central. Virtue, to Plato, is a form of knowledge, the knowledge of the Good, which guides individuals to act justly and wisely. Plato identified four primary virtues:

1. **Wisdom**: The ability to judge what is true and right.
2. **Courage**: The quality that allows individuals to face fear and uncertainty with resolve.
3. **Temperance** (or Moderation): Mastering desires helps maintain a balanced life.
4. **Justice**: Often seen as the supreme virtue, it represents the harmony of the soul, where all its parts function in unity to achieve the good.

The Soul and Its Tripartite Division: Plato perceived the soul as comprising three distinct parts:

1. **The Rational Part**: Focused on acquiring knowledge and understanding the Forms, particularly the Form of the Good.
2. **The Spirited Part**: Driven by honor and a sense of duty, it assists the rational part in controlling the desires.
3. **The Appetitive Part**: Concerned with basic desires and physical pleasures.

According to Plato, ethical living is achieved when the rational part governs the soul, keeping the spirited and appetitive parts in check. This harmony of the soul, orchestrated by reason, leads to virtuous living and justice.

Education and the Role of Philosopher-Kings: Plato envisioned an ideal society led by philosopher-kings who have a profound understanding of the Forms and are thus equipped to govern justly and wisely. Through their deep comprehension of the Good, these leaders guide society toward justice and harmony. Education plays a pivotal role in nurturing individuals capable of understanding the Forms. This philosophical education, depicted vividly in the "*Allegory of the Cave*," is a journey from the shadows of ignorance to the luminous world of knowledge and understanding, fostering a class of rulers who embody wisdom and justice.

Moral Psychology and Personal Development: Plato's moral psychology revolves around achieving harmony within the soul through cultivating virtues and pursuing knowledge. Individuals are expected to undergo personal development, a lifelong process of aligning closer with the Good. This alignment fosters moral development, where individuals gradually become more virtuous, achieving a harmonious soul-state that resonates with the Forms' eternal truths.

Ethical Implications of the Theory of Recollection: In Plato's philosophy, learning is a form of recollection, as the immortal soul already possesses knowledge of the Forms from prior existence. This theory implies that ethical knowledge is innate and that learning is a journey of rediscovering this inherent knowledge, guiding individuals to live ethically.

Conclusion: Plato's ethical theory, deeply integrated with his metaphysical and epistemological doctrines, offers a profound perspective on moral living. Through pursuing the Good and cultivating virtues, individuals can aspire to achieve a harmonious state of the soul, fostering a society that embodies wisdom and justice and nurturing the highest potentials of the human spirit.

Plato's philosophy beckons individuals on a transformative journey from the realm of shadows and illusions to the luminous world of eternal truths, where the ultimate goal is to achieve unity with the Good, the highest pinnacle of ethical and intellectual achievement.

**

REVIEW QUESTIONS

True/False Questions:

1. Plato believed that the physical world we perceive is the ultimate reality, and the Forms are merely abstract concepts with no true existence.
   True or False?

2. In Plato's ethical framework, justice is considered the harmony of the soul, where all its parts function in unity to achieve the good.
   True or False?

Multiple-Choice Questions:

3. Which of the following is NOT one of the four primary virtues identified by Plato?
   a) Wisdom
   b) Courage
   c) Temperance
   d) Honesty

4. According to Plato, the ultimate source of all truth and understanding is:
a) The Rational Part of the Soul
b) The Form of the Good
c) The Physical World
d) The Allegory of the Cave

Clinical Vignette:

5. A physician aims to treat not just the physical symptoms of a patient but also considers their mental and emotional well-being. This holistic approach reflects the influence of which philosophical concept from Plato?
a) The Allegory of the Cave
b) The Theory of Recollection
c) The Harmony of the Soul
d) The Theory of Forms

Basic Science Vignette:

6. Plato's emphasis on rational inquiry and systematic investigation laid the groundwork for early scientific thought. Which of the following best describes how Plato's philosophical approach influenced the development of the scientific method?
a) By relying solely on sensory experiences to understand the world
b) By emphasizing the importance of empirical observation without rational analysis
c) By promoting the use of deductive reasoning and logical consistency in investigations
d) By dismissing the need for structured inquiry and critical examination

Philosophy Vignette:

7. Plato's Theory of Forms posits that true reality consists of eternal and unchanging ideas, while the physical world is a mere shadow of these forms. How might this concept influence contemporary discussions on the nature of scientific laws and theories?
a) It suggests that scientific laws are merely subjective opinions with no objective basis
b) It implies that scientific theories are unchanging and eternal, reflecting a deeper reality
c) It encourages the idea that scientific laws are based solely on empirical observations without underlying principles
d) It promotes the view that scientific theories should be constantly revised and updated based on new evidence

Correct Answers:

1. False
2. True
3. d) Honesty
4. b) The Form of the Good
5. c) The Harmony of the Soul
6. c) By promoting the use of deductive reasoning and logical consistency in investigations
7. b) It implies that scientific theories are unchanging and eternal, reflecting a deeper reality

BEYOND THE CHAPTER

Plato (427-347 BCE)

- *Plato: Complete Works* edited by John M. Cooper
- *Plato: A Very Short Introduction* by Julia Annas
- *The Cambridge Companion to Plato* edited by Richard Kraut

Plato's Knowledge: Divided Line

- *Plato's Theory of Knowledge: The Theaetetus and the Sophist* by Francis M. Cornford
- *The Divided Line and the Structure of Plato's Republic* by Gerasimos Santas (Chapter in *History of Philosophy Quarterly*)
- *Plato's Divided Line: Essay on the Meaning of Plato's Thought* by Nestor-Luis Cordero

Plato's Knowledge: Physics

- *Plato's Natural Philosophy: A Study of the Timaeus-Critias* by Thomas Kjeller Johansen
- *Plato's Philosophy of Science* by Andrew Gregory
- *Plato's Cosmology: The Timaeus of Plato* translated by Francis M. Cornford

Plato's Scientific Method

- *Plato's Method of Dialectic* by Julius Stenzel
- *Plato's Philosophical Method: The Use of Dialectic in the Dialogues* by Nicholas D. Smith (Chapter in *The Cambridge Companion to Plato's Republic*)
- *Plato and the Method of Analysis* by John Palmer (Chapter in *Philosophy of Science*)

Plato's Ethics

- *Plato's Ethics* by Terence Irwin
- *Plato on Morality and the Good Life* by Gerasimos Santas (Chapter in *The Cambridge Companion to Plato*)
- *Plato's Moral Theory: The Early and Middle Dialogues* by Terence Irwin

***

Philosophy of Science — Medical Research

Chapter 65: Aristotle (384-322 BCE) — The Systematizer: Empiricism and Causality

Aristotle's integration of empirical observation and rational inquiry laid foundational principles for the scientific method, influencing medicine and ethics.

Abstract: Aristotle, one of history's quintessential polymaths, continues to shape various domains of human thought and understanding, with contributions spanning philosophy, science, medicine, and ethics. His groundbreaking blend of rationalist and empiricist philosophies fostered a harmonious approach to knowledge, intertwining empirical observation with rigorous, reasoned inquiry. He forged a precursor to the modern scientific method in science, uniting systematic observation and logical analysis in a continual dance of deduction and induction, laying foundational principles that have nurtured the evolution of scientific methodologies and epistemology. Moreover, his medical endeavors, particularly through meticulous biological investigations, shaped ancient conceptions of anatomy and the broader life sciences, indirectly facilitating methodological advancements in medical practice. Furthermore, Aristotle's ethical teachings, encapsulated predominantly in his "*Nicomachean Ethics*," have echoed through the centuries, indirectly fostering the development of modern bioethical principles rooted in human flourishing and moral virtue. Ultimately, Aristotle's legacy is a testimony to the enduring impact of a thinker who melded empirical observation with intellectual rigor, fostering a systematized approach to knowledge that resonated profoundly across disciplines and generations, fundamentally shaping Western intellectual tradition.

**

Introduction: Aristotle, a titan of ancient Greek philosophy, stands tall in the annals of history as one of the foremost thinkers who shaped Western intellectual thought. Born in Stagira in 384 BCE, he became a prominent figure at the Platonic Academy in Athens, later founding his institution, the *Lyceum*. Aristotle delved into various subjects, including metaphysics, ethics, politics, biology, and poetics. Unlike many thinkers of his time, he emphasized the importance of empirical observation and systematic categorization in understanding the world around him. His works, such as the "*Nicomachean Ethics*," "*Politics*," and "*Metaphysics*," continue to influence modern thought across disciplines, from the humanities to the natural sciences. Even today, more than two millennia after his death, Aristotle's profound insights and rigorous methodologies remind us of the interconnectedness of knowledge and the importance of seeking truth through reasoned inquiry.

Rationalism: Aristotle, in contrast to some of his predecessors, notably favored a rationalist approach to understanding the world, emphasizing the primacy of reason as the highest human faculty. While he recognized the significance of sensory experience as the starting point for knowledge, he believed true comprehension arose from the exercise of the intellect. Unlike the Platonic world of forms, which was accessible only through intellectual contemplation, Aristotle posited that the forms were embedded within the objects of the sensory world. For him, the act of abstraction by the mind — separating the form from its material embodiment — was a rational process that led to genuine knowledge. In his "*Posterior Analytics*," Aristotle detailed a method of deriving universal principles from particular observations, culminating in syllogistic reasoning. This deductive system stood a testament to his faith in rationality, underscoring that conclusions derived from valid premises were invariably true. By championing a philosophy rooted in empirical observations and rational deductions, Aristotle laid the groundwork for a harmonized approach to epistemology, reflecting his belief in the unparalleled capacity of human reason to discern and comprehend the truths of the universe.

Empiricism: While Aristotle is often celebrated for his rationalist tendencies, his empiricist orientation is vital to his philosophical framework. Aristotle firmly believed that all knowledge begins with sensory experience, a stance that starkly differentiates him from his teacher, Plato, who posited an innate realm of ideas. In Aristotle's perspective, the external world provides the raw material upon which the mind acts, shaping and categorizing experiences to form concepts and knowledge. This is evident in his work "*De Anima*," where he underscores the mind's blank-slate character, asserting that no innate ideas are present at birth. Instead, our cognitive understanding is progressively cultivated through interactions with the surrounding environment. In his biological treatises, Aristotle's meticulous observations and classifications of organisms underscore his commitment to empiricism, as he sought to derive knowledge from careful observation rather than mere speculation. Thus, while he recognized the indispensable role of reason in refining and understanding sensory experiences, the foundational bedrock of his epistemological approach was rooted in the empirical world, emphasizing the intricate dance between observation and intellect in the pursuit of truth.

The Scientific Method: Aristotle's deep-rooted dedication to systematic observation and logical reasoning is a precursor to the modern scientific method. While the methodologies of contemporary science have significantly evolved since his time, Aristotle's foundational principles laid an important groundwork. He championed gathering empirical data through observation and subjecting these observations to reasoned analysis. In his biological works, for instance, he meticulously classified organisms based on shared characteristics, exhibiting an early form of taxonomy. Further, in his "*Posterior Analytics*," he introduced moving from particular observations to universal truths, a rudimentary form of inductive reasoning. However, he also employed deductive reasoning, using general principles to make specific predictions. These intertwined approaches reflect the complementary induction and deduction processes central to today's scientific method. While Aristotle's approach was not experimental in the way modern science is, and some of his conclusions were later proven incorrect, his emphasis on systematic investigation, logical analysis, and the synthesis of observation and reason undeniably paved the way for future scientific endeavors and methodologies.

Medicine: Aristotle, although primarily known as a philosopher, played a significant role in shaping ancient conceptions of medicine through his biological and natural investigations. His inquiries into the natural world laid a foundational understanding of anatomy and the functions of various organisms, offering essential insights that subsequent medical practitioners utilized. While he wasn't a physician like Hippocrates, Aristotle's detailed examinations of animals in works like "*Historia Animalium*" provided comprehensive observations about their anatomy, reproduction, and behavior. These insights indirectly contributed to the broader understanding of physiology and life sciences, which are intrinsically linked to medicine. Furthermore, Aristotle's emphasis on empirical observation and systematic categorization encouraged a more structured and observational approach to medicine. His belief that nature operated according to certain rational principles led to the idea that illnesses, too, might have natural causes that could be understood and treated. At the same time, Aristotle's direct contributions to medicine were overshadowed by contemporaries like Hippocrates and later figures such as Galen; his overarching influence on systematic and empirical scientific approaches unquestionably laid the groundwork for methodological advances in the medical field.

Ethics: Aristotle's ethical teachings, primarily encapsulated in his "*Nicomachean Ethics*," have left a lasting impact on the conceptual framework of morality and have indirectly influenced principles like patient autonomy (informed consent), practitioner beneficence (do good), practitioner nonmaleficence (do no harm), and public justice (be fair). Central to his ethics is the idea of *eudaimonia*, often translated as "*flourishing*" or "*well-being*," which he believed to be the ultimate end of human life. Aristotle championed the "*golden mean*," balancing excess and deficiency in character traits to achieve this state. While he did not directly articulate modern bioethical principles, the foundations are evident. The idea of autonomy resonates with his emphasis on the significance of rational agency and self-directed virtues in determining one's good life. Beneficence aligns with his notion of active contribution to the polis and the good of others as essential for individual flourishing. Nonmaleficence echoes his caution against vices, actions that harm oneself or others, and the need for moral virtues to prevent such harm. Lastly, justice, extensively discussed in his "*Politics*," is viewed as a cardinal virtue crucial for the harmonious functioning of society. Aristotle's emphasis on character, virtue, and the communal aspects of human life deeply informs these ethical concepts, framing them within a broader vision of human flourishing and social cohesion.

Conclusion: Aristotle's enduring impact on human thought spans many domains, underscoring his status as a polymathic intellectual luminary. His contributions have etched indelible marks on philosophy, science, medicine, and ethics. While his rationalist inclinations and emphasis on empirical observation might seem dichotomous, they harmoniously shaped his approach to knowledge, fostering a blend of reasoned inquiry and systematic investigation. His pioneering notions on deduction and induction laid crucial groundwork for the scientific method's evolution, bridging the gap between observation and logical reasoning. His meticulous biological investigations in medicine set foundational principles for understanding anatomy and the natural world's intricacies. Ethically, Aristotle's focus on virtues and flourishing life contributed to the conceptual foundation of principles like patient autonomy (informed consent), practitioner beneficence (do good), practitioner nonmaleficence (do no harm), and public justice (be fair), resonating through time in the context of moral philosophy and bioethics. A beacon of wisdom and intellectual rigor, Aristotle's legacy is an enduring reminder of a single thinker's profound impact on shaping human understanding across diverse disciplines and generations.

Aristotle's Legacy: Systematized knowledge across diverse fields, championed empiricism, and introduced causality concepts, thereby laying the groundwork for much of Western science and philosophy.

**

A Closer Look

Aristotle's Knowledge

Abstract: In examining Aristotle's distinct approach to epistemology and his effort to reconcile the prevalent philosophical theories of his time, the narrative delves into the intricacies of Aristotle's metaphysical framework that harmoniously integrates the material with the moral and the sensory with the abstract. Tracing Aristotle's formative years and intellectual maturation at Plato's Academy, the text offers a nuanced analysis of his philosophical divergence from the materialists and Plato — his mentor. The focal point is Aristotle's ingenious concept of reality, where he conceptualizes the dual essence of every entity — the "*whatness*" or form and the "*thisness*" or matter, both intertwined and guiding the matter toward its fullest expression, a principle he coined as "*entelechy*." Furthermore, the text compares Aristotle's and John Locke's perspectives on human cognition, highlighting both their agreement on sensory experience as the basis of knowledge and their divergent views on the nature of universals. In stark contrast to Plato's quest for knowledge in an isolated world of forms, Aristotle's philosophy embarks on a grounded exploration of knowledge in the complex interconnectivity of "*formed matter*," establishing him as a figure who rooted the pursuit of knowledge firmly in the sensory and tangible world, offering a robust pathway to understanding the complexities of reality.

**

In the year 384 BCE, a historic event unfurled in the hamlet of Stagira in Thrace: the birth of a prodigious mind, Aristotle. His father, a respected physician to the Macedonian king, shaped the early years of the young Aristotle. At this juncture in history, Plato was forty-three years old, and Socrates' influential teachings were still echoing in the minds of thinkers despite his being gone from the world for fifteen years.

The seventeen-year-old Aristotle was then sent off on a journey to Plato's renowned Academy in Athens. This hallowed ground would become his intellectual home for the next two decades, with his departure only occasioned by the death of Plato in 347 BCE. The wheel of destiny then turned Aristotle toward tutoring the youthful Alexander, later known as Alexander the Great, son of King Philip of Macedonia.

How does Aristotle's approach to understanding knowledge or epistemology distinguish itself from his contemporaries and predecessors, specifically the materialists and Plato? The materialists, devoted to the physical world, struggled to account for humanity's moral dimension, as morality was independent of atoms and the void. As such, morality couldn't be seen, weighed, or measured. If morality wasn't an atom and a void, and if only atoms and the void exist, then what was morality?

On the other hand, Plato held an intriguing view where he placed the world of Ideas, or Forms, separate from the tangible world, a concept that posed a significant problem. By segregating moral Forms from our sensory experience, Plato found it challenging to tether morality to the world of spatial displacement perceived by our senses.

Amidst these philosophical tugs of war, Aristotle envisioned a theory of reality that could reconcile this dichotomy. His goal was to propose a metaphysical framework where both the sensory world and morality could coexist and be acknowledged as "*real*." Aristotle wished to confirm two key principles: firstly, he wanted to validate our sensory perception that reality is not static but ever-changing. This would reaffirm the value of reason as a tool for acquiring knowledge. Secondly, he sought to verify our innate intuition of humans as moral beings. In this manner, Aristotle aimed to carve a philosophical path that harmonized the material with the moral and the sensory with the abstract.

Examining the world through Aristotle's lens, every individual substance or particular comprises two intertwined aspects. First is the "*whatness*" or form — the common properties shared with other particulars that serve a specific purpose or function. Second, the "*thisness*" or matter — unique to the individual and made of physical "*stuff*." These concepts aren't separate but entwined.

Consider an acorn: An acorn's "*whatness*" lies in the characteristics it shares with other acorns, its "*thisness*" in its individuality. As the acorn grows into an oak tree, we see the material potentiality actualized, the acorn's purpose — its "*form*" — fulfilled. This development, this growth, is a process of coming-to-be, with the form or purpose guiding the matter toward its full expression — a concept Aristotle termed "*entelechy*."

However, this transformative journey isn't without its paradoxes. If we say that A changes to B, aren't we saying that A is itself and not itself? This apparent contradiction dissolves under Aristotle's metaphysical lens, where the enduring matter of A takes on a new form to become B, with growth seen as a systematic succession of smaller changes.

In the acorn's journey to becoming an oak, we encounter many stages of growth. However, the unity among these stages is in their collective progression toward a common purpose. Aristotle's resolution to the age-old puzzle of the one and the many lies in recognizing this unifying purpose.

To fully grasp any phenomenon, Aristotle proposed that we must understand its four causes:
1. the Material Cause (the material out of which it's made),
2. the Efficient Cause (motion by which it came to be),
3. the Formal Cause (into which it's made), and
4. the Final Cause (for the sake of which it's made).
These can be streamlined into the core concepts of matter (combining Material and Efficient Causes) and form (combining Formal and Final Causes).

Aristotle and John Locke, the 17th-century English philosopher, shared similarities in their views regarding human cognition, specifically the idea that all knowledge is derived from sensory experience. However, some key differences in their perspectives shouldn't be overlooked.

Locke is famously known for the theory of the "*tabula rasa*," or "*blank slate*." According to this view, humans are born with minds devoid of innate ideas or principles. Every bit of knowledge we gain, Locke argued, comes from our experience and interaction with the world.

Aristotle's view of the mind is somewhat similar. He, too, saw the mind as initially blank and capable of receiving impressions from the sensible forms of objects. Like wax takes the impression of a seal, the mind takes on the intelligible forms of things. Both philosophers agree that learning and understanding are impossible without sensory input.

But there's a nuanced difference in their understanding of forms and universals. For Locke, the mind abstracts general ideas from our particular sensory experiences. He, unlike Aristotle, emphasized that our knowledge of these abstractions could be flawed, and the universals we draw might not accurately represent the world.

On the other hand, Aristotle believed that universal forms are embedded within the particulars and that the mind can accurately perceive these forms. For Aristotle, the universals are real aspects of the world, not merely mental constructs, as Locke would argue.

So, while both Aristotle and Locke agree on the fundamental premise that sensory experience is crucial for knowledge, they diverge on the nature and reality of universals and the accuracy of our mental representations.

Plato sought knowledge in a separate world of Forms, with the Form of the Good as the pinnacle, a mysterious and incommunicable ideal beyond our tangible realm. On the other hand, Aristotle remained steadfastly grounded in our world. Plato sought abstract, general, static Knowledge, scrubbed clean of all particularity. Aristotle's quest, conversely, sought Knowledge in the multiple interconnectednesses of particulars — in "*formed matter*."

Aristotle's Scientific Method

Abstract: This discourse explores Aristotle's seminal contribution to the evolution of the scientific method, elucidating how his philosophies laid a foundation for empirical study and systematic inquiry. Aristotle's stance on empiricism heralded the significance of sensory experiences as conduits for acquiring knowledge, paralleling contemporary methods prioritizing data collection and observation. His proclivity for inductive reasoning, in which general conclusions are formulated from specific observations, is vital in contemporary scientific explorations, facilitating theory and hypothesis development. Furthermore, Aristotle's nuanced approach to causality offers a rich tapestry for investigating phenomena, advocating for a comprehensive understanding grounded in material, efficient, formal, and final causes. This approach, though transformed, still resonates within modern scientific endeavors that incessantly seek causal explanations. In addition, Aristotle's pioneering efforts in classification and systematization, notably in biology, presaged contemporary scientific taxonomy. Moreover, his strides in logic and argumentation, particularly in fostering syllogistic logic, have sculpted the frameworks within which scientific arguments are built and scrutinized today. The passage culminates by acknowledging Aristotle's forays into "*natural philosophy*," marking the infancy of the natural sciences through his meticulous observations and explicative endeavors concerning the natural world, underpinning the longevity of his influence in shaping scientific thought and methodology.

**

Aristotle's philosophy has influenced scientific thought, even though his approach differs from the modern scientific method:

Empiricism: Aristotle's belief that knowledge comes from sensory experience laid the groundwork for empiricism, a cornerstone of the scientific method. Empiricism emphasizes the role of evidence and experience in forming knowledge, aligning with the scientific method's data collection and observation stages.

Inductive Reasoning: Aristotle often used inductive reasoning, which involves broad generalizations based on specific observations. This form of reasoning is crucial in the scientific method, as it allows scientists to develop theories and hypotheses based on the data they collect.

Causality and Explanation: Aristotle's emphasis on understanding the causes of phenomena can be seen as an early form of scientific explanation. He proposed four types of causes (material, efficient, formal, and final) to explain why things are as they are comprehensive. While modern science doesn't use Aristotle's four causes, seeking causal explanations remains central to scientific inquiry.

Classification and Systematization: Aristotle's work in biology, where he classified animals into different species based on their characteristics, can be seen as a precursor to the systematic approach used in many scientific disciplines today.

Logic and Argumentation: Aristotle contributed significantly to logic, including developing syllogistic logic. His emphasis on clear argumentation and logical consistency has influenced how scientific arguments are constructed and evaluated.

Natural Philosophy: Aristotle's "*natural philosophy*" was an early form of the natural sciences. He made observations about the natural world and tried to explain them, a fundamental aspect of the scientific method.

While Aristotle's approach to knowledge differed from the experimental and hypothesis-driven approach of modern science, his emphasis on observation, explanation, and logical reasoning has had a lasting impact on the development of scientific thought.

Aristotle's Ethics

Abstract: In this discourse we explore Aristotle's approach to ethics, scrutinizing the foundational elements that define his concept of "*the good*" and how it serves as a lodestar in human actions and pursuits. Departing from Plato's theory of forms and the relativist stance of the Sophists, Aristotle proposes a realistic approach to ethics grounded in the tangible reality perceived through human experiences and actions. This realism manifests as an understanding that the ultimate goal of life is achieving happiness, characterized not by transient joy but by fulfilling one's inherent potential and purpose in harmony with one's true nature. Aristotle nuances this theory further, introducing the pivotal role of social relationships and community in nurturing virtue and realizing genuine happiness. This extends to delineating three distinct types of friendships, emphasizing the profound impact of virtuous friendships in fostering moral growth and contentment. Through a deep dive into Aristotle's perception of the contemplative life, the discourse illuminates the timeless relevance of his ethical philosophy in navigating the intricacies of human existence and the pursuit of a fulfilling life.

**

The grand quest of ethics is to unravel the true essence of the good. Ethics thrives not on cataloging the multitude of objects that humanity deems good but rather on discerning the very nature of goodness. In each unique situation where a person grapples with alternatives, there is an "*optimal*" course of action, a truth in ethical decisions.

Aristotle's ethical realism distinguished his philosophy from Plato and the Sophists. For Plato, Opinion was seen as an inadequate shadow of Knowledge, an imperfect precursor to being replaced by the pursuit of the "*Form of the Good.*" The Sophists also saw ethics as a realm of Opinions. However, they inferred this to mean ethical judgments were inherently subjective and relative, not truly in tune with reality. Aristotle, however, believed in the reality of perception. He argued that ethics, although different from physics and other sciences, still offered knowledge about reality.

Aristotle's ethics, much like a mirror, reflects his metaphysical stance. For him, the good is what ultimately satisfies an entity's purpose, its end, which is its form. In defining the human good, Aristotle puts forward a simple yet profound truth: every pursuit and action aims at some form of good. The good is the target at which all things aim. Yet, the chief good must be desirable for its own sake. This ultimate good, Aristotle believes, is happiness. *Happiness is the activity of the soul by virtue across a complete life*. It is the end of action, something final and self-sufficient.

Anything is happy when it fulfills its designated function, actualizing the form for which it was designed. The question of happiness for any entity aligns with the understanding of that entity's form — its purpose. Aristotle saw happiness as not merely fleeting joy or pleasure but a profound experience of living fully, realizing our potential without obstacles. He believed this is achieved when we're in harmony with our true nature. Amidst our complex lives and numerous activities, Aristotle saw contemplation — the cogitation of supreme truths about the universe — as the highest activity that truly expresses and actualizes our nature.

However, this contemplative life isn't accessible to everyone. Not everyone is born with the intellectual capacity or the peace and resources to engage in such a high level of thinking. Here, Aristotle acknowledges that wealth, while not happiness itself, is vital for attaining it.

Solitary existence, according to Aristotle, can't lead to happiness. Virtue, he argues, is not a solo act. It needs other people, not just for its cultivation but also for its practice. Can a person showcase justice, kindness, or courage in isolation? Hardly, Aristotle would say.

Aristotle also differentiates between three types of friendships: those based on utility, those driven by pleasure, and the rarest type, friendships between good people sharing similar virtues. The latter friendships require time and familiarity, and they offer mutual encouragement and example in the art of virtuous living.

**

REVIEW QUESTIONS

True/False Questions:

1. Aristotle believed that true knowledge begins with sensory experience and that abstract concepts are derived from these experiences through reason.
   True or False?

2. Aristotle's ethical theory is rooted in the idea that happiness is achieved by fulfilling one's potential and living in harmony with one's true nature.
   True or False?

Multiple-Choice Questions:

3. Which of the following is NOT one of Aristotle's four types of causes?
   a) Material Cause
   b) Efficient Cause
   c) Formal Cause
   d) Philosophical Cause

4. According to Aristotle, which type of friendship is considered the highest and most virtuous?
   a) Friendship based on utility
   b) Friendship based on pleasure
   c) Friendship based on virtue
   d) Friendship based on status

Clinical Vignette:

5. A medical practitioner emphasizes the importance of understanding the underlying causes of a patient's illness, considering various factors such as the material, efficient, formal, and final causes. This approach reflects the influence of which philosophical concept from Aristotle?
   a) The Theory of Forms
   b) The Four Causes
   c) The Golden Mean
   d) The Tripartite Soul

Basic Science Vignette:

6. Aristotle's emphasis on empirical observation and systematic categorization laid the groundwork for modern scientific inquiry. Which of the following best describes Aristotle's approach to understanding natural phenomena?
   a) By relying solely on divine explanations for natural events
   b) By emphasizing the importance of empirical observation and logical analysis
   c) By focusing exclusively on mathematical abstractions without sensory data
   d) By dismissing the need for systematic inquiry and categorization

Philosophy Vignette:

7. Aristotle's concept of the "*golden mean*" in ethics promotes:
   a) Extreme behaviors and actions as the ultimate path to happiness
   b) The avoidance of all physical pleasures to achieve moral virtue
   c) A balanced approach to behavior, avoiding both excess and deficiency
   d) The pursuit of material wealth as the highest form of good

Correct Answers:

1. True
2. True
3. d) Philosophical Cause
4. c) Friendship based on virtue
5. b) The Four Causes
6. b) By emphasizing the importance of empirical observation and logical analysis
7. c) A balanced approach to behavior, avoiding both excess and deficiency

BEYOND THE CHAPTER

Aristotle (384-322 BCE)

- *The Complete Works of Aristotle: The Revised Oxford Translation* edited by Jonathan Barnes
- *Aristotle: A Very Short Introduction* by Jonathan Barnes
- *The Cambridge Companion to Aristotle* edited by Jonathan Barnes

Aristotle's Knowledge

- *Aristotle on Knowledge and Learning: The Posterior Analytics* by David Bronstein
- *Aristotle's Theory of Knowledge* by Christoph Halbig (Chapter in *The Cambridge Companion to Aristotle's Nicomachean Ethics*)
- *Aristotle on Knowledge and Reality: A Philosophical Study* by Nicholas Rescher

Aristotle's Scientific Method

- *Aristotle's Scientific Method: The Analytics* by Paolo Crivelli
- *Aristotle's Theory of Scientific Method* by John W. I. Lee (Chapter in *A Companion to Aristotle*)
- *Aristotle's Scientific Method in the Nicomachean Ethics* by Michael V. Wedin (Chapter in *Philosophy and Phenomenological Research*)

Aristotle's Ethics

- *Aristotle's Ethics: Critical Essays* edited by Nancy Sherman
- *Aristotle's Nicomachean Ethics: An Introduction* by Michael Pakaluk
- *Aristotle's Ethics: Moral Development and Human Nature* by Hope May

***

Philosophy of Science

Medical Research

Chapter 66: Galen of Pergamum (GAY-len, 129-216 CE) — The Medical Synthesizer: Humoral Theory and Medical Ethics

Galen's blend of rationalism and empiricism in medical research established foundational principles for the scientific method and medical ethics. However, his contributions also included inaccuracies that influenced medical practice for over 1,500 years.

Abstract: Galen of Pergamum, a towering figure in ancient medicine and philosophy, profoundly shaped Western medical and scientific traditions. Championing a blend of rationalism and empiricism, he emphasized observational data and logical reasoning, laying the groundwork for modern scientific inquiry. His meticulous research, including animal dissections and vivisections, pioneered evidence-based inquiry. Galen's extensive writings on anatomy, physiology, and pharmacology became foundational texts, shaping medical discourse for centuries in Europe and the Islamic world. However, his adherence to the humoral theory and certain anatomical misconceptions perpetuated false beliefs that negatively impacted medical practice for over 1,500 years. His implicit adherence to medical ethics, evident in his respect for the Hippocratic Oath and emphasis on physicians' moral duty, set early ethical standards in medicine. Despite his contributions, modern ethical standards such as those outlined by the Institutional Animal Care and Use Committee (IACUC) would not condone many of his practices.

**

Introduction: Galen of Pergamum is one of antiquity's most influential figures in medicine and philosophy. Born in the vibrant city of Pergamum, in what is today modern-day Turkey, Galen's prodigious contributions fundamentally shaped medical understanding for nearly fifteen centuries. Trained in the rich medical traditions of Greece and the broader knowledge realms of the Roman Empire, he authored an astonishing array of works encompassing areas from anatomy and physiology to philosophy and ethics. His extensive dissections of animals, primarily monkeys and pigs, paved the way for breakthrough insights into the workings of various organs and systems within the body. While some of his conclusions were later revised with modern medical techniques and understanding, Galen's rigorous methodologies, holistic approach, and emphasis on empirical observation solidified his position as a cornerstone of Western medicine's foundation.

Rationalism: In the context of his time, Galen showcased a distinctively rationalist approach to medicine and science. Rather than solely relying on superstitions, divine intervention, or anecdotal accounts, as many of his contemporaries might, Galen emphasized the importance of empirical observation and logical reasoning. However, it is ironic that Galen, often celebrated as a critical thinker, fully embraced the Hippocratic corpus' humoral theory—a framework with no rational or empirical basis other than Galen's own rational and observational biases. The humoral theory posited that health depended on the balance of four bodily fluids (blood, phlegm, yellow bile, and black bile), was then continued to be accepted and perpetuated as medical doctrine, due to Galan's revered and unquestionable authority, for another 1,500 years. This irony highlights the complexities and limitations of even the most influential scientific minds.

Empiricism: Galen's commitment to empiricism, the philosophical doctrine that knowledge is derived from sensory experience, was a defining characteristic of his medical practice. Unlike those who might place unquestioned trust in ancient texts or inherited wisdom, such as was done with

Galen himself, Galen championed the direct observation and experiential study of the natural world. His extensive dissections of animals testify to this orientation, as he sought to gain firsthand knowledge of anatomy and physiology, even when such practices were controversial or challenging. For Galen, genuine understanding arose from tangible evidence rather than abstract theorization. His clinical methods also underscored this empiricist approach. As exemplified in the Hippocratic corpus, Galen carefully observed patients, recording symptoms, progressions of diseases, and the effects of various treatments, using these observations to refine his medical theories and practices. For example:

- **Fever Treatment**: Galen's observations led him to oppose the use of excessive cooling methods for treating fevers, advocating instead for balanced approaches that took into account the patient's overall condition.
- **Fracture Treatment**: Galen improved the methods for treating bone fractures, recommending the use of splints and proper alignment, which significantly reduced healing complications.

His magnum opus, "*On the Usefulness of the Parts of the Body*," exemplifies his empirical method, where he intricately detailed the functions and interrelations of bodily parts based on direct study. Through this unwavering commitment to empirical inquiry, Galen not only enhanced the medical knowledge of his time but also laid the groundwork for future generations to question, experiment, and learn through observation.

The Scientific Method: While the scientific method as we understand it today evolved much later, Galen's approach to medicine and inquiry can be seen as an embryonic form of this rigorous investigation process. He ardently advocated for direct observation, experimentation, and systematic documentation, principles that are fundamental to the contemporary scientific method. Galen's insistence on confirming or refuting established beliefs through hands-on study set a precedent for later scholars. His detailed animal dissections, for instance, were not mere academic exercises but rigorous attempts to understand the inner workings of the body, challenge prevailing notions, and correlate structure with function. When he made medical claims, they were often based on repeated observations and the outcomes of therapeutic interventions. By placing evidence at the heart of knowledge, as opposed to unchallenged acceptance of traditional beliefs or speculative reasoning, Galen sowed the seeds of an empirical approach that would, over time, evolve into the structured methodology of hypothesis testing and validation. Thus, while Galen himself did not conceptualize the "*scientific method*" in its modern form, his commitment to empirical inquiry and systematic experimentation undoubtedly influenced its eventual formation and the evolution of scientific inquiry.

Medicine: Galen's legacy in medicine is complex. As one of antiquity's most prolific medical writers, he penned numerous treatises encompassing various topics, from anatomy and physiology to therapeutics and philosophy. His groundbreaking animal dissections allowed him to elucidate and document aspects of the cardiovascular, nervous, and respiratory systems, offering insights that remained canonical until the Renaissance. However, these contributions must be viewed critically:

- **Pharmaceutical Preparations**: Galen developed complex formulas for drug compounding, many of which formed the basis of medicinal recipes for centuries. While innovative for his time, these formulas often lacked the scientific rigor and understanding of pharmacodynamics that modern medicine requires.
- **Surgical Techniques**: His detailed descriptions of surgical procedures, such as the removal of cataracts and treatment of spinal injuries, set standards for surgical practice that persisted until modern advancements. However, these were often based on limited anatomical understanding and lacked aseptic techniques.

Although Galen provided significant contributions, his adherence to certain erroneous beliefs had a huge negative impact on medical practice that would last for over 1,500 years. For example, his support of the humoral theory, which posited that health depended on the balance of four bodily fluids, led to treatments that were often ineffective and sometimes harmful. Practices such as bloodletting and the use of purgatives persisted because of Galen's authority, causing suffering and harm to countless patients. His incorrect theories about the circulatory system, believing that blood was continuously produced and consumed rather than circulated, delayed the correct understanding of human physiology until the work of William Harvey in the 17th century.

Ethics: Galen's influence on medical ethics, though less extensively documented than his clinical or anatomical work, offers glimpses into early thinking about the principles that guide the medical profession. Rooted in the Hippocratic oath and the ancient Greco-Roman tradition, his philosophy emphasized the physician's moral duty to prioritize the patient's well-being. While he didn't explicitly delineate the modern principles of patient autonomy (informed consent), practitioner beneficence (do good), practitioner nonmaleficence (do no harm), and public justice (be fair), these elements can be inferred from his writings. Beneficence and nonmaleficence are evident in his passionate advocacy for accurate knowledge and skill in practice, asserting that a physician's prime directive is to benefit the patient and avoid harm. His writings, emphasizing the importance of a doctor's moral character, hint at respect for patient autonomy, as the ethical physician would not deceive or coerce a patient. Galen's reverence for the Hippocratic Oath, which he believed every physician should uphold, showcases early considerations of the four principles of bioethics, patient autonomy (informed consent), practitioner beneficence (do good), practitioner nonmaleficence (do no harm), and public justice (be fair). While Galen's direct discourse on these ethical principles is scant, his broader philosophy and commitment to the Hippocratic tradition undeniably shaped the moral landscape of medicine in its formative stages.

Ethically, it is important to note that in Galen's research, he performed vivisections, or dissections of live animals, to understand the functions of various organs and systems. He primarily used animals such as pigs, monkeys, and dogs. These practices provided real-time observations of physiological processes but would be considered highly unethical by modern standards. Today, the Institutional Animal Care and Use Committee (IACUC) sets strict guidelines to minimize animal suffering and ensure ethical treatment in research. Galen's methods, which involved significant animal suffering, would not be permissible under these guidelines. The necessity of these vivisections is debatable; while they contributed to anatomical knowledge, they also reflect a darker aspect of early medical experimentation, where the suffering of animals was not a significant ethical consideration.

Conclusion: Galen of Pergamum's legacy is a testament to the complexities of historical progress in medicine and science. While his meticulous approach to observation, dissection, and documentation laid foundational stones for the scientific method and significantly advanced medical knowledge of his time, his adherence to the humoral theory and certain anatomical misconceptions perpetuated harmful practices for centuries. The irony of his rationalist and empiricist methods coexisting with unscientific doctrines underscores the limitations and biases inherent in all scientific inquiry. Furthermore, his use of vivisection highlights ethical challenges that remain relevant today. Galen's contributions, both positive and negative, serve as a critical reminder of the evolving nature of scientific understanding and the importance of continually reassessing medical practices and ethical standards. His work, therefore, remains a crucial chapter in the history of medicine, reflecting both the strides and the stumbles on the path to modern scientific and ethical principles.

Galen's Legacy: Consolidated and expanded upon the medical knowledge of his time, emphasizing the humoral theory, which became the definitive medical reference in Europe and the Islamic world for centuries, shaping medical practice and education. However, his adherence to incorrect theories also perpetuated false beliefs that negatively impacted patient care for many generations.

**

Review Questions:

True/False Questions:

1. Galen of Pergamum is known for pioneering the continuous blood circulation theory.
   True or False?

2. Galen's methodology included detailed animal dissections to understand human anatomy better.
   True or False?

Multiple-Choice Questions:

3. Which of the following works is primarily attributed to Galen of Pergamum?
   a) "De Materia Medica"
   b) "Enchiridion"
   c) "Natural History"
   d) "On the Usefulness of the Parts of the Body"

4. Galen's contributions to medicine included expanding which ancient medical theory?
   a) Germ theory
   b) Humorism
   c) Vitalism
   d) Evolution

Clinical Vignette:

5. A physician is consulting historical texts to understand ancient methods of diagnosing and treating imbalances in the body's humors. Which ancient medical practitioner's works would be most relevant to this research?
   a) Hippocrates
   b) Epictetus
   c) Galen
   d) Dioscorides

Basic Science Vignette:

6. A medical student is studying the contributions of Galen of Pergamum to the understanding of human anatomy. Which of the following practices best exemplifies Galen's empirical approach to medical science?
   a) Reliance on ancient texts without verification
   b) Performing animal dissections to observe internal organs
   c) Consulting astrological signs for diagnosis
   d) Using divination to predict patient outcomes

Philosophy Vignette:

7. In considering Galen's influence on medical ethics, which principle aligns most closely with his emphasis on a physician's moral duty to benefit the patient and avoid harm?
   a) Patient autonomy (informed consent)
   b) Practitioner beneficence (do good)
   c) Public justice (be fair)
   d) Practitioner nonmaleficence (do no harm)

Correct Answers:

1. False
2. True
3. d) "*On the Usefulness of the Parts of the Body*,"
4. b) Humorism
5. c) Galen
6. b) Performing animal dissections to observe internal organs
7. b) Practitioner beneficence (do good)

Beyond The Chapter
Galen of Pergamum (GAY-len, 129-216)

- *Galen: On the Constitution of the Art of Medicine* translated by Peter N. Singer
- *Galen: Method of Medicine* translated by Ian Johnston and G. H. R. Horsley (3 Volumes)
- *Galen and the World of Knowledge* edited by Christopher Gill, Tim Whitmarsh, and John Wilkins

***

Philosophy of Science

Medical Research

Chapter 67: Avicenna (Ibn Sina, 980-1037) — The Medical Synthesizer: The Canon of Medicine

Avicenna's synthesis of Greek philosophy and Islamic thought in 'The Canon of Medicine' revolutionized healthcare and scientific inquiry.

Abstract: Avicenna, also known as Ibn Sina, is a monumental figure in the Islamic Golden Age, a beacon of intellectual brilliance whose legacy has permeated centuries and across cultures. His multifaceted contributions spanned diverse domains, seamlessly integrating elements of rationalism and empiricism to forge novel pathways in philosophical epistemology and the scientific method. Avicenna's pioneering approaches in medicine, encapsulated in his seminal work "*The Canon of Medicine*," revolutionized healthcare practices, offering a meticulous amalgamation of empirical observations with theoretical insights, setting a benchmark for future scientific inquiries and healthcare delivery. Furthermore, his ethical frameworks, an amalgam of Aristotelian virtues and Islamic principles, have echoed through time, resonating significantly with contemporary bioethical considerations. Avicenna's profound synthesis of Greek philosophy and Islamic thought not only shaped the intellectual landscape of his time but continues to foster interdisciplinary dialogues, inspiring scholars and thinkers worldwide, thus solidifying his status as a timeless luminary in the annals of human knowledge and intellect.

**

Introduction: Avicenna, known as Ibn Sina in the Islamic world, was a polymath who significantly contributed to various fields during the Islamic Golden Age, especially medicine and philosophy. Born in 980 CE in the vicinity of Bukhara (present-day Uzbekistan), he demonstrated a prodigious intellect early in life, absorbing a vast range of knowledge that laid the foundation for his later scholarly pursuits. Arguably, his most renowned work, "*The Canon of Medicine*," served as the standard medical textbook for centuries in the East and the West. Beyond medicine, Avicenna's philosophical writings amalgamated Aristotelian thought and Neoplatonism with Islamic theology, profoundly influencing medieval scholasticism in Europe. He penned around 450 works throughout his lifetime, covering diverse topics from astronomy to music. Today, Avicenna stands as a symbol of the rich intellectual legacy of the Islamic world, his insights continuing to inspire and challenge thinkers across disciplines and cultures.

Rationalism: Avicenna (Ibn Sina) was a steadfast proponent of rationalism, deeply rooted in the belief that human reason can apprehend universal truths about reality. Drawing from the ancient Hellenistic traditions, especially Aristotelian thought and Neoplatonism, he postulated that the intellect could, through rigorous reasoning, arrive at definitive truths independent of empirical observation. In his major philosophical work, "*Kitab al-Shifa*" (The Book of Healing), Avicenna articulated his rationalist epistemology, emphasizing the role of the "*active intellect*" in actualizing potential knowledge within the human soul. He posited that certain fundamental truths, or "*first principles*," were self-evident and needed no external proof. The rational intellect could deduce further facts about the nature of existence from these axiomatic principles. His approach sharply diverged from the empirical methods favored by other scholars of his time, asserting the primacy of reason over sensory experience. While Avicenna did not dismiss empirical observation entirely, his philosophical methodology underscored that true and lasting knowledge came primarily through intellectual, deductive processes. This robust confidence in human reason cemented his legacy as one of the chief proponents of rationalist philosophy within the Islamic intellectual tradition.

Empiricism: While Avicenna (Ibn Sina) is predominantly hailed as a rationalist, disregarding his empiricist tendencies would be an oversimplification, especially in medicine and natural philosophy. Avicenna believed that knowledge begins with sensory perception, serving as the foundation for rational understanding. In "*The Canon of Medicine*," his magnum opus on medical science, he systematically amalgamated empirical observations with theoretical knowledge. Avicenna was among the first to introduce a systematic experimental approach, emphasizing the importance of repeated observation and testing of hypotheses. He recognized the importance of empirical data, especially in diagnosing and treating illnesses. In his philosophical works, while he upheld the superiority of intellectual knowledge for grasping metaphysical truths, he maintained that our senses inevitably mediate our initial access to the external world and its properties. Thus, for Avicenna, the interplay between empiricism and rationalism was crucial; sensory experience provided raw data which reason then refined, organized, and interpreted. In this sense, his epistemology elegantly wove together the empirical and the rational, signifying a nuanced understanding of knowledge acquisition that transcended strict categorization.

The Scientific Method: Avicenna (Ibn Sina) stands as a pivotal figure in the evolution of the scientific method, bridging classical traditions with emerging methodologies that would later influence the Renaissance. Rooted in his deep-seated belief in the interplay of empiricism and rationalism, Avicenna developed a systematic approach to investigation. In his "*Kitab al-Shifa*" (The Book of Healing), he outlined the importance of four essential steps for scientific inquiry: observation (emperical observation), hypothesis formation (rational analysis), experimentation (testing and validation), and verification (theoretical synthesis). By emphasizing the need for empirical data to substantiate theoretical claims, he championed a precursor to the modern experimental method. In his "*Canon of Medicine*," Avicenna underscored the importance of controlled experiments and the testing of remedies using a systematic approach, prioritizing evidence and repeatability. He often insisted on observational evidence and warned against blind acceptance of authority. This progressive attitude toward investigation and meticulous methodology laid foundational stones for subsequent scientific endeavors. Avicenna's approach not only enriched the scientific pursuits of the Islamic world but also permeated into Europe, shaping scholars' methodologies and epistemological underpinnings in the centuries that followed.

Medicine: Avicenna (Ibn Sina) is often dubbed the "*father of modern medicine*" due to his profound contributions to the field during the Islamic Golden Age. His magnum opus, "*The Canon of Medicine*" (al-Qanun fi'l-Tibb), is one of the most significant medical encyclopedias ever written, serving as a primary textbook in Europe and the Islamic world for centuries. This monumental work, spanning five volumes, meticulously cataloged known diseases, described their symptoms and provided a comprehensive list of pharmacological remedies derived from hundreds of medicinal plants and minerals. Avicenna adopted a holistic approach to medicine, emphasizing the interconnectedness of the body's organs and the importance of understanding the patient. He introduced innovative concepts, such as contagious diseases and the potential of soil and water to spread certain illnesses. Furthermore, he highlighted the significance of empirical observation and clinical trials, setting a precedent for methodical experimentation in medical sciences. By integrating Greco-Roman medical knowledge with contemporary discoveries and his insightful observations, Avicenna sculpted a medical tradition that melded empirical research with theoretical knowledge, indelibly shaping the trajectory of medical understanding and practice for generations.

Ethics: Avicenna's (Ibn Sina) profound philosophical forays encompassed ethics, where he wove Aristotelian principles with Islamic jurisprudential tenets. Although the terminologies of patient autonomy (informed consent), practitioner beneficence (do good), practitioner nonmaleficence (do no harm), and public justice (be fair) understood in contemporary bioethics were not explicitly framed in Avicenna's writings, their nascent concepts were evident. Rooted in Aristotle's virtue ethics, Avicenna asserted that the goal of human life was to achieve intellectual and moral perfection. This necessitated a life where individuals freely pursued knowledge and self-realization (hinting at the underpinnings of autonomy). Beneficence found echoes in his emphasis on achieving the greater good and promoting human well-being, especially in his medical treatises, where the physician's role as a healer was paramount. His medical ethics resonated with the principle of nonmaleficence; the imperative "*first, do no harm*" was vital in his therapeutic approach. Justice, though not articulated in modern terms, was imbued in his philosophy, advocating for the harmonious functioning of society based on fairness and the equitable distribution of resources. Avicenna's synthesis of Greek philosophical thought and Islamic principles contributed to an ethical framework that, while distinct in its historical and cultural context, laid the groundwork for later developments in moral philosophy and medical ethics.

Conclusion: Avicenna, or Ibn Sina, remains an indelible luminary in the pantheon of great thinkers, leaving an everlasting imprint on various disciplines during the Islamic Golden Age. As a polymath, he ingeniously synthesized Greek philosophy and Islamic thought, thereby shaping intellectual trajectories both in the East and the West. His influence spanned from philosophical epistemology, where he ardently championed rationalism while acknowledging empiricism's role, to the foundational principles of the scientific method, emphasizing empirical verification and systematic investigation. Avicenna's legacy in medicine is especially noteworthy; his "*Canon of Medicine*" defined medical practices for generations, merging empirical observations with theoretical rigor. Furthermore, his ethical deliberations, imbued with Aristotelian virtue and Islamic jurisprudence, crafted a moral framework that, although specific to its time, laid foundational principles echoing today's bioethical considerations. Thus, Avicenna stands as a testament to the enduring power of interdisciplinary synthesis, his insights bridging cultural epochs and continuing to resonate in contemporary academic and philosophical dialogues.

Avicenna's (Ibn Sina) Legacy: Synthesized and expanded upon Greek and Islamic knowledge, especially in medicine, with his "*Canon*," a seminal text in the West and the East for several centuries.

**

REVIEW QUESTIONS

True/False Questions:

1. Avicenna's (Ibn Sina) "*The Canon of Medicine*" served as a primary medical textbook for centuries in both the East and the West.
   True or False?

2. Avicenna solely relied on theoretical knowledge without emphasizing empirical observation in his medical practices.
   True or False?

Multiple-Choice Questions:

3. Which of the following works is Avicenna most famous for in the field of medicine?
   a) Kitāb al-Shifa (The Book of Healing)
   b) Kitāb al-Manāẓir (Book of Optics)
   c) Kitāb al-Ḥāwī
   d) The Canon of Medicine

4. Avicenna's approach to scientific inquiry emphasized which of the following methodologies?
   a) Sole reliance on sensory perception without rational analysis
   b) The combination of empirical observation and rational deduction
   c) Abstract theorization without empirical verification
   d) Acceptance of traditional knowledge without questioning

Clinical Vignette:

5. A medical student is studying the works of a historical figure who emphasized the importance of empirical observation, clinical trials, and holistic medicine. This figure's magnum opus served as a cornerstone in medical education for centuries. Which historical figure is the student likely studying?
   a) Rhazes (Al-Razi)
   b) Alhazen (Ibn al-Haytham)
   c) Avicenna (Ibn Sina)
   d) Al-Farabi

Basic Science Vignette:

6. A medical student is studying Avicenna's contributions to medicine, specifically his approach to understanding diseases. According to Avicenna's methodology in "*The Canon of Medicine*," what is the best way to confirm the effectiveness of a treatment?
   a) Relying solely on ancient texts and theories
   b) Using personal intuition and subjective experiences
   c) Conducting controlled experiments and observing patient outcomes
   d) Consulting astrological signs and omens

Philosophy Vignette:

7. Avicenna (Ibn Sina) is known for his rationalist philosophy, integrating Greek thought with Islamic principles. According to Avicenna, how can fundamental truths be known?
   a) Through sensory experiences alone
   b) By divine revelation without rational interpretation
   c) Through rigorous reasoning from self-evident first principles
   d) By accepting the consensus of ancient authoritie

Correct Answers:

1. True
2. False
3. d) The Canon of Medicine
4. b) The combination of empirical observation and rational deduction
5. c) Avicenna (Ibn Sina)
6. c) Conducting controlled experiments and observing patient outcomes
7. c) Through rigorous reasoning from self-evident first principles

BEYOND THE CHAPTER
Avicenna (Ibn Sina, 980-1037)

- *Avicenna: His Life and Works* by Soheil M. Afnan
- *The Life of Ibn Sina: A Critical Edition and Annotated Translation* by William E. Gohlman
- *Avicenna's Medicine: A New Translation of the 11th-Century Canon with Practical Applications for Integrative Health Care* by Mones Abu-Asab, Hakima Amri, Marc S. Micozzi

***

Chapter 68: Thomas Aquinas (1225-1274) — The Theological Synthesizer: Faith and Reason

Thomas Aquinas' integration of faith and reason fostered systematic inquiry and influenced ethical frameworks in medicine and science.

Abstract: In this analysis of Thomas Aquinas' legacy, a pivotal figure in Western intellectual history, we explore his profound influence across theology, philosophy, science, and ethics. Renowned as a "*Doctor of the Church*," Aquinas revolutionized theological discourse, blending faith and reason for nuanced explorations into creation and moral principles. His rationalist and empiricist inclinations, grounded in Aristotelian philosophy, harmonized sensory experiences with divine truths. Indirectly, he contributed to the birth of the scientific method, fostering empirical inquiries into the natural world. While not directly contributing to medicine, his philosophies laid a foundation for Christian medical ethics, linking the body and soul and guiding discussions on medical interventions' moral dimensions. Aquinas believed in human reason's capability to uncover terrestrial and divine truths, leaving a legacy that continues to guide discourse on the sacred and empirical realms.

**

Introduction: Thomas Aquinas (1225-1274), a Dominican friar and eminent theologian, is a monumental figure in the annals of Christian thought. Revered as a "*Doctor of the Church*," his writings have deeply influenced Catholic theology and Western philosophy for centuries. Rooted in the scholastic tradition, Aquinas' works — most notably the "*Summa Theologica*" and "*Summa contra Gentiles*" — represent a fusion of faith and reason, synthesizing the teachings of the ancient Greek philosopher Aristotle with the tenets of Christianity. Engaging in rigorous intellectual examination, Aquinas sought to understand the nature of God, the mysteries of creation, and the moral framework guiding human actions. His analytical approach to faith-based questions laid the foundation for the subsequent Catholic doctrine and contributed to the Western world's broader dialogue between reason and religion. His legacy continues to be a touchstone for scholars, theologians, and those seeking to understand the relationship between the divine and the rational.

Rationalism: While deeply rooted in religious orthodoxy, Thomas Aquinas exhibited significant rationalist inclinations in his approach to theology and philosophy. Though limited, he believed that human reason could grasp profound truths about God and the universe. Drawing inspiration from the works of Aristotle, whom he referred to as "*The Philosopher*," Aquinas posited that reason and faith are not mutually exclusive but complementary pathways to truth. In his magnum opus, the "*Summa Theologica*," he employed a systematic, logical structure to address theological questions, often presenting opposing viewpoints before concluding. This dialectical method showcased his respect for the power of human reason. For Aquinas, the natural world, God's creation, could be understood through empirical observation and rational inquiry. While he acknowledged that certain divine truths surpass human comprehension and must be accepted on faith, he also contended that aspects of divine revelation are amenable to rational analysis. This harmonization of faith and reason marked a significant shift from the prevailing Augustinian thought of his time, which leaned more toward mysticism. Aquinas' rationalist inclinations thus played a pivotal role in the development of scholasticism and cemented his position as a beacon of intellectual exploration in the Middle Ages.

Empiricism: Thomas Aquinas' empiricist orientation can be discerned in his belief that all knowledge begins with sensory experience. Embracing the Aristotelian adage "*nihil in intellectu nisi prius in sensu*" (nothing is in the intellect that was not first in the senses), Aquinas posited that our intellect is initially like a blank slate, devoid of concepts. It's through sensory experiences that it acquires knowledge. In this schema, the external world is apprehended by our senses, providing the raw material for our intellect to abstract and understand universal principles or forms. For instance, by encountering many individual dogs, we understand the universal concept of "*dogness*." This empiricist stance contrasted sharply with certain strands of Platonic and Augustinian thought, emphasizing innate ideas or truths independent of sensory experience. While Aquinas affirmed the transcendent nature of certain divine truths accessible only through revelation or faith, he simultaneously championed the idea that the natural world, a reflection of the Divine Creator, is a source of genuine knowledge that can be empirically and rationally investigated. This synthesis of empiricism and faith-oriented epistemology underscores Aquinas' profound commitment to the idea that the world is both knowable and a testament to divine wisdom.

The Scientific Method: Thomas Aquinas, though not a scientist in the modern sense, played a foundational role in shaping the intellectual environment from which the scientific method would eventually emerge. Grounded in his synthesis of Aristotelian philosophy and Christian theology, Aquinas emphasized the importance of empirical observation and rational analysis in understanding the natural world. He held that God's creation operates according to consistent, discoverable principles or laws and thus can be studied systematically. While the precise techniques of the scientific method — hypothesis testing, controlled experimentation, and repeated verification — were developed more fully in the centuries after Aquinas, his affirmation of the reliability of sensory experience and reason laid the crucial epistemological groundwork. Moreover, by arguing that reason and faith are harmonious, he created a space within Christian Europe to pursue natural knowledge without necessarily encroaching upon theological tenets. This framework, while not the scientific method per se, fostered an intellectual climate conducive to its eventual emergence. By championing reason's role in discerning truth, Aquinas indirectly contributed to the rise of a more empirical and systematic approach to studying the natural world — characteristics central to the scientific method.

Medicine: While Thomas Aquinas is primarily celebrated for his contributions to theology and philosophy, he did not directly contribute to medicine as we understand it today. However, his philosophical insights have implications for medical ethics and the understanding of the human being. Rooted in his holistic view of the person, Aquinas affirmed the union of body and soul, arguing against dualistic notions that devalue physical existence. This perspective upholds the intrinsic dignity of human life and has informed Catholic teachings on various medical and bioethical issues, such as the sanctity of life and the moral dimensions of medical interventions. Furthermore, Aquinas' emphasis on the natural law — the idea that there is a moral order ingrained in the nature of things — has had lasting repercussions in discussions about the ethics of medical practices and interventions. Though not a physician, his philosophical and theological musings, particularly those on human nature and ethics, have provided foundational principles for developing robust Christian medical ethics. Thus, while he didn't contribute to medicine in terms of clinical advancements or techniques, his influence is palpable in the realm of moral reasoning in medical contexts.

Ethics: Thomas Aquinas' expansive philosophical thought extends beyond the traditional boundaries of theology and philosophy, touching numerous disciplines, including medical ethics. His writings have been a beacon for many in the medical field, offering guidance on complex ethical dilemmas.

At the heart of his teachings is the profound respect for human dignity and the sanctity of life, which has shaped many positions on issues ranging from end-of-life decisions to reproductive technologies. Aquinas challenges purely materialistic views of human health and medicine by articulating the intrinsic connection between the body and soul. His emphasis on natural law has helped frame discussions on the inherent morality of certain medical interventions, urging professionals to consider the potential outcomes and the intrinsic moral dimensions of their actions. In an era where medical advancements often outpace ethical deliberations, the principles derived from Aquinas' thought provide a much-needed anchor, ensuring that care is rooted in a deep respect for the totality of the human person.

Conclusion: Thomas Aquinas is one of the most influential figures in the Western intellectual tradition, with a legacy spanning theology, philosophy, science, and ethics. His rigorous integration of faith and reason, drawing heavily from Aristotelian thought, laid foundational tenets for subsequent Catholic doctrine and spurred debates on the nature of truth in the wider Western world. From his rationalist persuasions to his empiricist leanings, Aquinas sought to elucidate that the divine and the knowable realm of human experience are not in conflict but offer mutual insights into the nature of reality. His views shaped the nascent inklings of the scientific method and anchored discussions on medical ethics, emphasizing the sanctity of life and the inherent morality of human actions. Beyond specific disciplines, the crux of Aquinas' impact lies in his profound commitment to the idea that while finite, human reason is a potent tool for understanding earthly phenomena and aspects of the divine. In the grand tapestry of history, his writings serve as a testament to the enduring quest to bridge the sacred and the empirical, guiding generations of thinkers, scholars, and believers.

Thomas Aquinas' Legacy: Harmonized Aristotelian philosophy with Christian theology in his "Summa Theologica," profoundly influencing Catholic doctrine and shaping the course of Western philosophy and theology.

**

REVIEW QUESTIONS

True/False Questions:

1. Thomas Aquinas believed that human reason and divine revelation are mutually exclusive pathways to truth.
   True or False?

2. Aquinas' philosophical insights significantly influenced the development of medical ethics, even though he did not directly contribute to clinical medicine.
   True or False?

Multiple-Choice Questions:

3. Which of the following best describes Thomas Aquinas' approach to understanding the natural world?
   a) He relied exclusively on religious texts for all knowledge.
   b) He emphasized the harmony between empirical observation and rational analysis.
   c) He believed mystical experiences were the primary source of knowledge.
   d) He rejected the use of sensory experience in gaining knowledge.

4. Thomas Aquinas' work "*Summa Theologica*" primarily focuses on:
   a) A compilation of medieval legal doctrines.
   b) Harmonizing Aristotelian philosophy with Christian theology.
   c) Outlining new medical techniques and treatments.
   d) Documenting mystical visions and experiences.

Clinical Vignette:

5. A medical ethicist is studying the historical foundations of medical ethics and comes across the works of Thomas Aquinas. How can Aquinas' contributions be best described in this context?
   a) He solely focused on religious dogma without considering human reason.
   b) He emphasized the intrinsic dignity of human life and the moral dimensions of medical interventions.
   c) He rejected the importance of empirical observation in understanding human health.
   d) He believed in the separation of body and soul, devaluing physical existence.

Basic Science Vignette:

6. Thomas Aquinas emphasized the importance of sensory experience as the foundation for acquiring knowledge. A researcher is studying a new plant species and wants to understand its medicinal properties. Based on Aquinas' empirical orientation, what would be the most appropriate first step in this study?
   a) Refer to ancient texts for information on similar plants
   b) Conduct a series of controlled experiments to observe the plant's effects
   c) Meditate on the possible divine properties of the plant
   d) Use the plant in human subjects immediately without prior testing

Philosophy Vignette:

7. Thomas Aquinas harmonized faith and reason, arguing that human reason could uncover truths about God and the universe. How does Aquinas' approach to theology and philosophy reflect this harmonization?
   a) By asserting that reason alone is sufficient to understand all divine truths
   b) By dismissing the relevance of sensory experience in acquiring knowledge
   c) By employing logical structure and rational analysis to explore theological questions
   d) By rejecting the teachings of Aristotle in favor of purely mystical experiences

Correct Answers:

1. False
2. True
3. b) He emphasized the harmony between empirical observation and rational analysis
4. b) Harmonizing Aristotelian philosophy with Christian theology
5. b) He emphasized the intrinsic dignity of human life and the moral dimensions of medical interventions
6. b) Conduct a series of controlled experiments to observe the plant's effects
7. c) By employing logical structure and rational analysis to explore theological questions

BEYOND THE CHAPTER
Thomas Aquinas (1225-1274)

- *Aquinas: A Beginner's Guide* by Edward Feser
- *Thomas Aquinas: Selected Writings* edited by Ralph McInerny
- *Aquinas's Shorter Summa: Saint Thomas's Own Concise Version of His Summa Theologica* by Thomas Aquinas

***

Chapter 69: Nicolaus Copernicus (1473-1543) — The Heliocentric Revolutionary: Shifting the Cosmic Paradigm

Nicolaus Copernicus' heliocentric model, grounded in rationalism and empirical observation, revolutionized our understanding of the cosmos.

Abstract: Nicolaus Copernicus (1473-1543) is a transformative figure in scientific history, catalyzing a paradigm shift in our understanding of the cosmos. An astronomer, mathematician, physician, and canon lawyer from Poland, he challenged the geocentric model with his heliocentric theory, detailed in "*De revolutionibus orbium coelestium*." This radical proposal, based on rationalism and empirical observations, sought to align celestial mechanics with logical coherence and empirical truth, breaking from the Ptolemaic system. His methodological approach, intertwining observation with theory, prefigured the scientific method, laying the groundwork for the scientific revolution. Though his contributions to medicine were minor, his interdisciplinary expertise embodied the Renaissance spirit. His commitment to intellectual truth and the pursuit of knowledge for humanity's betterment marked a seismic shift in our understanding of the cosmos. Copernicus' legacy endures as a testament to the power of intellectual rigor and fearless inquiry.

**

Introduction: Nicolaus Copernicus (1473–1543) is a pivotal figure in the annals of scientific history, heralding a profound shift in how humanity understood its place in the cosmos. Born in the Kingdom of Poland, Copernicus was not merely an astronomer but also a mathematician, physician, and canon lawyer. Yet, it was his groundbreaking work in astronomy for which he is most renowned. Challenging the long-held geocentric model, which posited the Earth as the unmoving center of the universe, Copernicus courageously proposed a heliocentric system where Earth and other planets orbited the Sun. His magnum opus, "*De revolutionibus orbium coelestium*" ("*On the Revolutions of the Celestial Spheres*"), published just before his death, laid the foundation for the scientific revolution. Copernicus's ideas did not emerge in isolation; they were part of a broader continuum of astronomical thought. Influences from earlier scholars, including Avicenna (Ibn Sina), played a crucial role in shaping the intellectual landscape. Avicenna, a Persian polymath, had critiqued the prevailing Ptolemaic system and integrated astronomy with physics and philosophy, providing a foundation that would indirectly support the heliocentric model. By introducing a sun-centered model, Copernicus made a chain of discoveries that would reshape the fabric of scientific and philosophical thought, paving the way for future luminaries like Galileo, Kepler, and Newton.

Rationalism: Nicolaus Copernicus, while deeply rooted in the scholarly traditions of his time, exhibited a distinct rationalist inclination that would eventually drive the seismic shifts in astronomical thought. As a philosophical doctrine, rationalism emphasizes reason as the primary source of knowledge, often over empirical or sensory experiences. In Copernicus' case, his dissatisfaction with the geocentric model was not merely due to its complexity but primarily because it failed to align with the principles of pure reason. The Ptolemaic system, riddled with epicycles and deferents, seemed unwieldy and counterintuitive. Copernicus sought a more intellectually satisfying explanation for the observed motions of celestial bodies. Believing in the harmonious nature of the universe and drawing from ancient Greek sources, he revisited the heliocentric idea first proposed 1,800 years earlier by Aristarchus of Samos (c. 310 – c. 230 BCE). Aristarchus's student, Seleucus of Seleucia (c. 190 – c. 150 BCE), provided empirical and mathematical arguments in its favor. Despite their efforts, these ideas were overshadowed by Aristotle's geocentric worldview. However, Copernicus revived and illuminated these ancient ideas with his own observations and mathematical formulations. While his model still contained epicycles and was not entirely free from complexities, it was motivated by a rationalist's quest for an elegant, coherent system aligned more intuitively with human reason. His heliocentric theory, thus, was not just an alternative model to what was currently held, but a manifestation of the rationalist spirit — prioritizing logical coherence and simplicity over the strict adherence to scholastic Aristotelian doctrines of the medieval universitites, or over-reliance on imperfect observations.

Empiricism: While Nicolaus Copernicus is often hailed for his groundbreaking heliocentric model, it's essential to recognize the empirical foundation upon which his work rested. Empiricism emphasizes knowledge derived from experience and sensory observations as the cornerstone of understanding. Copernicus, though driven by a desire for a more rationally coherent model, was deeply influenced by empirical observations of celestial bodies. He was not working in an evidential vacuum. Instead, he carefully examined and reinterpreted existing astronomical data, identifying inconsistencies in the geocentric model's predictions and the actual observed positions of planets. While the Ptolemaic system attempted to account for these discrepancies with intricate epicycles, Copernicus saw an opportunity to align theory more closely with empirical reality through the heliocentric model. His work became an early testament to the empirical method's power, intertwining observation with theory. Even though Copernicus lacked the advanced observational tools that later scientists like Galileo would employ, his empiricist orientation marked a significant step toward the evolving scientific methodology that would come to rely heavily on observation, experimentation, and evidence as the bedrock of knowledge.

The Scientific Method: Nicolaus Copernicus' transformative work in astronomy can be seen as a precursor to the formalization of the scientific method, a systematic approach to inquiry that relies on observation, hypothesis formation, experimentation, and theory development. While the scientific method as it's understood today was more explicitly shaped by later figures like Galileo, Kepler, and Bacon, Copernicus played a pivotal role in its evolution. By questioning the established geocentric model, he demonstrated the importance of skepticism toward accepted knowledge. His heliocentric proposal wasn't just a bold hypothesis; it was rooted in scrutiny of available astronomical data, illustrating the interplay between observation and theory. Moreover, his willingness to publish "*De revolutionibus orbium coelestium*," even in the face of potential controversy, highlighted the significance of open discourse and peer review in scientific progress. Though Copernicus did not conduct "*experiments*" in the manner later associated with the scientific method, his systematic approach to problem-solving, reliance on empirical data, and commitment to logical coherence laid essential groundwork for the methodological advancements shaping the scientific revolution and the centuries to follow.

Medicine: Although primarily known for his revolutionary contributions to astronomy, Nicolaus Copernicus was also a trained physician, a lesser-known facet of his multifaceted career. Studying medicine at the University of Padua, one of the leading medical schools of the Renaissance, Copernicus acquired knowledge and skills that he later applied in his service to various ecclesiastical figures, including his uncle, the Bishop of Warmia. While his direct contributions to medicine do not parallel the magnitude of his astronomical work, his medical training informed his meticulous, systematic approach to problem-solving, a trait evident in "*De revolutionibus orbium coelestium*." Furthermore, Copernicus lived when the boundaries between different fields of study were more porous, allowing scholars like him to draw interdisciplinary connections. For instance, the Renaissance era saw a renewed interest in the works of ancient Greek physicians, wherein philosophical ideas about cosmic harmony and order might intersect with medical theories. While groundbreaking discoveries or treatments don't mark Copernicus' legacy in medicine, his broader commitment to scholarly rigor and the synthesis of knowledge across disciplines exemplifies the Renaissance's integrative spirit.

Ethics: hile Nicolaus Copernicus is celebrated for his revolutionary contributions to astronomy and, to a lesser extent, medicine, he is not directly associated with the formal development of core ethical principles like patient autonomy (informed consent), practitioner beneficence (do good), practitioner nonmaleficence (do no harm), and public justice (be fair) as they are understood in modern bioethics. Although these principles have their origins in ancient texts like the Hippocratic corpus (460 - unknown BCE), which Copernicus would have known, they were more explicitly articulated and codified in the context of modern medical ethics centuries after his time. However, Copernicus' dedication to truth and courage in challenging established paradigms can be seen as an implicit endorsement of intellectual autonomy, the freedom to pursue knowledge unencumbered by dogma or societal pressures. His commitment to advancing an accurate understanding of the universe, which he likely believed would benefit humanity's broader knowledge base, resonates with the principle of beneficence. Yet, drawing direct links between Copernicus and the ethical tenets that govern fields like medicine or research today would be speculative. Instead, one might interpret Copernicus' legacy as a broader testament to the ethical imperatives of intellectual honesty, courage in the face of potential controversy, and the pursuit of truth for the greater good.

Conclusion: Nicolaus Copernicus' multifaceted contributions spanned the realms of astronomy, medicine, and beyond, reshaping humanity's perception of its position within the universe and laying the foundations for centuries of scientific inquiry. Through his heliocentric model, Copernicus did not merely challenge geocentrism but catalyzed a paradigm shift that would lead to the broader scientific revolution. His rationalist inclinations and empirical orientation epitomize the harmonious interplay between reason and observation, integral to scientific progress. Beyond astronomy, his work in medicine and his interdisciplinary approach, emblematic of the Renaissance, emphasized the interconnectedness of knowledge. While not directly contributing to modern ethical principles, his intellectual bravery and commitment to truth are lasting reminders of the ethical responsibilities inherent in scholarly pursuits. Collectively, Copernicus' legacy underscores the transformative power of curiosity, intellectual rigor, and the fearless pursuit of knowledge, regardless of the dominant paradigms of one's time.

Nicolaus Copernicus' Legacy: Proposed the heliocentric model of the solar system, challenging geocentric views and revolutionizing the understanding of the cosmos.

**

REVIEW QUESTIONS

True/False Questions:

1. Nicolaus Copernicus' heliocentric theory was primarily driven by his rationalist quest for an elegant, coherent system aligned more intuitively with human reason.
   True or False?

2. Nicolaus Copernicus conducted numerous experiments similar to the modern scientific method to prove his heliocentric theory.
   True or False?

Multiple-Choice Questions:

3. Which field did Copernicus not significantly contribute to during his lifetime?
   a) Astronomy
   b) Medicine
   c) Canon law
   d) Chemistry

4. What motivated Copernicus to propose the heliocentric model?
   a) A desire to align celestial mechanics with religious doctrines
   b) The need to simplify the Ptolemaic system's complexities
   c) Empirical evidence from advanced telescopic observations
   d) Influence from contemporary scientists' discoveries

Clinical Vignette:

5. A scholar is studying the influence of Copernicus' heliocentric theory on the scientific method. The scholar notes that Copernicus emphasized the importance of aligning theory with empirical observations and logical coherence. Which principle of Copernicus' methodology is being highlighted?
   a) Reliance on religious doctrines
   b) Acceptance of established paradigms
   c) Integration of observation with theory
   d) Use of advanced experimental tools

Basic Science Vignette:

6. Nicolaus Copernicus' heliocentric model fundamentally altered our understanding of the solar system. A student learning about planetary motions wants to understand why Mars appears to move backward in the sky at certain times, a phenomenon known as retrograde motion. Based on Copernicus' heliocentric theory, what is the most accurate explanation for this observation?
   a) Mars changes its direction of orbit around the Sun periodically
   b) Earth moves faster in its orbit around the Sun, overtaking Mars, which creates the illusion of Mars moving backward
   c) Mars' orbit is highly elliptical, causing it to periodically reverse direction
   d) The gravitational pull of Jupiter causes Mars to move backward temporarily

Philosophy Vignette:

7. Nicolaus Copernicus emphasized the importance of rationalism and empirical observation in developing his heliocentric model. How does Copernicus' approach to understanding the cosmos exemplify the principles of rationalism and empiricism?
   a) By relying solely on religious texts to explain celestial movements
   b) By using logical reasoning and empirical data to challenge the geocentric model
   c) By accepting the geocentric model without question
   d) By rejecting all forms of observational data in favor of pure speculation

Correct Answers:

1. True
2. False
3. d) Chemistry
4. b) The need to simplify the Ptolemaic system's complexities
5. c) Integration of observation with theory
6. b) Earth moves faster in its orbit around the Sun, overtaking Mars, which creates the illusion of Mars moving backward
7. b) By using logical reasoning and empirical data to challenge the geocentric model

BEYOND THE CHAPTER
Nicolaus Copernicus (1473-1543)

- *Nicolaus Copernicus: Making the Earth a Planet* by Owen Gingerich
- *On the Revolutions of Heavenly Spheres* by Nicolaus Copernicus, translated by Charles Glenn Wallis
- *Copernicus: A Very Short Introduction* by Owen Gingerich

***

Philosophy of Science

Medical Research

WEEK THIRTEEN: EMPIRICISM AND INQUIRY

WEEK THIRTEEN

Class Hour 1

- Chapter 70: Girolamo Fracastoro (1478-1553) — The Germ Theorist: Seed-like Entities and Contagion
- Chapter 71: Paracelsus (1493-1541) — The Chemical Healer: Experimenting with Medicine
- Chapter 72: Ambroise Paré (1510-1590) — The Surgical Innovator: Compassionate Techniques

Class Hour 2

- Chapter 73: Andreas Vesalius (1514-1564) — The Anatomical Innovator: Revealing the Human Body
- Chapter 74: Francis Bacon (1561-1626) — The Empirical Philosopher: Method and Experimentation
- Chapter 75: Galileo Galilei (1564-1642) — The Observational Pioneer: Telescopic Revelations

EXAM IV: 50Q - online

Introduction

In the thirteenth week of our course, we explore the transformative period during which the foundations of modern science and medicine were solidified through groundbreaking discoveries and innovative methodologies. This era witnessed the convergence of rationalist and empirical approaches, fundamentally altering our understanding of disease, anatomy, and the natural world.

Girolamo Fracastoro, known as the Germ Theorist, introduced the revolutionary idea that diseases spread through "*seed-like entities*." His rationalist and empirical approach challenged prevailing notions and laid the groundwork for the germ theory of disease. By emphasizing the importance of contagion and systematic investigation, Fracastoro's work marked a pivotal shift toward evidence-based medical practice, influencing subsequent methodologies and fostering a deeper understanding of infectious diseases.

Paracelsus, the Chemical Healer, departed from traditional Galenic medicine by advocating the use of chemicals and minerals in treatments. His empirical approach emphasized experimentation and the practical efficacy of treatments, which laid rudimentary principles for the scientific method. Paracelsus' focus on hands-on experimentation and patient-centric care significantly influenced ethical considerations in healthcare, promoting a more individualized and effective approach to medical treatment.

Ambroise Paré, the Surgical Innovator, transformed surgical practices with his compassionate techniques. Rejecting the brutal methods of his time, Paré modernized surgery by treating wounds without cauterization and developing more humane methods. His empirical approach, characterized by critical thinking and practical application, led to improved patient outcomes and fostered an evidence-based approach to medicine. Paré's emphasis on compassionate care and systematic inquiry established ethical foundations that continue to resonate in modern medical practice.

Andreas Vesalius, the Anatomical Innovator, revolutionized the field of anatomy with his seminal work "*De humani corporis fabrica*." By challenging Galenic orthodoxy with detailed human anatomical studies, Vesalius' blend of rationalist and empirical methodologies provided a deeper, evidence-based understanding of the human body. His meticulous dissections and comprehensive illustrations laid foundational principles for modern medical practices and ethical considerations in dissection, transforming anatomical science.

Francis Bacon, the Empirical Philosopher, advocated for the inductive method in his influential work "*Novum Organum*." Bacon's rationalist and empirical approach laid the philosophical groundwork for the scientific method, emphasizing the importance of observation, experimentation, and systematic inquiry. His ideas revolutionized scientific and medical practices, promoting evidence-based methods and fostering ethical considerations in scientific research, thus paving the way for modern empirical investigation.

Galileo Galilei, the Observational Pioneer, forever changed our understanding of the cosmos with his telescopic revelations. By providing empirical evidence for the Copernican heliocentric model, Galileo championed the use of observation and experimentation in astronomy. His rationalist and empirical methodology detached science from traditional constraints, influencing modern scientific inquiry and ethical principles in medical practices. Galileo's insistence on empirical evidence and rigorous methodology set new standards for scientific investigation.

This week, we witness the remarkable evolution of scientific thought and medical practice through the pioneering efforts of these visionary figures. Their contributions not only advanced our understanding of disease, anatomy, and the universe but also established enduring principles of empirical investigation, systematic inquiry, and ethical consideration. As we explore their legacies, we gain a deeper appreciation for the transformative power of rationalist and empirical approaches in shaping the course of scientific and medical history.

***

Chapter 70: Girolamo Fracastoro (1478-1553) — The Germ Theorist: Seed-like Entities and Contagion

Girolamo Fracastoro's germ theory and empirical approach to disease transmission marked a pivotal shift towards rational and observational science.

Abstract: In the Renaissance, Girolamo Fracastoro epitomized the blend of art and science, pushing boundaries in medicine, astronomy, and literature. Hailed as a harbinger of germ theory, he revolutionized the understanding of infectious diseases, proposing they were transmitted through "*seed-like entities*" in his treatise "*De Contagione et Contagiosis Morbis*." His poetic discourse on syphilis coined the term and showcased the intersection of narrative and scientific observation. Beyond medicine, his contributions to astronomy reflected the empirical methods he used in his medical inquiries, paving the way for future discoveries. Fracastoro's curiosity and meticulous approach marked a departure from superstitions, catalyzing a shift toward rationalism and empiricism, foundational principles of the scientific method. His legacy encapsulates the Renaissance spirit—a fervent quest for knowledge and reverence for the empirical world, offering a foundation for modern science and medicine.

**

Introduction: Girolamo Fracastoro, a multifaceted intellectual of the Renaissance era, was born in 1478 in Verona, Italy, and wore several hats in his lifetime — physician, poet, astronomer, and geologist, to name a few. Most famously remembered for his groundbreaking contributions to the medical field, Fracastoro first proposed the theory that epidemic diseases were caused by transferable tiny particles or "*spores*." This revolutionary idea was articulated in his seminal work "*De Contagione et Contagiosis Morbis*." He also introduced the term "*syphilis*" — named after the protagonist of his poetic work that chronicled the disease's origins. Given the Renaissance era's emphasis on classical knowledge, it is likely that Fracastoro was aware of the Hippocratic Corpus, a collection of medical texts from ancient Greece (460 - unknown BCE). The Hippocratic Corpus, though lacking a modern understanding of germs, included significant observations on contagious diseases and preventive measures. Adhering to the miasma theory, it attributed diseases to "*bad air*" and environmental factors while noting the contagious nature of certain diseases and advocating for sanitation and hygiene. These early insights would have provided Fracastoro with a foundational understanding, influencing his development of early germ theory concepts. In addition to the classical influences, Fracastoro's ideas were also built upon the earlier work of Islamic scholars, particularly Avicenna (Ibn Sina, 980-1037). In his seminal work "*The Canon of Medicine*," Avicenna discussed the idea of contagion and described how certain diseases could spread through contact, air, and water. He noted that diseases could be transmitted through airborne particles and recommended quarantine measures for patients with infectious diseases. These concepts were critical in shaping the intellectual environment that allowed Fracastoro to articulate his germ theory which was primarily based on epidemiological evidence. Venturing beyond medicine, his insights on the movement of planets laid foundational principles for future astronomers. Bridging the divide between science and art, Fracastoro stands as a testament to the Renaissance spirit, with his myriad achievements epitomizing the era's genuine quest for knowledge and understanding.

Rationalism: During a period of superstition and dogma, Girolamo Fracastoro stood out as a beacon of rationalist thought. At a time when mystical explanations for natural phenomena were pervasive, he championed the empirical examination of the world. His treatise, "*De Contagione et Contagiosis Morbis*," is emblematic of his rationalist inclinations. Rather than attributing the spread of diseases like syphilis to divine wrath or astral influences, Fracastoro posited that these maladies were transmitted by minuscule, infectious particles or "*semina*." This groundbreaking theory bore the hallmarks of Renaissance rationalism: a reliance on observation, a move away from religious or superstitious explanations, and a commitment to systematic reasoning. It's worth noting that while his particle theory was a precursor to the modern germ theory of disease, his hypothesis was rooted in reason and observation rather than pure evidence from experiments. Fracastoro's approach heralded a budding age of scientific inquiry, where the thirst for knowledge veered away from age-old beliefs and toward rational, evidence-based explanations. Through his work and writings, he embodies the transformative spirit of the Renaissance, where reason began to emerge as the touchstone of understanding the natural world.

Empiricism: Girolamo Fracastoro's commitment to empirical methodologies marked a paradigm shift from the prevailing speculative thought of his era. His empiricist orientation was evident in his emphasis on observation and direct experience as foundations for gaining knowledge. While his contemporaries often leaned on ancient texts and dogmatic principles to understand the world, Fracastoro sought tangible evidence. In "*De Contagione et Contagiosis Morbis*," he postulated that invisible, transferable entities caused diseases. Instead of merely theorizing, he based his conclusions on careful observation of disease patterns, transmission, and progression. For instance, observing the way syphilis spread among populations, he discerned that it followed a discernible and consistent pattern that could not merely be ascribed to superstition or divine decree. His pioneering approach sought to correlate causes with effects using systematic observation — a cornerstone of empiricism. While Fracastoro did not have access to the advanced experimental tools that would later cement the germ theory of disease, his unwavering dedication to grounding his hypotheses in observed realities was a crucial stride toward the empirical traditions that modern science now holds sacrosanct.

The Scientific Method: Girolamo Fracastoro, while not directly responsible for the modern formulation of the scientific method, undeniably played a pivotal role in its embryonic stages during the Renaissance. His insistence on observation and experience as the foundation for understanding disease laid a crucial groundwork for the iterative process of hypothesis, experimentation, and conclusion that characterizes the scientific method. In "*De Contagione et Contagiosis Morbis*," Fracastoro departed from purely speculative or theological explanations and instead championed an approach rooted in observable patterns and systematic reasoning. He exemplified an early form of hypothesis testing by postulating that diseases were spread by unseen particles based on observed phenomena. Although he lacked the experimental apparatus and controlled testing environments that later scientists would utilize, his dedication to grounding theories in empirical evidence significantly departed from contemporary modes of thought. In essence, Fracastoro's work represents a transitional moment in the evolution of scientific inquiry, bridging the gap between medieval scholasticism and the more rigorous, systematic approaches that would define the scientific method in subsequent centuries.

Medicine: In the annals of medical history, Girolamo Fracastoro occupies a unique position for his visionary contributions during the Renaissance. Challenging the dominant paradigms of his era, he authored "*De Contagione et Contagiosis Morbis*," a groundbreaking treatise that introduced the revolutionary concept that diseases were spread by minute, transferable particles, or "*semina.*" This early precursor to the modern germ theory of disease was a profound departure from commonly held beliefs that attributed the illness to divine intervention, miasmas, or imbalances of humors. Furthermore, Fracastoro coined the term "*syphilis*" in his poetic work "*Syphilis sive morbus gallicus*," intertwining medical observation with literature to recount the tale of the disease's presumed origins. His holistic approach to understanding illness, encompassing empirical observation and systematic reasoning, laid essential groundwork for subsequent advancements in epidemiology and infectious disease control. While he practiced in an era devoid of advanced microscopic tools or refined laboratory techniques, Fracastoro's perceptiveness and unwavering commitment to rational inquiry carved a path for future generations, solidifying his legacy as one of the seminal figures in the evolution of modern medicine.

Ethics: Girolamo Fracastoro's work primarily revolved around the domains of medicine, astronomy, and literature, with his most prominent contributions lying in the realm of epidemiology. However, directly linking Fracastoro to the formulation of modern bioethical principles such as patient autonomy (informed consent), practitioner beneficence (do good), practitioner nonmaleficence (do no harm), and public justice (be fair) can be challenging, as these concepts were more explicitly codified much later. That said, his emphasis on empirical evidence and rational understanding in the face of diseases provided an indirect influence. By championing a model of disease transmission rooted in observable phenomena rather than moral failings or divine punishment, Fracastoro indirectly promoted a more humane and rational approach to patient care. This shift away from moralistic judgments could be seen as a nascent form of beneficence and nonmaleficence, emphasizing the importance of understanding and treating diseases without bias. Although it would be anachronistic to attribute modern ethical formulations directly to Fracastoro, his groundbreaking perspectives on disease undeniably contributed to a broader shift in medicine toward a more evidence-based and patient-centric approach, which would later become intertwined with the core ethical principles of medical practice.

Conclusion: Girolamo Fracastoro's legacy in the tapestry of scientific and medical history is both profound and multifaceted. Born amidst the intellectual ferment of the Renaissance, Fracastoro epitomized the spirit of the age: a voracious curiosity combined with a rigorous approach to understanding the natural world. His pioneering ideas, particularly in medicine, laid the groundwork for future generations to build upon. From his early insights on the transmission of diseases to his foundational role in shaping empirical methodologics, he served as a bridge between ancient beliefs and modern scientific inquiry. Moreover, his departure from moralistic judgments in understanding diseases heralded a more compassionate and ethically driven approach to medicine. While only sometimes recognized alongside the more celebrated figures of the Renaissance, Fracastoro's contributions were instrumental in moving forward the ideals of rationalism, empiricism, and evidence-based understanding. His enduring legacy is a testament to the transformative power of curiosity, observation, and rigorous reasoning in reshaping our understanding of the world.

Girolamo Fracastoro's Legacy: Introduced the idea of contagion, positing that diseases were spread by "*seed-like entities*," laying foundational insights for studying infectious diseases.

**

REVIEW QUESTIONS

True/False Questions:

1. Girolamo Fracastoro's treatise "*De Contagione et Contagiosis Morbis*" introduced the concept that diseases were transmitted through "*seed-like entities*."
   True or False?

2. Fracastoro's work in medicine was heavily reliant on divine and mystical explanations for diseases.
   True or False?

Multiple-Choice Questions:

3. Which of the following fields did Girolamo Fracastoro significantly contribute to during his lifetime?
   a) Physics
   b) Chemistry
   c) Epidemiology
   d) Sociology

4. What was a key methodological approach used by Fracastoro in his study of diseases?
   a) Relying solely on ancient texts
   b) Observing and recording disease patterns
   c) Conducting advanced laboratory experiments
   d) Utilizing mystical and astrological explanations

Clinical Vignette:

5. A physician in the Renaissance period is examining the spread of a new infectious disease and decides to adopt Fracastoro's approach by emphasizing empirical observation and rational analysis. Which principle of Fracastoro's methodology is the physician most closely following?
   a) Reliance on mystical explanations
   b) Use of ancient texts without questioning
   c) Empirical evidence and observation
   d) Acceptance of divine intervention

Basic Science Vignette:

6. A Renaissance physician is studying the spread of a mysterious disease following the principles outlined by Girolamo Fracastoro. Which concept best illustrates Fracastoro's approach to understanding contagion?
   a) Diseases are caused by imbalances in bodily humors
   b) Illnesses are spread through divine intervention
   c) Diseases are transmitted by minute, transferable particles
   d) Epidemics result from astrological alignments

Philosophy Vignette:

7. Girolamo Fracastoro's work on disease transmission exemplifies a shift towards which philosophical approach in understanding natural phenomena?
   a) Mysticism, focusing on supernatural explanations
   b) Rationalism, emphasizing logical deduction and reasoning
   c) Empiricism, relying on observation and evidence
   d) Scholasticism, integrating faith and reason

Correct Answers:

1. True
2. False
3. c) Epidemiology
4. b) Observing and recording disease patterns
5. c) Empirical evidence and observation
6. c) Diseases are transmitted by minute, transferable particles
7. c) Empiricism, relying on observation and evidence

BEYOND THE CHAPTER
Girolamo Fracastoro (1478-1553)

- *Fracastoro's Syphilis* by Geoffrey Eatough
- *Girolamo Fracastoro: Latin Poetry* edited by James Gardner
- *The Life and Works of Fracastoro* by Jerome Bylebyl

***

Philosophy of Science™ — Medical Research

Chapter 71: Paracelsus (1493-1541) — The Chemical Healer: Experimenting with Medicine

Paracelsus' empirical exploration of chemicals and minerals for remedies laid foundational principles for modern pharmacology and toxicology.

Abstract: Paracelsus, a seminal figure of the medieval and Renaissance eras, transformed medical science and ethics during a period reliant on ancient doctrines. Born Philippus Aureolus Theophrastus Bombastus von Hohenheim, he emphasized empirical observation and experiential learning, steering medicine away from Galenic practices. Paracelsus pioneered pharmacology and toxicology, exploring chemicals and minerals for remedies and laying the principles for the modern scientific method through hands-on experimentation. Despite his roots in alchemy, he adopted a rational approach, prioritizing efficacy and direct engagement with the natural world over ancient authorities. His patient-centric ethos infused ethical considerations into medical practice, advocating personalized care and prioritizing patient wellbeing. This shift fostered an environment integrating science, medicine, and ethics, cementing Paracelsus' legacy as a transformative figure who bridged the mystical and the empirical in his relentless pursuit of knowledge and advancement.

**

Introduction: Paracelsus, born Philippus Aureolus Theophrastus Bombastus von Hohenheim in the late 15th century, is a cornerstone figure in the annals of medical and alchemical history. Straddling the medieval and Renaissance periods, his unorthodox approach to medicine was underpinned by the belief that insights could be gained from ancient texts, personal experience, and direct observation of the natural world. Challenging the traditional Galenic medicine of his time, Paracelsus introduced revolutionary concepts, emphasizing the importance of "*signatures*" from nature in drug discovery and promoting the idea that agents outside the body caused diseases — his inclination toward chemical remedies led to the foundation of modern pharmacology. Often polarizing for his unorthodox methods and sharp critiques of academic medicine, Paracelsus nonetheless paved the way for significant medical advancements, anchoring his insights in both the mystical traditions of alchemy and the empirical rigor of early modern science.

Rationalism: While often remembered for his alchemical pursuits, Paracelsus exhibited rationalist inclinations ahead of his time, challenging the dogmatic adherence to ancient authorities that dominated the medical landscape of the Renaissance. For Paracelsus, the rational study of nature and empirical observation held greater value than blind adherence to age-old texts. He believed in"*knowledge through experience,*" positing that true understanding of the world, especially in medicine, came from reading and accepting classical sources and direct observation, experimentation, and personal experience. This emphasis on empirical data collection laid the groundwork for the scientific method that would be fully realized in subsequent centuries. Paracelsus' commitment to practical experimentation meant he often favored chemical and mineral remedies over traditional herbal medicines, underscoring a shift toward more reproducible and quantifiable therapeutic agents. His approach was not merely speculative or mystical; it was rooted in the belief that the world operated on discernible principles that could be understood and harnessed through reasoned investigation, making him a pivotal figure in the transition from medieval alchemy to modern chemistry and medicine.

Empiricism: Paracelsus, defying the rigid scholasticism of his era, staunchly championed an empiricist orientation toward medical knowledge and practice. Contrary to the prevailing reliance on ancient Greco-Roman texts, he believed that genuine knowledge sprang from direct engagement with the natural world. To Paracelsus, understanding medicine and the human body required firsthand observation, experimentation, and practical experience rather than mere theoretical contemplation. His journeys across Europe, where he learned from diverse folk traditions and witnessed varied medical practices, reinforced his empiricist convictions. These experiences led him to introduce treatments and remedies based on what worked in practice, even if they did not conform to established doctrines. For instance, his pioneering work in toxicology stemmed from keen observations of the effects of substances on the human body, formulating the foundational idea that "*the dose makes the poison.*" By emphasizing a hands-on, observational approach, Paracelsus laid the groundwork for future scientists and physicians to prioritize empirical evidence over unfounded tradition, paving the way for a more scientific and systematic approach to medicine.

The Scientific Method: While not directly credited with formulating the scientific method as it's recognized today, Paracelsus played an instrumental role in fostering its foundational principles. Paracelsus championed the importance of empirical observation and hands-on experimentation in an age of reliance on ancient authorities like Galen and Aristotle. He argued against blind adherence to ancient texts, insisting instead on the importance of firsthand knowledge derived from real-world observations. His commitment to experiential learning and the application of knowledge marked a shift from speculative philosophy to evidence-based practice. Paracelsus laid the groundwork for the systematic inquiry that would later crystallize into the scientific method by advocating for testing hypotheses through practical experiments, especially in medicine and alchemy. His insistent call for the verification of knowledge and the importance he placed on tangible results challenged the traditional modes of learning and set the stage for the subsequent scientific revolution of the 17th century. In essence, Paracelsus' approach to knowledge and investigation was an early precursor to the empirical rigor and skepticism of modern scientific inquiry.

Medicine: Paracelsus, often dubbed the "*father of toxicology*," revolutionized Renaissance medicine through his groundbreaking approaches and theories. Departing sharply from the prevailing Galenic humoral theory, which posited that diseases arose from imbalances in the body's four humors, Paracelsus introduced the novel concept that illnesses were caused by external agents invading the body. He proposed that these agents could be combated using specific chemical remedies, laying the foundation for targeted pharmacological treatments. His famous maxim, "*All things are poison, and nothing is without poison; the dosage alone makes it so a thing is not a poison,*" underscored the significance of dosage in therapeutic interventions, birthing modern toxicology's fundamental principles. Moreover, Paracelsus' reliance on empirical observation led him to prioritize chemical and mineral remedies over traditional herbal ones, recognizing their consistent efficacy and potency. By championing the idea of harnessing nature's "*signatures*" or hints, he bridged mysticism and early modern science in the quest for effective treatments. His holistic approach, which considered both the spiritual and physical dimensions of health, combined with his emphasis on evidence-based remedies, profoundly reshaped medical paradigms and heralded a new era of therapeutic practice.

Ethics: Paracelsus, while primarily known for his contributions to medicine and alchemy, also indirectly influenced the ethical principles that form the bedrock of modern medical practice. Rooted in his patient-centric approach, he emphasized the importance of tailoring treatments to individual needs, thus touching on autonomy or respecting patients' unique conditions and choices. His relentless pursuit of effective treatments and belief in the physician's sacred duty to heal resonates with the principle of beneficence, the commitment to act in the patient's best interests. Concurrently, his foundational work in toxicology, encapsulated in his aphorism that the dose determines the poison, aligns with the principle of nonmaleficence, highlighting the necessity to avoid harm. And though Paracelsus may not have explicitly discussed justice in modern terms, his travels, encounters with diverse cultures, and willingness to learn from various folk traditions suggest an openness to equity in healthcare and knowledge. While not a direct architect of modern medical ethics, Paracelsus' pioneering approach to medicine and his profound respect for the sanctity of life certainly laid the groundwork for later discussions about ethical patient care.

Conclusion: Paracelsus' indelible mark on the evolution of science, medicine, and ethics is a testament to his visionary thinking during the transformative era between the medieval period and the Renaissance. As a fervent advocate for empirical observation and experiential learning, he championed a radical departure from the established dogmas of his time, emphasizing the value of firsthand knowledge over the mere deference to ancient authorities. His pioneering work in toxicology, combined with his revolutionary approaches to medical treatments, reshaped the very paradigms of healthcare and laid foundational principles for modern pharmacology. Beyond his technical contributions, Paracelsus' patient-centric ethos subtly pushed forward ethical considerations that would later become central to medical practice. In essence, his holistic vision bridged the mystical and the empirical and forged new pathways for integrating science, medicine, and ethics, securing his legacy as a transformative figure in the annals of human knowledge and advancement.

Paracelsus' Legacy: Advocated for using chemicals and minerals in medicine, pushing medicine away from traditional Galenic practices and toward a more experimental and empirical approach.

**

Review Questions:

True/False Questions:

1. Paracelsus emphasized empirical observation and experiential learning, steering medicine away from age-old Galenic practices.
   True or False?

2. Paracelsus' medical approach heavily relied on theoretical speculation rather than hands-on experimentation.
   True or False?

Multiple-Choice Questions:

3. Which field did Paracelsus significantly contribute to during his lifetime?
   a) Physics
   b) Toxicology
   c) Astronomy
   d) Sociology

4. What was a key methodological approach used by Paracelsus in his study of medicine?
   a) Reliance on ancient texts without questioning
   b) Observing and experimenting with chemicals and minerals
   c) Conducting advanced astronomical observations
   d) Utilizing mystical and astrological explanations

Clinical Vignette:

5. A medical researcher, inspired by Paracelsus, decides to explore new remedies for a disease by focusing on the chemical properties of various minerals. The researcher emphasizes practical experimentation and careful observation of effects. Which principle of Paracelsus' methodology is the researcher most closely following?
   a) Reliance on mystical explanations
   b) Use of ancient texts without questioning
   c) Empirical evidence and observation
   d) Acceptance of divine intervention

Basic Science Vignette:

6. A Renaissance physician is studying the effects of a new chemical remedy on a disease, following the principles of Paracelsus. Which statement best captures Paracelsus' approach to this study?
   a) Using ancient texts exclusively to understand the remedy's effects
   b) Relying on mystical and astrological interpretations
   c) Observing the remedy's effects directly through experimentation
   d) Combining the remedy with traditional herbal medicines without testing

Philosophy Vignette:

7. Paracelsus' shift towards chemical remedies in medicine exemplifies a move away from which traditional medical approach?
   a) Ayurvedic practices
   b) Chinese traditional medicine
   c) Galenic humoral theory
   d) Native American herbalism

Correct Answers:

1. True
2. False
3. b) Toxicology
4. b) Observing and experimenting with chemicals and minerals
5. c) Empirical evidence and observation
6. c) Observing the remedy's effects directly through experimentation
7. c) Galenic humoral theory

Beyond The Chapter
Paracelsus (1493-1541)

- *Paracelsus: An Introduction to Philosophical Medicine in the Era of the Renaissance* by Walter Pagel
- *The Life and the Doctrines of Philippus Theophrastus, Bombast of Hohenheim, Known by the Name of Paracelsus*by Franz Hartmann
- *Paracelsus: Medicine, Magic and Mission at the End of Time* by Charles Webster

***

Philosophy of Science™ — Medical Research

Chapter 72: Ambroise Paré (1510-1590) — The Surgical Innovator: Compassionate Techniques

Ambroise Paré's compassionate and empirical approach to surgery revolutionized battlefield medicine and evidence-based practice.

Abstract: In the annals of medical history, Ambroise Paré (1510-1590) emerges as a beacon of innovation, profoundly transforming surgery and battlefield medicine during the Renaissance era, characterized by an explosive expansion in scientific and artistic exploration. Paré eschewed traditional brutal practices, such as treating gunshot wounds with boiling oil, in favor of more compassionate and empirically effective methods, revolutionizing the survival rates of wounded soldiers. His relentless curiosity and hands-on approach to medicine fostered the development of new surgical instruments. They heralded a shift toward evidence-based practice, epitomizing the rationalist and empirical impulses of the time. Through a confluence of advanced surgical techniques, ethical considerations, and an unyielding commitment to enhancing patient welfare, Paré laid a robust foundation for modern medicine, ushering in a new era characterized by critical thinking and an empirical approach to healing. His enduring legacy is a testament to the transformative power of integrating compassion, scientific rigor, and humanism, forever altering the trajectory of medical science and ethics.

**

Introduction: Ambroise Paré (1510-1590) stands as one of the most transformative figures in the annals of medical history, primarily for his pioneering work in surgery and battlefield medicine. Born in France during the Renaissance, a period rich in scientific and artistic exploration, Paré's innovative techniques and insights vastly improved the survival rates of soldiers and reshaped surgical practices of the time. He is perhaps best remembered for challenging the then-conventional treatment of gunshot wounds with boiling oil, introducing more humane and effective treatments instead. Equally significant, Paré developed groundbreaking surgical instruments, wrote seminal medical texts, and advocated for a hands-on, empirical approach to medicine, relying on observation and experience rather than age-old doctrines. His legacy not only underscores the evolution of surgery and trauma care but also exemplifies the spirit of curiosity and humanism that the Renaissance era celebrated, forever changing the trajectory of medical science.

Rationalism: Ambroise Paré's contributions to medicine were deeply rooted in a rationalist approach that prioritized empirical observation and direct experience over traditional doctrine. At a time when ancient texts and often unchallenged beliefs guided many surgical practices, Paré consistently sought to understand the underlying causes of ailments and identify treatments based on observed outcomes. His famous shift from using boiling oil — a painful and largely ineffective treatment for gunshot wounds based on traditional teachings — to a more compassionate and efficacious method stemmed from his direct observations of the detrimental effects of the former and the surprising success of alternatives he devised. Moreover, Paré's meticulous documentation of his methods and findings in his numerous writings displayed scientific rigor ahead of its time. By favoring evidence and reason over dogma, Paré revolutionized specific treatments and exemplified the broader rationalist impulse of the Renaissance, paving the way for future medical practitioners to prioritize direct observation, experimentation, and logical reasoning in their pursuit of healing.

Empiricism: Ambroise Paré's work in medicine bore the distinct mark of empiricism, an orientation that values knowledge derived from direct and tangible experience. Distancing himself from the dogmatic adherence to ancient texts and inherited practices that dominated his era, Paré turned to firsthand observation and experimentation to inform his medical decisions. This empiricist inclination was most evident in his treatment of gunshot wounds. When confronted with the prevailing method of cauterizing such wounds with boiling oil — a treatment recommended by ancient authorities but not grounded in evidence — Paré decided to use a gentler salve, partly out of necessity. The unexpectedly positive results led him to adopt this new treatment permanently, driven by his direct observations rather than tradition. Furthermore, he meticulously documented his clinical encounters, ensuring that his practices were rooted in and could be evaluated based on empirical evidence. In doing so, Paré championed a more humane approach to treatment and laid foundational stones for the scientific methodology in medicine, foregrounding the importance of experiential evidence over mere theoretical conjecture.

The Scientific Method: While Ambroise Paré is not conventionally credited with formally developing the scientific method, his approach to medicine undeniably prefigured its core tenets and significantly influenced its maturation in medical science. Paré's emphasis on direct observation, systematic experimentation, and meticulous documentation acted as a beacon when many medical treatments were determined by long-held traditions or ancient texts, often untested against empirical reality. His decision to abandon the cauterization of gunshot wounds with boiling oil in favor of a milder and more effective salve, based on firsthand observations and comparative outcomes, mirrors the scientific method's cyclical observation, hypothesis, experimentation, and refinement process. By grounding his practice in evidence and challenging the status quo when results dictated, Paré became a harbinger of a more systematic and empirical approach to knowledge acquisition. Through his writings and teachings, Paré advanced specific medical practices and exemplified a methodological shift toward evidence-based reasoning, foreshadowing the more formalized scientific methodologies that would gain prominence in subsequent centuries.

Medicine: Ambroise Paré, a luminary of Renaissance medicine, dramatically revolutionized surgical practices and trauma care with his trailblazing techniques and relentless quest for knowledge. Often dubbed the "*father of modern surgery*," Paré's most noteworthy contribution was his radical departure from the painful and harmful practice of treating gunshot wounds with boiling oil. After accidentally running out of the scalding liquid during a battlefield surgery, Paré improvised with a more benign mixture of egg yolk, rose oil, and turpentine, observing a remarkably improved patient recovery. This empirical discovery led him to advocate for gentler and more effective wound treatments, drastically improving soldiers' survival rates. Additionally, Paré designed numerous surgical instruments, enhancing the precision and efficiency of various procedures. His extensive writings, including his magnum opus "*Les Oeuvres*," served as seminal texts for surgeons across Europe, ushering in a new era of evidence-based medical practice. Beyond mere technical contributions, Paré's commitment to the well-being of his patients, coupled with his empirical approach, solidified his enduring legacy as a cornerstone figure in the evolution of medical science and ethics.

Ethics: Ambroise Paré's contributions to medicine extended beyond mere techniques to encompass core ethical principles that would shape the very ethos of medical practice. His opposition to the then-standard use of boiling oil for gunshot wounds encapsulates the principle of nonmaleficence, or "*do no harm*." Where traditional methods caused immense suffering, Paré championed more effective and humane treatments, promoting beneficence by actively seeking the well-being of his patients. His empirical approach, which valued patient outcomes and direct observations over established dogma, also indirectly bolstered patient autonomy. By pioneering evidence-based practices, he empowered patients and their caregivers to make informed decisions based on actual results rather than unchallenged traditions. Furthermore, Paré's commitment to documenting and sharing his findings democratized medical knowledge, subtly weaving the principle of justice into his work by ensuring that the benefits of his discoveries were accessible to many rather than being reserved for an elite few. Though he might not have articulated these ethical principles as we know them today, Paré's practice and philosophy undeniably laid the groundwork for their emergence and consolidation in subsequent medical ethics discourse.

Conclusion: Ambroise Paré's indelible mark on the landscape of medical history is a testament to the synergy of compassion, curiosity, and scientific rigor. As a pioneering figure of the Renaissance, Paré, with his innovative surgical techniques and transformative treatments, not only elevated the standards of care but also redefined the ethos of medical practice. By marrying rationalism and empiricism, he emphasized the importance of evidence and experience over untested tradition. His contributions went beyond procedural advancements, laying the foundation for the scientific method's principles in medicine, advocating for evidence-based practices, and upholding ethical standards prioritizing patient well-being and autonomy. Paré's legacy is not just about the specific methods or instruments he developed but about the holistic shift he championed in the approach to medicine — a shift toward critical thinking, evidence-based practice, and unwavering commitment to patient welfare. He remains an enduring beacon, illuminating the path for future generations of medical professionals and researchers.

Ambroise Paré's Legacy: Modernized surgical methods and emphasized the importance of treating wounds without cauterization, transforming surgical practices.

**

Review Questions:

True/False Questions:

1. Ambroise Paré introduced the practice of treating gunshot wounds with boiling oil to improve patient outcomes.
   True or False?

2. Paré's approach to medicine was characterized by empirical observation and direct experience, rather than reliance on traditional doctrines.
   True or False?

Multiple-Choice Questions:

3. Which significant contribution is Ambroise Paré known for in the field of medicine?
   a) Developing new astronomical models
   b) Introducing advanced chemical theories
   c) Revolutionizing surgical techniques and wound care
   d) Establishing principles of physics

4. What principle did Ambroise Paré emphasize in his medical practice?
   a) Adherence to ancient texts without questioning
   b) Use of mystical and astrological explanations
   c) Observation, experimentation, and patient well-being
   d) Reliance on religious doctrines for treatment

Clinical Vignette:

5. A modern surgeon is inspired by Ambroise Paré's methods and decides to abandon traditional but painful treatment methods in favor of more compassionate and empirically effective alternatives. Which principle of Paré's methodology is the surgeon most closely following?
   a) Reliance on ancient texts
   b) Use of mystical explanations
   c) Empirical evidence and observation
   d) Acceptance of divine intervention

Basic Science Vignette:

6. A Renaissance surgeon, inspired by Ambroise Paré, is considering alternatives to cauterizing gunshot wounds with boiling oil. Which approach best aligns with Paré's innovations?
   a) Continuing to use boiling oil as the primary treatment
   b) Applying a milder salve based on empirical observations of improved recovery
   c) Using herbal remedies with no prior testing
   d) Relying solely on astrological signs to guide treatment

Philosophy Vignette:

7. Ambroise Paré's approach to surgery exemplified a move away from which of the following traditional medical practices?
   a) Treating diseases with mystical rituals
   b) Using bloodletting as a primary treatment
   c) Relying on boiling oil to cauterize wounds
   d) Emphasizing the importance of diet and exercise exclusively

Correct Answers:

1. False
2. True
3. c) Revolutionizing surgical techniques and wound care
4. c) Observation, experimentation, and patient well-being
5. c) Empirical evidence and observation
6, b) Applying a milder salve based on empirical observations of improved recovery
7. c) Relying on boiling oil to cauterize wounds

Beyond The Chapter
Ambroise Paré (1510-1590)

- *Ambroise Paré: Surgeon of the Renaissance* by Wallace Hamby
- *The Apologie and Treatise of Ambroise Paré* edited by Geoffrey Keynes
- *Ambroise Paré and His Times, 1510-1590* by Stephen Paget

***

Chapter 73: Andreas Vesalius (1514-1564) — The Anatomical Innovator: Revealing the Human Body

Andreas Vesalius' empirical dissections and systematic study of human anatomy laid the groundwork for modern anatomical science.

Abstract: Amid the Renaissance's fervent reevaluation of ancient doctrines, Andreas Vesalius emerged as a beacon of enlightenment, transforming medical science. Breaking from the reliance on Galen's texts, Vesalius pioneered a radical approach to studying human anatomy, embodying the era's rationalist and empiricist tendencies. His meticulous dissections of human cadavers revealed numerous inaccuracies in Galenic teachings. In his monumental work "*De humani corporis fabrica*," Vesalius corrected misconceptions and introduced a detailed representation of the human body, fostering an evidence-based understanding of anatomy. His direct and systematic study emphasized observation and experimentation, laying the groundwork for the modern scientific method. Vesalius's influence transcended science, igniting ethical discussions on human dissection and shaping contemporary medical ethics. Andreas Vesalius stands as a luminary in medical history, embodying empirical rigor and critical thinking, and establishing a legacy of relentless curiosity and pursuit of truth in the medical sciences.

**

Introduction: Andreas Vesalius, born in 1514 in Brussels, stands as a towering figure in the annals of medical history. Often hailed as the father of modern anatomy, Vesalius radically transformed the study and understanding of the human body during the Renaissance. Before his groundbreaking work, medical education largely depended on ancient texts, particularly those by Galen, and often involved the dissection of animals rather than human corpses. However, Vesalius' insatiable curiosity and meticulous observations led him to challenge these traditional notions. His magnum opus, "*De humani corporis fabrica*" (On the Fabric of the Human Body), published in 1543, corrected not only many of Galen's inaccuracies but also introduced detailed illustrations that offered a vivid window into the inner workings of the human body. Vesalius' emphasis on direct observation and empirical evidence set the stage for the evolution of the modern scientific method in medicine. It cemented his legacy as a revolutionary figure in the annals of medical science.

Some of Andreas Vesalius corrections to the anatomical teachings of Galen:

1. Structure of the Human Heart:
   - Vesalius corrected Galen's assertion that the human heart had tiny pores in the septum allowing blood to flow between the left and right ventricles. Vesalius demonstrated that no such pores existed and that blood passed from the right side of the heart to the left via the lungs (through pulmonary circulation).
2. Jawbone Structure:
   - Vesalius found that the human jawbone (mandible) was a single bone, contrary to Galen's belief that it was composed of two separate bones. This correction highlighted significant differ-

ences between human and animal anatomy, as Galen's observations were based primarily on animal dissections.

3. Structure of the Liver:
   - Vesalius corrected Galen's description of the liver, particularly the number and arrangement of its lobes. Galen described the liver based on animal anatomy, which did not accurately reflect the human liver's structure.
4. Vascular System:
   - Vesalius disproved Galen's description of the rete mirabile, a network of blood vessels at the base of the brain, which Galen had described based on animal studies. Vesalius showed that this structure did not exist in humans.
5. Sternum Structure:
   - Vesalius corrected Galen's description of the sternum (breastbone). Galen had described the sternum as being composed of seven segments based on animal dissections, while Vesalius demonstrated that the human sternum typically has three segments.
6. Musculature:
   - Vesalius corrected many of Galen's descriptions of muscles, including their locations, attachments, and functions. His detailed observations and illustrations provided a more accurate understanding of human musculature.
7. Nervous System:
   - Vesalius provided more accurate descriptions of the brain and spinal cord, including the structure of the ventricles in the brain, correcting several of Galen's misconceptions based on animal anatomy.
8. Reproductive System:
   - Vesalius corrected several of Galen's errors regarding the reproductive organs. He provided detailed descriptions of the male and female reproductive systems, highlighting differences that Galen had not accurately described.

Rationalism: Andreas Vesalius' approach to anatomy epitomized the burgeoning rationalist tendencies of the Renaissance era, emphasizing empirical observation and firsthand knowledge over long-standing dogma. While the medical establishment of his day heavily relied on ancient authorities like Galen, whose texts were revered as almost infallible, Vesalius was unafraid to challenge the orthodoxy. His commitment to rational inquiry led him to undertake direct dissections of human cadavers — a departure from the norm where animal dissections were used as proxies for human anatomy. He made numerous discoveries through these dissections, highlighting inaccuracies in Galenic teachings. His insistence on seeing to believe, rather than accepting truths merely on authority, was a hallmark of rationalist thought. Vesalius believed that understanding the human body required a direct and systematic study of its structure, free from the distortions of age-old misconceptions. His publication, "*De humani corporis fabrica*," underscored this rationalist ethos, combining meticulous descriptions with detailed illustrations, thereby providing an evidence-based foundation for anatomy. In an age poised between medieval scholasticism and modern science, Vesalius' rationalist leanings paved the way for a more empirical and skeptical approach to studying the natural world.

Empiricism: Andreas Vesalius is often heralded as a pioneering figure of empiricism in medical science. In stark contrast to the prevailing reliance on ancient texts, Vesalius championed the cause of direct observation and experiential learning. Rooted in the empiricist conviction that knowledge can be gained from sensory experience, he prioritized hands-on anatomical dissections of human bodies over merely adhering to the teachings of ancient authorities like Galen. This hands-on approach led him to uncover numerous anatomical discrepancies between the revered texts and actual human anatomy. His magnum opus, "*De humani corporis fabrica*," is a testament to his empiricist orientation, merging meticulously detailed descriptions with precise illustrations based on his dissections. Through this work, Vesalius emphasized the indispensability of empirical evidence in advancing anatomical knowledge, thereby challenging the traditional deference to ancient authorities. His unwavering commitment to observation, experimentation, and evidence-based conclusions positioned Vesalius at the forefront of a paradigm shift, catalyzing a transformation from scholastic deference to the empirical rigor that would characterize modern science.

The Scientific Method: Andreas Vesalius, through his revolutionary work in anatomy, played a seminal role in laying the foundational principles that would later be integral to the scientific method. Before Vesalius, the study of the human body was shackled by unquestioning adherence to ancient authorities, chiefly Galen. Vesalius, however, championed a rigorous approach to scientific inquiry that prioritized observation, hypothesis, experimentation, and validation. His methodical dissections of human cadavers and the subsequent comparison of his findings to traditional texts exemplified a nascent form of hypothesis testing. By highlighting discrepancies between established teachings and his empirical observations, Vesalius showcased the importance of evidence over dogma. "*De humani corporis fabrica*," his magisterial work, wasn't just a compendium of anatomy and a methodological manifesto, illustrating how systematic observation could unearth truths about the natural world. While Vesalius might not have formulated the scientific method as we understand it today, his insistence on empirical evidence, skepticism toward established beliefs without validation, and commitment to rigorous documentation undeniably paved the way for the empirical rigor and iterative inquiry that underpin the modern scientific method.

Medicine: Andreas Vesalius, with his audacious inquiries into human anatomy, ushered in a new era of medicine, revolutionizing its understanding and practice. Before Vesalius, medical wisdom was largely derived from ancient texts, particularly those of Galen, seldom subjected to critical scrutiny. Vesalius, however, undertook direct dissections of human cadavers, unveiling a myriad of anatomical structures in unprecedented detail. His findings, meticulously cataloged in "*De humani corporis fabrica*," debunked several Galenic misconceptions, redefining the landscape of anatomical knowledge. This work was not merely an anatomical treatise but also a pedagogical tool enriched with intricate illustrations, facilitating the medical community's understanding of the human body's intricacies. Vesalius' dedication to firsthand observation and evidence-based conclusions emphasized the importance of empirical study in medicine, challenging the blind allegiance to ancient authorities. His methods and discoveries set new standards for medical education, pushing for hands-on learning and fostering a culture of critical thinking and skepticism. In doing so, Vesalius laid the groundwork for modern medicine, emphasizing the role of direct observation, rigorous documentation, and continuous questioning in the pursuit of medical truths.

Ethics: Andreas Vesalius' groundbreaking ventures into anatomical studies, particularly his direct dissections of human cadavers, had far-reaching implications, both scientifically and ethically. His work indirectly prompted the medical community to grapple with the ethical dimensions of using human bodies for scientific inquiry. While Vesalius sourced cadavers largely from executed criminals — a practice considered acceptable in his time — his high-profile dissections ignited discussions on the dignity and rights of the deceased. This early discourse laid the groundwork for modern principles like autonomy and respect for individuals, even in death. Furthermore, by correcting long-standing misconceptions in anatomy, Vesalius' meticulous research embodied the principles of beneficence and nonmaleficence. His accurate depictions of the human body equipped physicians with better knowledge, enabling them to provide beneficial interventions and avoid harm. Lastly, although Vesalius might not have explicitly addressed justice in modern terms, his commitment to accurate knowledge can be seen as an endeavor toward equal and fair treatment of patients through improved medical understanding. While Vesalius operated in a different ethical milieu, his work undeniably set in motion a cascade of reflections on the ethical dimensions of medical practice and research.

Conclusion: Andreas Vesalius, a Renaissance luminary, revolutionized the medical landscape with his unparalleled dedication to empirical observation, challenging the unquestioned dogmas of his time. His audacious endeavors in dissecting human cadavers recalibrated our understanding of human anatomy. They established foundational principles that shaped the evolution of the scientific method, integrating the ideals of rationalism, empiricism, and ethical consideration. Vesalius' seminal work, "*De humani corporis fabrica*," is a beacon of his commitment to evidence-based knowledge, toppling misconceptions rooted in ancient authority. His emphasis on direct observation laid the bedrock for modern medicine's empirical orientation, emphasizing the necessity for firsthand knowledge and continuous inquiry. Moreover, Vesalius inadvertently ignited essential ethical conversations, emphasizing the dignity of the deceased and the profound responsibility of the medical profession. Collectively, Vesalius' contributions remain a testament to the transformative power of critical thinking, curiosity, and empirical rigor, underscoring the enduring importance of challenging established norms in the pursuit of truth.

Andreas Vesalius' Legacy: Published "*De humani corporis fabrica,*" a groundbreaking anatomical work that challenged Galenic orthodoxy and advanced the detailed study of the human body.

**

Review Questions:

True/False Questions:

1. Andreas Vesalius' work "*De humani corporis fabrica*" corrected many inaccuracies in Galenic teachings about human anatomy.
   True or False?

2. Vesalius relied heavily on animal dissections to study human anatomy, following the traditional methods of his time.
   True or False?

Multiple-Choice Questions:

3. Which key methodological approach did Andreas Vesalius pioneer in the study of human anatomy?
   a) The use of mystical and astrological explanations
   b) Reliance on ancient texts without questioning
   c) Direct observation and dissection of human cadavers
   d) Conducting advanced chemical experiments

4. What significant impact did Vesalius' work have on the field of medicine?
   a) It introduced new astronomical models
   b) It debunked several Galenic misconceptions about human anatomy
   c) It established principles of physics
   d) It promoted the use of traditional herbal remedies

Clinical Vignette:

5. A modern medical student is inspired by Vesalius' approach to learning and decides to verify the accuracy of anatomical drawings in their textbook by conducting their own dissections and observations. Which principle of Vesalius' methodology is the student most closely following?
   a) Reliance on ancient texts
   b) Use of mystical explanations
   c) Empirical evidence and observation
   d) Acceptance of unverified authority

Basic Science Vignette:

6. A medical student in the Renaissance era is studying the human body following Andreas Vesalius' principles. Which practice best reflects Vesalius' approach to anatomy?
   a) Relying solely on ancient texts for anatomical knowledge
   b) Conducting dissections of animal bodies to infer human anatomy
   c) Performing dissections of human cadavers to observe anatomical structures directly
   d) Using astrological signs to understand human anatomy

Philosophy Vignette:

7. Andreas Vesalius' work in anatomy exemplified a departure from which traditional reliance in the medical field?
   a) Empirical observation and experimentation
   b) Galenic texts and ancient authorities
   c) Direct dissection and observation of human bodies
   d) Ethical considerations in medical practice

Correct Answers:

1. True
2. False
3. c) Direct observation and dissection of human cadavers
4. b) It debunked several Galenic misconceptions about human anatomy
5. c) Empirical evidence and observation
6. c) Performing dissections of human cadavers to observe anatomical structures directly
7. b) Galenic texts and ancient authorities

Beyond The Chapter
Andreas Vesalius (1514-1564)

- *Andreas Vesalius: The Making, The Madman, and the Myth* by Joffe SN
- *Vesalius: The China Root Epistle: A New Translation and Critical Edition* by Daniel H. Garrison
- *Andreas Vesalius Bruxellensis: The Bloodletting Letter of 1539* by Andreas Vesalius, translated by Roger Kenneth French

***

Chapter 74: Francis Bacon (1561-1626) — The Empirical Philosopher: Method and Experimentation

Francis Bacon's systematic approach to knowledge acquisition laid the cornerstone for the modern scientific method.

Abstract: Francis Bacon, a prodigious thinker of the Renaissance, reshaped the trajectory of scientific and philosophical inquiry. An advocate for empirical and rationalist traditions, Bacon criticized deference to established authorities and urged a systematic approach to knowledge acquisition. Through works like "*Novum Organum*," he emphasized marrying empirical observation with reasoned interpretation, pioneering a methodological approach that became the cornerstone of the scientific method. This empirical ethos catalyzed the Scientific Revolution and profoundly influenced modern medicine, steering it toward evidence-based practices. Although not explicitly outlined, his philosophical tenets laid the foundations of contemporary bioethics, emphasizing truth-seeking, beneficence, and a rational, transparent approach to science. Bacon's legacy is a beacon in the intellectual landscape, embodying the shift from traditional scholastic methods to a modern era marked by empirical scrutiny, systematic reasoning, and ethical integrity in pursuing knowledge.

**

Introduction: Francis Bacon, born in 1561, was a multifaceted figure in the annals of English history — a philosopher, statesman, scientist, jurist, orator, and author. Often hailed as the father of the scientific method, Bacon's advocacy for empirical inquiry and inductive reasoning laid foundational tenets for the Scientific Revolution. He believed knowledge was power, urging a departure from the traditional scholastic methods of the medieval academy, which prioritized deduction and reliance on established authorities. In addition to his scientific contributions, Bacon held political posts, serving both Queen Elizabeth I and King James I. His written works, such as "*The Advancement of Learning*" and "*Novum Organum*," sought to reimagine human understanding of the natural world and science, challenging contemporaries to explore knowledge through systematic experimentation and observation. While his life was fraught with controversies and political intrigues, Francis Bacon's intellectual legacy solidified his place as a pivotal figure in the development of modern scientific thought.

Rationalism: Francis Bacon, while often associated with the empirical tradition of scientific inquiry, also embodied elements of rationalism, the philosophy that reason is the primary source of knowledge. Bacon's rationalist tendencies are especially evident in his disdain for the unquestioning acceptance of ancient authorities, which had long dominated European intellectual thought. He argued that reason, not mere adherence to tradition, should guide scientific and philosophical pursuits. In his seminal work "*Novum Organum*," Bacon highlighted the biases and fallacies (which he termed "*idols*") that cloud human judgment, emphasizing the need for a method of reasoning that minimizes these errors. Although his vision of the scientific method stressed the importance of empirical observations, it also required systematic reasoning to derive broader truths from those observations. For Bacon, reason and empirical evidence went hand in hand: while observations provided data about the world, reason was essential to interpret, understand, and apply these data effectively. Thus, while he was a staunch advocate for empiricism, Bacon's insistence on the corrective and guiding power of reason also planted him within the rationalist tradition.

Empiricism: Francis Bacon is celebrated as a pivotal figure in the empiricist tradition, which posits that knowledge can be gained from sensory experience. Opposing the scholastic reverence for ancient texts and deductive reasoning, Bacon championed the direct observation of the natural world as the most reliable pathway to genuine understanding. He posited that the human mind is prone to errors when left to its devices, a notion he expounded upon through the concept of the "*idols*" in his work "*Novum Organum*." These "*idols*' are systematic errors in thinking that can mislead judgments. To counteract them, Bacon proposed a systematic approach to investigation — what would come to be known as the scientific method. This method emphasizes experimentation and collecting data from the external environment to derive conclusions. For Bacon, true knowledge was not to be found in age-old manuscripts or solely in reasoning but in the tangible and methodical exploration of the world. Francis Bacon laid the philosophical groundwork for centuries of scientific advancement and discovery by emphasizing the importance of empirical evidence and the systematic observation of nature.

The Scientific Method: Francis Bacon's contributions to the development of the scientific method are foundational and profound. Bacon championed a new approach to understanding the natural world by rejecting the dominant scholasticism of his era, which was deeply rooted in Aristotelian philosophy and deference to ancient authorities. In his pivotal work, "*Novum Organum*," he outlined a method of inquiry based on inductive reasoning, where general principles are derived from specific observations. He proposed that investigations commence without preconceived notions, allowing data and evidence to guide conclusions rather than fitting observations into established frameworks. Moreover, Bacon emphasized the role of experimentation in confirming or refuting hypotheses. He warned against the "*idols*" of the mind or inherent biases that could distort understanding and believed these biases could be minimized only through a systematic, empirical approach. By emphasizing observation, experimentation, and inductive reasoning, Bacon laid down the intellectual architecture for the scientific method to shape the Scientific Revolution and the subsequent course of modern scientific inquiry. His philosophy underscored that knowledge is acquired not merely through reason or tradition alone but through meticulous interaction with and study of the world.

Medicine: While not a physician, Francis Bacon left an indelible mark on medicine by promoting empiricism and the scientific method. His philosophy profoundly influenced the way medical research and practices evolved. Bacon advocated for direct observation and systematic experimentation, emphasizing the need to abandon reliance solely on ancient medical texts, which had guided medical practices for centuries. His call for rigorous investigation into the natural world extended to studying diseases, treatments, and human anatomy. By promoting a shift from dogmatic adherence to ancient authorities like Galen and Hippocrates to a more empirical approach, Bacon indirectly paved the way for numerous medical advancements. For instance, his emphasis on experimentation resonated with the work of contemporaries like William Harvey, who used empirical methods to uncover the circulatory system's workings. While Bacon did not directly contribute specific medical knowledge or innovations, his philosophical framework provided a fertile ground for the evolution of modern medicine, emphasizing evidence-based practices and the continuous quest for knowledge through direct observation and systematic inquiry.

Ethics: Francis Bacon's influence on ethics, especially as they intersect with scientific and medical fields, can be viewed through his broader philosophical tenets. While he did not directly outline principles like patient autonomy (informed consent), practitioner beneficence (do good), practitioner nonmaleficence (do no harm), and public justice (be fair) as they are understood in contemporary bioethics, Bacon's emphasis on rigorous inquiry, truth-seeking, and skepticism toward untested assumptions has indirect implications for ethical considerations. By promoting a rigorous scientific method rooted in empiricism, Bacon implicitly championed the ethical principle of beneficence, advocating for practices that better humanity through genuine knowledge. His warnings against the "*idols*" of the mind, which are biases and false notions, echo the ethical principle of nonmaleficence by urging against actions or beliefs that could harm or mislead. Furthermore, his call for broad-based education and the dissemination of knowledge is an early endorsement of autonomy, empowering individuals with information to make informed choices. Although not explicitly framed in terms of modern bioethical discourse, Bacon's advocacy for a transparent, evidence-based approach to knowledge and his deep respect for truth can be considered foundational to the ethical principles that guide contemporary science and medicine.

Conclusion: Francis Bacon's legacy spans many intellectual domains, from philosophy to the sciences, reflecting a transformative approach to knowledge acquisition and application. He stands as a testament to the evolution of thought from the age of deference to ancient texts to an era marked by empirical observation and systematic reasoning. Bacon's emphasis on inductive methods shifted the trajectory of scientific inquiry, anchoring it in tangible evidence and observable phenomena. Moreover, he recognized the limitations and biases inherent in human cognition and proposed methods to counteract these "*idols*" to arrive at genuine truths. While he did not directly delve into specialized fields like medicine, his philosophies influenced their advancement, particularly in transitioning from traditional practices to evidence-based methodologies. His ethical inclinations, though not explicitly articulated in modern terms, laid the groundwork for principles underpinning responsible and informed practice in various disciplines. Through his writings and ideas, Francis Bacon catalyzed a paradigm shift that has continued to shape modern thought, reaffirming the value of critical inquiry, empirical evidence, and ethical integrity in pursuing knowledge.

Francis Bacon's Legacy: Advocated for the inductive method and empiricism in the "*Novum Organum*," laying philosophical groundwork for the scientific method and its emphasis on observation and experimentation.

**

REVIEW QUESTIONS

True/False Questions:

1. Francis Bacon is often hailed as the father of the scientific method due to his advocacy for empirical inquiry and inductive reasoning.
   True or False?

2. Bacon's approach to knowledge acquisition relied heavily on the unquestioning acceptance of ancient texts and authorities.
   True or False?

Multiple-Choice Questions:

3. Which work of Francis Bacon is most associated with outlining his vision of the scientific method?
   a) The Advancement of Learning
   b) Novum Organum
   c) De Humani Corporis Fabrica
   d) The Almagest

4. What was a key methodological approach advocated by Francis Bacon in his scientific inquiries?
   a) Reliance on mystical and astrological explanations
   b) Deductive reasoning based on ancient texts
   c) Inductive reasoning based on empirical observations
   d) Use of divine revelations to guide research

Clinical Vignette:

5. A modern researcher, inspired by Francis Bacon, decides to conduct a series of experiments to test a new medical hypothesis, emphasizing observation and empirical data collection rather than relying on existing theoretical frameworks. Which principle of Bacon's methodology is the researcher most closely following?
   a) Reliance on ancient texts
   b) Use of mystical explanations
   c) Empirical evidence and inductive reasoning
   d) Acceptance of unverified authority

Basic Science Vignette:

6. A scientist inspired by Francis Bacon's methodologies is conducting an experiment to understand a new plant species. Which practice best aligns with Bacon's approach to scientific inquiry?
   a) Relying on ancient texts to classify the plant without further investigation
   b) Using deductive reasoning based on theoretical assumptions about plants
   c) Systematically observing and experimenting with the plant to gather data and derive conclusions
   d) Consulting astrological charts to understand the plant's properties

Philosophy Vignette:

7. Francis Bacon's advocacy for the scientific method emphasized a departure from which traditional approach to acquiring knowledge?
   a) Empirical observation and experimentation
   b) Reliance on established authorities and ancient texts
   c) Systematic reasoning and hypothesis testing
   d) Inductive reasoning and evidence-based inquiry

Correct Answers:

1. True
2. False
3. b) Novum Organum
4. c) Inductive reasoning based on empirical observations
5. c) Empirical evidence and inductive reasoning
6. c) Systematically observing and experimenting with the plant to gather data and derive conclusions
7. b) Reliance on established authorities and ancient texts

BEYOND THE CHAPTER
Francis Bacon (1561-1626)

- *Francis Bacon: The Temper of a Man* by Catherine Drinker Bowen
- *The Advancement of Learning* by Francis Bacon
- *Francis Bacon: A Life* by Jane Stevenson

***

Philosophy of Science

Medical Research

Chapter 75: Galileo Galilei (1564-1642) — The Observational Pioneer: Telescopic Revelations

Galileo Galilei's use of the telescope and commitment to rationalism and empiricism revolutionized our understanding of the cosmos.

Abstract: Galileo Galilei is a cornerstone figure in the transition from ancient dogmatic approaches to the systematic methodology of modern science. Revered as the "*father of modern science*," Galileo revolutionized our understanding of the cosmos with his pioneering use of the telescope, challenging geocentric doctrines. His commitment to rationalism and empiricism elevated observation and logical reasoning above traditional authority. By meticulously applying these principles, he laid the foundation for the scientific method, fostering rigorous experimentation and empirical validation. Galileo's influence extended beyond astronomy and physics, subtly impacting medicine by promoting evidence-based practice and inspiring the development of tools like the microscope. His commitment to intellectual autonomy and truth-seeking, despite personal cost, highlighted the ethical dilemmas in pursuing knowledge. Galileo's multifaceted legacy echoes through time, marking numerous domains of human thought and practice, epitomizing the transformative power of curiosity, evidence-based inquiry, and intellectual courage.

**

Introduction: Galileo Galilei, often referred to as the "*father of modern science*," stands as a monumental figure in the annals of scientific discovery. Born in Pisa, Italy, in 1564, he pioneered the use of the telescope for astronomical observation, unveiling the mysteries of the universe in ways previously inconceivable. Galileo's discoveries, such as the moons of Jupiter and the phases of Venus, not only challenged the prevailing geocentric model of the cosmos but also sparked a profound shift in humanity's understanding of its place in the vast expanse of space. However, his unyielding advocacy for the heliocentric model, which posited the Sun at the center of the solar system, brought him into conflict with the Catholic Church, leading to his infamous trial and subsequent house arrest. Beyond his astronomical endeavors, Galileo made significant contributions to physics, particularly the study of motion, further solidifying his legacy as one of history's preeminent thinkers.

Rationalism: Galileo Galilei's commitment to the principles of rationalism was foundational to his approach to the scientific method and his quest to understand the natural world. Embracing the belief that reason and logic, rather than traditional authority or superstition, should be the primary means of acquiring knowledge, Galileo often found himself at odds with the established doctrines of his time. He was adamant that empirical observations, when interpreted through reason, were the clearest path to truth. His reliance on the telescope, a revolutionary direct observation tool, underscored this rationalist perspective. Challenging centuries of Aristotelian beliefs and Church doctrines, he argued for a heliocentric universe model based on what he could see and deduce rather than on accepted dogma. His rigorous application of mathematics and experimentation in physics also illustrated his rationalist inclinations, as he sought objective and quantifiable truths about the behavior of objects in motion. Despite the immense societal and institutional pressures he faced, this commitment to reason and empirical evidence makes Galileo a quintessential figure in the history of rationalist thought.

Empiricism: Galileo Galilei, with his firm dedication to the principles of empiricism, revolutionized the scientific approach by emphasizing the importance of direct observation and experience in acquiring knowledge. He believed that to understand the universe's intricacies truly, one had to observe, measure, and test phenomena rather than rely solely on pre-existing beliefs or theoretical conjectures. Galileo's commitment to this empirical approach was exemplified in his use of the telescope, through which he made groundbreaking discoveries, such as the craters on the Moon, the phases of Venus, and the moons orbiting Jupiter. These direct observations challenged the conventional wisdom of the time and laid the groundwork for a more evidence-based understanding of the cosmos. Furthermore, in his studies of motion and mechanics, Galileo conducted a series of experiments, relying on practical measurements and tests to understand the principles governing the behavior of moving objects. Through these endeavors, Galileo underscored the paramount importance of evidence and firsthand experience in the quest for knowledge, firmly establishing empiricism as a cornerstone of the scientific method.

The Scientific Method: Galileo Galilei, an emblematic figure in the evolution of the scientific method, fundamentally shifted the way humanity pursued knowledge by integrating systematic observation, experimentation, and analytical reasoning. Before Galileo, many scientific beliefs were rooted in ancient philosophical traditions or interpreted through a religious lens. Galileo championed that hypotheses about the natural world should be subjected to rigorous testing and empirical verification rather than accepted on authority alone. His studies, particularly in kinematics, involved carefully designed experiments, often repeated under various conditions, to isolate and understand fundamental principles of motion. His insistence on empirical evidence was groundbreaking; for instance, his telescopic observations provided tangible evidence for heliocentrism, challenging the longstanding geocentric model. Moreover, Galileo's meticulous documentation and his commitment to mathematical description set precedents for recording and communicating experiments. Through his practices and challenges to the status quo, Galileo sowed the seeds for the modern scientific method, emphasizing the iterative process of observation, hypothesis formation, experimentation, and conclusion. His legacy thus extends beyond his specific discoveries to the very way in which science is conducted today.

Medicine: While Galileo Galilei is most renowned for his groundbreaking contributions to astronomy, physics, and the scientific method, his influence indirectly touched the realm of medicine. Although Galileo did not directly delve into medical research or practice, his emphasis on empirical observation and experimentation paved the way for a more systematic and evidence-based approach in all sciences, including medicine. His championed principles urged scholars to question traditional beliefs and seek demonstrable evidence, leading to a broader scientific revolution. Moreover, the technological advancements associated with Galileo, particularly his improvements to the telescope, had parallels in medical tools: the development and refinement of instruments like the microscope soon after Galileo's time allowed for detailed observations of the microscopic world, transforming areas such as bacteriology and pathology. His work on blood circulation, while less extensive than that of contemporaries like William Harvey, reflected his interest in physiological processes. Thus, while Galileo's direct contributions to medicine might not be as pronounced as in other fields, the spirit of inquiry and innovation he embodied undeniably influenced the medical advancements of subsequent generations.

Ethics: Galileo Galilei, though primarily celebrated for his scientific achievements, also inadvertently influenced the evolution of ethical thought, especially as it pertained to the pursuit of knowledge and its impact on society. His clashes with the establishment, particularly the Church, over his defense of heliocentrism highlighted the ethical principle of autonomy — the right to intellectual freedom and self-determination in the pursuit of truth. Galileo's unwavering commitment to what he believed was demonstrable evidence illustrated the ethical stance of beneficence, emphasizing the intrinsic good in advancing human understanding. At the same time, his story serves as a cautionary tale on nonmaleficence, revealing the potential harm of suppressing knowledge or curtailing intellectual freedoms. Furthermore, his eventual trial and house arrest demonstrate the complexities of justice in knowledge dissemination, where power dynamics can often overshadow objective truth. While Galileo did not directly systematically address these ethical principles, his life and struggles underscored the inherent ethical challenges that arise when new knowledge confronts established beliefs and power structures, thereby indirectly shaping future discourse on the ethics of knowledge, discovery, and its dissemination.

Conclusion: Galileo Galilei's multifaceted legacy is a testament to his profound influence on the trajectory of human knowledge. A staunch proponent of rationalism, he championed the power of reasoned thought, challenging deep-seated beliefs and traditions. Through his unwavering commitment to empiricism, he laid the groundwork for evidence-based investigation, forever altering the methodologies of various scientific disciplines. Central to this was his role in refining and promoting the scientific method, ensuring that the pursuit of knowledge was systematic, rigorous, and grounded in verifiable observations. Beyond the physical sciences, Galileo's intellectual ethos resonated in medicine, heralding a shift toward more empirical approaches and the development of crucial observation tools. Yet, his impact was not limited to scientific realms; his confrontations with the Church brought to the fore pressing ethical considerations concerning intellectual freedom, the benefits and harms of knowledge dissemination, and the dynamics of power and truth. Galileo's contributions extend far beyond the telescopic revelations for which he is most famed, marking him as a transformative figure whose ideas have resonated through centuries, shaping diverse fields and prompting pivotal shifts in thought and practice.

Galileo Galilei's Legacy: Championed the use of telescopes in astronomy and provided empirical evidence for the Copernican heliocentric model, further detaching science from traditional religious and philosophical constraints.

**

REVIEW QUESTIONS

True/False Questions:

1. Galileo Galilei's use of the telescope provided empirical evidence that supported the heliocentric model of the solar system.
   True or False?

2. Galileo's scientific approach was characterized by reliance on traditional authority and ancient texts over empirical observation.
   True or False?

Multiple-Choice Questions:

3. Which of the following discoveries by Galileo challenged the geocentric model of the universe?
   a) The existence of Saturn's rings
   b) The moons orbiting Jupiter
   c) The discovery of Neptune
   d) The phases of the Moon

4. What key principle did Galileo emphasize in his scientific methodology?
   a) Reliance on mystical explanations
   b) Deductive reasoning based on ancient texts
   c) Empirical observation and experimentation
   d) Use of divine revelations to guide research

Clinical Vignette:

5. A modern scientist, inspired by Galileo's methodology, decides to challenge a widely accepted theory by designing experiments to test its validity through direct observation and empirical data collection. Which principle of Galileo's methodology is the scientist most closely following?
   a) Reliance on ancient texts
   b) Use of mystical explanations
   c) Empirical evidence and experimentation
   d) Acceptance of unverified authority

Basic Science Vignette:

6. A student inspired by Galileo Galilei is conducting experiments to understand the motion of objects. Which practice best reflects Galileo's approach to studying motion?
   a) Relying on ancient texts without questioning their accuracy
   b) Conducting systematic experiments and precise measurements of objects in motion
   c) Using astrological charts to explain the behavior of moving objects
   d) Assuming that all motion is controlled by divine intervention without further investigation

Philosophy Vignette:

7. Galileo Galilei's advocacy for the heliocentric model and empirical evidence represented a departure from which traditional cosmological belief?
   a) The heliocentric model, where the Sun is at the center
   b) The geocentric model, where Earth is at the center
   c) The theory of an expanding universe
   d) The Newtonian model of universal gravitation

Correct Answers:

1. True
2. False
3. b) The moons orbiting Jupiter
4. c) Empirical observation and experimentation
5. c) Empirical evidence and experimentation
6. b) Conducting systematic experiments and precise measurements of objects in motion
7. b) The geocentric model, where Earth is at the center

BEYOND THE CHAPTER
Galileo Galilei (1564-1642)

- *Galileo's Daughter: A Historical Memoir of Science, Faith, and Love* by Dava Sobel
- *Galileo: Watcher of the Skies* by David Wootton
- *Dialogue Concerning the Two Chief World Systems* by Galileo Galilei, translated by Stillman Drake

***

Philosophy of Science

Medical Research

WEEK FOURTEEN: THINKERS AND PIONEERS

WEEK FOURTEEN

Class Hour 1

- Chapter 76: William Harvey (1578-1657) — The Circulatory Pioneer: Heart and Blood
- Chapter 77: René Descartes (1596-1650) — The Rationalist Thinker: Cogito, ergo sum
- Chapter 78: Antonie van Leeuwenhoek (1632-1723) — The Microscopic Discoverer: Unveiling the Invisible

Class Hour 2

- Chapter 79: Isaac Newton (1643-1727) — The Universal Theorist: Laws of Motion and Gravitation
- Chapter 80: David Hume (1711-1776) — The Empirical Skeptic: Questioning Causality
- Chapter 81: Immanuel Kant (1724-1804) — The Synthesizer: Rationalism and Empiricism

---

Introduction

In the fourteenth week of our journey through Philosophy of Science, we encounter a fascinating array of thinkers whose groundbreaking ideas and discoveries revolutionized our understanding of the natural world, human physiology, and the principles guiding scientific inquiry and medical ethics.

William Harvey, the Circulatory Pioneer, challenged centuries-old beliefs about the vascular system by demonstrating the circulation of blood. Through meticulous observation and rational analysis, Harvey refuted longstanding misconceptions and provided an evidence-based understanding of physiology. His work not only transformed cardiology and medical practices but also emphasized the importance of empirical research and ethical principles in medical investigation.

René Descartes, the Rationalist Thinker, introduced a new era in philosophy with his famous declaration "*Cogito, ergo sum*" ("*I think, therefore I am*"). Emphasizing reason and rationalism, Descartes' approach influenced scientific methodologies, promoting a mechanistic understanding of the human body. His ideas laid the foundations for modern medical ethics and patient care, highlighting the role of reason and systematic thought in understanding human physiology and health.

Antonie van Leeuwenhoek, the Microscopic Discoverer, revolutionized biology with his detailed observations using microscopes. By unveiling the hidden world of microorganisms, Leeuwenhoek laid the foundation for microbiology, profoundly impacting medical science. His rationalist and empirical approach emphasized the importance of transparency and reproducibility in scientific research, fostering ethical principles that continue to guide scientific inquiry today.

Isaac Newton, the Universal Theorist, united the heavens and earth under a single mechanistic framework with his laws of motion and universal gravitation. Newton's rationalist and empirical approach revolutionized physics, providing a comprehensive understanding of natural phenomena. His work indirectly advanced medical technologies and emphasized ethical principles in scientific inquiry, underscoring the importance of systematic investigation and rational analysis.

David Hume, the Empirical Skeptic, presented a radical form of empiricism that questioned causality and the concept of the self. Hume's skepticism reshaped Western philosophy and influenced scientific methodology, medicine, and ethical principles. By challenging assumptions and promoting critical inquiry, Hume's ideas highlighted the importance of evidence and reason in patient care and research.

Immanuel Kant, the Synthesizer, bridged the gap between rationalism and empiricism in his "*Critique of Pure Reason*." Kant proposed that human knowledge arises from the interaction between innate structures and experience, influencing scientific methodology, medicine, and ethical principles. His emphasis on autonomy, beneficence, nonmaleficence, and justice provided a robust ethical framework that continues to guide medical practice and scientific research.

This week, we explore the profound contributions of these thinkers, whose ideas and discoveries have left an indelible mark on the fields of science, medicine, and ethics. Through their rationalist and empirical approaches, they not only advanced our understanding of the natural world and human physiology but also established enduring principles that continue to shape scientific inquiry and ethical practices. As we delve into their legacies, we gain a deeper appreciation for the intricate interplay between reason, observation, and ethical considerations in the pursuit of knowledge and the betterment of human health.

***

Philosophy of Science — Medical Research

Chapter 76: **William Harvey** (1578-1657) — The Circulatory Pioneer: Heart and Blood

William Harvey's discovery of blood circulation, through empirical and rational methods, revolutionized medical science.

Abstract: William Harvey stands as a titan in the history of medical science, fundamentally altering our comprehension of human physiology with his groundbreaking revelation of the circulatory system and the heart's central role therein. His unwavering dedication to rationalism and empiricism propelled medicine from a field steeped in ancient beliefs to a discipline grounded in observation and logic. Through meticulous experimentation and logical reasoning, outlined in his seminal work "*De Motu Cordis*," Harvey not only debunked the ancient, accepted theories about blood circulation but also firmly established the principles of the scientific method, emphasizing a systematic approach based on empirical evidence and continual refinement of hypotheses. Moreover, his approach indirectly fostered the emergence of ethical considerations in medical research and practice, heralding a shift toward evidence-based knowledge that prioritized patient well-being. In revolutionizing the understanding of the cardiovascular system and pioneering an empirical approach to medical science, Harvey laid the bedrock for modern cardiology and a legacy that continues to shape the realms of medicine, research, and ethics today.

**

Introduction: William Harvey, a pioneering figure in the annals of medical history, was the English physician responsible for the monumental discovery of blood circulation and the heart's central role in this process. Born in 1578, in the Tudor period of England, Harvey's work laid the foundation for modern cardiology and profoundly revolutionized our understanding of the human body's inner workings. Before his groundbreaking revelation, the widely accepted beliefs about blood circulation were largely derived from the teachings of the ancient Greek physician Galen. However, through meticulous experimentation and observation, Harvey demonstrated that blood moves continuously through the body, pumped by the heart. This monumental insight debunked centuries of medical dogma and set a precedent for empirical medical methodology, ushering in a new medical research and understanding era.

Rationalism: William Harvey's approach to understanding the human body's physiology bears the hallmark of rationalist thought. At a time when medical knowledge heavily relied on ancient texts, particularly those of Galen, Harvey was not content with mere acceptance. Instead, he embodied the spirit of Renaissance rationalism by prioritizing direct observation, systematic experimentation, and logical reasoning. Harvey's determination to unravel the mysteries of blood circulation is a perfect example. By meticulously dissecting numerous animal species and conducting various experiments, he concluded that blood must circulate in a closed system powered by the heart — a stark contrast to the prevailing Galenic belief of blood's ebb and flow. Harvey's insistence on empirical evidence and reliance on reason over tradition showcased his deep-seated rationalist inclinations. His work exemplified the transition from the medieval reverence for authoritative texts to the modern emphasis on observation and logic. By embracing this rationalist approach, Harvey revolutionized our understanding of the cardiovascular system and reaffirmed that reason and direct experience are essential tools in pursuing knowledge.

Empiricism: William Harvey's empiricist orientation is most vividly reflected in his rigorous commitment to observation and experimentation as the cornerstones of scientific inquiry. In an age when medical wisdom was largely anchored to ancient texts, Harvey championed the belief that knowledge should be derived from direct experience and tangible evidence. His magnum opus, "*De Motu Cordis*" (On the Motion of the Heart and Blood), is a testament to his empiricist methodology. Here, Harvey meticulously detailed his experiments, emphasizing the role of direct observation in dissecting animals and noting the actual movements and capacities of the heart and blood vessels. He challenged the age-old blood circulation beliefs, largely based on speculation and textual interpretation, and replaced them with findings rooted in empirical evidence. Harvey's emphasis on firsthand observation and systematic experimentation represented a significant shift from speculative philosophy to evidence-based science. By adhering to these empiricist principles, he reshaped the understanding of human physiology and laid down a foundational approach that would become central to the following scientific endeavors.

The Scientific Method: William Harvey's contributions to medical science transcended the specific domain of cardiology and blood circulation; he played an integral role in solidifying the foundations of the modern scientific method. Before Harvey, reliance on ancient texts and dogmatic adherence to traditional beliefs often overshadowed the need for direct observation and experimentation. Harvey, in his quest to understand the cardiovascular system, epitomized the essence of the scientific method by consistently combining observation with rigorous experimentation. His meticulous investigations in "*De Motu Cordis*" demonstrated a structured approach: posing a question about blood movement, forming hypotheses, conducting experiments to test these hypotheses, and drawing conclusions based on empirical evidence. Through this systematic approach, Harvey effectively overturned centuries-old misconceptions, showcasing the power and necessity of empirical inquiry. His work underscored the fundamental importance of challenging established knowledge through evidence-based examination. It reinforced the idea that hypotheses and theories must be continually tested and refined in light of new data. In essence, William Harvey's endeavors crystallized the principles of the scientific method, setting a precedent for future scientists and researchers across various disciplines.

Medicine: William Harvey's monumental contributions to medicine forever altered its trajectory, setting a new standard for research and understanding the human body. Before Harvey, the comprehension of human physiology was clouded by misconceptions rooted in ancient texts, particularly the works of Galen. Harvey's pioneering research into the cardiovascular system, detailed in his seminal work "*De Motu Cordis*," unveiled the true nature of blood circulation: a closed-loop system where the heart functions as the central pump, continuously propelling blood throughout the body. This revelation debunked the long-held belief in the ebb and flow of blood and the idea that the body's tissues consumed it. Beyond this groundbreaking discovery, Harvey's emphasis on direct observation and systematic experimentation heralded a new era of empirical medicine, emphasizing the importance of evidence-based practice. His insistence on firsthand inquiry over mere deference to authoritative texts promoted a culture of skepticism and rigorous investigation in medical science. Consequently, William Harvey's legacy in medicine is twofold: he reshaped the understanding of a fundamental physiological process. He underscored the necessity of empirical research, laying the groundwork for future advancements in the field.

Ethics: While William Harvey is primarily celebrated for his groundbreaking work on blood circulation, his broader impact on the medical world indirectly influenced the evolution of ethical principles in research and practice. Harvey's meticulous and transparent methodologies, grounded in observation and experimentation, underscored the importance of genuine inquiry over adherence to dogma. This commitment to truth and rigorous methodology is an early nod to the principle of beneficence: acting in the patient's best interest by ensuring that medical knowledge and practices are evidence-based. Moreover, by challenging prevailing misconceptions, Harvey highlighted the dangers of unchallenged beliefs and practices, resonating with the principle of nonmaleficence, which stresses the importance of "*doing no harm*." While direct links between Harvey's work and the principles of autonomy and justice are more tenuous, his enduring legacy of promoting evidence-based medicine undoubtedly created a foundation upon which patient rights and equitable care could be built. While Harvey may not have directly addressed these principles, his dedication to accurate, evidence-based science set the stage for a more ethically informed medical profession.

Conclusion: William Harvey's indelible imprint on the annals of scientific and medical history is a testament to his unwavering commitment to observation, reason, and empirical research. As the luminary who redefined our understanding of blood circulation, he dismantled centuries of misconceptions, propelling medicine into a new enlightenment era. Embracing rationalist and empiricist approaches, Harvey championed the value of direct experience and logical reasoning, leading to his monumental discoveries. His rigorous methodology underscored the essence of the scientific method, emphasizing the importance of evidence-based inquiry. Beyond his direct contributions to cardiology, Harvey set the stage for future medical advancements, underscoring the importance of empirical research and an evidence-based approach. Furthermore, his emphasis on genuine inquiry and accurate understanding indirectly paved the way for the evolution of ethical considerations in medical practice and research. William Harvey's legacy is vast and multifaceted; he was a pioneering physician and a trailblazer whose methodologies and principles continue to influence modern medicine, research, and ethics.

William Harvey's Legacy: Demonstrated the circulation of blood and the role of the heart as a pump, refuting longstanding beliefs about the body's vascular system and laying the foundation for modern cardiology.

**

REVIEW QUESTIONS

True/False Questions:

1. William Harvey's discovery of blood circulation fundamentally changed the understanding of human physiology and refuted ancient beliefs about the body's vascular system.
   True or False?

2. Harvey's approach to studying the human body was characterized by strict adherence to ancient texts rather than direct observation and experimentation.
   True or False?

Multiple-Choice Questions:

3. Which seminal work did William Harvey write to detail his findings on blood circulation?
   a) De Humani Corporis Fabrica
   b) Novum Organum
   c) De Motu Cordis
   d) Principia Mathematica

4. What methodological approach did William Harvey emphasize in his scientific research?
   a) Reliance on ancient texts without questioning
   b) Empirical observation and systematic experimentation
   c) Mystical and astrological interpretations
   d) Use of divine revelations to guide research

Clinical Vignette:

5. A contemporary medical researcher, inspired by William Harvey, decides to investigate a new hypothesis about heart function by conducting detailed experiments and observing the physiological processes in various animal species. Which principle of Harvey's methodology is the researcher most closely following?
   a) Reliance on ancient texts
   b) Use of mystical explanations
   c) Empirical evidence and experimentation
   d) Acceptance of unverified authority

Basic Science Vignette:

6. A medical researcher inspired by William Harvey is studying the circulatory system. Which practice best reflects Harvey's approach to understanding blood circulation?
   a) Relying on ancient texts without conducting new experiments
   b) Using speculative theories without empirical evidence
   c) Performing detailed dissections and experiments to observe blood flow directly
   d) Consulting astrological charts to explain the body's functions

Philosophy Vignette:

7. William Harvey's discovery of blood circulation challenged which traditional belief about the vascular system?
   a) The idea that blood circulates in a closed loop system
   b) The belief in the ebb and flow of blood without continuous circulation
   c) The concept of the heart as the central pump of blood
   d) The understanding of blood being consumed by tissues

Correct Answers:

1. True
2. False
3. c) De Motu Cordis
4. b) Empirical observation and systematic experimentation
5. c) Empirical evidence and experimentation
6. c) Performing detailed dissections and experiments to observe blood flow directly
7. b) The belief in the ebb and flow of blood without continuous circulation

BEYOND THE CHAPTER
William Harvey (1578-1657)

- *William Harvey: A Life in Circulation* by Thomas Wright
- *William Harvey and the Mechanics of the Heart* by Jole Shackelford
- *The Works of William Harvey* translated by Robert Willis

***

Philosophy of Science — Medical Research

Chapter 77: René Descartes (1596-1650) — The Rationalist Thinker: *Cogito, ergo sum*

René Descartes' rationalist philosophy and systematic doubt laid the foundations for modern epistemology and scientific inquiry.

Abstract: René Descartes (1596-1650) stands as a pivotal figure whose multidisciplinary contributions catalyzed a paradigm shift, heralding the Enlightenment. Known as the "*father of modern philosophy*," Descartes employed systematic doubt to forge a new epistemological pathway, establishing rationalism as central to knowledge acquisition. While primarily a rationalist, he acknowledged sensory experiences' role in understanding the external world, creating a nuanced stance between rationalism and empiricism. Descartes' analytical methodologies revolutionized the scientific method, promoting logical and clear exploration of the natural world. His mechanistic view of the human body transformed medicine, advocating for rational analysis of physiological processes over mystical interpretations. Although not central to ethical discourse, his focus on individual consciousness and reason influenced Enlightenment principles of personal autonomy and rational moral judgment. Descartes' legacy, epitomized by "*Cogito, ergo sum,*" underscores the transformative power of reason in advancing human thought, leaving an enduring influence across centuries.

**

Introduction: René Descartes (1596–1650), often dubbed the "*father of modern philosophy*," was a French mathematician, scientist, and philosopher whose transformative ideas laid the groundwork for the Enlightenment. A key figure in the Scientific Revolution, Descartes is best known for his philosophical declaration, "*Cogito, ergo sum*" (I think, therefore I am), a foundational statement in Western philosophy emphasizing the primacy of individual consciousness and reason. Rejecting the scholastic Aristotelianism of the medieval universities, Descartes embarked on a journey to establish a new knowledge system, proposing that only certain knowledge is based on self-evident principles, much like mathematical truths. His dualistic view separated the mind from the body, proposing two distinct substances: the immaterial mind and the material body. Furthermore, Descartes' pioneering work in mathematics, especially Cartesian coordinate systems, has been instrumental in the development of modern calculus. Descartes profoundly influenced many fields through his interdisciplinary contributions, reshaping history and human thought.

Rationalism: René Descartes is a pivotal figure in the development of rationalism, a philosophical tradition that holds that the primary source of knowledge is reason rather than sensory experience. For Descartes, doubting was central to his method, which involved discarding any belief that could be challenged and then reconstructing knowledge on an indubitable foundation. This approach led him to his famous assertion, "*Cogito, ergo sum*" (I think, therefore I am). Here, Descartes identified the act of thinking as the most certain, as even doubt presupposed a thinking subject. This prioritization of reason and introspection over empirical observation sharply contrasted with the empirical traditions of the time. While he didn't dismiss the value of the senses entirely, Descartes believed they were prone to deception and illusion. Instead, he sought a foundation for knowledge analogous to the axiomatic foundations of mathematics. He thought that, through pure reason, one could arrive at clear and distinct self-evident and irrefutable ideas. This rationalist perspective reshaped epistemology and provided a robust counterpoint to empiricism, leading to rich philosophical dialogues that have persisted through the centuries.

Empiricism: It is important to clarify that René Descartes is primarily associated with rationalism, not empiricism. However, to present a nuanced understanding of his philosophy, one can argue that Descartes did not entirely dismiss sensory experience's role in acquiring knowledge. While he was skeptical of the senses due to their potential for deception (as in the case of optical illusions or dreams), he believed that once foundational truths were established through reason, sensory information could be interpreted in their light. In his work, "*Meditations on First Philosophy*," Descartes underwent a methodical process of doubt, questioning the veracity of everything, including sensory experiences. Yet, by the end of his meditations, he recognized that the external world, apprehended through the senses, does have a form of reality, albeit one contingent upon the existence of a benevolent God who ensures that our senses, generally speaking, do not deceive us. This acknowledgment of sensory-derived knowledge places Descartes in a unique position between pure rationalism and empiricism. While he never granted the senses the same foundational role that empirical philosophers like John Locke or David Hume did, Descartes' philosophy did leave room for sensory experience within the broader epistemological framework.

The Scientific Method: René Descartes' influence on the evolution of the scientific method is profound and enduring. In an age when scientific inquiry emerged from its entanglement with scholastic and theological traditions, Descartes championed an approach anchored in systematic doubt and analytical reasoning. In his "*Discourse on the Method*," he proposed a systematic procedure for scientific and philosophical inquiry: begin by doubting everything to identify what is certain, break problems down into their simplest components, solve the simpler problems first, and then systematically list complete solutions. This method emphasized the importance of clear, logical reasoning and rejecting ambiguity. Descartes' emphasis on deduction and mathematical reasoning also encouraged the idea that the natural world operates according to immutable laws, which can be discerned through careful, rational investigation. While the modern scientific method has evolved to emphasize empirical observation and hypothesis testing, Descartes' call for clarity, precision, and systematic reasoning remains foundational. Through his writings, he laid the groundwork for the systematic exploration of the natural world, helping to pave the way for subsequent scientific thinkers and their rigorous methodologies.

Medicine: René Descartes' contributions to medicine, though not as central as his roles in philosophy or mathematics, reflect his overarching desire to understand the nature of existence through reason. Main to this was his mechanistic view of the human body. Descartes perceived the human body as an intricate machine operating according to physical principles. In this framework, he sought to explain physiological processes, such as blood circulation, in mechanical terms. His dualistic philosophy posited a clear distinction between the mind (a non-material entity) and the body (a material, mechanical entity), which had significant implications for medicine. While he acknowledged the interaction between mind and body, his emphasis on the body's mechanistic nature encouraged a more objective, systematic approach to understanding human physiology and disease. Furthermore, by framing the body as a machine subject to physical laws, Descartes indirectly promoted that malfunctions (diseases) could be studied, understood, and rectified through rational intervention. While his specific medical theories might be considered rudimentary by modern standards, Descartes' mechanistic view laid the groundwork for the subsequent development of biomedical science and the shift away from more mystical or humoral explanations of disease.

Ethics: René Descartes, while renowned for his contributions to epistemology and metaphysics, was not a principal figure in developing ethical principles, especially those specifically outlined in biomedical ethics like patient autonomy (informed consent), practitioner beneficence (do good), practitioner nonmaleficence (do no harm), and public justice (be fair). However, his broader philosophical ideas indirectly informed the Enlightenment ethos, which paved the way for the rise of individual rights and the importance of reason in moral judgment. Descartes' emphasis on the individual's thinking as the foundation of knowledge can be extrapolated to underline the value of personal autonomy, the right of individuals to make decisions about their own lives without coercion. His dualistic separation of mind and body arguably elevates the realm of the mind, suggesting a sanctity and inviolability of personal consciousness and, by extension, personal choice. While he did not delve deeply into moral frameworks in the manner of contemporaries like Thomas Hobbes or later philosophers like Immanuel Kant, the Cartesian shift toward a rational, individual-centered worldview helped establish a philosophical landscape in which the concepts of autonomy, individual rights, and the application of reason to moral dilemmas became paramount.

Conclusion: René Descartes' legacy is vast, permeating numerous fields and setting the stage for much modern thought. As the "*father of modern philosophy*," Descartes shifted the epistemological lens toward individual consciousness, ushering in an era where reason stood paramount. His signature statement, "*Cogito, ergo sum*," became a testament to the primacy of thought and the indubitable nature of self-awareness. His rationalist inclinations fortified the foundations of knowledge, demanding clarity akin to mathematical axioms. Yet, he also cautiously acknowledged the role of empirical experiences. In science, Descartes' systematic methodologies emphasized analytical reasoning, influencing future investigations and shaping the trajectory of the Scientific Revolution. His perspective on the human body as a mechanistic entity drove a more objective approach to medicine, sidelining ancient mystical explanations. While not a major ethical figure, the Cartesian focus on the individual's rational mind indirectly fueled Enlightenment values, accentuating personal autonomy and the power of reason in moral judgments. In totality, Descartes' multifaceted contributions have left an indelible mark on human thought, reshaping disciplines and forging new pathways in our ceaseless quest for understanding.

René Descartes' Legacy: Pioneered modern philosophy with "*Cogito, ergo sum*" (I think, therefore I am), emphasizing the primacy of reason and laying the foundation for rationalism.

**

REVIEW QUESTIONS

True/False Questions:

1. René Descartes is known for his philosophical declaration "*Cogito, ergo sum*" (I think, therefore I am), which emphasizes the primacy of individual consciousness and reason.
   True or False?

2. Descartes' approach to acquiring knowledge was heavily based on empirical observation and sensory experiences without relying on rational analysis.
   True or False?

Multiple-Choice Questions:

3. Which seminal work of Descartes outlines his systematic procedure for scientific and philosophical inquiry?
   a) Principia Mathematica
   b) Discourse on the Method
   c) De Humani Corporis Fabrica
   d) Novum Organum

4. What was a key methodological approach advocated by René Descartes in his scientific and philosophical inquiries?
   a) Reliance on mystical and astrological explanations
   b) Deductive reasoning and systematic doubt
   c) Empirical observation without rational analysis
   d) Use of divine revelations to guide research

Clinical Vignette:

5. A contemporary philosopher, inspired by René Descartes, decides to reexamine long-standing philosophical beliefs by systematically doubting them and seeking knowledge based on clear and distinct principles. Which principle of Descartes' methodology is the philosopher most closely following?
   a) Reliance on ancient texts
   b) Use of mystical explanations
   c) Systematic doubt and rational analysis
   d) Acceptance of unverified authority

Basic Science Vignette:

6. A scientist inspired by René Descartes is developing a new method to study the human body. Which practice best reflects Descartes' approach to understanding human physiology?
   a) Relying solely on traditional beliefs without questioning them
   b) Using astrological signs to explain bodily functions
   c) Employing rational analysis and mechanistic principles to study physiological processes
   d) Assuming that diseases are caused by supernatural forces without investigation

Philosophy Vignette:

7. René Descartes' declaration "*Cogito, ergo sum*" (I think, therefore I am) marked a departure from which traditional philosophical belief?
   a) The belief in the primacy of empirical observation over reason
   b) The reliance on ancient texts as the primary source of knowledge
   c) The focus on collective consciousness over individual thought
   d) The acceptance of sensory experience as infallible

Correct Answers:

1. True
2. False
3. b) Discourse on the Method
4. b) Deductive reasoning and systematic doubt
5. c) Systematic doubt and rational analysis
6. c) Employing rational analysis and mechanistic principles to study physiological processes
7. b) The reliance on ancient texts as the primary source of knowledge

BEYOND THE CHAPTER
René Descartes (1596-1650)

- *Descartes: A Biography* by Desmond M. Clarke
- *Meditations on First Philosophy* by René Descartes, translated by Donald A. Cress
- *Descartes: Selected Philosophical Writings* translated by John Cottingham, Robert Stoothoff, and Dugald Murdoch

***

Chapter 78: Antonie van Leeuwenhoek (1632-1723) — The Microscopic Discoverer: Unveiling the Invisible

Antonie van Leeuwenhoek's pioneering use of the microscope unveiled a microscopic universe, transforming biological sciences.

Abstract: Antonie van Leeuwenhoek, a pioneering figure from the 17th century, radically reshaped the trajectory of biological sciences, laying a robust foundation for microbiology. A self-taught lens-maker and meticulous observer, Leeuwenhoek unlocked an entirely unseen microscopic universe, redefining humanity's comprehension of life and biology. His approach, a synthesis of rationalist and empiricist philosophies, emphasized objective investigation and firsthand experience, echoing the scientific revolution's spirit. Leeuwenhoek's meticulous methodologies significantly contributed to the evolution of the scientific method, emphasizing reproducibility and peer review. His discoveries notably impacted medicine, ushering in a new understanding of diseases and human physiology and setting the stage for future developments in the field. While not explicitly a contributor to the discourse on scientific ethics, Leeuwenhoek's commitment to truth and knowledge dissemination can be seen as an early representation of responsible scientific conduct. Leeuwenhoek's remarkable legacy remains a vivid testament to the transformative power of curiosity-driven exploration and diligent observation in shaping human history and knowledge.

**

Introduction: Antonie van Leeuwenhoek, a self-taught Dutch tradesman born in the 17th century, holds an illustrious position in the annals of scientific history as the "*Father of Microbiology*." With no formal academic training, Leeuwenhoek's insatiable curiosity drove him to perfect the art of lens-making, crafting microscopes of unparalleled magnification for his time. Through these microscopes, he unveiled an unseen world teeming with microscopic organisms, or "*little animals*," as he affectionately termed them. His meticulous observations and descriptions of bacteria, protozoa, and even human cells charted unexplored territories, transforming our understanding of the biological realm. His groundbreaking discoveries laid the foundation for microbiology and challenged established perceptions of life, emphasizing the minuscule's significance and reshaping biological sciences' trajectory. Through his unwavering dedication, Leeuwenhoek embodied the potential of self-guided learning and the profound impact of individual inquiry on the progression of science.

Rationalism: Antonie van Leeuwenhoek, in many ways, exemplified the rationalist spirit of the Scientific Revolution of the 17th century. While not a philosopher like Descartes or Spinoza, his approach to studying the microscopic realm underscored key principles of rationalism. Leeuwenhoek relied heavily on empirical observation, grounding his assertions in what he could perceive and measure rather than relying on established doctrine or speculative reasoning. His rigorous documentation of microscopic observations, often accompanied by detailed sketches, demonstrated a deep commitment to the idea that knowledge should be derived from objective and systematic investigation. Furthermore, Leeuwenhoek's correspondence with the Royal Society of London underscores his belief in the universal nature of reason and the communicability of scientific findings. His work championed the principle that understanding the natural world required a departure from tradition, superstition, or mere acceptance of established thought. Instead, Leeuwenhoek's investigations highlighted the importance of direct observation, empirical evidence, and logical reasoning in discerning the intricacies of the world around us, embodying the essence of the scientific rationalist approach.

Empiricism: Antonie van Leeuwenhoek's endeavors in the realm of microscopy epitomize the empiricist orientation, a foundational pillar of the Scientific Revolution. Rather than relying on the established knowledge of the time or purely deductive reasoning, Leeuwenhoek anchored his explorations in direct, sensory experience. With each specimen he studied under his self-made microscopes, he meticulously recorded what he observed, even when those observations clashed with his era's prevailing scientific and philosophical notions. For Leeuwenhoek, the unseen microscopic world revealed truths that could only be ascertained through firsthand observation. His detailed descriptions, whether of bacterial motion or the intricate structure of tissue, emphasized data drawn directly from experience. In essence, Leeuwenhoek's methodological approach was grounded in the empiricist belief that sensory experiences, rather than innate ideas or established doctrines, were the primary source of knowledge. His relentless pursuit of observational data and commitment to sharing his findings with the broader scientific community marked a significant stride for empiricism, underscoring the idea that understanding the natural world demands a genuine engagement with it through direct experience and observation.

The Scientific Method: While Antonie van Leeuwenhoek did not lay out a formalized version of the scientific method as later thinkers would, his practices contributed to its evolution and wider acceptance. A methodical combination of observation, description, and verification characterized his approach to microscopy. Leeuwenhoek would observe minute organisms under his meticulously crafted lenses, diligently record his findings with accompanying sketches, and then reexamine his observations to ensure their accuracy. His commitment to repeatability — often reconfirming his discoveries with multiple specimens and different instruments — echoed the scientific method's emphasis on reproducibility. Furthermore, Leeuwenhoek welcomed external verification by communicating his findings in detailed letters to the Royal Society of London, inviting peers to challenge or confirm his discoveries. This open approach to scientific inquiry, where conclusions were derived from empirical evidence and then exposed to external scrutiny, showcased foundational elements of the scientific method. Through his rigorous and transparent practices, Leeuwenhoek advanced our knowledge of the microscopic world and reinforced the virtues of systematic inquiry, empirical evidence, and peer review that would become hallmarks of modern scientific exploration.

Medicine: Antonie van Leeuwenhoek's forays into the microscopic realm had profound implications for the field of medicine, laying the groundwork for future discoveries that would revolutionize medical science. His detailed observations were the first to introduce scientists and physicians to a vast universe of microscopic organisms, some of which would later be identified as pathogens responsible for various diseases. By illustrating the existence of these microorganisms, which he often referred to as "*animalcules*," Leeuwenhoek inadvertently paved the way for the germ theory of disease, a foundational concept in modern medicine. His studies on human blood, where he described red blood cells, enhanced our understanding of its composition and function. Furthermore, his examinations of plaque from his teeth, which revealed many living organisms, gestured toward the importance of oral hygiene and its connection to health. While he did not possess the knowledge to link specific microbes to specific diseases directly, Leeuwenhoek's pioneering work essentially set the stage for later medical researchers to explore, identify, and understand the microscopic agents of disease, ultimately catalyzing advancements in diagnostics, treatments, and preventive measures in the field of medicine.

Ethics: While Antonie van Leeuwenhoek is primarily celebrated for his revolutionary advancements in microscopy and biology, his direct influence on developing the core ethical principles commonly cited in modern bioethics — patient autonomy (informed consent), practitioner beneficence (do good), practitioner nonmaleficence (do no harm), and public justice (be fair) — is less evident. However, his scientific approach offers indirect insights. Leeuwenhoek's rigorous commitment to truth-telling, as he faithfully reported his observations even when they countered prevailing beliefs, indirectly speaks to the principle of beneficence, as disseminating accurate information is vital for the well-being of society. His correspondence with the Royal Society and his transparent sharing of discoveries could be interpreted as promoting autonomy by equipping others with knowledge allowing informed decisions. While Leeuwenhoek's work did not overtly address nonmaleficence or justice, his practices illuminated the need for integrity in scientific endeavors, hinting at the broader responsibility of researchers to do no harm and to ensure that the benefits and burdens of scientific advancements are equitably distributed. Although Leeuwenhoek did not explicitly shape these ethical principles, his dedication to unbiased observation, transparency, and sharing knowledge can be seen as an early embodiment of the spirit of ethical inquiry and responsibility in scientific research.

Conclusion: Antonie van Leeuwenhoek's monumental contributions to science and medicine, encapsulated in his in-depth examinations of the microscopic world, ushered in a new era of understanding and discovery. With an uncanny ability to bridge rationalism and empiricism, Leeuwenhoek illuminated the uncharted realms of the microscopic, radically altering perceptions of life and biology. By steadfastly adhering to an empirical approach, emphasizing the importance of direct observation, and fostering transparency in his work, he played an instrumental role in the evolution of the scientific method. Furthermore, his revelations in microbiology underpinned significant advancements in medicine, particularly in our comprehension of disease etiology and human physiology. While he may not have directly crafted ethical doctrines as we know them today, Leeuwenhoek's commitment to truth, transparency, and knowledge-sharing set a precedent for the ethical conduct of scientific research. His enduring legacy, built upon curiosity and detailed observation, is a testament to the indomitable spirit of human inquiry and its capacity to reshape history.

Antonie van Leeuwenhoek's Legacy: Revolutionized biology with his detailed observations using microscopes, introducing the world to the existence of microorganisms.

**

Review Questions:

True/False Questions:

1. Antonie van Leeuwenhoek is known as the "*Father of Microbiology*" due to his detailed observations of microscopic organisms using self-made microscopes.
   True or False?

2. Leeuwenhoek's work heavily relied on established doctrines rather than direct observation and empirical evidence.
   True or False?

Multiple-Choice Questions:

3. Which significant contribution did Antonie van Leeuwenhoek make to the field of biology?
   a) Discovering the laws of planetary motion
   b) Unveiling the existence of microorganisms
   c) Formulating the theory of evolution
   d) Proposing the germ theory of disease

4. What was a central element of Leeuwenhoek's scientific methodology?
   a) Relying on ancient texts
   b) Conducting repeatable and verifiable experiments
   c) Using mystical explanations
   d) Following astrological predictions

Clinical Vignette:

5. A modern biologist is inspired by Leeuwenhoek and decides to study the microbial composition of dental plaque using advanced microscopy techniques. Which principle of Leeuwenhoek's approach is this biologist applying?
   a) Acceptance of traditional beliefs
   b) Reliance on theoretical models without evidence
   c) Empirical observation and detailed documentation
   d) Ignoring direct sensory experience

Basic Science Vignette:

6. A student is replicating Antonie van Leeuwenhoek's experiments by observing microorganisms in a drop of pond water. Which of the following best reflects Leeuwenhoek's approach to scientific discovery?
   a) Ignoring unexpected observations and focusing only on what is known
   b) Systematically observing, documenting, and verifying the findings through repeated experiments
   c) Relying on second-hand descriptions of microorganisms without personal observation
   d) Dismissing findings that contradict established scientific beliefs

Philosophy Vignette:

7. Antonie van Leeuwenhoek's contribution to the scientific method is best exemplified by which practice?
   a) Developing theoretical models without experimental evidence
   b) Relying solely on ancient texts to understand microscopic life
   c) Emphasizing empirical observation, meticulous documentation, and inviting peer review
   d) Rejecting the need for reproducibility in scientific experiments

Correct Answers:

1. True
2. False
3. b) Unveiling the existence of microorganisms
4. b) Conducting repeatable and verifiable experiments
5. c) Empirical observation and detailed documentation
6. b) Systematically observing, documenting, and verifying the findings through repeated experiments
7. c) Emphasizing empirical observation, meticulous documentation, and inviting peer review

Beyond The Chapter
Antonie van Leeuwenhoek (1632-1723)

- *Antoni van Leeuwenhoek: Master of the Microscope* by Clifford Dobell
- *The Collected Letters of Antoni van Leeuwenhoek* (19 Volumes) edited by L. C. Palm
- *Leeuwenhoek's Legatees and Beijerinck's Beneficiaries: A History of Medical Virology in The Netherlands* by Ton van Helvoort

***

Philosophy of Science

Medical Research

Chapter 79: Isaac Newton (1643-1727) — The Universal Theorist: Laws of Motion and Gravitation

Newton's laws of motion and gravitation transformed our understanding of the universe, blending rationalism and empiricism to lay the foundations for modern science.

Abstract: In the history of scientific exploration and intellectual revolution, Isaac Newton towers as a seminal figure, orchestrating a seismic shift in humanity's comprehension of the universe. Born in 1643 in Woolsthorpe, England, Newton not only formulated the laws that govern motion and universal gravitation but also laid the mathematical foundations that propel a multitude of scientific disciplines today. This account encapsulates Newton's unparalleled contributions, marked by a meticulous blend of rationalism and empiricism, which spearheaded the scientific revolution of the 17th century and crystallized the methodological framework for modern scientific inquiry. His tireless pursuits ventured beyond mathematics and physics, indirectly influencing fields like medicine while underlining an implicit ethos of integrity and truth in academic endeavors. Newton's indelible legacy, therefore, embodies not just his groundbreaking discoveries but also the spirit of relentless, unbiased exploration, which has profoundly shaped the evolution of scientific study and continues to fuel humanity's insatiable quest for knowledge.

**

Introduction: Isaac Newton, born in 1643 in Woolsthorpe, England, stands as one of the most influential figures in the annals of science. A polymath whose insights spanned the fields of mathematics, physics, and astronomy, Newton's groundbreaking work laid the foundation for much of modern scientific understanding. He most famously formulated the laws of motion and universal gravitation, transforming how humanity perceived the universe. These principles, penned in his magnum opus, "*Philosophiæ Naturalis Principia Mathematica*," not only elucidated the mechanics of the cosmos but also inspired subsequent generations of scientists, including Albert Einstein. Beyond physics, Newton's contributions to calculus, though developed contemporaneously with the mathematician Gottfried Wilhelm Leibniz, have become fundamental to a wide spectrum of scientific disciplines. In the history of human thought, few have left a legacy as profound and enduring as Newton, making his works a cornerstone in the evolution of scientific study.

Rationalism: Isaac Newton, often celebrated for his unparalleled contributions to science, embodied the spirit of rationalism that marked the scientific revolution of the 17th century. Rooted in the belief that reason and logical deduction were primary sources of knowledge, rationalism posited that the universe, in all its vast complexity, could be understood by applying intellect and systematic thought. Newton's methodological approach to the study of nature exemplified this conviction. In his "*Philosophiæ Naturalis Principia Mathematica*," he employed a rigorous, deductive framework, beginning with self-evident axioms or "*definitions*" and "*laws of motion*" and deriving from them a multitude of propositions, a method reminiscent of Euclid's geometrical proofs. Newton's belief in the capacity of reason to decode the mysteries of the universe was epitomized in his famous assertion, "*I frame no hypotheses*." Instead of resorting to untested assumptions, he sought to ground his conclusions firmly in empirical evidence and logical derivation. In doing so, Newton advanced scientific knowledge and championed the rationalist ethos, asserting the power of human reason to unveil the laws governing the cosmos.

Empiricism: Though Isaac Newton is often associated with the rationalist tradition due to his methodological approach to mathematics, he equally exemplifies the empiricist orientation in his scientific endeavors. Empiricism emphasizes knowledge derived from sensory experience and experimentation, and Newton's work, especially in physics, was deeply rooted in this philosophy. For instance, his investigations into the nature of light were grounded in meticulous experiments using prisms, leading him to foundational insights about the spectrum of colors. Likewise, while his laws of motion and gravitation were presented within a logical framework, they were inherently empirical, as they were drawn from and tested against observable phenomena. Newton's famous adage, "*Hypotheses non fingo*" or "*I frame no hypotheses*," underscores his reluctance to speculate beyond what could be directly inferred from the data. His commitment to empirical methods and meticulous observations set a gold standard for scientific inquiry. In essence, Newton's genius lay not just in his capacity for abstract thought but also in his empirical rigor, emphasizing that a genuine understanding of the universe required a delicate balance between reason and observation.

The Scientific Method: Isaac Newton's contributions to science transcends the content of his discoveries; he played a pivotal role in crystallizing the methodology that underpins modern scientific inquiry, known as the scientific method. Before Newton, the line between philosophy, alchemy, and science was often blurred. Through his meticulous approach in "*Philosophiæ Naturalis Principia Mathematica*" and other works, Newton demonstrated the effectiveness of a systematic method that begins with defining basic principles or axioms, proceeds with logical derivations, and seeks empirical verification. His rigorous commitment to uniting theory with experiments underscored the necessity of grounding abstract ideas in observable phenomena. For instance, his three laws of motion weren't mere philosophical postulates; they were derived from and validated against real-world observations. Moreover, Newton emphasized repeatability, a crucial tenet of the scientific method, asserting that for findings to be universally accepted, they must consistently be replicable under the same conditions. In shaping the scientific method, Newton effectively set the standard for how science would be conducted for centuries, emphasizing an iterative process of hypothesis, experiment, observation, and refinement that remains at the heart of scientific endeavors today.

Medicine: Unlike his monumental contributions to physics, mathematics, and astronomy, Isaac Newton's direct medical contributions were limited. However, his work indirectly impacted the scientific understanding that would later shape medical advancements. For instance, Newton's exploration of optics and the nature of light paved the way for future developments in fields like ophthalmology. His inquiries into alchemy, a precursor to modern chemistry, involved meticulous experiments with various substances, some of which had medicinal properties. While he did not produce medical cures or groundbreaking medical theories, his rigorous approach to experimentation and emphasis on empirical evidence indirectly influenced how medicine began to evolve like other scientific disciplines. Instead of relying solely on ancient texts or speculative theories, medical researchers started giving more weight to direct observation and methodical experimentation, mirroring Newton's approach in his scientific pursuits. Thus, while Newton was not a medical pioneer immediately, his overarching scientific methodologies and principles would eventually become foundational in medicine's systematic study and advancement.

Ethics: Isaac Newton, while primarily known for his scientific and mathematical innovations, was not a prominent figure in the domain of ethical philosophy, especially when it comes to principles like patient autonomy (informed consent), practitioner beneficence (do good), practitioner nonmaleficence (do no harm), and public justice (be fair), which are central to modern bioethics. However,

Newton's commitment to truth, rigorous inquiry, and intellectual honesty indirectly emphasized the importance of integrity and sincerity in scholarly pursuits. This adherence to truth can be perceived as a broader ethical stance, wherein knowledge and its advancements are pursued for the greater good, aligning with the principle of beneficence. Moreover, his reluctance to publish speculative or unverified claims can be seen as an early recognition of nonmaleficence in the sense of not misleading or causing potential harm to the academic community. Newton's overarching ethos of rigorous inquiry, free from personal bias and grounded in empirical evidence, is an implicit endorsement of justice in that it values objective truth over individual prejudice. Nonetheless, linking Newton directly to the aforementioned ethical principles is a stretch; his influence on them is more inferred from his general approach to scholarship than any direct commentary or focus on ethical theory.

Conclusion: In the annals of human history, Isaac Newton emerged as a colossal figure, wielding an influence that reshaped the very fabric of scientific thought and inquiry. His groundbreaking work in mathematics, physics, and astronomy redefined humanity's understanding of the cosmos and set rigorous standards for empirical study and logical deduction. Newton's delicate balance between rationalism and empiricism emphasized the synthesis of reason with sensory experience, providing a robust framework for the scientific method that governs research today. While his direct contributions to fields like medicine and ethics might appear limited, the ripple effects of his methodological approach, emphasizing truth, empirical evidence, and intellectual integrity, have permeated diverse disciplines. Newton's legacy, therefore, isn't confined merely to his discoveries; it's found in the enduring ethos of rigorous, unbiased inquiry he championed — an ethos that has, over centuries, propelled humanity's ceaseless quest for knowledge.

Isaac Newton's Legacy: Codified the laws of motion and universal gravitation, uniting the heavens and earth under a single mechanistic framework and heralding the era of classical physics.

**

REVIEW QUESTIONS

True/False Questions:

1. Isaac Newton formulated the laws of motion and universal gravitation.
   True or False?

2. Newton's approach to science emphasized reason over empirical observation.
   True or False?

Multiple-Choice Questions:

3. Which of the following works by Newton laid the foundation for modern physics?
   a) "*The Sceptical Chymist*"
   b) "*Philosophiæ Naturalis Principia Mathematica*"
   c) "*On the Origin of Species*"
   d) "*De Revolutionibus Orbium Coelestium*"

4. What was Newton's famous adage that underscored his reluctance to speculate beyond observable phenomena?
   a) "*I think, therefore I am.*"
   b) "*Hypotheses non fingo.*"
   c) "*To be, or not to be.*"
   d) "*The unexamined life is not worth living.*"

Clinical Vignette:

5. A researcher in the 17th century conducts meticulous experiments and emphasizes empirical evidence to validate his findings. His approach influences future generations of scientists in various fields, including medicine. Which scientist's methodological rigor does this scenario most closely resemble?
   a) René Descartes
   b) Isaac Newton
   c) John Locke
   d) Baruch Spinoza

Basic Science Vignette:

6. A physics student is studying the principles that govern planetary motion and decides to apply Newton's laws to understand the orbits of planets around the sun. According to Newton's laws, what is the primary force that keeps the planets in orbit?
   a) Electromagnetic force
   b) Gravitational force
   c) Nuclear force
   d) Frictional force

Philosophy Vignette:

7. In a discussion about the foundations of scientific inquiry, a student refers to Newton's famous statement, "*I frame no hypotheses*." What does this statement best illustrate about Newton's approach to science?
   a) His rejection of the need for empirical evidence
   b) His commitment to not speculating beyond observable data
   c) His preference for philosophical discourse over experimentation
   d) His belief in the supernatural explanations for natural phenomena

Correct Answers:

1. True
2. False
3. b) "*Philosophiæ Naturalis Principia Mathematica*"
4. b) "*Hypotheses non fingo*"
5. b) Isaac Newton
6. b) Gravitational force
7. b) His commitment to not speculating beyond observable data

BEYOND THE CHAPTER
Isaac Newton (1643-1727)

- *Isaac Newton* by James Gleick
- *The Principia: Mathematical Principles of Natural Philosophy* by Isaac Newton, translated by I. Bernard Cohen and Anne Whitman
- *Never at Rest: A Biography of Isaac Newton* by Richard S. Westfall

***

Chapter 80: David Hume (1711-1776) — The Empirical Skeptic: Questioning Causality

Hume's skepticism of causality and emphasis on empirical evidence reshaped epistemology and ethics, promoting a rigorous, evidence-based approach to knowledge.

Abstract: David Hume, a luminary of the Scottish Enlightenment, profoundly shaped Western thought, particularly in epistemology and ethics. His commitment to empiricism, notably in "*A Treatise of Human Nature*," shifted philosophical inquiry toward sensory experiences, reevaluating beliefs about knowledge and existence. By scrutinizing causal reasoning and the transient self, Hume influenced theological discourse and scientific methodology, promoting an evidence-based approach. His explorations of the human mind balanced empiricism and rationalism, recognizing the role of experience while logically structuring knowledge. Hume's ethical discourses, grounded in sentiment and human nature, created a nuanced landscape for moral deliberations. Even in fields like medicine, his philosophical foundations indirectly guided epistemological and ethical considerations. Hume's legacy lies in his critique of foundational concepts, explorations of human understanding through sensory experience, and efforts to bridge empirical observations with rational thought. His enduring influence offers insights that continue to inform and challenge contemporary thinkers, solidifying his status as a timeless figure in intellectual history.

**

Introduction: David Hume, a seminal figure in the Scottish Enlightenment, is widely acknowledged as one of the most influential philosophers in the Western tradition. Born in Edinburgh, Hume's contributions spanned the realms of epistemology, metaphysics, ethics, and history. In his revolutionary work, "*A Treatise of Human Nature*," he posited that human understanding is fundamentally rooted in sensory experience, challenging longstanding assumptions about the nature of knowledge. This empiricist approach questioned the basis of causal reasoning and the existence of an unchanging self, prompting thinkers of his time and subsequent generations to reconsider how we understand the world. Moreover, Hume's skepticism about religious beliefs and his exploration of the limitations of human reason has profoundly impacted philosophy, theology, and the sciences. His groundbreaking thoughts, lucid prose, and incisive critiques have cemented his place as a cornerstone of Western philosophy's pantheon.

Rationalism: It's important to note that David Hume is typically categorized as an empiricist rather than a rationalist. However, to understand the nuances of his thought, one could argue that while he heavily leaned toward empiricism, he did have some intersections with rationalist thinking. Hume believed that all ideas stem from sensory experiences or "*impressions*," which is a distinctly empiricist view. Yet, he also delved into the operations of the human mind, considering how we process and organize these impressions into coherent thought. In this, Hume acknowledged the mind's ability to connect ideas through principles of association such as resemblance, contiguity, and cause and effect. Even though he famously critiqued the concept of causation, asserting that we cannot derive a necessary connection between cause and effect through experience alone, this inquiry into mental operations demonstrates a certain rationalist inclination. However, this should not be mistaken for an endorsement of innate knowledge or a rejection of empiricism. Instead, Hume's approach can be seen as an attempt to bridge empirical observations with the logical structures of human reason, illustrating his philosophy's intricate dance between empiricist and rationalist inclinations.

Empiricism: David Hume is prominently recognized as one of the foremost figures in the empiricist tradition. This philosophical standpoint stresses the primacy of sensory experience as the foundation for all knowledge. In opposition to the rationalist claim that reason and innate ideas are the primary sources of knowledge, Hume contended that everything we know originates from our experiences or "*impressions*," and the "*ideas*" we form are merely fainter copies or representations of these impressions. His influential work, "*A Treatise of Human Nature*," delves deeply into this empiricist perspective, positing that our mind categorizes and links these ideas mainly through principles of association, such as resemblance, contiguity in time or place, and cause and effect. However, Hume's empiricism also led him to profound skepticism. Notably, he challenged the certainty of causal reasoning, arguing that our belief in cause and effect arises from habit and experience rather than any inherent logical necessity. For Hume, our knowledge isn't grounded in objective certainties discerned by reason but is constructed from the regularities and patterns we observe in our sensory experiences. His stringent empiricism has had lasting implications, compelling future philosophers and scientists to rigorously examine the foundations of their beliefs and methodologies.

The Scientific Method: David Hume's contributions to philosophy have significantly affected the development and understanding of the scientific method. While he did not lay out a specific scientific methodology, his exploration of empiricism and causality has profoundly impacted how scientists conceptualize and interpret empirical evidence. Hume's scrutiny of causation, where he argued that cause and effect relationships are not directly observable but are inferred from habituated patterns of experience, has instilled caution in the scientific community. He emphasized that even though two events are observed in succession, it does not necessarily mean one causes the other; this challenges scientists to seek rigorous evidence and not to make hasty conclusions. Furthermore, his skepticism prompted an emphasis on the tentative nature of scientific theories, suggesting that no idea can be proven conclusively but can only be corroborated through consistent empirical observations. Hume's insights into the limitations of human reason and observation have paved the way for a more humble, iterative, and evidence-based scientific approach. While he may not have been a scientist, his philosophical inquiries have shaped the epistemological foundations upon which modern scientific methodology rests.

Medicine: David Hume, primarily celebrated for his work in philosophy, did not directly contribute to medicine through research, treatments, or medical methodologies. However, his philosophical inquiries have tangentially influenced medical epistemology and the ethical considerations surrounding medical practice. His empiricist leanings, which emphasize the significance of experience-based knowledge, align well with the empirical methodologies fundamental to medical research. Clinical observations, experimentation, and evidence-based approaches echo Hume's emphasis on deriving understanding from direct experience. Moreover, Hume's examination of human nature, emotions, and moral judgments offers valuable insights into the patient-practitioner relationship and the ethical complexities of medical decisions. His assertion that moral judgments often stem from feelings and sentiments rather than pure reason can be particularly relevant in medical ethics, where emotional and personal factors frequently intertwine with objective decision-making. While Hume's direct impact on medicine might be limited, the ripple effects of his philosophical explorations certainly resonate within the realm of medical thinking, research, and ethics.

Ethics: David Hume's exploration into ethics presented a shift from traditional moral rationalism to an understanding of morality rooted in sentiment and human nature. While he didn't directly categorize his thoughts under the principles of patient autonomy (informed consent), practitioner beneficence (do good), practitioner nonmaleficence (do no harm), and public justice (be fair), his insights indirectly touch upon these ideas. Hume argued that moral judgments arise not from reason alone but primarily from human sentiments or feelings. He believed that our approval or disapproval of actions is largely based on the pleasure or displeasure they evoke. When considering autonomy, Hume's emphasis on individual sentiment and experience underscores the importance of respecting individual choices and agency. Regarding beneficence and nonmaleficence, his ideas highlight the intrinsic human inclination toward acts that promote societal harmony and avoid harm. As for justice, Hume perceived it as an artificial virtue necessary for societal cohesion but not inherently rooted in human nature. For Hume, justice arises from conventions that ensure cooperation and prevent harm. Though Hume did not frame his ethical discussions in the terms used in modern bioethics, his emphasis on sentiment as a basis for moral judgment has enduring implications for how we perceive and prioritize these principles in ethical deliberations.

Conclusion: David Hume's indelible mark on Western thought is evident in his comprehensive examinations of human understanding, knowledge, and morality. While firmly grounded in the empiricist tradition, Hume navigated through the terrains of rationalism, presenting a nuanced perspective on how humans derive and process knowledge. He highlighted the intricate relationship between experience and reason, challenging conventional notions and prompting a reevaluation of foundational beliefs. He profoundly influenced various fields beyond philosophy, from the scientific method's rigor to the ethical considerations inherent in medical practice. Central to his philosophical narrative is the emphasis on sentiment as a primary driver of human judgment, an idea that has reshaped ethical discourse. Hume's ability to interweave skepticism with empirical investigation has positioned him as a central figure in the annals of philosophical thought, and his teachings continue to inform, challenge, and inspire thinkers across disciplines. Through his writings, Hume invites perpetual reflection on the nature of human understanding, urging us to approach knowledge with humility and critical rigor.

David Hume's Legacy: Presented a radical empiricism that questioned causality and the concept of the self, reshaping Western philosophy and paving the way for later thinkers like Kant.

**

REVIEW QUESTIONS

True/False Questions:

1. David Hume's commitment to empiricism is most notably detailed in his work "*A Treatise of Human Nature*."
   True or False?

2. Hume argued that moral judgments are derived primarily from human reason rather than sentiments or feelings.
   True or False?

Multiple-Choice Questions:

3. Which of the following best captures Hume's view on causality?
   a) Causality is directly observable through sensory experience.
   b) Causality is an inherent logical necessity.
   c) Causality is inferred from habituated patterns of experience.
   d) Causality does not exist in any form.

4. How did Hume's empiricism influence the development of the scientific method?
   a) It led to the invention of new scientific instruments.
   b) It emphasized the need for rigorous evidence and empirical observation.
   c) It promoted the idea that all scientific theories are conclusively proven.
   d) It discouraged skepticism about scientific claims.

Clinical Vignette:

5. A medical ethicist is examining the influence of David Hume on contemporary medical ethics. Which aspect of Hume's philosophy would be most relevant in understanding the emotional dimensions of patient care?
   a) His critique of rationalist epistemology.
   b) His emphasis on the role of sentiments in moral judgments.
   c) His theories on causal reasoning.
   d) His views on the unchanging self.

Basic Science Vignette:

6. David Hume's skepticism about causality influenced scientific methodology. Which of the following best reflects Hume's view on causal relationships?
   a) Causal relationships are directly observable and can be proven with certainty
   b) Causal relationships are inferred from patterns of experience but cannot be proven with certainty
   c) Causal relationships are innate ideas present in the human mind
   d) Causal relationships are irrelevant to scientific inquiry

Philosophy Vignette:

7. In his ethical explorations, David Hume emphasized the role of sentiment in moral judgments. Which of the following best captures Hume's view on the source of moral judgments?
   a) Moral judgments are based solely on logical reasoning and factual analysis
   b) Moral judgments arise from divine commandments and religious teachings
   c) Moral judgments stem from human sentiments and feelings of pleasure and displeasure
   d) Moral judgments are determined by societal laws and regulations

Correct Answers:

1. True
2. False
3. c) Causality is inferred from habituated patterns of experience
4. b) It emphasized the need for rigorous evidence and empirical observation
5. b) His emphasis on the role of sentiments in moral judgments
6. b) Causal relationships are inferred from patterns of experience but cannot be proven with certainty
7. c) Moral judgments stem from human sentiments and feelings of pleasure and displeasure

BEYOND THE CHAPTER
David Hume (1711-1776)

- *David Hume: The Philosopher as Historian* by Nicholas Phillipson
- *An Enquiry Concerning Human Understanding* by David Hume
- *Hume's True Scepticism* by Donald C. Ainslie

***

Chapter 81: Immanuel Kant (1724-1804) — The Synthesizer: Rationalism and Empiricism

Kant's critical philosophy bridged rationalism and empiricism, grounding knowledge in both reason and experience while laying a foundation for modern ethics.

Abstract: Immanuel Kant (1724-1804) is a pivotal figure in philosophy, bridging rationalism and empiricism and revolutionizing Western thought. In the 18th century, Kant navigated between Descartes' rationalism and Hume's empiricism. His seminal work, "*Critique of Pure Reason*," proposed that inherent mental structures and sensory experiences together shape human knowledge. He argued that the mind, with "*a priori*" concepts, actively shapes our perception of reality. Kant's ethical philosophy, articulated in the "*categorical imperative*," established a universal foundation for moral actions, emphasizing the intrinsic worth of rational beings. This philosophy influenced modern ethics, supporting principles like patient autonomy (informed consent), practitioner beneficence (do good), nonmaleficence (do no harm), and public justice (be fair). Kant's ideas laid a profound foundation for scientific inquiry and medical ethics, highlighting autonomy and individual dignity. His legacy continues to inspire philosophical discourse and deepen our understanding of human existence.

**

Introduction: Immanuel Kant, an 18th-century German philosopher, is often considered one of the most influential thinkers in modern Western philosophy. Born in Königsberg, now Kaliningrad, Russia, Kant's extensive body of work touched on many subjects, but he is perhaps most renowned for his contributions to metaphysics, epistemology, and ethics. Central to his thought is the "*Critique of Pure Reason*," where he sought to address the limitations and capacities of human knowledge, introducing groundbreaking ideas such as the distinction between phenomena (things as they appear to us) and noumena (things in themselves). His "*categorical imperative*" notion in ethics presents a universal basis for moral action, emphasizing duty and the intrinsic worth of rational beings. Additionally, Kant's thoughts on aesthetics, anthropology, and political philosophy have left an indelible mark on subsequent generations of thinkers, and his philosophies continue to inspire debates and discussions today. Whether challenging or supporting his views, scholars and enthusiasts alike cannot deny Kant's pivotal role in shaping the philosophical landscape.

Rationalism: Immanuel Kant's philosophical endeavors represented a synthesis between the empirical traditions associated with British philosophers like David Hume and the rationalist inclinations of the continental tradition epitomized by figures such as René Descartes. While Kant acknowledged the importance of sensory experience in acquiring knowledge, he also contended that pure reason has capacities beyond mere empirical observation. For Kant, certain fundamental concepts — like space, time, and causality — are not derived from experience but are inherent structures of the human mind that shape our perception of the world. This "*transcendental*" form of rationalism posits that while our knowledge begins with experience, it doesn't necessarily arise from experience. Instead, our mind organizes sensory data according to these "*a priori*" concepts, enabling empirical knowledge. In his "*Critique of Pure Reason*," Kant extensively explores these ideas, aiming to demonstrate how pure reason can deduce certain synthetic "*a priori*" propositions. By navigating a middle path between empiricism and rationalism, Kant sought to carve out a space for metaphysics in the modern era, securing a role for reason in determining the bounds of possible knowledge.

Empiricism: Immanuel Kant's engagement with empiricism was primarily a response to the ideas of British empiricists, especially David Hume, who contended that all knowledge is rooted in sensory experience. While Kant held rationalist convictions, he recognized the crucial role of experience in forming knowledge. He famously noted that "*thoughts without content are empty, intuitions without concepts are blind.*" By this, he meant that while our rational capacities structure our perception and understanding, sensory experience (intuition) provides the raw material for knowledge. Without this empirical input, our rational faculties would have nothing to process. Kant's unique contribution was his synthesis of these views: while we can't have knowledge without sensory experience, our mind isn't just a passive recipient of sensory data. Instead, it actively structures these data according to inherent, *"a priori"* concepts, like space and time. Thus, for Kant, empirical observations are vital, but they're always filtered and shaped by the mind's intrinsic conceptual framework. In this way, Kant tried to rescue the possibility of objective knowledge by showing how empiricism and rationalism, seemingly at odds, work hand in hand in our cognitive processes.

The Scientific Method: Immanuel Kant's influence on the scientific method is less about altering its procedural nuances and more about providing a philosophical foundation for scientific inquiry. Responding to the empiricist claims of David Hume, who questioned the certainty of cause and effect, Kant sought to justify how scientific knowledge could be both empirical and certain. He argued that while our sensory experiences provide the raw data for knowledge, the human mind inherently structures these data consistently and regularly, using "*a priori*" concepts like causality, space, and time. This means that while we can't directly know the "thing-in-itself" (noumenon), we can have confidence in our observations and causal inferences (phenomena) because the mind's innate conceptual framework consistently structures them. Thus, Kant's work provided a philosophical grounding for the predictability and reliability of scientific observations, reinforcing the idea that the natural world operates according to consistent, knowable laws. While Kant didn't modify the scientific method per se, his philosophy bolstered confidence in its capability to yield objective and universal knowledge about the natural world.

Medicine: Immanuel Kant is primarily celebrated for his contributions to metaphysics, epistemology, and ethics rather than direct contributions to medicine. However, his philosophy has indirect implications for the field. Kant's emphasis on individuals' autonomy and intrinsic worth has been influential in bioethics, particularly in discussions about patient consent, autonomy, and medical ethics. His assertion that individuals should always be treated as ends in themselves and never merely as a means to an end underpins modern notions of patient rights and dignity. While Kant himself did not delve into medical practices or theories, his broader philosophical principles have been invoked in medical debates to argue for the inviolability of individual rights, the importance of informed consent, and the ethical considerations of medical interventions and research. While Kant did not contribute directly to the progression of medical knowledge or techniques, his philosophical ideas provide a foundational ethical framework from which many contemporary medical practices and principles derive their moral grounding.

Ethics: Immanuel Kant's ethical philosophy has left a profound mark on the foundations of contemporary moral principles. Central to Kantian ethics is the idea of autonomy and the intrinsic worth of rational agents. He asserted that individuals should always be treated as ends in themselves and never merely as means to an end. This principle underscores the modern emphasis on autonomy in ethical considerations, emphasizing the importance of respecting individual choices and upholding patient rights, especially in medical contexts. Beneficence and nonmaleficence can also be traced back to Kant's "*categorical imperative*," which demands actions to be universally applicable without contradiction. In essence, we should act to promote good (beneficence) and avoid harm (nonmaleficence) if we expect the same actions to be universalized. Finally, the principle of justice, ensuring fairness and equality, resonates with Kant's insistence on universal laws and the equal worth of all rational beings. Although Kant did not delineate these ethical principles explicitly as they're understood in contemporary discourse, his emphasis on duty, universalizability, and the intrinsic dignity of individuals laid the philosophical groundwork that shaped the evolution of these central tenets of modern ethics.

Conclusion: Immanuel Kant's indelible mark on philosophy is evidenced in his meticulous exploration of the nature of human knowledge, morality, and experience. Born in 18th-century Königsberg, his synthesis of empiricism and rationalism in the "*Critique of Pure Reason*" offered a revolutionary perspective on how the human mind perceives and processes the world. In bridging the gulf between the empirical traditions of figures like Hume and the rationalist doctrines of philosophers such as Descartes, Kant elucidated the inherent structures of the mind that shape our understanding of reality. Beyond epistemology, his ethical treatises, prominently the concept of the "*categorical imperative*," set the stage for modern moral philosophy, underscoring individual beings' inherent value and autonomy. Moreover, his philosophical groundwork supported scientific inquiry and imbued the medical field with an ethical compass, emphasizing patient rights and dignity. Kant's vast contributions extend from aesthetics to political thought, and his ideas remain foundational in contemporary philosophical dialogues. His philosophical *oeuvre* continues to shape, challenge, and inspire modern thought across various disciplines.

Immanuel Kant's Legacy: Synthesized rationalism and empiricism in his "*Critique of Pure Reason*," proposing that human knowledge arises from our innate structures interacting with experience.

**

REVIEW QUESTIONS

True/False Questions:

1. Immanuel Kant's "*Critique of Pure Reason*" attempts to reconcile rationalism and empiricism by asserting that human knowledge arises from the interaction between inherent mental structures and sensory experiences.
   True or False?

2. Kant's ethical principle of the "*categorical imperative*" emphasizes that individuals should be treated as a means to an end rather than as ends in themselves.
   True or False?

Multiple-Choice Questions:

3. What is a central tenet of Kant's epistemology?
   a) All knowledge is derived solely from sensory experience.
   b) Reason alone, without sensory input, can provide all knowledge.
   c) Knowledge is formed by the mind structuring sensory experiences through inherent concepts.
   d) Knowledge of the world is completely unknowable.

4. How does Kant's philosophy influence modern medical ethics?
   a) By providing specific medical procedures.
   b) By emphasizing patient autonomy and the intrinsic worth of individuals.
   c) By promoting the use of traditional medical practices.
   d) By rejecting the importance of informed consent.

Clinical Vignette:

5. A healthcare professional is reflecting on the ethical implications of patient autonomy in medical practice. Which aspect of Kant's philosophy would most support the idea that patients should be fully informed and allowed to make their own healthcare decisions?
   a) Kant's theory of the noumenon.
   b) Kant's emphasis on the categorical imperative.
   c) Kant's critique of pure reason.
   d) Kant's views on aesthetics.

Basic Science Vignette:

6. Immanuel Kant argued that human understanding arises from the interaction between innate mental structures and sensory experiences. According to Kant, what concept is not derived from experience but is instead an inherent structure of the human mind?
   a) Gravity
   b) Time
   c) Electricity
   d) Photosynthesis

Philosophy Vignette:

7. Immanuel Kant's "*categorical imperative*" is central to his ethical philosophy. Which of the following best describes one aspect of the categorical imperative?
   a) Act only according to maxims that can be universalized without contradiction
   b) Act based on the greatest happiness for the greatest number
   c) Act according to the dictates of divine command
   d) Act according to societal norms and customs

Correct Answers:

1. True
2. False
3. c) Knowledge is formed by the mind structuring sensory experiences through inherent concepts
4. b) By emphasizing patient autonomy and the intrinsic worth of individuals
5. b) Kant's emphasis on the categorical imperative
6. b) Time
7. a) Act only according to maxims that can be universalized without contradiction

BEYOND THE CHAPTER
Immanuel Kant (1724-1804)

- *Kant: A Biography* by Manfred Kuehn
- *Critique of Pure Reason* by Immanuel Kant, translated by Paul Guyer and Allen W. Wood
- *Kant: Political Writings* edited by Hans Reiss

***

Philosophy of Science

Medical Research

WEEK FIFTEEN: INNOVATORS IN SCIENCE

WEEK FIFTEEN

Class Hour 1

- Chapter 82: Edward Jenner (1749-1823) — The Vaccine Innovator: Conquering Smallpox
- Chapter 83: Thomas Malthus (1766-1834) — The Population Theorist: Limits of Growth
- Chapter 84: John Stuart Mill (1806-1873) — The Utilitarian Philosopher: Greatest Good for the Greatest Number

Class Hour 2

- Chapter 85: Charles Darwin (1809-1882) — The Evolutionary Theorist: Natural Selection
- Chapter 86: Ignaz Semmelweis (1818-1865) — The Hygiene Pioneer: Handwashing in Medicine
- Chapter 87: Gregor Mendel (1822-1884) — The Geneticist: Principles of Inheritance

---

Introduction

In the fifteenth week of our course, we delve into a period of extraordinary advancements in science and medicine, marked by groundbreaking discoveries and the establishment of new ethical principles guiding medical practice and scientific inquiry.

Edward Jenner, known as the Vaccine Innovator, made a monumental contribution to medicine by pioneering the concept of vaccines. His work on smallpox laid the foundation for immunology, effectively revolutionizing the field. Jenner's rationalist and empirical approach not only conquered one of humanity's most devastating diseases but also underscored the ethical principles of beneficence, nonmaleficence, and public justice. His commitment to improving public health through vaccination set a precedent for ethical medical research and practice.

Thomas Malthus, the Population Theorist, recognized the inherent tension between exponential population growth and the linear expansion of resources. Malthus' rationalist and empirical approach profoundly influenced economics, environmental studies, and ethical discussions surrounding sustainability, reproductive autonomy, and social justice. His insights into population dynamics spurred important debates on the limits of growth and the ethical implications of resource distribution.

John Stuart Mill, the Utilitarian Philosopher, championed the principles of utilitarianism, advocating for the greatest good for the greatest number. Emphasizing empirical experience as the source of knowledge, Mill's rationalist and empirical approach had a significant impact on scientific methodology, medicine, and ethics. His promotion of beneficence, nonmaleficence, and justice in public health policy continues to resonate, guiding decisions that affect the well-being of society.

Charles Darwin, the Evolutionary Theorist, reshaped our understanding of biology with his theory of evolution by natural selection. Darwin's rationalist and empirical approach challenged established views on the origins of species and profoundly influenced scientific methodology, medicine, and ethical principles. His work fostered a deeper understanding of human health and disease, emphasizing the interconnectedness of all life forms and the evolutionary processes that shape our existence.

Ignaz Semmelweis, the Hygiene Pioneer, transformed medical practices with his advocacy for hand hygiene in medical settings. By significantly reducing hospital-acquired infections through simple yet effective measures, Semmelweis' rationalist and empirical approach highlighted the importance of cleanliness in patient care. His emphasis on beneficence, nonmaleficence, and justice in medical practice underscored the ethical imperative to protect patients from preventable harm.

Gregor Mendel, the Geneticist, laid the foundation for the field of genetics with his meticulous work on pea plants. Uncovering the principles of inheritance, Mendel's rationalist and empirical approach revolutionized biology and medicine. His discoveries prompted new ethical considerations in genetic research and patient care, emphasizing the responsible application of genetic knowledge and the ethical implications of manipulating hereditary traits.

This week, we explore the transformative contributions of these visionaries, whose work not only advanced scientific knowledge but also established enduring ethical principles that continue to guide contemporary science and medicine. Through their rationalist and empirical approaches, they addressed some of the most pressing challenges of their time, fostering a legacy of innovation and ethical responsibility that remains relevant today. As we reflect on their achievements, we gain a deeper appreciation for the profound impact of their discoveries on our understanding of health, disease, and the ethical practice of medicine.

***

Chapter 82: Edward Jenner (1749-1823) — The Vaccine Innovator: Conquering Smallpox

Jenner's development of the smallpox vaccine exemplified the power of empirical science in medicine, marking a milestone in public health and preventive care.

Abstract: In a pivotal moment in medical history, Edward Jenner revolutionized preventive medicine by developing the world's first vaccine targeting smallpox, a disease that once ravaged populations globally. His methodical approach, deeply rooted in rationalism and empiricism, exemplified the scientific method in medical research, paving the way for immunology. Jenner's relentless pursuit of evidence-based knowledge and commitment to observational science underscored empiricism's transformative power in understanding and combating diseases. Moreover, his work delineated early contours of ethical considerations in medical experiments, sowing seeds for modern principles of patient autonomy (informed consent), practitioner beneficence (do good), practitioner nonmaleficence (do no harm), and public justice (be fair). His groundbreaking initiative has saved countless lives and established a precedent for scientific inquiry that is both innovative and ethically conscious, cementing his legacy as a towering figure in medicine and beyond.

**

Introduction: Edward Jenner, an English physician and scientist, is celebrated as the pioneer of the smallpox vaccine, the world's first vaccine, marking a turning point in the annals of medicine. Born in the 18th century, his groundbreaking work laid the foundation for immunology, ushering in a new epoch of preventive medicine. Although variolation, an early method of inoculating against smallpox using the virus itself, was practiced before Jenner's discovery, it came with significant risk. Jenner's innovative approach involved using the relatively harmless cowpox virus to confer immunity against the deadly smallpox. This realization stemmed from an observation that milkmaids who contracted cowpox didn't subsequently fall ill with smallpox. By 1796, he successfully demonstrated his hypothesis through a series of experiments, forever changing the trajectory of public health. Edward Jenner's contributions have saved countless lives from the scourge of smallpox and paved the way for modern-day vaccination practices, making him a towering figure in the history of medicine.

Rationalism: Edward Jenner's approach to medical science embodied the principles of rationalism, emphasizing reason and evidence over superstition or unverified beliefs. As an Enlightenment thinker, Jenner believed in systematically investigating the natural world and applying reasoned thinking to derive truths. This perspective was evident in his observations and experiments surrounding smallpox and cowpox. Rather than merely accepting his day's prevalent theories or myths, Jenner sought a logical and empirical foundation for his work. His curiosity about the anecdotal link between cowpox in milkmaids and their immunity to smallpox led him to conduct controlled experiments. His groundbreaking vaccination with the cowpox material on young James Phipps demonstrated his commitment to empirical testing. Rather than relying on hearsay, Jenner rigorously documented his findings and subjected them to scrutiny by the broader scientific community. This adherence to a structured methodology and reliance on observational data over preconceived notions were hallmarks of his rationalist approach. Edward Jenner's methodical exploration and application of evidence-based conclusions symbolized the Enlightenment's rationalist spirit, transforming the scientific landscape and societal perceptions of disease prevention.

Empiricism: Edward Jenner's work in developing the smallpox vaccine was deeply rooted in empiricism, emphasizing the importance of direct observation and experimental evidence in understanding the world. Rather than solely relying on theoretical constructs or existing knowledge, Jenner turned to tangible, firsthand evidence to guide his scientific inquiry. His empiricist orientation was highlighted by his detailed observations of the relationship between cowpox and smallpox. The widely circulated folklore that milkmaids who had contracted cowpox seemed immune to smallpox was not merely accepted by Jenner as truth; instead, he sought concrete evidence to validate or refute this claim. His deliberate experiment, where he inoculated James Phipps with material from a cowpox sore and later exposed him to smallpox, clearly manifested his commitment to empirical methods. When Phipps remained healthy, Jenner had direct evidence supporting the protective properties of cowpox against the more deadly disease. By consistently favoring experience and experimental results as the primary sources of knowledge, Edward Jenner epitomized the empiricist approach, ensuring that his discoveries were grounded in verifiable reality rather than mere speculation.

The Scientific Method: Edward Jenner's work on the smallpox vaccine exemplifies and reinforces the significance of the scientific method in pioneering discoveries. While he didn't invent the scientific method, his approach to the correlation between cowpox and smallpox immunity demonstrates its practical application. Beginning with an observation that milkmaids who contracted cowpox appeared resistant to smallpox, he hypothesized that cowpox could offer protection against the deadlier disease. His commitment to testing this hypothesis under controlled conditions set Jenner apart. He conducted an experiment by inoculating James Phipps with cowpox and subsequently exposing him to smallpox, ensuring a systematic method of collecting evidence. Upon observing the results, where Phipps showed no sign of developing the disease, Jenner gathered data that would inform the broader scientific community. Through documentation and peer review, his findings underwent scrutiny, further solidifying their validity. Jenner's meticulous approach validated his hypothesis and underscored the importance of structured, evidence-based investigation in scientific pursuits. In this way, Edward Jenner's landmark research on vaccination bolstered the principles of the scientific method, reminding future generations of its enduring value in the quest for knowledge.

Medicine: Edward Jenner's monumental contributions to medicine reshaped medical history and public health. At the forefront of his legacy is the development of the world's first vaccine, targeting the devastating disease of smallpox. Before Jenner's research, smallpox decimated populations, causing severe morbidity and mortality. By keenly observing that milkmaids who had contracted cowpox seemed resistant to smallpox, Jenner formulated a hypothesis about potential cross-immunity. In a groundbreaking experiment, he inoculated James Phipps with cowpox and exposed him to smallpox, demonstrating that the boy was protected against the latter. This pioneering approach led to the advent of the smallpox vaccine, marking the dawn of immunization as a preventive medical measure. As a result, the once-feared disease was eventually eradicated worldwide by 1980, showcasing the immeasurable impact of Jenner's work. Furthermore, his efforts set the stage for the broader field of immunology, influencing generations of scientists and laying the groundwork for subsequent vaccine developments. Through this singular contribution, Edward Jenner saved countless lives and transformed the disease prevention paradigm, establishing himself as a luminary in the annals of medicine.

Ethics: Edward Jenner's pioneering work in developing the smallpox vaccine inadvertently raised ethical considerations that continue to influence medical practice and research. His experiment on young boy James Phipps can be viewed through the ethical principles of patient autonomy (informed consent), practitioner beneficence (do good), practitioner nonmaleficence (do no harm), and public justice (be fair). On the one hand, Jenner's vaccine represented beneficence, as it sought to benefit society by potentially eradicating a deadly disease. The underlying motivation was to do good and improve public health. Regarding nonmaleficence or the principle of "*do no harm*," Jenner's actions become more contentious. While Phipps did not develop smallpox post-inoculation, exposing a child to potential harm without informed consent is ethically questionable by today's standards. The concept of autonomy, which emphasizes the patient's right to make decisions about their own body, needed to be explicitly adhered to, especially considering the age of Phipps and the absence of informed consent processes. Lastly, the principle of justice, ensuring that benefits and burdens are distributed fairly, becomes relevant in the broader application of Jenner's vaccine, protecting many regardless of socioeconomic status. Jenner's work, while groundbreaking, provides an early context for the evolving nature of ethical considerations in medical research and practice.

Conclusion: In reflection, Edward Jenner's multifaceted legacy is a testament to the power of scientific rigor, observation, and innovation. A forerunner in the annals of medicine, Jenner's dedication to empirical methods and commitment to the principles of rationalism and the scientific method reshaped our understanding of disease prevention. Through developing the first smallpox vaccine, he catalyzed a transformative shift in public health and pioneered the broader realm of immunology. Beyond the technical contributions, Jenner's work prompted future generations to grapple with ethical considerations that remain at the forefront of medical research today. By balancing the benefits of scientific advancement with the moral imperatives of patient autonomy (informed consent), practitioner beneficence (do good), practitioner nonmaleficence (do no harm), and public justice (be fair), Jenner's legacy invites a continuous examination of our responsibility in pursuing knowledge. His work serves as both an inspiration and a reminder of an individual's profound impact, driving humanity forward while challenging it to reflect upon its actions and intentions.

Edward Jenner's Legacy: He pioneered the concept of vaccines with his work on smallpox, laying the foundation for immunology and changing the course of medical history.

**

REVIEW QUESTIONS

True/False Questions:

1. Edward Jenner's development of the smallpox vaccine marked the first successful use of vaccination to prevent disease.
   True or False?

2. Jenner's experiments on James Phipps strictly followed the modern principles of informed consent and patient autonomy.
   True or False?

Multiple-Choice Questions:

3. What was the primary observation that led Edward Jenner to develop the smallpox vaccine?
   a) Smallpox could be cured with herbal remedies.
   b) Milkmaids who contracted cowpox did not get smallpox.
   c) Smallpox could be treated with bed rest and hydration.
   d) Smallpox was spread by contaminated water.

4. How did Edward Jenner's work influence the field of medicine?
   a) By developing antibiotics.
   b) By introducing the concept of vaccination.
   c) By inventing the stethoscope.
   d) By discovering penicillin.

Clinical Vignette:

5. A medical historian is studying the ethical implications of early vaccination practices. Which aspect of Edward Jenner's work would most highlight the evolution of ethical standards in medical research?
   a) His development of the polio vaccine.
   b) His use of cowpox material for vaccination.
   c) His experiment on James Phipps without informed consent.
   d) His invention of antiseptic techniques.

Basic Science Vignette:

6. Edward Jenner's development of the smallpox vaccine was based on which observation?
   a) Smallpox survivors were immune to cowpox
   b) Milkmaids who had contracted cowpox did not get smallpox
   c) Cowpox and smallpox were identical viruses
   d) Variolation was ineffective in preventing smallpox

Philosophy Vignette:

7. Edward Jenner's experiments on James Phipps can be evaluated through the ethical principle of patient autonomy. Which of the following best reflects the challenge Jenner faced regarding patient autonomy?
   a) Ensuring that the vaccine was effective
   b) Balancing the risk of side effects with potential benefits
   c) Conducting experiments without informed consent
   d) Providing the vaccine to as many people as possible

Correct Answers:

1. True
2. False
3. b) Milkmaids who contracted cowpox did not get smallpox
4. b) By introducing the concept of vaccination
5. c) His experiment on James Phipps without informed consent
6. b) Milkmaids who had contracted cowpox did not get smallpox
7. c) Conducting experiments without informed consent

BEYOND THE CHAPTER
Edward Jenner (1749-1823)

- *Edward Jenner: The Cheltenham Years, 1795-1823 (Being a Chronicle of the Vaccination Campaign)* by Brian Norman Ford
- *The Life of Edward Jenner M.D.: With Illustrations of His Doctrines, and Selections from His Correspondence* by John Baron
- *Edward Jenner: Conqueror of Smallpox* by Steven Parker

***

Chapter 83: Thomas Malthus (1766-1834) — The Population Theorist: Limits of Growth

Malthus's analysis of population growth and resources highlighted the empirical and rational study of socio-economic issues, influencing environmental and economic theories.

Abstract: Thomas Malthus's work in the late 18th and early 19th centuries marked a defining moment in demography, political economy, and environmental studies. Known for his analysis of population growth and resource availability, Malthus's contributions extend deeply into rationalist and empiricist paradigms, fostering an enduring dialogue across various fields, including medicine and ethics. His rationalist approach provided a reasoned analysis of societal trends, opposing the utopian views of his time. His empiricist stance relied on observable phenomena, embedding his theories with lasting relevance and promoting evidence-based exploration of socio-economic issues. Although not directly involved in medicine, Malthus highlighted the connection between population dynamics and health, emphasizing the role of medical interventions in societal well-being. His theories also engage with ethical dilemmas, linking reproductive autonomy, social justice, and environmental sustainability. Malthus's legacy continues to influence contemporary discussions, urging society to balance human aspiration with ecological limits amid growing environmental challenges.

**

Introduction: Thomas Malthus, an influential 18th-century British scholar, is best known for his groundbreaking contributions to demography and political economy. Born in 1766, Malthus articulated theories that fundamentally shaped discussions about population dynamics and their interplay with resources. His most seminal work, "*An Essay on the Principle of Population*," posited that while populations grow geometrically (exponential growth), resources, particularly food, increase arithmetically (linear growth). This discrepancy, Malthus argued, would inevitably lead to overpopulation, resulting in societal challenges such as famine, disease, and war, which would then act as natural checks to curb population growth. While his predictions were contentious and have been the subject of extensive debate, they laid the groundwork for subsequent population economics and ecology studies. The "*Malthusian*" perspective, as it came to be known, catalyzed crucial discussions on sustainable growth, resource management, and the complex relationship between human societies and their environments.

Rationalism: Thomas Malthus, though primarily remembered for his theories on population and resources, was deeply rooted in the rationalist tradition that typified Enlightenment thought. Rationalism, with its emphasis on reason as the chief source and test of knowledge, found its way into Malthus's approach to understanding the complex dynamics between population growth and the availability of resources. Instead of relying on religious or mystical explanations, Malthus applied a reasoned, systematic analysis of demographic trends. In "*An Essay on the Principle of Population*," he synthesized empirical observations with deductive reasoning to articulate his theory. While the population grows at a geometric rate, food supply increases only arithmetically, leading inevitably to a crisis point. Malthus's approach underscored the importance of logical analysis, drawing conclusions from observable phenomena and extrapolating potential future outcomes. By leveraging reason as his primary tool, Malthus not only critiqued the overly optimistic views of his contemporaries who believed in the indefinite perfectibility of society but also established a foundational framework for future studies in demography, economics, and ecology. His rationalist inclinations thus played a pivotal role in crafting an argument that has continued to shape debates on population and sustainability for over two centuries.

Empiricism: Alongside his rationalist leanings, Thomas Malthus also displayed a distinct empiricist orientation, emphasizing the role of sensory experience and observation in shaping knowledge. His thoughts on population dynamics and resource constraints were not just grounded in abstract reasoning; empirical observations of the world around him deeply influenced them. Malthus meticulously studied historical and contemporary data, analyzing birth rate trends, death rate, and food production. In "*An Essay on the Principle of Population*," he frequently drew upon real-world examples of famines, diseases, and societal challenges as evidence to bolster his claims about the impending dangers of unchecked population growth. His empiricist approach is particularly evident when he discusses the various "*checks*" to population — both preventive and positive — which he derived from actual societal phenomena rather than mere theoretical constructs. By marrying empirical evidence with rational analysis, Malthus crafted a compelling argument that has stood the test of time, even though it has been debated and refined. His commitment to empiricism underscored the importance of grounding theoretical postulations in observable reality. This methodological stance has greatly influenced subsequent demography, economics, and environmental studies work.

The Scientific Method: While Thomas Malthus is not typically lauded as a direct architect of the scientific method, his demography and political economy work exhibited methodological principles that resonate with scientific inquiry. Malthus's investigation into population dynamics and resource availability was systematic and rigorous. He began with a hypothesis — specifically, the geometric growth of populations versus the arithmetic growth of resources — and sought empirical evidence, both historical and contemporary, to test his theory. By juxtaposing his reasoned deductions with real-world observations, he exemplified the iterative process of hypothesis formation, data collection, and theory refinement. Moreover, his readiness to engage with critics, refine his arguments, and adapt his theories in subsequent editions of his essay mirrors the scientific ethos of continual revision in light of new evidence. While the realm Malthus operated in was more social science than natural science, his rigorous, evidence-based approach paved the way for future scholars to apply similar methodologies to diverse fields of inquiry. As such, Malthus's contributions can be seen as a subtle yet significant influence in the broader tapestry of the evolution of systematic, empirical investigation — cornerstones of the scientific method.

Medicine: Thomas Malthus, primarily recognized for his theories on population dynamics and resource constraints, did not directly contribute to the field of medicine as contemporaneous scientists or doctors might have. However, the implications of his work indirectly intersected with medical and public health considerations. Malthus postulated that unchecked population growth would inevitably lead to natural checks, such as famine, disease, and mortality. By highlighting disease as a potential consequence of overpopulation, his theories underscored the significance of medical interventions in mitigating population pressures. While Malthus viewed diseases as inevitable outcomes of population surges, his insights inadvertently stressed the role of medicine and public health in altering demographic trajectories. In the larger discourse, Malthus's work can be seen as a catalyst emphasizing the importance of understanding disease patterns in population dynamics. His emphasis on preventive checks, like moral restraint, can also be extrapolated to broader public health strategies that preemptively address health crises. While not a medical scholar per se, Malthus's insights on the interplay between population and disease certainly provided a backdrop against which subsequent public health policies and medical strategies were discussed and developed.

Ethics: Thomas Malthus's theories, while primarily demographic and economic, carry profound ethical implications, particularly when viewed through the lens of principles such as patient autonomy (informed consent), practitioner beneficence (do good), practitioner nonmaleficence (do no harm), and public justice (be fair). Malthus's emphasis on "*moral restraint*" as a preventive check to population growth touches upon the principle of autonomy, suggesting individuals have both the freedom and responsibility to make informed reproductive choices. This idea indirectly champions the importance of education and empowerment, especially in reproductive health. The principle of beneficence can be discerned in Malthus's underlying desire for societal well-being by preventing the catastrophic consequences of overpopulation. Conversely, his stark portrayal of "*positive checks*" like famine and disease can be seen as a dire warning against nonmaleficence, or the ethical commitment to do no harm, by allowing populations to swell unchecked. Lastly, the principle of justice, which concerns the fair distribution of resources and opportunities, lies at the heart of Malthus's thesis. His concerns about the disproportionate impact of population growth on society's most vulnerable highlighted the ethical dilemmas surrounding equitable resource allocation. Though Malthus may not have explicitly framed his work in modern bioethical terms, his theories undeniably grapple with foundational ethical concerns that continue to shape discourse on sustainability, equity, and public health.

Conclusion: Thomas Malthus's intellectual legacy extends far beyond his primary work on population dynamics, encompassing areas of rationalist and empiricist thought, methodological rigor echoing the scientific method, indirect intersections with medicine, and profound ethical considerations. By highlighting the potential pitfalls of unchecked population growth and the inevitable constraints of limited resources, Malthus challenged prevailing notions of human progress and instigated meaningful discussions on sustainable development, environmental concerns, and ethical governance. His marriage of reason with empirical evidence showcased an approach that would be instrumental in shaping future academic pursuits, from economics and sociology to public health and ethics. While his conclusions stirred controversy and debate, their foundational importance cannot be understated. Malthus's insights provide a cautionary tale of the delicate balance between human ambition and ecological constraints, and his lasting influence serves as a testament to the enduring power of interdisciplinary thought in addressing society's most pressing challenges.

Thomas Malthus's Legacy: Malthus is best recognized for his pioneering insights into the tensions between exponential population growth and linear resource expansion.

**

Review Questions:

True/False Questions:

1. Thomas Malthus's "*An Essay on the Principle of Population*" posits that population growth occurs at an arithmetic rate while food supply increases at a geometric rate.
   True or False?

2. Malthus believed that moral restraint, such as delaying marriage, could serve as a preventive check to population growth.
   True or False?

Multiple-Choice Questions:

3. What was a central premise of Thomas Malthus's theory on population growth?
   a) Population growth and resource availability always maintain a perfect balance.
   b) Population growth is exponential, while food production increases linearly.
   c) Population growth has no impact on resource availability.
   d) Food production can always keep pace with population growth through technological innovation.

4. Which of the following best describes the influence of Malthus's work on medical and public health considerations?
   a) It directly led to the development of vaccines.
   b) It highlighted the role of diseases as natural checks on population growth.
   c) It promoted the use of antibiotics to control population.
   d) It advocated for herbal remedies to combat overpopulation.

Clinical Vignette:

5. A public health policy maker is analyzing the ethical implications of population control measures. Which aspect of Thomas Malthus's theories would be most relevant to their considerations?
   a) His development of economic theories.
   b) His emphasis on moral restraint and preventive checks to population growth.
   c) His advocacy for agricultural innovation.
   d) His views on the use of natural resources for industrial growth.

Basic Science Vignette:

6. Thomas Malthus observed that populations grow at a geometric rate, while resources grow at an arithmetic rate. What is the primary consequence of this discrepancy according to Malthus?
   a) Unlimited resources will always be available
   b) Population growth will slow down naturally without any consequences
   c) Population growth will outpace resource availability, leading to famine, disease, and other checks
   d) Resources will eventually grow at the same rate as populations

Philosophy Vignette:

7. Thomas Malthus's theories on population growth and resource limitations can be evaluated through the ethical principle of justice. Which of the following best reflects the challenge Malthus highlighted regarding justice?
   a) Ensuring that everyone has access to unlimited resources
   b) Balancing the fair distribution of limited resources among a growing population
   c) Encouraging unlimited population growth without regard for resources
   d) Promoting the idea that population growth will naturally equalize resources

Correct Answers:

1. False
2. True
3. b) Population growth is exponential, while food production increases linearly
4. b) It highlighted the role of diseases as natural checks on population growth
5. b) His emphasis on moral restraint and preventive checks to population growth
6. c) Population growth will outpace resource availability, leading to famine, disease, and other checks
7. b) Balancing the fair distribution of limited resources among a growing population

Beyond The Chapter
Thomas Malthus (1766-1834)

- *An Essay on the Principle of Population* by Thomas Malthus
- *Malthus: The Life and Legacies of an Untimely Prophet* by Robert J. Mayhew
- *God and the Secular: A Philosophical Assessment of the Works of Thomas Malthus* by D. Pullen

***

Chapter 84: John Stuart Mill (1806-1873) — The Utilitarian Philosopher: Greatest Good for the Greatest Number

Mill's utilitarianism, advocating for the greatest happiness principle, merged rationalist and empirical approaches to ethics, influencing modern moral philosophy and public policy.

Abstract: John Stuart Mill, a key figure in 19th-century intellectual discourse, significantly shaped liberal and ethical thought. His contributions span politics, epistemology, and ethics, blending rationalist and empiricist perspectives. Mill's doctrine, grounded in cognitive faculties and empirical experience, offered a harmonized approach to knowledge, challenging strict dichotomies. His formulation of "*Mill's Methods*" profoundly influenced scientific inquiry by fostering a logical approach to discerning causal relationships from data. While not directly contributing to medicine, Mill's philosophical underpinnings resonate in medical ethics, bioethics, patient autonomy, and public health policy. Central to Mill's legacy is his advocacy for utilitarianism, an ethical theory promoting the greatest happiness for the greatest number. This theory continues to shape contemporary moral philosophy and applied ethics, influencing views on individual rights, societal responsibilities, and human welfare. Through his work, Mill remains a beacon of intellectual rigor and moral philosophy, catalyzing reflections on human nature and societal constructs, and leaving an enduring imprint on modern thought.

**

Introduction: John Stuart Mill (1806-1873), a towering figure in 19th-century philosophy and political thought, remains one of the most influential thinkers in liberalism, utilitarianism, and women's rights. Born in London to a devoutly utilitarian father, Mill was educated rigorously from a young age, laying the foundation for his later contributions to philosophy and political economy. Mill championed the cause of individual freedom and autonomy against societal interference throughout his life, crystallized in his seminal work "*On Liberty*." He also developed and refined the ethical theory of utilitarianism, advocating for the "*greatest good for the greatest number*," as presented in "*Utilitarianism*." Furthermore, Mill's progressive views on gender equality, as articulated in "*The Subjection of Women*," set the stage for modern feminist thought. His vast work addresses ethics, political philosophy, economics, epistemology, and metaphysics, solidifying Mill's enduring legacy in shaping modern liberal thought.

Rationalism: While often associated with empiricism due to his defense of inductive reasoning, John Stuart Mill also displayed marked rationalist tendencies throughout his philosophical explorations. Mill held that human understanding was rooted in experience and innate cognitive faculties. While he acknowledged the fundamental role of sensory experience in knowledge acquisition, evident in his work "*A System of Logic*," he diverged from strict empiricists in emphasizing the crucial part of the human mind in organizing and interpreting those experiences. Mill believed that some basic principles of logic and mathematics were not derived merely from experience but rather from the mind's inherent structures. This idea echoes rationalist sentiments, suggesting that there exist intrinsic ways of understanding that the mind brings to its experiences. While he did not embrace the innate ideas doctrine often linked with classical rationalism, Mill's insistence on the interplay between inherent mental faculties and empirical evidence placed him in a unique position between the two predominant schools of thought: rationalism and empiricism. His work reflects an attempt to harmonize these often conflicting philosophical approaches, foregrounding the importance of reason in interpreting and understanding our sensory experiences.

Empiricism: John Stuart Mill is a prominent figure in the empiricist tradition, which posits that knowledge primarily arises from sensory experience. Rooted in this belief, Mill's empiricist orientation is profoundly evident in his work "*A System of Logic*," where he elaborates on the principles of inductive reasoning. Contrary to the rationalists who contend the existence of innate ideas or principles independent of experience, Mill argued that even our most fundamental concepts and beliefs have their origins in experience. For instance, he suggested that mathematical and logical truths, often seen by rationalists as purely *a priori*, gain their meaning and justification through our empirical encounters with the world. The "*canons of induction*" he formulated became foundational in scientific inquiry, underlining the methodological processes of gathering empirical evidence and deriving general principles from specific observations. Furthermore, Mill's empiricism extended to his understanding of human nature and morality, as he believed that moral principles, like scientific ones, could be discerned and refined through observation and experience. By championing empiricism, Mill emphasized the primacy of observation, experimentation, and experience in acquiring genuine knowledge, positioning them as the bedrock of scientific and philosophical inquiry.

The Scientific Method: John Stuart Mill's influence on the scientific method is deeply rooted in his rigorous exploration of inductive reasoning, primarily encapsulated in his work "*A System of Logic*." Mill was deeply concerned with how general truths or principles could be derived from specific observations. To this end, he formulated the "*Methods of Induction*" or "*Mill's Methods*," a set of five principles to discern the causal relationships between phenomena. These methods — the Method of Agreement, the Method of Difference, the Joint Method of Agreement and Difference, the Method of Residues, and the Method of Concomitant Variations — are heuristic tools to infer causality from empirical observations. While these methods didn't replace the experimental approach, they provided a structured way to analyze observational data, especially in areas where controlled experiments were challenging. By establishing these methods, Mill played a pivotal role in formalizing and enriching the scientific method, emphasizing the importance of systematic observation and rigorous reasoning. His contributions served as a bridge, ensuring that empirical science maintained a strict, logical foundation and underscoring the necessity of systematic approaches in pursuing knowledge.

Medicine: John Stuart Mill, primarily known for his work in philosophy, political economy, and ethics, did not directly contribute to the field of medicine in the way that a physician or medical researcher might. However, his intellectual underpinnings have indirectly affected medical ethics and methodology. Mill's principle of utility, advocating for the "*greatest good for the greatest number*," has been invoked in bioethical discussions, particularly when considering the balance between individual rights and collective welfare in public health decisions. Moreover, his staunch defense of individual liberty, articulated in "*On Liberty*," has been foundational in discussions about patient autonomy, informed consent, and the right to refuse treatment. His induction methods, designed to discern causal relationships, have also been referenced in epidemiological studies and the analysis of observational data in medicine. While Mill wasn't a direct contributor to medical science, his philosophical principles, and methodological insights provide an essential backdrop for various medical and bioethical deliberations, underscoring the interdisciplinary nature of philosophical thought and its impact on diverse fields, including medicine.

Ethics: John Stuart Mill's imprint on ethical principles is particularly evident in his development and defense of utilitarianism, which posits that actions are right in proportion as they promote happiness and wrong as they tend to produce the reverse. Mill's "*greatest good for the greatest number*" doctrine has profoundly influenced the principle of beneficence, which seeks to promote the well-being of individuals and the general public. This might translate to actions or decisions that contribute to patient health and well-being within medical ethics. Mill's robust defense of individual rights and autonomy in "*On Liberty*" directly aligns with the principle of autonomy, emphasizing the individual's right to self-determination and freedom from external coercion. The focus of nonmaleficence, which dictates that one should do no harm, can be indirectly related to Mill's harm principle, where he asserts that the only justification for interfering with someone's liberty is to prevent harm to others. Lastly, the principle of justice, ensuring fair distribution of benefits, risks, and costs, resonates with Mill's utilitarian framework, which aims for the greatest benefit distributed widely. Mill's ethical insights support these principles, solidifying his lasting influence on moral philosophy and applied ethics.

Conclusion: John Stuart Mill, an emblematic philosopher of the 19th century, has carved an indelible niche in the annals of intellectual history. His multi-faceted contributions span areas as diverse as politics, ethics, epistemology, and even tangentially, medicine. A strong proponent of both rationalist and empiricist thinking, Mill championed a nuanced approach to knowledge acquisition, advocating for a harmonious interplay between inherent cognitive faculties and sensory experiences. His contributions to the scientific method, especially "*Mill's Methods,*" have further cemented his reputation as a meticulous thinker, emphasizing rigorous reasoning and observation. Even in realms where he wasn't a primary contributor, such as medicine, Mill's foundational ideas, especially concerning ethics, have proven influential, guiding deliberations about patient rights and public health. At the core of his ethical philosophy lies utilitarianism, which has shaped modern interpretations of beneficence, autonomy, nonmaleficence, and justice. Mill's intellectual odyssey, marked by his seminal works and methodological insights, has sculpted philosophical discourse and provided guiding principles for subsequent generations across diverse disciplines.

John Stuart Mill's Legacy: Championed utilitarianism and the idea of "*the greatest good for the greatest number*," advocating for individual rights and freedom in a societal context.

**

REVIEW QUESTIONS

True/False Questions:

1. John Stuart Mill is known for his contributions to liberalism, utilitarianism, and women's rights.
   True or False?

2. Mill's principle of utility advocates for the greatest happiness for the greatest number.
   True or False?

Multiple-Choice Questions:

3. What is John Stuart Mill's significant work that outlines his views on individual freedom?
   a) A System of Logic
   b) The Subjection of Women
   c) On Liberty
   d) Utilitarianism

4. Mill's Methods are primarily concerned with:
   a) Establishing metaphysical truths
   b) Discerning causal relationships through empirical observations
   c) Formulating moral principles
   d) Critiquing political systems

Clinical Vignette:

5. A bioethicist is applying Mill's utilitarian principles to a public health policy decision. They must balance individual rights and the collective welfare. Which of the following best aligns with Mill's utilitarian approach?
   a) Prioritizing individual freedoms above all else
   b) Implementing a policy that benefits the majority, even if it restricts some individual freedoms
   c) Ignoring the consequences of the policy on the population
   d) Focusing solely on the rights of the minority group

Basic Science Vignette:

6. John Stuart Mill's "*Methods of Induction*" are primarily designed to:
   a) Prove metaphysical theories
   b) Establish causal relationships from empirical observations
   c) Confirm religious doctrines
   d) Critique rationalist philosophies

Philosophy Vignette:

7. Mill's principle of utilitarianism advocates that actions are right if they:
   a) Conform to societal norms
   b) Promote the greatest happiness for the greatest number
   c) Uphold traditional values
   d) Follow divine commandments

Correct Answers:

1. True
2. True
3. c) On Liberty
4. b) Discerning causal relationships through empirical observations
5. b) Implementing a policy that benefits the majority, even if it restricts some individual freedoms
6. b) Establish causal relationships from empirical observations
7. b) Promote the greatest happiness for the greatest number

BEYOND THE CHAPTER
John Stuart Mill (1806-1873)

- *John Stuart Mill: A Biography* by Nicholas Capaldi
- *On Liberty* by John Stuart Mill
- *Utilitarianism* by John Stuart Mill

***

Chapter 85: Charles Darwin (1809-1882) — The Evolutionary Theorist: Natural Selection

Darwin's theory of natural selection, grounded in empirical observation and rational analysis, revolutionized biology and our understanding of life's diversity.

Abstract: Charles Darwin, the eminent 19th-century British naturalist, fundamentally transformed our understanding of life with his theory of evolution by natural selection. His journey aboard the HMS Beagle exemplified his use of rationalist principles and an empirical approach, challenging prevailing creationist doctrines and promoting evidence-based reasoning. Darwin's meticulous application of the scientific method set a gold standard for academic inquiry, fostering rigorous scrutiny and peer review. Although not directly involved in medicine, his insights indirectly shaped evolutionary medicine, revealing the genetic and evolutionary foundations of human health and disease. Darwin's theories also prompted profound ethical reflections, compelling society to reconsider deeply held beliefs about human nature and our relationship with the natural world. His legacy, marked by intellectual courage and curiosity, continues to inspire humanity to explore, understand, and appreciate the interconnected wonders of life on Earth. Darwin remains a beacon in the annals of intellectual history, urging ongoing exploration and understanding of life's intricate tapestry.

**

Introduction: Charles Darwin, a 19th-century British naturalist, revolutionized the scientific understanding of the origin and evolution of life with his groundbreaking work, "*On the Origin of Species using Natural Selection*." Darwin, born in 1809 in Shrewsbury, England, embarked on the HMS Beagle's voyage in 1831, allowing him to observe diverse species across different environments. The observations and insights he gathered during this journey became the foundation for his theory of natural selection, proposing that species evolve over long periods through adaptation to their environments. This theory contested the prevailing views of creationism and introduced a new paradigm for understanding the interconnectedness of life on Earth. Darwin's pioneering ideas sparked debate, influencing many scientific disciplines, theology, and philosophy. Today, his legacy is celebrated as the cornerstone of modern evolutionary biology.

Rationalism: Charles Darwin, while often associated primarily with the realm of biology, was deeply rooted in the tradition of rationalism. His scientific approach to understanding the natural world was characterized by a rigorous reliance on observation, evidence, and reason, as opposed to relying on traditional or dogmatic beliefs. Contrary to the predominantly creationist worldview of his time, which was rooted in religious interpretations of the origins of life, Darwin sought naturalistic explanations based on empirical evidence. His detailed observations, particularly during the voyage of the HMS Beagle, were followed by years of systematic analysis, experimentation, and the logical synthesis of vast amounts of data. This commitment to reason and evidence led him to formulate the theory of evolution by natural selection. Through this lens, Darwin's work can be seen as a quintessential example of rationalist thought; he valued evidence over entrenched belief and endeavored to understand the world through a framework built upon logical inference and critical examination.

This dedication to rationalism revolutionized biology and underscored the broader cultural shift toward evidence-based reasoning in the modern era.

Empiricism: Charles Darwin, beyond his other intellectual orientations, was fundamentally an empiricist, valuing knowledge derived from sensory experience and observation. The foundation of his monumental work on evolution was built upon meticulous observations he made during his travels aboard the HMS Beagle, where he extensively cataloged diverse species and noted variations in similar species across different environments. This rich tapestry of empirical data from real-world environments like the Galápagos Islands formed the bedrock of his evolutionary theory. Instead of speculating in an armchair manner about the origins and adaptations of species, Darwin directly engaged with the natural world, accumulating a wealth of tangible evidence. This deep commitment to empiricism drove him to conclude based on what he could observe and verify rather than relying solely on abstract reasoning or pre-existing beliefs. Darwin's theory of natural selection emerged from a synthesis of myriad empirical observations, solidifying his stance as one of history's most influential empiricists and highlighting the power of observation in shaping our understanding of the natural world.

The Scientific Method: Charles Darwin's work is a testament to the scientific method's enduring power in furthering human understanding. Although the principles of the scientific method had been established before his time, Darwin's approach to his research on evolution underscored and enhanced its importance. Beginning with observations made during his travels aboard the HMS Beagle, he formulated hypotheses about the interrelatedness of species and their adaptations over time. Instead of hastily publishing these findings, Darwin meticulously tested and refined his hypotheses over two decades, collecting additional data, seeking contrary evidence, and conducting experiments. In doing so, he exemplified the iterative nature of the scientific process: posing questions, collecting evidence, and refining theories in light of new findings. His eventual publication of "*On the Origin of Species*" was a model of careful argumentation based on extensive empirical evidence. The rigorous scrutiny and debate his work underwent after publication further underscored the method's critical aspect of peer review. While Darwin did not redefine the scientific method, his application in the face of potential controversy solidified its role as the gold standard for scientific inquiry and investigation.

Medicine: Charles Darwin, primarily known for his groundbreaking work on evolution and natural selection, did not directly contribute to medicine as pioneering physicians or researchers of his time did. However, his ideas and theories have indirectly shaped medical understanding and practice. Darwin's emphasis on variation and adaptation offers insights into why certain traits or susceptibilities might exist in populations and how they have evolved. His work laid the foundation for evolutionary medicine, which seeks to understand why diseases and disorders occur by considering evolutionary processes. For instance, understanding our evolutionary history can shed light on vulnerabilities in the human body, like the persistence of certain genetic disorders or our susceptibility to certain illnesses. Moreover, Darwin's work on the "*Expression of the Emotions in Man and Animals*" provided early insights into the universality of emotional expressions across cultures, which has implications for psychiatric and neurological studies. While Darwin might not have directly revolutionized medical practice or treatment, his evolutionary lens has enriched and deepened the medical community's understanding of human health, disease, and the broader biological context in which they exist.

Ethics: Charles Darwin's revolutionary work on evolution, while primarily biological, had profound implications for how humans view themselves and, by extension, ethical considerations. His idea that all species, including humans, are interconnected through the *Tree of Life* challenged traditional beliefs about human superiority and our distinct place in the universe. This newfound kinship with the natural world indirectly emphasized the ethical principle of autonomy, urging a reconsideration of humans' dominion over nature and other beings. Darwin's depiction of nature's often harsh, competitive, and indifferent mechanisms also sparked debates about beneficence and nonmaleficence. If natural selection was nature's way, should society mirror this or strive for a higher, compassionate ethic? Furthermore, the realization that variations and adaptations arise due to environmental pressures led to reflections on justice, specifically ensuring that all individuals, regardless of their natural endowments, have equitable societal opportunities. While Darwin did not set out to redefine ethical norms, the implications of his work unavoidably permeated moral philosophy, urging society to grapple with its values and principles in light of an evolved understanding of human nature and our place in the natural world.

Conclusion: Charles Darwin's monumental contributions have resonated through various fields, marking an indelible footprint on our collective intellectual landscape. His theories on evolution, grounded in meticulous observation and the synthesis of vast data, not only reshaped biological sciences but also influenced our perspectives on reason, observation, scientific methodology, medicine, and ethics. Darwin's insistence on empirical research and evidence-based conclusions ushered in a new age of scientific rigor and inquiry. While his direct contributions to medicine were subtle, the ripple effects of his ideas provided deeper insights into human health and the evolutionary underpinnings of diseases. Moreover, Darwin catalyzed profound reflections on our ethical responsibilities toward ourselves, other species, and the planet by unraveling the interconnected tapestry of life and challenging human-centric worldviews. Darwin's legacy is not confined to a singular domain but permeates multiple facets of human thought and understanding, urging continual introspection, curiosity, and growth.

Charles Darwin's Legacy: Proposed the theory of evolution by natural selection, radically reshaping biology and challenging established views on species' origins.

**

REVIEW QUESTIONS

True/False Questions:

1. Charles Darwin's theory of evolution by natural selection emphasized that species evolve over time through adaptation to their environments.
   True or False?

2. Darwin's work had no significant influence on the medical field or medical ethics.
   True or False?

Multiple-Choice Questions:

3. What was the name of Darwin's groundbreaking work that introduced the theory of natural selection?
   a) The Descent of Man
   b) On the Origin of Species
   c) The Voyage of the Beagle
   d) The Expression of the Emotions in Man and Animals

4. Which scientific method did Darwin extensively use to develop his theory of evolution?
   a) Hypothetical speculation
   b) Empirical observation and experimentation
   c) Philosophical debate
   d) Mathematical modeling

Clinical Vignette:

5. A physician is researching the genetic basis of a particular disease and is inspired by Darwin's approach to understanding variations and adaptations in species. Which of the following practices would best align with Darwin's empirical approach?
   a) Theorizing about the disease without collecting data
   b) Conducting experiments and gathering genetic data from different populations
   c) Ignoring genetic variations and focusing solely on environmental factors
   d) Relying on traditional beliefs about the disease

Basic Science Vignette:

6. A population of beetles living on a forest floor exhibits a variety of colorations ranging from light brown to dark brown. Over several generations, the forest experiences frequent fires that char the ground, making it darker. Observing this change, a scientist notices an increase in the proportion of dark brown beetles compared to light brown beetles. What principle does this scenario best illustrate?
   a) Genetic drift
   b) Artificial selection
   c) Natural selection
   d) Inheritance of acquired characteristics

Philosophy Vignette:

7. A philosopher is examining Darwin's influence on existential questions about human purpose. Based on Darwin's evolutionary theory, what is a key takeaway regarding human existence?a) Human purpose is defined by a preordained divine plan
   b) Human purpose is to dominate other species and exploit natural resources
   c) Human existence is part of a broader continuum of life, emphasizing adaptation and survival
   d) Human purpose is to transcend natural laws through technological advancement

Correct Answers:

1. True
2. False
3. b) On the Origin of Species
4. b) Empirical observation and experimentation
5. b) Conducting experiments and gathering genetic data from different populations
6. c) Natural selection
7. c) Human existence is part of a broader continuum of life, emphasizing adaptation and survival

BEYOND THE CHAPTER
Charles Darwin (1809-1882)

- *Charles Darwin: The Power of Place* by Janet Browne
- *On the Origin of Species* by Charles Darwin
- *The Autobiography of Charles Darwin* edited by Nora Barlow

***

Chapter 86: Ignaz Semmelweis (1818-1865) — The Hygiene Pioneer: Hand-washing in Medicine

Semmelweis's advocacy for hand hygiene in medical practice, based on empirical observation, drastically reduced mortality rates and transformed hospital hygiene.

Abstract: In the mid-19th century, Ignaz Semmelweis instituted a paradigm shift in medical hygiene and patient care, asserting the vital significance of antiseptic practices in hospitals. Semmelweis, hailed as the "*savior of mothers*," undertook meticulous observational studies, embodying the empirical and rationalist approaches to scientific inquiry, drastically reducing mortality rates associated with puerperal fever. Through his revolutionary advocacy for hand hygiene in medical settings, he challenged and reshaped existing medical doctrines and echoed the core principles of modern bioethics, emphasizing beneficence, nonmaleficence, and justice in patient care. Semmelweis' steadfast dedication to enhancing medical protocols, grounded in the systematic application of the scientific method, carved out a lasting legacy, demonstrating the profound impact of evidence-based reasoning and ethical conviction in the ongoing advancement of medicine. His efforts ushered in a new era of patient safety. They highlighted the enduring importance of antiseptic measures, forging a path for developing modern germ theory and fostering equitable standards in medical practice.

**

Introduction: Ignaz Semmelweis, a 19th-century Hungarian physician, is often lauded as the pioneer of antiseptic procedures in modern medicine. Born in 1818 in Buda, today's Budapest, Semmelweis made groundbreaking observations that dramatically reduced the incidence of puerperal fever, a fatal condition once common among women who had recently given birth. While working at the Vienna General Hospital in the 1840s, he noticed stark disparities in mortality rates between two maternity clinics: one run by male doctors and medical students and the other by midwives. Through diligent observation and investigative spirit, Semmelweis deduced that doctors, who often moved directly from autopsies to delivering babies, were inadvertently transmitting lethal pathogens. This led him to introduce a simple but revolutionary protocol: handwashing with a chlorinated lime solution before attending to patients. Despite the profound positive impact of this practice, Semmelweis faced resistance and skepticism from many contemporaries. Nonetheless, his contributions set the foundation for adopting antiseptic procedures and recognizing the critical importance of hygiene in medical practice.

Rationalism: Ignaz Semmelweis, throughout his medical career, exemplified the principles of rationalism, which emphasizes reason and evidence over tradition or dogma. When confronted with the mystery of the high maternal mortality rate due to puerperal fever at the Vienna General Hospital, Semmelweis didn't resort to the prevailing notions or superstitions of his time. Instead, he systematically investigated, relying on empirical evidence and logical reasoning. His astute observations of the disparate death rates between two maternity wards – one overseen by doctors and another by midwives – led him to probe deeper into the practices unique to each community. His rationalist approach enabled him to draw a connection between the doctors who conducted autopsies before attending childbirths and the spread of the fever. Undeterred by the established medical beliefs, Semmelweis deduced that some "*cadaveric particles*" were responsible for the spread. He introduced a handwashing regime using a chlorinated lime solution to test his theory, drastically reducing the mortality rate. In an era when germ theory was yet to be established, Semmelweis' unwavering trust in observation, experimentation, and logical deduction made him a beacon of rationalism in the medical field.

Empiricism: Ignaz Semmelweis stands as a paragon of empiricism in the annals of medical history. Empiricism, which stresses knowledge derived from sensory experience and observable phenomena, was deeply ingrained in Semmelweis' approach to the perplexing issue of puerperal fever at the Vienna General Hospital. Rather than accepting prevalent beliefs or theoretical explanations, Semmelweis sought tangible evidence. His empirical journey began with observing a stark difference in mortality rates between two maternity clinics. His subsequent inquiry was driven by meticulous data collection, charting the patterns and commonalities of the disease's occurrence. When a colleague's death from a similar infection provided a crucial clue, Semmelweis furthered his empirical approach, hypothesizing that "*cadaveric particles*" transferred from autopsies were responsible for the disease. But rather than stopping at a theoretical assertion, he put his hypothesis to the test. Instituting a regime of handwashing with a chlorinated lime solution, he closely monitored the outcomes, witnessing a dramatic decline in the mortality rate. Semmelweis exemplified the principles of empiricism in medicine through rigorous observation, data collection, hypothesis testing, and drawing conclusions based on tangible results.

The Scientific Method: Ignaz Semmelweis' work, though not directly shaping the development of the scientific method itself, stands as a powerful testament to its application in medicine. His approach to the problem of puerperal fever at the Vienna General Hospital followed the essential steps of the scientific method. First, Semmelweis observed an unsettling discrepancy in maternal mortality rates between the two clinics. From this observation, he formulated a question: Why was the death rate so much higher in one clinic than the other? His investigative journey led to the hypothesis that contamination from cadavers, handled by doctors and medical students during autopsies, was a potential source of the deadly infections. Semmelweis instituted a procedural change to test this theory, introducing a handwashing regimen with chlorinated lime before doctors attended births. Subsequent observation revealed a dramatic reduction in mortality rates, confirming his hypothesis. While facing resistance from the medical community, Semmelweis' perseverance and systematic approach demonstrated the potency of empirical observation, hypothesis testing, and iterative learning in medicine, embodying the principles of the scientific method and setting a precedent for future medical research.

Medicine: Ignaz Semmelweis' invaluable medical contributions reverberate strongly even in contemporary medical practices. Known colloquially as the "*savior of mothers*," Semmelweis fundamentally shifted the understanding of hospital-acquired infections in the mid-19th century. Working in Vienna's maternity wards, he discerned a disturbingly high incidence of puerperal fever, a deadly condition afflicting postpartum women. His keen observations and systematic approach unveiled a connection between doctors transitioning from autopsies to childbirth without adequate sanitation and the rampant spread of the disease. Challenging prevailing medical orthodoxy, Semmelweis posited that "*cadaveric particles*" were being transferred, causing the fever. His solution was strikingly simple yet profoundly transformative: he mandated handwashing with a chlorinated lime solution before attending to childbirth. This measure resulted in a steep drop in mortality rates, underscoring the critical role of hygiene in patient care. While his contemporaries largely rebuffed his findings, often due to the unsettling implications for their practices, Semmelweis' work laid the groundwork for the later development of the germ theory of disease. His emphasis on sanitation has since become a cornerstone of modern medicine, saving countless lives and setting a standard for rigorous patient care protocols.

Ethics: While Ignaz Semmelweis did not explicitly articulate modern bioethical principles such as patient autonomy (informed consent), practitioner beneficence (do good), practitioner nonmaleficence (do no harm), and public justice (be fair), his work and struggles undeniably reverberated with these tenets. Semmelweis' insistent push for hand hygiene fundamentally revolved around beneficence, the duty to act to benefit patients, and nonmaleficence, the commitment to do no harm. By championing handwashing, he sought to curtail the inadvertent harm medical practitioners inflicted upon mothers by transmitting lethal infections, advocating for their safety and well-being. His dedication to lowering maternal mortality can also be viewed through the lens of justice, as he strived for all patients, irrespective of their status or the clinic they attended, to have an equal chance at surviving childbirth. Though the concept of autonomy, which emphasizes patient rights and informed consent, was not widely recognized in Semmelweis' time, his advocacy for improved hygiene standards indirectly catered to this principle by seeking a standard of care that would better respect patients' well-being and dignity. Thus, while Semmelweis may have yet to be directly engaged in the discourse on medical ethics as we understand it today, his legacy undeniably resonates with these foundational principles, highlighting the inherent ethical responsibilities of healthcare providers.

Conclusion: Ignaz Semmelweis' groundbreaking endeavors in the realm of medicine not only advanced the fields of medical hygiene and patient care but also reverberated through the broader spectrum of scientific investigation, ethics, and epistemology. Drawing from empiricism and rationalism, Semmelweis exemplified a steadfast commitment to observation and evidence-based reasoning, paving the way for later proponents of the germ theory of disease. By courageously challenging the entrenched medical doctrines of his time and introducing antiseptic procedures, he bolstered the pillars of modern medicine. His insistence on rigorous hygiene protocols underscored an implicit emphasis on bioethical principles, advocating for patient safety and equitable care standards. While he faced significant resistance and skepticism during his lifetime, Semmelweis' legacy has cemented him as a seminal figure in medical history. His unwavering dedication to improving patient outcomes and setting higher medical standards remains a testament to the transformative power of scientific inquiry, perseverance, and ethical conviction.

Ignaz Semmelweis' Legacy: Advocated for hand hygiene in medical settings, significantly reducing hospital-acquired infections and highlighting the importance of antiseptic practices.

**

Review Questions:

True/False Questions:

1. Ignaz Semmelweis is known for his groundbreaking work in reducing puerperal fever through the introduction of hand hygiene practices.
   True or False?

2. Semmelweis' theories were widely accepted and implemented without resistance during his lifetime.
   True or False?

Multiple-Choice Questions:

3. What key practice did Ignaz Semmelweis introduce to reduce the incidence of puerperal fever?
   a) Sterilizing medical instruments
   b) Handwashing with a chlorinated lime solution
   c) Using antibiotics
   d) Isolating infected patients

4. Semmelweis' approach to solving the problem of puerperal fever is best described as:
   a) Reliance on traditional medical beliefs
   b) Empirical observation and systematic experimentation
   c) Theoretical speculation without practical application
   d) Blind adherence to contemporary medical practices

Clinical Vignette:

5. A hospital is experiencing a high rate of postoperative infections. Inspired by Ignaz Semmelweis' work, which of the following measures would be most effective in reducing infection rates?
   a) Increasing the use of antibiotics for all patients
   b) Mandating handwashing with an antiseptic solution before and after all surgical procedures
   c) Requiring all staff to wear masks at all times
   d) Performing more frequent sterilization of operating rooms

Basic Science Vignette:

6. Dr. Green is working in a maternity ward and notices a high incidence of postpartum infections among patients. Inspired by historical practices, she decides to implement a handwashing protocol with a chlorinated solution before any examination or delivery procedures. This measure leads to a significant drop in infection rates. What principle does this scenario best illustrate?
   a) Antibiotic resistance
   b) Hand hygiene and antiseptic practices
   c) Vaccination protocols
   d) Genetic predisposition

Philosophy Vignette:

7. A medical ethics committee is discussing the historical importance of hand hygiene in preventing infections. Reflecting on Ignaz Semmelweis' contributions, they consider his advocacy for handwashing as a foundational practice. Which ethical principle is most emphasized in this context according to Semmelweis' approach?
   a) Autonomy
   b) Beneficence
   c) Nonmaleficence
   d) Justice

Correct Answers:

1. True
2. False
3. b) Handwashing with a chlorinated lime solution
4. b) Empirical observation and systematic experimentation
5. b) Mandating handwashing with an antiseptic solution before and after all surgical procedures
6. b) Hand hygiene and antiseptic practices
7. b) Beneficence

Beyond The Chapter
Ignaz Semmelweis (1818-1865)

- *The Doctors' Plague: Germs, Childbed Fever, and the Strange Story of Ignac Semmelweis* by Sherwin B. Nuland
- *Etiology, Concept and Prophylaxis of Childbed Fever* by Ignaz Semmelweis, translated by K. Codell Carter
- *Semmelweis: His Life and Work* by Andrija Štampar

***

Philosophy of Science

Medical Research

Chapter 87: Gregor Mendel (1822-1884) — The Geneticist: Principles of Inheritance

Mendel's experiments with pea plants, uncovering the principles of inheritance, laid the groundwork for modern genetics through empirical research and logical analysis.

Abstract: In the mid-19th century, Augustinian monk Gregor Mendel revolutionized our understanding of heredity through meticulous pea plant experiments at St. Thomas Abbey. Mendel's work embodied rationalism and empiricism, emphasizing logical deduction and evidence-based methodology. His systematic use of the scientific method, characterized by precise data collection and analysis, established a standard for inquiry in the natural sciences. While not directly related to medicine, Mendel's insights became fundamental for diagnosing and treating genetic disorders, fostering the development of personalized medicine. His findings also sparked ethical discussions about genetic advancements, emphasizing patient autonomy (informed consent), practitioner beneficence (do good), nonmaleficence (do no harm), and public justice (be fair) in biomedicine. Mendel's legacy extends beyond his genetic discoveries, significantly impacting medical science and ethical considerations in modern genetics. His contributions mark a transformative epoch in scientific exploration and human understanding, underscoring the profound influence of his pioneering work on genetics and biomedicine.

**

Introduction: Gregor Mendel, an Augustinian monk born in 1822 in what is now the Czech Republic, laid the foundation for genetics through his meticulous experiments on pea plants. Working in the garden of St. Thomas Abbey during the mid-19th century, Mendel demonstrated that certain traits followed particular patterns in successive generations, leading to his formulation of the fundamental laws of inheritance. Though initially overlooked, his discoveries would later revolutionize the understanding of heredity, challenging established notions and providing the basis for modern genetic studies. Mendel's pioneering work, encapsulated in his paper "*Experiments on Plant Hybridization,*" published in 1866, established him as the "f*ather of genetics*" and underscored the significance of systematic scientific investigation in unveiling nature's mysteries.

Rationalism: Gregor Mendel, in his scientific pursuits, exhibited strong rationalist inclinations, exemplified by his systematic approach to studying inheritance in pea plants. Rooted in the belief that the universe operates based on consistent, discoverable principles, rationalism posits that reason and logic are primary sources of knowledge. Mendel's experiments exemplify this tenet. Rather than relying solely on observations or accepting prevailing notions of heredity, Mendel devised systematic cross-breeding experiments to decipher patterns of inheritance. His meticulous data collection and analysis led him to postulate clear, logical laws — the Law of Segregation and the Law of Independent Assortment — describing the inheritance of discrete traits. Mendel's rational approach was a divergence from the more speculative theories of his contemporaries. In the face of skepticism and limited immediate recognition, he trusted his empirical evidence and the logical deductions drawn from them. Thus, Mendel's work not only contributed foundational concepts to genetics but also served as a testament to the power of rationalist thought in scientific discovery.

Empiricism: Gregor Mendel's scientific endeavors were deeply entrenched in empiricism, a philosophical approach emphasizing the role of sensory experience and observable evidence in acquiring knowledge. Rather than deriving his hypotheses from abstract theories or pre-existing dogmas, Mendel looked directly to nature, crafting systematic experiments with pea plants to understand hereditary processes. Over eight years, he meticulously tracked the appearance of various traits across thousands of plant generations, amassing a comprehensive set of data. This vast empirical evidence allowed Mendel to discern consistent patterns of inheritance, leading to his formulation of the foundational laws of genetics. By placing observation and experimentation at the heart of his research, Mendel exemplified the empiricist tradition, underpinning his groundbreaking conclusions with verifiable data. This rigorous, evidence-based methodology not only set a precedent for future genetic studies but also reinforced the essential role of empiricism in driving scientific discovery and challenging established paradigms.

The Scientific Method: Gregor Mendel's groundbreaking work with pea plants is often heralded not just for its revelations in the realm of genetics but also for its adherence to and reinforcement of the principles of the scientific method. Operating in a period when the lines between science, speculation, and philosophy were often blurred, Mendel's experiments stood out for their systematic approach. He began with clear, testable hypotheses about heredity. He then designed controlled experiments, ensuring that variables were isolated and potential confounding factors minimized. As he meticulously bred and cross-bred thousands of plants, he collected robust data, which he analyzed precisely. He drew logical conclusions from these data, formulating the fundamental inheritance laws. When others doubted or misunderstood his findings, Mendel's reliance on the scientific method ensured his results were replicable and verifiable. Through his rigorous application of this method, Mendel advanced the field of genetics. He bolstered the scientific method's standing as the gold standard for inquiry and discovery in the natural sciences.

Medicine: While Gregor Mendel did not directly delve into medicine, the implications of his pioneering work on genetics have had profound ramifications for medical science. By elucidating the fundamental laws of inheritance, Mendel laid the groundwork for our understanding of how genetic traits are passed down through generations. This knowledge proved indispensable in the 20th century as researchers began to unravel the complexities of human genetics and its relationship to disease. Mendelian inheritance patterns have become a cornerstone in diagnosing, understanding, and treating various genetic disorders, from cystic fibrosis to sickle cell anemia. His foundational principles have enabled medical professionals to predict the likelihood of an individual inheriting specific genetic conditions, guiding critical decisions in fields like genetic counseling. Moreover, Mendel's contributions indirectly paved the way for the development of personalized medicine, where treatments can be tailored based on an individual's genetic makeup. While Mendel may have never fathomed the vast medical applications of his pea plant experiments, his legacy is firmly embedded in the annals of medical genetics, transforming how we approach, diagnose, and treat genetic diseases.

Ethics: Though Gregor Mendel's direct work was centered on the principles of genetics rather than ethics, the implications of his discoveries have had far-reaching consequences in the ethical considerations of modern genetics and biomedicine. As the field of genetics grew, so did the ethical dilemmas surrounding genetic testing, cloning, gene editing, and other associated technologies. Autonomy, the right of an individual to make informed, voluntary decisions, became paramount when considering genetic testing and its implications for future health, family planning, or even personal identity. Beneficence, the duty to act in the patient's best interest, is tested when determining the right course of action based on genetic findings. Nonmaleficence, or the principle of "*do no harm*," is intertwined with the potential risks of genetic interventions and the unforeseen consequences they might bring.

Finally, justice, which underscores the equitable distribution of benefits, risks, and costs, plays a role in debates over access to genetic technologies and treatments. While Mendel may not have directly influenced these principles, his foundational work in genetics set the stage for the complex web of ethical considerations that have since emerged, underscoring the intricate relationship between scientific advancements and ethical responsibility.

Conclusion: Gregor Mendel's indelible mark on the scientific landscape reaches far beyond his meticulously plotted pea garden in St. Thomas Abbey. Through his pioneering work, Mendel decoded the basic tenets of genetic inheritance and reshaped how scientific inquiry was approached. His confluence of rationalism and empiricism underscored the importance of logical reasoning and observable data in scientific discovery. Further, Mendel's strict adherence to the scientific method emphasized the essence of rigorous, systematic experimentation, setting a gold standard for subsequent scientific endeavors. Though not directly medical, his legacy has reverberated through the annals of medical genetics, offering tools to diagnose, understand, and treat various genetic disorders. Additionally, as the world grappled with the profound implications of genetic understanding, Mendel's work inadvertently birthed myriad ethical dilemmas in biomedicine, nudging society to reflect on the responsibilities accompanying knowledge. Through his unwavering curiosity and meticulous approach, Mendel, the humble monk from the Czech Republic, ushered in a new epoch of scientific and ethical introspection that continues to shape our world today.

Gregor Mendel's Legacy: He laid the foundation for genetics with his work on pea plants, uncovering the principles of inheritance and traits' predictability.

**

Review Questions:

True/False Questions:

1. Gregor Mendel's experiments on pea plants led to the formulation of the fundamental laws of inheritance.
   True or False?

2. Mendel's work had no significant impact on the field of medicine.
   True or False?

Multiple-Choice Questions:

3. What are the key principles that Gregor Mendel formulated from his experiments on pea plants?
   a) The Law of Gravity and the Law of Thermodynamics
   b) The Law of Segregation and the Law of Independent Assortment
   c) The Law of Evolution and the Law of Natural Selection
   d) The Law of Conservation and the Law of Motion

4. Which approach best describes Mendel's methodology in his experiments?
   a) Theoretical speculation without empirical data
   b) Random observation without systematic analysis
   c) Systematic experimentation with precise data collection and analysis
   d) Reliance on anecdotal evidence and tradition

Clinical Vignette:

5. A geneticist is researching the inheritance pattern of a genetic disorder in a family. Inspired by Mendel's work, which of the following practices would best align with Mendel's scientific approach?
a) Formulating hypotheses based on family history and conducting controlled breeding experiments
b) Using theoretical models without collecting family data
c) Ignoring patterns and focusing on individual cases without broader analysis
d) Relying solely on anecdotal reports from family members

Basic Science Vignette:

6. A genetics researcher is studying the inheritance of a particular trait in pea plants, following Gregor Mendel's principles. If a plant with yellow seeds (dominant) is crossed with a plant with green seeds (recessive), what would be the expected phenotype of the offspring in the F1 generation?
a) All green seeds
b) A mix of yellow and green seeds
c) All yellow seeds
d) An equal number of yellow and green seeds

Philosophy Vignette:

7. Considering the ethical implications of Mendel's discoveries, which bioethical principle is most directly related to ensuring that genetic information is used to benefit patients and improve their health outcomes?
a) Autonomy
b) Beneficence
c) Nonmaleficence
d) Justice

Correct Answers:

1. True
2. False
3. b) The Law of Segregation and the Law of Independent Assortment
4. c) Systematic experimentation with precise data collection and analysis
5. a) Formulating hypotheses based on family history and conducting controlled breeding experiments
6. c) All yellow seeds
7. b) Beneficence

Beyond The Chapter
Gregor Mendel (1822-1884)

- *Gregor Mendel: The First Geneticist* by Vítězslav Orel
- *Mendel in the Kitchen: A Scientist's View of Genetically Modified Foods* by Nina Fedoroff and Nancy Marie Brown
- *Mendel's Legacy: The Origin of Classical Genetics* by Elof Axel Carlson

***

Philosophy of Science

Medical Research

WEEK SIXTEEN: MODERN MEDICAL INNOVATORS

WEEK SIXTEEN

Class Hour 1

- Chapter 88: Louis Pasteur (1822-1895) — The Germ Theory Pioneer: Microbial Revolution
- Chapter 89: Alexander Fleming (1881-1955) — The Antibiotic Pioneer: Penicillin
- Chapter 90: Albert Sabin (1906-1993) and Jonas Salk (1914-1995) — The Polio Warriors: Vaccine Pioneers

Class Hour 2

- Chapter 91: Hans Jonas (1903-1993) — The Ethical Philosopher: Responsibility in Science
- Chapter 92: Henry K. Beecher (1904-1976) — The Ethical Clinician: Informed Consent
- Chapter 93: Joseph Murray (1919-2012) — The Transplant Pioneer: Organ Transplants

EXAM V: 50Q - online

Introduction

In week sixteen, we reach the culmination of our exploration into Philosophy of Science, focusing on key figures whose groundbreaking contributions have shaped modern medicine and bioethics. These pioneers, through their rationalist and empirical approaches, have revolutionized our understanding of disease, treatment, and ethical considerations in scientific practice.

Louis Pasteur, the Germ Theory Pioneer, revolutionized medical microbiology and public health with his development of the germ theory of disease and the process of pasteurization. Pasteur's meticulous experiments and observations provided compelling evidence that microorganisms cause disease, fundamentally altering medical science. His work not only advanced public health but also underscored the ethical principles of beneficence and the duty to protect communities from infectious diseases.

Alexander Fleming, the Antibiotic Pioneer, made a monumental discovery with penicillin, the first true antibiotic. This breakthrough ushered in the antibiotic era, dramatically reducing deaths from bacterial infections and transforming medical practice. Fleming's rationalist and empirical methods highlighted the importance of patient care, beneficence, and nonmaleficence, setting new standards for medical treatment and ethical responsibility in healthcare.

Albert Sabin and Jonas Salk, known as the Polio Warriors, developed oral and injectable polio vaccines, respectively. Their efforts nearly eradicated a crippling and often deadly disease, demonstrating the profound impact of vaccines on public health. Sabin's and Salk's approaches exemplified the principles of beneficence and nonmaleficence, emphasizing the ethical duty to prevent disease and promote health on a global scale.

Hans Jonas, the Ethical Philosopher, introduced a bioethical framework that emphasized responsibility toward future generations and the natural world. His work profoundly shaped the ethics of science, medicine, and technology, encouraging a deeper moral engagement with the long-term consequences of scientific advancements. Jonas' philosophy of responsibility remains a cornerstone in contemporary bioethical discourse, advocating for the ethical consideration of our actions' impact on future generations.

Henry K. Beecher, the Ethical Clinician, transformed medical research methodologies by championing informed consent and patient autonomy. Beecher's rationalist and empirical approach emphasized the importance of methodological rigor and moral responsibility in clinical research. His work significantly impacted healthcare ethics, ensuring that patients' rights and autonomy are respected in medical practices and research protocols.

Joseph Murray, the Transplant Pioneer, conducted the first successful human organ transplant, opening a new frontier in medical science. Murray's work not only revolutionized transplant medicine but also emphasized the importance of bioethical principles in patient care. His pioneering efforts underscored the necessity of ethical considerations in transplantation practices, ensuring that patient welfare and ethical standards guide medical advancements.

As we conclude our course, we reflect on the transformative contributions of these extraordinary individuals. Their rationalist and empirical approaches have not only advanced scientific knowledge and medical practice but have also established enduring ethical principles that continue to guide contemporary science and medicine. Through their pioneering work, we gain a deeper appreciation for the intricate interplay between scientific discovery and ethical responsibility, understanding how each breakthrough shapes our collective commitment to improving human health and well-being.

***

Chapter 88: Louis Pasteur (1822-1895) — The Germ Theory Pioneer: Microbial Revolution

Pasteur's germ theory, established through meticulous experiments, revolutionized medicine and public health, emphasizing the importance of empirical science.

Abstract: Louis Pasteur, a visionary French chemist and microbiologist of the 19th century, revolutionized multiple disciplines, including biology, medicine, and the food industry, heralding a new era of scientific advancement and public health improvement. His systematic integration of rationalism and empiricism fostered a robust application of the scientific method, dismantling the prevalent theory of spontaneous generation and establishing the germ theory of disease. This seminal shift in understanding, supported by meticulous and innovative experiments such as the swan-neck flask experiment, not only set new standards for scientific inquiry but also paved the way for transformative medical interventions, including the inception of vaccines for rabies and anthrax and the development of the pasteurization process. Beyond his tangible contributions, Pasteur's unwavering commitment to societal well-being reflected seminal ethical tenets that continue to influence modern medical and scientific practices, emphasizing beneficence, safety, and inclusivity. Hence, Pasteur's lasting legacy transcends his groundbreaking discoveries, embodying a holistic approach that synergizes inquiry, evidence, and ethics, significantly elevating human knowledge and health trajectory.

**

Introduction: Louis Pasteur, a 19th-century French chemist and microbiologist, stands as one of the most impactful figures in the annals of science. Born in 1822 in Dole, France, Pasteur's pioneering contributions fundamentally altered our understanding of biology and medicine. He debunked the widely accepted myth of spontaneous generation, asserting that microorganisms come from other organisms, laying the groundwork for the germ theory of disease. Pasteur's innovations did not end there: his pasteurization process, where heating liquids destroy bacteria and then allowing them to cool, revolutionized the food industry by significantly extending the shelf life of perishable products like milk. His persistent investigations into the causes and prevention of diseases led to the development of the first vaccines for rabies and anthrax, saving countless lives. Throughout his illustrious career, Pasteur's insights and inventions dramatically reduced the prevalence of dangerous illnesses and set the foundation for many of modern medicine's best practices.

Rationalism: Though deeply rooted in empirical observation and experimentation, Louis Pasteur's rationalist inclination shaped his scientific inquiries. Rationalism emphasizes the role of reason and logic over sensory experience, and Pasteur frequently approached scientific problems by first questioning prevailing assumptions and dogmas. For instance, in an era when the notion of spontaneous generation — the idea that life could arise spontaneously from non-living matter — was widely accepted, Pasteur was skeptical. Rather than merely taking this belief or basing his opinion solely on direct observations, he designed meticulous experiments to test the hypothesis. His work culminated in the famous swan-neck flask experiment, where he demonstrated that microorganisms didn't appear in a growth medium unless exposed to air contaminants. This disproved spontaneous generation and underscored the importance of logical and rational problem-solving in the scientific method.

Pasteur's insistence on questioning and challenging established beliefs, utilizing reason to guide hypotheses, and then rigorously testing these hypotheses with empirical methods embodies the marriage of rationalism with the scientific method, forever changing the landscape of microbiology and medicine.

Empiricism: Louis Pasteur's scientific endeavors vividly exemplify an empiricist orientation, which champions knowledge acquisition through sensory experience, particularly observation and experimentation. This empiricist bent was most evident in the meticulousness with which Pasteur approached scientific problems. When confronting the prevalent belief in spontaneous generation, Pasteur designed and executed a series of groundbreaking experiments rather than relying solely on theoretical conjecture. His swan-neck flask experiment, for instance, was the epitome of empirical methodology. By creating a flask with a long, curved neck, he showed that while air could enter, dust and microbes could not, preventing the broth inside from spoiling. When the neck was broken and the broth exposed directly to the air, it rapidly became contaminated, visually demonstrating the presence and role of microorganisms in decay. Through hands-on investigations, Pasteur validated or debunked theories based on tangible, replicable evidence rather than mere speculation. His relentless pursuit of empirical evidence, combined with a deep trust in the observatory power of experimentation, fortified the role of empiricism in the sciences, showcasing its indispensable value in discerning the truths of the natural world.

The Scientific Method: Louis Pasteur's work stands as a monumental testament to the application and furtherance of the scientific method, a systematic procedure for investigating phenomena, acquiring knowledge, and integrating previous knowledge. While the scientific method had its roots well before Pasteur's time, his contributions exemplify its principles in action, reinforcing its significance to contemporary science. Pasteur consistently began with keen observations, from which he formulated testable hypotheses, particularly evident when challenging the doctrine of spontaneous generation. The rigorous experiments he designed, such as the swan-neck flask experiment, showcased a methodological approach to validation or refutation. By ensuring that his experiments were replicable, Pasteur underscored the necessity of reproducibility in scientific inquiry, a key tenet of the method. Furthermore, he was transparent about his results, inviting peer review and critique, and was willing to amend his conclusions in the face of new evidence. In doing so, Pasteur applied the scientific method to produce groundbreaking discoveries and exemplified its role as a self-correcting and ever-evolving process, reinforcing its central position in the progress of modern scientific endeavors.

Medicine: Louis Pasteur's indelible mark on medicine can hardly be overstated, having pioneered discoveries that radically transformed the field and saved innumerable lives. Perhaps most notably, Pasteur's work laid the foundation for the germ theory of disease, which postulated that microorganisms cause many diseases. This revelation overturned centuries of medical misunderstanding and set the stage for advances in hygiene, sterilization, and, eventually, the development of antibiotics. His investigative prowess was further showcased when he turned his attention to vaccine development. Through careful attenuation of harmful agents, Pasteur developed the first vaccine for rabies, a fatal disease that had plagued humanity for ages. This seminal work opened the door for creating other vaccines, acting as a blueprint for immunization techniques that prevent countless illnesses today. Moreover, his innovation of pasteurization, though often associated with the food industry, had profound implications for medicine by establishing a method to curtail the spread of pathogens. Through these contributions, Louis Pasteur not only reshaped the trajectory of medical science but also considerably enhanced the quality and longevity of human life in the face of once-deadly afflictions.

Ethics: While Louis Pasteur is primarily celebrated for his scientific achievements, his work and ethos have implicit bearings on the ethical principles that guide modern medical and scientific endeavors. Pasteur's unwavering dedication to combating diseases speaks to the principle of beneficence, the commitment to act in the patient's or community's best interest. His relentless pursuit of vaccines for rabies and anthrax underscored a drive to achieve greater societal good. The principle of nonmaleficence, or "*do no harm*," can be observed in his meticulous approach to experimentation, ensuring that the techniques he developed, such as pasteurization, were safe and beneficial. Autonomy, the right of an individual to make decisions, while not directly addressed in Pasteur's work, is implicitly respected in his drive to provide tools and knowledge that allow individuals and societies to make informed health choices. Finally, his endeavors touch upon justice, or the fair distribution of benefits and burdens. By working on widespread health threats of his time, Pasteur sought solutions that had the potential to benefit all, irrespective of their socioeconomic status. Although he might not have explicitly verbalized these principles, Pasteur's legacy undeniably reverberates through the ethical foundations of modern medicine and science.

Conclusion: Louis Pasteur's multifaceted impact on the world of science and medicine is both profound and enduring. His explorations seamlessly blended the principles of rationalism and empiricism, driving the evolution and affirmation of the scientific method. Through careful observation, hypothesis formation, and rigorous testing, he dismantled long-held beliefs and illuminated the truths of the microbial world, reshaping our understanding of disease causation and prevention. In medicine, his revelations regarding the germ theory and the subsequent development of vaccines served as critical turning points, ushering in an era where once-fatal diseases became preventable. Beyond his tangible discoveries, Pasteur's influence extended into the ethical sphere, reflecting the core principles that guide today's scientific and medical practices. His dedication to the welfare of humanity, exemplified by his tireless work and innovations, set a benchmark for subsequent generations of researchers. In essence, Pasteur's legacy is not just about the specifics of his discoveries but the holistic manner in which he approached and transformed various domains, reinforcing the importance of inquiry, evidence, and ethics in advancing human knowledge and wellbeing.

Louis Pasteur's Legacy: Developed the germ theory of disease and pasteurization, greatly advancing medical microbiology and public health.

**

REVIEW QUESTIONS

True/False Questions:

1. Louis Pasteur's experiments debunked the theory of spontaneous generation and established the germ theory of disease.
   True or False?

2. Pasteur's work had little impact on the development of vaccines and modern medical practices.
   True or False?

Multiple-Choice Questions:

3. Which experiment by Louis Pasteur demonstrated that microorganisms do not spontaneously generate in sterile broth unless exposed to air contaminants?
   a) Swan-neck flask experiment
   b) Pasteurization experiment
   c) Anthrax vaccine experiment
   d) Rabies vaccine experiment

4. What process did Louis Pasteur develop to extend the shelf life of perishable products like milk by destroying bacteria through heating?
   a) Sterilization
   b) Pasteurization
   c) Fermentation
   d) Filtration

Clinical Vignette:

5. A public health official is working to reduce the incidence of a deadly bacterial infection in a rural community. Inspired by Pasteur's principles, which of the following strategies would best align with his approach?
   a) Ignoring sanitation practices and focusing solely on treating infected individuals
   b) Implementing widespread education on hygiene and sanitation practices, and ensuring access to clean water
   c) Encouraging traditional medicine practices without scientific validation
   d) Waiting for the infection rates to decrease naturally without intervention

Basic Science Vignette:

6. A patient presents with symptoms of a bacterial infection. The attending physician decides to apply a method to ensure the liquid medication is free of harmful bacteria by heating it and then allowing it to cool. Which of the following processes is the physician using?
   a) Fermentation
   b) Distillation
   c) Pasteurization
   d) Filtration

Philosophy Vignette:

7. Louis Pasteur's approach to scientific inquiry involved questioning prevailing assumptions and designing experiments to test these hypotheses. How does this reflect the philosophical principle of rationalism?
   a) It emphasizes the collection of sensory data over theoretical reasoning
   b) It highlights the importance of empirical evidence without any pre-existing hypothesis
   c) It underscores the use of reason and logic to challenge established beliefs and form hypotheses
   d) It focuses solely on the practical applications of scientific discoveries

Correct Answers:

1. True
2. False
3. a) Swan-neck flask experiment
4. b) Pasteurization
5. b) Implementing widespread education on hygiene and sanitation practices, and ensuring access to clean water
6. c) Pasteurization
7. c) It underscores the use of reason and logic to challenge established beliefs and form hypotheses

BEYOND THE CHAPTER
Louis Pasteur (1822-1895)

- *Louis Pasteur: A Life* by Patrice Debré
- *Pasteur's Quadrant: Basic Science and Technological Innovation* by Donald E. Stokes
- *The Private Science of Louis Pasteur* by Gerald L. Geison

***

Philosophy of Science

Medical Research

Chapter 89: Alexander Fleming (1881-1955) — The Antibiotic Pioneer: Penicillin

Fleming's discovery of penicillin marked a breakthrough in medical science, showcasing the life-saving potential of empirical research and scientific innovation.

Abstract: Alexander Fleming, the prodigious bacteriologist who discovered penicillin, left an indelible mark on scientific history and modern medicine. Grounded in rationalism and empiricism, Fleming's pioneering efforts reshaped disease management and unlocked unprecedented potential in medicine. His keen observation of Penicillium notatum's antibacterial properties heralded the antibiotic era. Fleming's influence extended beyond his discoveries; he fortified the scientific method, emphasizing observation, hypothesis formulation, and empirical validation in biomedical research. His guidance on the potential misuse of antibiotics highlighted ethical engagement, stressing beneficence and caution against nonmaleficence in medical advancements. Fleming's legacy transcends the laboratory, epitomizing the symbiosis of rigorous scientific inquiry and ethical considerations, setting a gold standard for researchers and clinicians. His work stands as a testament to the transformative power of curiosity, diligence, and foresight in fostering discoveries that save countless lives and shape healthcare's trajectory.

**

Introduction: Alexander Fleming, born in 1881 in rural Scotland, was a groundbreaking bacteriologist whose discoveries irrevocably shifted the trajectory of modern medicine. While serving in World War I, he witnessed the limitations of antiseptics in treating deep wound infections, sparking his interest in antibacterial agents. This curiosity eventually culminated in his seminal discovery in 1928: the mold Penicillium notatum produced a substance that killed many bacteria. This substance, which he named penicillin, became the world's first true antibiotic. Beyond its immediate life-saving capabilities, penicillin paved the way for developing other antibiotics, fundamentally changing our approach to bacterial infections and disease management. Fleming's fortuitous finding and subsequent work heralded a new era in medicine, where once-deadly infections became treatable, and surgeries and medical interventions could be undertaken with significantly reduced risk.

Rationalism: Alexander Fleming's scientific approach was deeply rooted in rationalist principles, favoring empirical evidence, observation, and reasoned analysis over mere speculation or unverified beliefs. While some might view the discovery of penicillin as serendipitous, Fleming's keen observational skills and deductive reasoning allowed him to recognize the mold's antibacterial potential. Rather than dismissing the contaminated petri dish, he rationally hypothesized that the mold was producing a substance inhibiting bacterial growth. Employing meticulous experimentation and systematic investigation, he isolated and studied this substance, eventually identifying it as penicillin. Fleming's rigorous approach to science is a testament to his rationalist inclinations. He believed in drawing conclusions based on evidence and consistently subjected his findings to tests and validations, ensuring they could be replicated and stood up to scrutiny. In a world where many early 20th-century medical treatments were based on tradition or anecdotal evidence, Fleming's steadfast commitment to rational inquiry and empirical research set a gold standard for scientific exploration and laid the groundwork for subsequent breakthroughs in the field of medicine.

Empiricism: Alexander Fleming's work embodied the essence of empiricism, which emphasizes knowledge derived from sensory experience and observation. His keen observation of the "*halo*" of inhibited bacterial growth around the mold in a petri dish wasn't just a chance encounter but an empiricist's acknowledgment of an unusual phenomenon demanding further investigation. Fleming didn't rely on pre-existing theoretical constructs or abstract reasoning alone. Instead, he embarked on a series of practical experiments to test and understand the properties of the substance produced by the mold, later identified as penicillin. Each step of his research was anchored in tangible observations, measurements, and experiments that provided concrete evidence of penicillin's antibacterial properties. His reliance on direct experience and experimental verification, rather than mere theoretical speculation, exemplifies the empiricist approach. In a nutshell, Fleming's monumental discovery and subsequent contributions to medicine were grounded in his commitment to empiricism, emphasizing the importance of evidence-based inquiry and the tangible, measurable aspects of the natural world as primary sources of knowledge.

The Scientific Method: While Alexander Fleming is more directly celebrated for his discovery of penicillin than for reshaping the scientific method itself, his work exemplified and reinforced the importance of this method in biomedical research. Fleming's approach to his findings emphasized observation, hypothesis formulation, experimentation, and analysis — the foundational pillars of the scientific method. When confronted with an anomaly, the clear zone around the Penicillium mold on a bacterial culture, he didn't dismiss it as mere contamination. Instead, he formulated hypotheses about its cause and systematically tested them. His meticulous experiments with the mold extract provided reproducible results that other scientists could verify independently. Moreover, Fleming's transparency in documenting and sharing his findings, including the limitations and potential of penicillin, underscored the importance of peer review and replication in the scientific process. Though he didn't redefine the scientific method per se, Fleming's rigorous adherence to it in his groundbreaking discovery showcased its value in producing reliable, transformative knowledge, reinforcing its centrality in modern scientific endeavors.

Medicine: Alexander Fleming's contributions to medicine are monumental, with his discovery of penicillin marking a pivotal moment in medical history. Before Fleming's epoch-making find in 1928, humanity had limited defenses against bacterial infections, leading to countless deaths from diseases that are easily treatable today. Fleming's astute observation of a mold inhibiting bacterial growth led to the identification of penicillin, the world's first antibiotic. This antibiotic revolutionized the treatment of bacterial infections, transforming them from potential death sentences into manageable conditions. Beyond penicillin, his discovery catalyzed further research into antibiotics, broadening the spectrum of drugs available to combat various pathogens. Additionally, introducing antibiotics like penicillin made surgical procedures far safer, drastically reducing post-operative infections and associated mortality rates. Through this singular discovery, Fleming ushered in the era of antibiotics, fundamentally altering the landscape of medicine. His legacy includes the countless lives saved directly through penicillin's use and the broader antibiotic research it inspired, setting the stage for modern medicine's continued battle against infectious diseases.

Ethics: Alexander Fleming's contributions to medicine, while primarily scientific, also indirectly influenced the realm of medical ethics, particularly when viewed through the lens of principles like patient autonomy (informed consent), practitioner beneficence (do good), practitioner nonmaleficence (do no harm), and public justice (be fair). Fleming's discovery of penicillin emphasized beneficence, as the antibiotic has since saved millions of lives, significantly reducing suffering from bacterial infections. In advocating for its appropriate use, he implicitly championed nonmaleficence, warning against the under-dosing of penicillin, which might lead to antibiotic-resistant strains — a prescient concern given today's challenges with antibiotic resistance. Fleming's work also touches upon justice: while he was central to the discovery of penicillin, the subsequent mass production and global distribution efforts ensured that it was not just a luxury for the few but a life-saving treatment accessible to many, striving toward equitable healthcare. However, Fleming's work didn't address the principle of autonomy; this would become more central in later medical discussions, focusing on informed consent and patient rights. While Fleming might not have directly engaged with modern ethical terminologies, the ramifications of his discovery and his foresight regarding its misuse highlight a deep ethical underpinning in his approach to medicine.

Conclusion: Alexander Fleming's lasting imprint on the annals of science and medicine is undeniable, serving as a beacon of innovation, diligence, and foresight. Rooted in the principles of rationalism and empiricism, his meticulous approach underscored the indispensable nature of observation, evidence-based inquiry, and the scientific method in driving breakthroughs. The advent of penicillin, his most celebrated discovery, dramatically reshaped medical landscapes, offering humanity a potent weapon against previously indomitable bacterial foes. Beyond the immediate implications for healthcare, Fleming's conscientious warnings about antibiotic misuse exhibit an inherent ethical consciousness, illuminating potential pitfalls in medical advancements. Fleming's legacy is multifaceted: a testament to scientific rigor, a touchstone for the medical revolution, and a gentle reminder of the ethical considerations intrinsic to progress. Through his life's work, Fleming stands as a paragon of how curiosity, coupled with discipline and foresight, can lead to discoveries that echo through generations.

Alexander Fleming's Legacy: He discovered penicillin, ushering in the antibiotic era and dramatically reducing deaths from bacterial infections.

**

REVIEW QUESTIONS

True/False Questions:

1. Alexander Fleming's discovery of penicillin was entirely serendipitous and did not involve any rational analysis or empirical research.
   True or False?

2. Fleming's discovery of penicillin led to the first true antibiotic, which fundamentally changed our approach to bacterial infections and disease management.
   True or False?

Multiple-Choice Questions:

3. What principle was central to Alexander Fleming's scientific approach, as highlighted in the chapter?
   a) Speculation
   b) Anecdotal evidence
   c) Rationalism
   d) Tradition

4. Which of the following best describes Fleming's concern regarding the potential misuse of antibiotics?
   a) It could lead to higher costs for medical treatments.
   b) It might cause more bacterial infections.
   c) It could result in the development of antibiotic-resistant strains.
   d) It would reduce the effectiveness of vaccines.

Clinical Vignette:

5. A patient presents with a severe bacterial infection that has not responded to standard treatments. The medical team decides to use a new antibiotic developed based on principles similar to those used by Alexander Fleming. What should be the primary consideration to ensure the effectiveness and ethical use of this new antibiotic?
   a) Ensuring it is used widely regardless of resistance patterns
   b) Using it in low doses to avoid side effects
   c) Administering it with informed consent and monitoring for resistance
   d) Prescribing it to all patients regardless of their condition

Basic Science Vignette:

6. A microbiologist notices that bacteria do not grow around a particular mold in a petri dish. Following Alexander Fleming's approach, what should the microbiologist do next to potentially discover a new antibiotic?
   a) Discard the contaminated petri dish
   b) Document and further investigate the properties of the mold
   c) Assume the mold is harmful and avoid further experiments
   d) Compare the mold to existing molds known to be non-antibiotic

Philosophy Vignette:

7. Alexander Fleming's discovery of penicillin emphasizes the importance of empirical observation in scientific research. How does this approach contrast with purely theoretical scientific methods?
   a) By focusing on hypothesis generation without the need for data
   b) By valuing direct experimental evidence over theoretical speculation
   c) By relying solely on mathematical models and simulations
   d) By avoiding the formulation of any hypotheses

Correct Answers:

1. False
2. True
3. c) Rationalism
4. c) It could result in the development of antibiotic-resistant strains
5. c) Administering it with informed consent and monitoring for resistance
6. b) Document and further investigate the properties of the mold
7. b) By valuing direct experimental evidence over theoretical speculation

BEYOND THE CHAPTER
Alexander Fleming (1881-1955)

- *Penicillin Man: Alexander Fleming and the Antibiotic Revolution* by Kevin Brown
- *The Life of Sir Alexander Fleming: Discoverer of Penicillin* by André Maurois
- *Alexander Fleming: The Man and the Myth* by Gwyn Macfarlane

***

Chapter 90: Albert Sabin (1906-1993) and **Jonas Salk** (1914-1995) — The Polio Warriors: Vaccine Pioneers

Sabin and Salk's development of polio vaccines exemplified the transformative power of empirical science in public health, eradicating a crippling disease.

Abstract: Amid the 20th-century polio epidemic, the concerted efforts of Dr. Albert Sabin and Dr. Jonas Salk heralded a revolutionary era in medical science and global health initiatives. Embarking on divergent paths but united in purpose, these venerable scientists unveiled groundbreaking vaccines that drastically reduced the global incidence of poliomyelitis. Salk introduced the inactivated polio vaccine in 1955, utilizing a killed virus. Sabin later pioneered an oral vaccine using a weakened form of the virus, facilitating easier administration during mass immunization campaigns. Together, their empirical and rationalistic approaches to scientific inquiry, grounded in the rigors of the scientific method, epitomized the transformative power of evidence-based research. Beyond their technical contributions, Salk and Sabin showcased an enduring commitment to ethical principles, prioritizing public welfare over personal gain. Their combined legacies are a beacon of hope in the relentless pursuit of medical advancements and a testament to the monumental societal upliftment achievable through collaborative scientific endeavors and global cooperation.

**

Introduction: In the mid-20th century, amidst the crippling fear and rising incidence of polio, two pioneering scientists emerged, forever altering the trajectory of this global health crisis. Dr. Jonas Salk and Dr. Albert Sabin, striving for the same objective but following divergent paths, developed vaccines to combat poliomyelitis (polio). Salk introduced the inactivated polio vaccine (IPV) in 1955, which required an injection and offered immunity by using a killed form of the virus. Sabin, in contrast, introduced the oral polio vaccine (OPV) in the early 1960s, utilizing a weakened form of the virus to stimulate immunity. Their contributions precipitated a dramatic decline in polio cases, shaping public health campaigns and laying the groundwork for a global commitment to eradicate the disease. While their methods differed, Salk and Sabin's names are forever intertwined in medical history, symbolizing hope, resilience, and the tenacious spirit of scientific inquiry.

Rationalism: Albert Sabin and Jonas Salk, in their pursuit to combat poliomyelitis, exemplified the rationalist approach to scientific inquiry. Grounded in empirical evidence and the belief that reason and logic are central to understanding and influencing the natural world, both scientists harnessed systematic methodologies to develop their respective vaccines. Utilizing an empirical process, Salk meticulously developed the inactivated polio vaccine (IPV) based on a killed version of the virus, ensuring its safety through rigorous testing. Similarly, Sabin's work on the oral polio vaccine (OPV) stemmed from systematically evaluating weakened virus strains to determine their efficacy in eliciting immunity without causing the disease. Their endeavors were rooted in hope or serendipity and staunch belief in the scientific method. Their rationalist inclinations didn't merely rest on generating results but were further reflected in their commitment to wide-scale vaccination campaigns and public health education. Sabin and Salk epitomized the rationalist ethos by prioritizing evidence over superstition or prejudice, demonstrating how structured, reasoned approaches can yield transformative solutions to humanity's most pressing challenges.

Empiricism: In their quest to defeat poliomyelitis, Albert Sabin and Jonas Salk showcased a profound empiricist orientation in their scientific endeavors. Empiricism, the philosophy that knowledge arises primarily from sensory experience and observational evidence, was at the core of their vaccine development processes. Salk's development of the inactivated polio vaccine (IPV) was rooted in meticulous experimentation and iterative testing, ensuring its efficacy and safety through tangible evidence rather than just theoretical propositions. Similarly, Sabin's oral polio vaccine (OPV) resulted from systematic observations and methodical trials with weakened virus strains to demonstrate its ability to confer immunity without triggering the disease. For both scientists, the credibility of their vaccines was established not on untested hypotheses but on tangible, reproducible results drawn from the real world. This empiricist approach was instrumental in garnering public trust and facilitating the wide adoption of their vaccines. Through their unwavering commitment to evidence-based research, Sabin and Salk illuminated the power of empiricism in advancing medical science and safeguarding public health.

The Scientific Method: Albert Sabin and Jonas Salk, though not the inventors of the scientific method, certainly stand as exemplars of its rigorous application in medical research. Their relentless pursuit of effective vaccines against polio underlined the necessity of a systematic approach to scientific inquiry, which involves formulating hypotheses, conducting controlled experiments, and drawing evidence-based conclusions. Salk's efforts to create the inactivated polio vaccine (IPV) saw him conducting meticulous experiments and iterative testing, ensuring that findings were based on reproducible results, not mere conjecture. Similarly, Sabin's journey with the oral polio vaccine (OPV) was defined by controlled observations, rigorous trials, and repeated validations with weakened virus strains. Their work underscored the indispensability of peer reviews, reproducibility, and transparency in research findings. While they did not redefine the scientific method per se, their high-profile successes in the face of a global health crisis demonstrated its value. By embodying its principles in their research, Sabin and Salk not only advanced the fight against polio but also reinforced the foundational importance of the scientific method in advancing medical knowledge.

Medicine: Albert Sabin and Jonas Salk, two luminaries in the annals of medical history, made monumental contributions to the field, particularly in the fight against poliomyelitis, a disease that had, for decades, plagued societies with paralysis and death. In the 1950s, Salk developed the inactivated polio vaccine (IPV), the first to offer protection against the debilitating disease. Using a killed virus, the IPV, administered via injection, quickly became a beacon of hope, drastically reducing polio's prevalence. Parallel, Sabin took a different approach, perfecting an oral vaccine in the early 1960s that employed a live attenuated virus. The oral polio vaccine (OPV) was effective and easier to administer, especially in large-scale immunization campaigns. Beyond the immediate impact of nearly eradicating a dreaded disease, their work set benchmarks in vaccine research, clinical testing, and public health collaboration. Their legacies are not just tethered to polio; they stand as a testament to what focused scientific inquiry, coupled with a commitment to humanity, can achieve in medicine. Through their endeavors, Sabin and Salk have indelibly shaped the course of medical research and global health initiatives.

Ethics: While primarily celebrated for their groundbreaking work on polio vaccines, Albert Sabin and Jonas Salk indirectly cast light on the core ethical principles underpinning medical research. In their rigorous quest for effective vaccines, both showcased a commitment to beneficence — actively working for the good of individuals and society. Salk notably refused to patent his vaccine, emphasizing its role as a public good. This resonates with the principle of justice — ensuring that benefits, risks, and costs are fairly distributed across society. Their extensive clinical trials and careful monitoring sought to uphold nonmaleficence, providing their interventions did no harm. The meticulous testing and rigorous validation they applied before the mass deployment of their vaccines exemplify this dedication. Moreover, their transparent communication with the public and efforts to educate about the vaccines can be seen as acknowledging autonomy by ensuring individuals are informed and can make health choices. Though the landscape of medical ethics has evolved since their time, Sabin's and Salk's work offers foundational lessons on balancing the rapid progression of science with a deep-seated commitment to ethical integrity.

Conclusion: With their remarkable efforts against poliomyelitis, Albert Sabin and Jonas Salk have left an indelible mark on medical history and broader societal frameworks of understanding and approaching health challenges. Their dedicated pursuit of vaccines underscored the paramount importance of evidence-based methodologies in medical science. Their work wasn't just a testament to technical prowess but also to ethical imperatives that guide the scientific community. Salk's decision to forgo patenting his vaccine for the greater public good speaks to a deep-seated sense of justice and beneficence. On the other hand, Sabin's rigorous trials and public health campaigns highlighted the intertwining of medical advancements with public education and autonomy. Both scientists, in their distinct capacities, reaffirmed the role of research, not just as a tool for discovery but as an instrument of societal upliftment and global cooperation. In essence, the legacies of Sabin and Salk go beyond polio; they resonate as beacons of hope, integrity, and dedication in the ever-evolving landscape of medical science and public health.

Albert Sabin's and Jonas Salk's Legacy: Developed oral and injectable polio vaccines, respectively, nearly eradicating a crippling and often deadly disease from large parts of the world.

**

REVIEW QUESTIONS

True/False Questions:

1. Jonas Salk developed the inactivated polio vaccine (IPV) using a killed virus, and Albert Sabin developed the oral polio vaccine (OPV) using a weakened form of the virus.
   True or False?

2. Both Salk and Sabin prioritized personal gain over public welfare in the development of their polio vaccines.
   True or False?

Multiple-Choice Questions:

3. What ethical principle did Jonas Salk demonstrate by refusing to patent his polio vaccine?
   a) Practitioner beneficence (do good)
   b) Patient autonomy (informed consent)
   c) Practitioner nonmaleficence (do no harm)
   d) Public justice (be fair)

4. Which method was employed by Albert Sabin to make his polio vaccine easier to administer during mass immunization campaigns?
   a) Injection of a killed virus
   b) Oral administration of a weakened virus
   c) Nasal spray of a live virus
   d) Topical application of a weakened virus

Clinical Vignette:

5. A public health official is planning a large-scale immunization campaign to eradicate polio in a developing country. Considering logistical challenges and the need for widespread coverage, which polio vaccine, developed by one of these scientists, would likely be chosen for ease of administration and effective mass immunization?
   a) Inactivated polio vaccine (IPV) developed by Jonas Salk
   b) Oral polio vaccine (OPV) developed by Albert Sabin
   c) Both IPV and OPV combined
   d) Neither, a new vaccine should be developed

Basic Science Vignette:

6. During a medical conference, a speaker discusses the different types of polio vaccines and mentions one that uses a killed virus to confer immunity. Which vaccine is the speaker referring to?
   a) Oral Polio Vaccine (OPV)
   b) Live Attenuated Vaccine
   c) Inactivated Polio Vaccine (IPV)
   d) Recombinant Vaccine

Philosophy Vignette:

7. In an ethics seminar, a student discusses how a scientist decided not to patent a vaccine to ensure it would be freely accessible to the public, emphasizing the principle of beneficence. This decision was made by which scientist?
   a) Albert Sabin
   b) Jonas Salk
   c) Louis Pasteur
   d) Edward Jenner

Correct Answers:

1. True
2. False
3. d) Public justice (be fair)
4. b) Oral administration of a weakened virus
5. b) Oral polio vaccine (OPV) developed by Albert Sabin
6. c) Inactivated Polio Vaccine (IPV)
7. b) Jonas Salk

BEYOND THE CHAPTER
Albert Sabin (1906-1993) and Jonas Salk (1914-1995)

- *Jonas Salk: A Life* by Charlotte DeCroes Jacobs
- *Albert Sabin: A Life of Triumph* by Karin L. McGowan
- *Polio: An American Story* by David M. Oshinsky

***

Philosophy of Science

Medical Research

Chapter 91: Hans Jonas (1903-1993) — The Ethical Philosopher: Responsibility in Science

Jonas's ethical philosophy emphasized responsibility in science and technology, grounding bioethics in both rational and empirical traditions.

Abstract: Hans Jonas emerges as a seminal figure in 20th-century philosophical discourse, expertly navigating the complex interplay between ethics, technology, and science. Gaining prominence in the latter half of the century, Jonas forged a revolutionary path by integrating rationalist and empiricist elements into a rich, nuanced approach to ethics, laying the groundwork for modern bioethics. Rooted in the principle of responsibility, Jonas urged society to broaden its ethical perspective to include present and future generations and the natural world, challenging traditional frameworks. His philosophies emphasize patient autonomy (informed consent), practitioner beneficence (do good), practitioner nonmaleficence (do no harm), and public justice (be fair). These principles have guided critical discourse on the moral implications of scientific and technological advancements, fostering robust ethical decision-making in contemporary science and medicine. Jonas' enduring legacy highlights a visionary approach to ethics, encouraging deeper engagement with pressing moral questions and promoting responsible stewardship in science, medicine, and technology, honoring human dignity and the natural world.

**

Introduction: Hans Jonas, a German-born philosopher and bioethicist, is a pivotal figure in the discourse on ethics, particularly regarding technology, science, and medicine. Born in 1903, his work gained prominence in the latter half of the 20th century, countering his time's existentialist and phenomenological philosophies. At the core of Jonas' ethical inquiry is the exploration of the relationship between humanity and nature, especially in the context of modern scientific advancements. He raised critical questions about the moral implications of technological innovations, including genetic engineering and environmental degradation. His seminal work, "*The Imperative of Responsibility*," challenged traditional ethical frameworks, arguing for an expanded sphere of ethical responsibility that considers present individuals, future generations, and the natural world. This expansion of moral purview formed the basis for modern bioethics, providing scholars and practitioners with tools for ethical decision-making in increasingly complex scientific landscapes. In the following analysis, we will delve deeper into Jonas' contributions to modern thought by evaluating their rationalist foundations, empiricist orientation, impact on the scientific method, influence on medicine, and shaping of the four principal tenets of bioethics: patient autonomy (informed consent), practitioner beneficence (do good), practitioner nonmaleficence (do no harm), and public justice (be fair).

Rationalism: Hans Jonas' philosophical outlook, while incorporating various influences, can have a rationalist underpinning, particularly when understanding ethics in the face of technological and scientific advancements. Unlike strict empiricists, who focus solely on sensory experience as the source of knowledge, Jonas appreciated the role of innate cognitive faculties in moral reasoning. He asserted that ethical principles, to some extent, are hardwired into human consciousness and can be deduced through rational thought. In Jonas' view, our logical, "*a priori*" understanding of responsibility obliges us to extend our ethical concerns beyond immediate contexts to encompass future generations and the natural world. His approach is rooted in the idea that human cognition comes with built-in capabilities for moral reasoning — faculties like foresight, conceptual abstraction, and the capacity for value judgment — that enable us to understand complex ethical issues even without empirical data. This rationale influenced his articulation of principles like responsibility toward future generations. This notion cannot be entirely arrived at through practical methods but requires deductive reasoning based on hypotheses about human dignity and the value of life. Jonas' emphasis on the moral imperative as a rational construct serves as a testament to the influence of rationalist thought in his philosophical framework.

Empiricism: While Hans Jonas is often considered a rationalist in his philosophical approach, particularly in ethics, it would be reductive to overlook the empiricist elements that inform his work. Jonas deeply engaged with the empirical realm of science and technology, using real-world advancements and challenges as the backdrop against which ethical questions should be posed and answered. In doing so, he acknowledged that humans obtain valuable and necessary knowledge through sensory experience and interaction with the external world. Particularly in his explorations of bioethics and the ethical implications of technological innovation, Jonas employed inductive reasoning, drawing from specific instances or scenarios to make broader moral claims. For example, his critiques of genetic engineering and environmental degradation were often rooted in empirical data, highlighting how these practices have concrete consequences for human and non-human life. Although he insisted on universal ethical principles, these principles were often elucidated and refined through empirical examination of real-world issues. Thus, Jonas offers a nuanced approach, synthesizing rationalist and empiricist perspectives to provide a robust framework for ethical analysis, one that accounts for both the universal and the particular, the "*a priori*" and the experiential.

The Scientific Method: Hans Jonas didn't directly contribute to the formation or application of the scientific method; his work serves as a philosophical commentary on the ethical implications of scientific and technological pursuits. Nonetheless, he played a pivotal role in shaping the discourse around the scientific method by questioning the ethical boundaries and responsibilities that scientific inquiry should respect. Jonas warned of the "*God-like*" powers science bestows upon humanity, insisting that the scientific method's standard procedures for establishing empirical truth are insufficient for tackling moral and ethical truths. His critiques introduce a level of ethical reflexivity that adds a layer of complexity to the scientific method's traditional steps. By doing so, Jonas argued for a more inclusive scientific approach that accounts for long-term implications and broader responsibilities. His perspective prompts scientists to consider ethical dimensions before formulating hypotheses, effectively incorporating a moral "*step*" into the scientific process. While Jonas did not alter the core elements of the scientific method, he did challenge its limitations, encouraging a more holistic approach that recognizes the impact of scientific inquiries on society, the environment, and future generations. Thus, he contributed to shaping a more ethically aware science practice, adding a dimension especially crucial in medicine, bioengineering, and environmental science.

Medicine: Hans Jonas' contributions to the medical healthcare field have been primarily ethical, offering frameworks for decision-making that resonate with contemporary challenges in healthcare. While he didn't directly contribute to clinical practices or medical healthcare standards of care in the way a clinician or researcher might, his ethical considerations have influenced bioethical standards and thereby indirectly shaped the medical healthcare standard of care from a moral perspective. Jonas' concept of "*responsibility*" has had a significant impact, urging medical healthcare professionals to consider the broader ethical implications of their actions, including long-term consequences and effects on future generations. This is particularly relevant in genetic engineering, reproductive technologies, and end-of-life care. His philosophy calls for greater scrutiny of procedures and technologies that might be empirically sound but ethically problematic, pushing the medical community to integrate ethical deliberations into their standard protocols. Jonas adds a layer of moral rigor to the evidence-based practices traditionally used to establish medical healthcare standards of care. His work calls on the medical community to go beyond peer-reviewed, evidence-based practices and consider ethical principles when determining the adequacy and appropriateness of care, impacting how medical ethics are integrated into healthcare decision-making.

Ethics: Hans Jonas' ethical philosophy has had a profound influence on modern bioethics, especially on the four cardinal principles: patient autonomy (informed consent), practitioner beneficence (do good), practitioner nonmaleficence (do no harm), and public justice (be fair). Regarding autonomy and informed consent, Jonas emphasized the moral importance of recognizing individual agency, especially when advanced medical technologies can potentially compromise it. His focus on "*responsibility*" adds depth to the principle of beneficence by extending the obligation to do good to individual patients, future generations, and the ecosystem. This broadened scope of responsibility also strengthens the focus on nonmaleficence, or "*do no harm*," by urging medical healthcare professionals to consider the long-term and far-reaching consequences of their actions, especially in emerging fields like genetic engineering. Similarly, his ethical inquiries have enriched the principle of justice, urging us to think about fairness in allocating medical healthcare resources and distributing risks and benefits across society and time. By introducing the concept of extended responsibility into bioethics, Jonas adds a temporal and ecological dimension that urges us to think beyond the immediate clinical context. His work serves as an ethical compass that guides practitioners and policy-makers alike in navigating the complex moral landscapes that technological advancements in healthcare present, thereby leaving an indelible mark on bioethical thought.

Conclusion: Hans Jonas' contributions to philosophy, bioethics, and our understanding of the responsibilities entailed by scientific and technological advancements cannot be overstated. Born in 1903 and achieving prominence in the latter half of the 20th century, he provided thoughtful counterpoints to his era's existentialist and phenomenological philosophies. Jonas pioneered a framework for understanding ethics that emphasized our responsibilities to the present, future generations, and the natural world. This ethical longview has applications in diverse realms, from bioethical principles in medical healthcare to environmental sustainability considerations, rationalist moral reasoning, and empiricist evaluations of real-world technologies. His nuanced approach has effectively served as a cornerstone for ethical decision-making in complex scientific landscapes, urging the integration of ethical considerations into the scientific method, medical practices, and broader social policies. By synthesizing rationalist and empiricist elements, Jonas offers a comprehensive ethical model that balances universal principles with the particularities of real-world challenges. His philosophies have had a lasting impact on modern thought, especially in the moral dimensions of science, medicine, and technology, challenging us to become more accountable stewards of the intellectual and material powers we wield.

Hans Jonas' Legacy: His groundbreaking work in bioethics, particularly the concept of "*responsibility*," has forever shaped ethical considerations in science, medicine, and technology, challenging us to extend our moral concern to future generations and the natural world.

**

Review Questions:

True/False Questions:

1. Hans Jonas emphasized the principle of responsibility, urging ethical consideration for future generations and the natural world.
   True or False?

2. Hans Jonas contributed directly to the clinical practices and medical healthcare standards of care in the way a clinician might.
   True or False?

Multiple-Choice Questions:

3. Which philosophical approach is most closely associated with Hans Jonas' view on ethics and technology?
   a) Utilitarianism
   b) Existentialism
   c) Rationalism
   d) Empiricism

4. How did Hans Jonas' concept of responsibility influence the field of bioethics?
   a) By focusing solely on the well-being of current patients
   b) By emphasizing immediate, short-term outcomes
   c) By extending ethical consideration to future generations and the environment
   d) By disregarding the long-term consequences of medical interventions

Clinical Vignette:

5. A medical professional is considering a new genetic engineering procedure that could have significant long-term impacts. Inspired by Hans Jonas' philosophy, what should be the primary ethical consideration in their decision-making process?
   a) The immediate financial benefits of the procedure
   b) The long-term consequences and responsibilities toward future generations
   c) The popularity of the procedure among patients
   d) The ease and convenience of the procedure

Basic Science Vignette:

6. A genetic engineering company is developing a new technology to alter human DNA to prevent hereditary diseases. While the technology shows promise, there are significant unknowns about its long-term effects on future generations. Based on Hans Jonas' philosophy, what should the company prioritize in its decision-making process?
   a) Immediate market profitability and shareholder returns
   b) Speed of development to stay ahead of competitors
   c) Long-term responsibility to future generations and ethical implications
   d) Minimizing costs and maximizing efficiency

Philosophy Vignette:

7. In a medical ethics debate, a practitioner argues that implementing a new life-saving technology without thorough long-term testing is justified because it could save many lives immediately. How would Hans Jonas likely respond to this argument?
   a) He would agree, prioritizing immediate benefits over potential future risks
   b) He would emphasize the necessity of informed consent from patients
   c) He would argue for the importance of long-term testing to ensure no harm to future generations
   d) He would prioritize cost-effectiveness and efficiency in medical interventions

Correct Answers:

1. True
2. False
3. c) Rationalism
4. c) By extending ethical consideration to future generations and the environment
5. b) The long-term consequences and responsibilities toward future generations
6. c) Long-term responsibility to future generations and ethical implications
7. c) He would argue for the importance of long-term testing to ensure no harm to future generations

Beyond The Chapter
Hans Jonas (1903-1993)

- *The Legacy of Hans Jonas: Judaism and the Phenomenon of Life* edited by Hava Tirosh-Samuelson and Christian Wiese
- *The Imperative of Responsibility: In Search of an Ethics for the Technological Age* by Hans Jonas
- *The Phenomenon of Life: Toward a Philosophical Biology* by Hans Jonas

***

Chapter 92: Henry K. Beecher (1904-1976) — The Ethical Clinician: Informed Consent

Beecher's advocacy for informed consent in clinical research transformed medical ethics, emphasizing patient autonomy and empirical integrity.

Abstract: Henry Knowles Beecher's indelible legacy in the medical fraternity is marked by his pivotal roles in anesthesiology and bioethics. Born in 1904, Beecher advanced empirical foundations in anesthesiology and introduced controlled clinical trials, embedding rationalist principles into medical ethics and transforming the ethical landscape of clinical research and healthcare. By harmonizing empiricism and rationalism, Beecher cultivated methodological rigor and ethical sanctity within medical research, emphasizing empirical effectiveness and the moral responsibilities of medical interventions. His groundbreaking 1966 exposé on unethical medical experiments highlighted the necessity of informed consent and ethical oversight, marking a watershed moment in bioethics. Beecher's rationalist approach reinforced the four cornerstones of bioethics: patient autonomy (informed consent), practitioner beneficence (do good), nonmaleficence (do no harm), and public justice (be fair). His contributions to anesthesiology revolutionized pain management protocols, elevating standards of care. As a harbinger of change, Beecher's legacy stands as a testament to modern medicine's intellectual and moral progress, shaping the evolving realms of medical practice and ethical consideration.

**

Introduction: Henry Knowles Beecher, born in 1904, was a pioneering anesthesiologist and medical ethicist whose work has left an indelible mark on the field of medicine, both in its clinical and ethical dimensions. Beecher gained prominence as a faculty member at Harvard Medical School and was instrumental in developing controlled clinical trials as a foundational practice in medical research. His 1966 paper, "*Ethics and Clinical Research*," served as a watershed moment in the evolution of bioethics, exposing unethical medical experiments and calling for greater scrutiny and ethical oversight in research. While he initially embarked on a career deeply rooted in the empiricist tradition — advocating for evidence-based approaches in anesthesiology — his focus shifted increasingly toward ethical considerations, encapsulating a rationalist orientation that sought to apply reasoned ethical principles to complex medical issues. His work has profoundly impacted the four pillars of bioethics — patient autonomy (informed consent), practitioner beneficence (do good), practitioner nonmaleficence (do no harm), and public justice (be fair) — by promoting a framework that guides physicians and researchers in their moral obligations to patients and subjects. Thus, Henry K. Beecher remains a towering figure whose contributions continue to shape modern medicine's ethical and methodological landscape.

Rationalism: Henry K. Beecher's contributions to medical ethics and anesthesiology were deeply rooted in rationalist principles, advocating for a reasoned approach to problem-solving and applying "*a priori*" ethical considerations. Beecher's call for controlled clinical trials was not just for empirical rigor but also an assertion of the necessity of logical frameworks and conceptual clarity in medical research. He relied on deductive reasoning to construct ethical principles that could guide the inherently complex and variable field of medicine. For instance, his exploration of informed consent and vulnerability in clinical research subjects invoked innate reasoning and moral obligation faculties. The core ideas behind patient autonomy (informed consent), practitioner beneficence (do good), practitioner nonmaleficence (do no harm), and public justice (be fair), which he strongly influenced, are, at their root, principles that can be logically deduced to apply across a variety of medical scenarios. Beecher utilized these "*a priori*" principles to develop a more humane and ethical approach to medical research and patient care. His commitment to these principles guided him toward his landmark 1966 article, "*Ethics and Clinical Research*," published in the New England Journal of Medicine. In this article, Beecher highlighted 22 examples of unethical clinical research practices, significantly influencing the development of ethical guidelines and standards for research involving human subjects, highlighting the importance of applying rationalist ethical frameworks to safeguard human dignity and welfare. Thus, Beecher's rationalist foundation was crucial in developing ethical guidelines that both stem from and guide our evolved reasoning faculties and moral sensibility.

Empiricism: Henry K. Beecher's commitment to empiricism was most vividly evident in his pioneering work in anesthesiology and the design of controlled clinical trials. He firmly believed that medical practices and theories should be grounded in observable phenomena and verifiable data, aligning closely with the empiricist view that knowledge comes primarily from sensory experiences. This empiricist orientation led Beecher to advocate for methodologically rigorous trials that could produce reliable and replicable results, thereby moving the medical field closer to evidence-based practice. His approach relied on inductive reasoning, using specific observations from carefully designed experiments to make general conclusions about anesthetic agents, pain management, and other medical interventions. His insistence on empirical validation also extended to ethical considerations; for example, his landmark 1966 paper, "*Ethics and Clinical Research*," didn't just philosophize about moral failings but provided 22 concrete cases of unethical conduct, offering an empirically grounded critique that was both compelling and irrefutable. By melding empiricism with his broader ethical framework, Beecher brought a level of scientific rigor to the practice and ethics of medicine that was largely unprecedented, thereby shaping the field into its modern form, where empirically verified effectiveness and ethical considerations are deeply intertwined.

The Scientific Method: Henry K. Beecher shaped the scientific method as it applies to clinical research and medicine. Before his interventions, the field of medicine needed to be more rigorous in adopting scientific methodologies. Beecher advocated including controlled clinical trials as a standard procedure for assessing the efficacy of medical interventions. He introduced the scientific method into a domain where anecdotal evidence, untested procedures, and tradition often held sway. Beecher's insistence on controlled trials allowed the formation of testable hypotheses about treatment efficacy, which could be verified or falsified through meticulous data collection and analysis. Moreover, he emphasized the ethical obligations involved in medical research, including the need for informed consent and safeguards against harm, thus ensuring that the scientific method was implemented rigorously and ethically. His influence led to more standardized protocols for data collection and more robust peer review processes, raising the standards of credibility and replicability in medical research. Through his work, Beecher cemented the scientific method as an indispensable tool in pursuing medical knowledge, elevating clinical research's rigor and ethical integrity. His contributions provided a robust framework that has deeply influenced medical research's methodological standards and ethical practices, setting a precedent for the systematic, replicable, and ethically sound scientific inquiry we expect today.

Medicine: Henry K. Beecher's seminal contributions to the field of medicine were twofold: he significantly advanced the science of anesthesiology and laid the ethical groundwork for clinical research. As an anesthesiologist, Beecher's pioneering research helped establish more effective and humane pain management methods during surgery. His rigorous empirical studies on anesthetic agents led to safer and more effective protocols, raising the healthcare standards of care in anesthesia. Beecher's advocacy for controlled clinical trials fundamentally changed how medical research was conducted, emphasizing the need for peer-reviewed, evidence-based practices. His work thus played an instrumental role in establishing the scientific and ethical benchmarks that have since become integral to the medical healthcare community's standards of care. In highlighting ethical lapses in existing research through his groundbreaking 1966 paper "*Ethics and Clinical Research*," Beecher further impacted these standards by insisting on including ethical considerations alongside scientific rigor. His attention to informed consent, patient safety, and ethical oversight have become cornerstones in the guidelines and practices that govern clinical research and patient care today. By elevating the scientific and ethical aspects of medicine, Henry K. Beecher has left an indelible imprint on medical healthcare standards of care, shaping them into the continually evolving, evidence-based, and ethically grounded practices we recognize today.

Ethics: Henry K. Beecher's influence on bioethics is particularly evident in conceptualizing and mainstream acceptance of its four foundational pillars: patient autonomy (informed consent), practitioner beneficence (do good), practitioner nonmaleficence (do no harm), and public justice (be fair). In his landmark 1966 article, "*Ethics and Clinical Research*," Henry K. Beecher highlighted 22 examples of unethical research practices:

1. Investigations with children involving the administration of live viruses or bacteria without proper informed consent:
   - Willowbrook Hepatitis Study: Mentally disabled children at the Willowbrook State School were deliberately infected with hepatitis to study the disease.
2. Studies on institutionalized individuals where hepatitis was deliberately induced:
   - Willowbrook Hepatitis Study: Mentally disabled children at the Willowbrook State School were deliberately infected with hepatitis to study the disease.
3. Research on mentally ill patients involving the administration of drugs that induced dangerous side effects:
   - Fernald State School Radiation Studies: Mentally ill patients were given radioactive substances without their knowledge.Studies involving withholding effective treatments to compare outcomes with untreated groups.
4. Studies involving withholding effective treatments to compare outcomes with untreated groups:
   - Tuskegee Syphilis Study: African American men with syphilis were denied effective treatment to study the disease's natural progression.
5. Investigations on patients undergoing surgery where procedures were carried out without their knowledge or consent:
   - Nuremberg War Crimes: Many cases during WWII where prisoners were subjected to surgeries without consent, including some referenced in the Nuremberg Trials.
6. Studies on prisoners involving the administration of toxic substances:
   - Holmesburg Prison Experiments: Prisoners were exposed to various toxic substances to study their effects.
7. Research involving the deliberate induction of illness in healthy volunteers without clear therapeutic intent:
   - Guatemala Syphilis Experiment: Healthy individuals were deliberately infected with syphilis to study the effects.

8. Experiments on terminally ill patients without their consent:
   - Specific case not cited by name but referred to terminally ill patients given experimental treatments without consent.
9. Studies involving administration of experimental drugs without proper informed consent:
   - Specific case not cited by name but included studies where experimental drugs were given without proper consent.
10. Investigations where patients were subjected to high levels of radiation without adequate safety measures:
    - Fernald State School Radiation Studies: Students were exposed to radioactive substances.
11. Experiments involving the withholding of anesthesia or pain relief to study pain responses:
    - Specific case not cited by name but included studies where anesthesia was withheld.
12. Studies on poor and minority populations involving risky procedures without adequate informed consent:
    - Tuskegee Syphilis Study: Poor African American men were subjected to the study without informed consent.
13. Research involving the administration of known harmful substances to study their effects:
    - Holmesburg Prison Experiments: Prisoners exposed to harmful substances.
14. Investigations where participants were not fully informed about the potential risks and benefits of the study:
    - Specific case not cited by name but referenced studies with incomplete disclosure.
15. Experiments on vulnerable populations such as children or mentally incapacitated individuals without proper safeguards:
    - Willowbrook Hepatitis Study: Mentally disabled children infected with hepatitis.
16. Studies involving the use of placebos where effective treatments were available and withheld:
    - Tuskegee Syphilis Study: Effective treatment was withheld to observe the disease progression.
17. Research where participants were deceived about the true nature of the study:
    - Guatemala Syphilis Experiment: Participants were deceived about the nature of the study.
18. Investigations involving the induction of psychotic states in patients without therapeutic justification:
    - Specific case not cited by name but referenced studies where psychotic states were induced.
19. Studies where patients were subjected to unnecessary procedures for research purposes:
    - Specific case not cited by name but included unnecessary medical procedures for research
20. Research involving the deliberate exposure of participants to harmful substances to study their effects:
    - Holmesburg Prison Experiments: Deliberate exposure to harmful substances.
21. Experiments where participants were not properly monitored for adverse effects:
    - Specific case not cited by name but referenced studies with inadequate monitoring.
22. Investigations where researchers had conflicts of interest that were not disclosed to participants:
    - Specific case not cited by name but referenced studies with inadequate monitoring.
23. Investigations where researchers had conflicts of interest that were not disclosed to participants:
    - Specific case not cited by name but included undisclosed conflicts of interest in research studies.

Beecher's article was instrumental in highlighting these unethical practices and advocating for stricter ethical standards and oversight in clinical research. His paper catalyzed urgent ethical discussions, particularly emphasizing the importance of informed consent. This reinforced the principle of autonomy, stressing that patients and research subjects have the right to control their bodies and make informed decisions about their medical care. Beecher's insistence on methodological rigor and evidence-based medicine further contributed to beneficence, emphasizing that medical interventions should be founded on reliable data to ensure they are effective and beneficial. Simultaneously, his call for ethical oversight in clinical research laid the groundwork for the principle of nonmaleficence, advocating that medical experiments should "*do no harm*" and that mechanisms must be in place to protect against unethical treatment of patients and research subjects. Lastly, Beecher's focus on ethical standards implicitly and explicitly touched on justice by demanding fairness in how medical research is conducted, including how participants are chosen and how burdens and benefits are distributed. Thus, Beecher's work has been instrumental in shaping the ethical landscape of modern medicine, embedding these four principles into the practices, guidelines, and norms that govern medical research and patient care today.

Conclusion: Henry K. Beecher is a monumental figure whose influence reverberates through anesthesiology, medical research, and bioethics. With his deeply rooted rationalist and empiricist foundations, he masterfully blended "*a priori*" ethical considerations with empirical rigor, elevating medical practice and research's scientific and ethical standards. Beecher pioneered the scientific method within clinical medicine, introducing rigorous controlled clinical trials and ethical scrutiny as non-negotiable elements of medical research. His seminal contributions in anesthesiology revolutionized pain management and safety protocols, lifting healthcare standards of care to unprecedented levels. On the ethical front, his groundbreaking work in 1966 ignited a seismic shift in the moral foundations of medicine, indelibly shaping the principles of patient autonomy (informed consent), practitioner beneficence (do good), practitioner nonmaleficence (do no harm), and public justice (be fair) that guide today's medical community. Beecher's multifaceted impact is a testament to his intellectual breadth and moral commitment. The principles, methodologies, and ethical frameworks he championed have become the bedrock upon which modern medicine stands, a legacy of enduring importance that continues to guide and challenge today's medical and moral landscapes.

Henry K. Beecher's Legacy: A transformative figure in anesthesiology and bioethics, Beecher revolutionized medical research methodologies and established the ethical foundations — especially informed consent and the principles of patient autonomy (informed consent), practitioner beneficence (do good), practitioner nonmaleficence (do no harm), and public justice (be fair) — that have become integral to the ethical conduct of modern medicine.

**

REVIEW QUESTIONS

True/False Questions:

1. Henry K. Beecher was instrumental in developing controlled clinical trials and elevating ethical standards in medical research.
   True or False?

2. Beecher's 1966 paper, "*Ethics and Clinical Research*," had minimal impact on the field of bioethics.
   True or False?

Multiple-Choice Questions:

3. Which of the following best describes Henry K. Beecher's approach to medical research?
   a) Theoretical speculation without empirical evidence
   b) Rigorous empirical studies combined with ethical considerations
   c) Focus solely on empirical data without ethical concerns
   d) Reliance on traditional methods without innovative changes

4. Henry K. Beecher's work emphasized the importance of which bioethical principle in medical research?
   a) Autonomy (informed consent)
   b) Beneficence (do good)
   c) Nonmaleficence (do no harm)
   d) All of the above

Clinical Vignette:

5. A researcher is designing a clinical trial for a new drug. Inspired by Henry K. Beecher's principles, what key ethical consideration should be included in the trial design?
   a) Maximizing the number of participants
   b) Ensuring thorough documentation for publication
   c) Obtaining informed consent from all participants
   d) Prioritizing cost-effectiveness of the trial

Basic Science Vignette:

6. A 52-year-old patient is scheduled for a major surgery. The anesthesiologist discusses the procedure and potential risks with the patient, who then consents to the surgery. During the operation, the anesthesiologist follows rigorous protocols to ensure the patient's safety and pain management. What principle of bioethics is primarily being demonstrated by obtaining the patient's consent?
   a) Practitioner beneficence
   b) Public justice
   c) Patient autonomy
   d) Practitioner nonmaleficence

Philosophy Vignette:

7. A clinical researcher is designing a new study to test the effectiveness of a novel pain management drug. In the planning phase, the researcher emphasizes the importance of informed consent, rigorous data collection, and ethical oversight to protect participants. Which ethical principle is the researcher primarily addressing by ensuring informed consent and ethical oversight?
   a) Practitioner nonmaleficence
   b) Practitioner beneficence
   c) Patient autonomy
   d) Public justice

Correct Answers:

1. True
2. False
3. b) Rigorous empirical studies combined with ethical considerations
4. d) All of the above
5. c) Obtaining informed consent from all participants
6. c) Patient autonomy
7. c) Patient autonomy

BEYOND THE CHAPTER
Henry K. Beecher (1904-1976)

- *The Development of Ethics: A Historical and Critical Study Volume III: From Kant to Rawls* by Henry Sidgwick
- *Research and the Individual: Human Studies* by Henry K. Beecher
- *Henry Knowles Beecher and the Development of Informed Consent in Anesthesia Research* by Mark A. Rockoff

***

Chapter 93: Joseph Murray (1919-2012) — The Transplant Pioneer: Organ Transplants

Murray's success in organ transplantation, grounded in empirical research and ethical consideration, transformed medical practice and patient care.

Abstract: In the mid-20th century, Dr. Joseph Murray emerged as a towering figure who bridged medicine and bioethics with unparalleled expertise. Murray, a beacon of rationalist thought and empiricist approach, advanced organ transplantation from skepticism to groundbreaking success. His theoretical frameworks and strategies to overcome organ rejection propelled transplantation medicine forward. Murray's adherence to empirical methods, characterized by rigorous clinical trials and hands-on experiences, paved the way for successful kidney transplants. His meticulous approach fostered medical innovations and cemented the foundations of the scientific method. Murray's work deeply impacted the four principles of bioethics: patient autonomy (informed consent), practitioner beneficence (do good), nonmaleficence (do no harm), and public justice (be fair). His holistic approach to patient care illuminated the ethical complexities in medical advancements. As a precursor of a revolutionary era in healthcare, Murray's life and work represent the pinnacle of human intellect and compassion, reshaping medicine and granting countless individuals a second chance at life.

**

Introduction: Joseph Murray, a pioneering figure in the annals of medical history, was a surgeon who navigated the intricate channels of science and ethics to alter the trajectory of organ transplantation. Born in the early 20th century, his life intersected with an epoch of vast scientific curiosity and accelerated medical advancements. An amalgamation of rationalist and empiricist approaches, Murray dared to venture into the unknown territories of transplantation when the field was fraught with skepticism and inchoate understanding. His relentless dedication to clinical observation and empirical data collection, combined with a well-grounded theoretical framework, resulted in groundbreaking work, particularly in kidney transplantation. As we delve deeper into his life and contributions, we will discern his pivotal role in shaping the scientific method and his profound influence on the four cardinal ethical principles of bioethics: patient autonomy (informed consent), practitioner beneficence (do good), practitioner nonmaleficence (do no harm), and public justice (be fair). Such a comprehensive analysis of Murray offers a panoramic view of a titan who sculpted the landscapes of medicine and bioethics.

Rationalism: Joseph Murray's rationalist inclinations were prominently evident in medical science and practice. While the empirical evidence provided through experimentation and clinical trials was undeniably essential in his work, the conceptual foundation from which he approached the complexities of organ transplantation exhibited traces of rationalist thought. Murray was among the few who believed, even before tangible evidence emerged that it was fundamentally possible for one human being to accept an organ from another without outright rejection. This belief stemmed from his inherent reasoning that if the body could be made to "*accept*" the foreign organ, transplantation could revolutionize medicine. By relying on theoretical frameworks and logical deductions, he hypothesized strategies to overcome organ rejection. It's worth noting that while others remained confined to the empirical setbacks of early transplantations, Murray's rationalist convictions paved the way for him to envisage solutions that might not have been immediately evident from empirical data alone. Thus, through a blend of reason, logic, and a steadfast belief in the "*a priori*" understanding of human biology, Murray embarked on a journey that changed the face of transplantation medicine.

Empiricism: Joseph Murray's empiricist orientation manifested robustly in his methodical approach to transplantation medicine. Grounded in the belief that knowledge is derived predominantly from sensory experiences, Murray was acutely aware that successful transplantation required a systematic understanding of empirical data obtained from direct observation and experimentation. His revolutionary strides in kidney transplantation were not solely based on theoretical constructs but were deeply rooted in extensive clinical trials and hands-on experiences. For instance, his experimentation with irradiation on patients before transplantation, while ethically controversial by today's standards, was an empirical attempt to address organ rejection. This dedication to direct observation was further exemplified when he and his team monitored post-operative patients, amassing invaluable data on organ rejection and refining their techniques based on these observations. Through such meticulous scrutiny of sensory data — watching, touching, and evaluating the living body in its dynamic response to transplants — Murray solidified his understanding and enhanced the methodologies in transplantation medicine. While his rationalist inclinations provided a conceptual direction, his empiricist orientation gave the tangible means and methods to make the previously unimaginable feat of successful organ transplantation a reality.

The Scientific Method: Joseph Murray, in his relentless pursuit of knowledge within the realm of transplantation medicine, invariably upheld and contributed to the sanctity of the scientific method. Beginning with his observations of organ rejection, Murray's curiosity birthed pivotal questions about the underlying causes and potential solutions. His hypotheses — primarily centered on suppressing the body's immune response to accept foreign organs — were crafted with rigorous precision and testability in mind. His experiments, from studying twin transplants to implementing irradiation, were systematic attempts to validate these hypotheses. Murray's meticulous data collection and analysis, especially his close observation of post-transplant patients, provided insights that either validated or refuted his initial propositions. Moreover, the replicability of his techniques was evident when, after his groundbreaking successes, organ transplantation began to be practiced more widely, following the methodologies he helped pioneer. Finally, Murray's pioneering work was not exempt from the scrutiny of the larger medical community. Through peer reviews, debates, and discussions, his methodologies and findings were continually refined. By adhering to and fortifying the principles of the scientific method, Joseph Murray not only advanced transplantation medicine but also exemplified the rigorous standards that scientific exploration demands.

Medicine: Dr. Joseph Murray's monumental contributions to medicine, particularly in organ transplantation, forever altered the trajectory of healthcare. Before Murray's pioneering work, successfully transplanting organs between non-twin individuals was a distant dream, fraught with myriad challenges, chiefly the body's natural inclination to reject foreign organs. In 1954, Murray etched his name in medical history by completing the first kidney transplant between identical twins, a landmark achievement that opened the doors for further exploration. Yet, his relentless pursuit to understand and combat organ rejection solidified his stature. By innovating techniques, such as irradiation and immunosuppressive drugs, Murray expanded the horizons of transplantation medicine. These groundbreaking methodologies, informed by rigorous research and clinical observation, became foundational pillars in establishing new standards of care in transplantation. They adhered to and raised the bar for evidence-based, peer-reviewed practices within the healthcare community. Through his endeavors, Murray played an instrumental role in recalibrating and elevating medical healthcare standards in transplantation. He ensured that countless patients would benefit from life-saving procedures based on his pioneering work.

Ethics: Joseph Murray's groundbreaking work in transplantation inevitably intersected with the cardinal ethical pillars of bioethics: patient autonomy (informed consent), practitioner beneficence (do good), practitioner nonmaleficence (do no harm), and public justice (be fair). As he navigated the developing terrain of organ transplantation, the principle of autonomy, especially informed consent, became pivotal. Recognizing the profound ethical implications of transplanting organs, Murray championed the need for comprehensive patient understanding, ensuring that recipients and donors were fully aware of potential risks and outcomes. His commitment to beneficence was evident in his unyielding pursuit to improve and save lives, pioneering techniques that would increase transplantation success rates. Yet, Murray's work also grappled with nonmaleficence, as early experimental procedures, such as pre-transplant irradiation, carried inherent risks. While he constantly endeavored to minimize harm, the evolving nature of transplantation medicine invariably presented challenges. Lastly, the principle of justice — ensuring fairness in organ allocation and accessibility — gained prominence as transplantation became more feasible. While Murray may not have singlehandedly addressed all the complex ethical dilemmas associated with organ distribution, his groundbreaking work undeniably laid the foundation for these essential conversations. In essence, Murray's legacy in transplantation is intertwined with his profound influence on shaping and navigating the ethical dimensions of modern medicine.

Conclusion: Dr. Joseph Murray's odyssey through the complex labyrinths of medicine and bioethics stands as an emblem of human endeavor, combining the art of healing with the rigor of science and the nuances of ethical considerations. As the maestro behind successful organ transplantation, Murray didn't merely reshape a facet of medicine; he redefined life's possibilities, offering countless individuals a second chance at life. By harmoniously interweaving rationalist thought with empirical observation, he manifested the ideal interplay of theory and practice in medical science. His commitment to the scientific method set new precedents in research, while his pioneering methodologies in transplantation ushered in a new era of medical care. Yet, his legacy isn't limited to just scientific advancements. Murray's conscientious approach to the ethical facets of transplantation underscored the imperative balance between innovation and morality, highlighting the quintessential role of patient autonomy (informed consent), practitioner beneficence (do good), practitioner nonmaleficence (do no harm), and public justice (be fair) in patient care. In totality, Dr. Joseph Murray's life and work serve as an enduring beacon, illuminating the boundless potential of human intellect, compassion, and ethical consideration in serving humanity.

Joseph Murray's Legacy: Conducted the first successful human organ transplant, pioneering a new medical frontier that has since saved countless lives.

**

Review Questions:

True/False Questions:

1. Joseph Murray was primarily known for his contributions to cardiac surgery. True or False?
2. Murray's pioneering work led to the first successful human kidney transplant between identical twins. True or False?

Multiple-Choice Questions:

3. What was one of the primary challenges Joseph Murray faced in his early transplantation work?
   a) Finding suitable donors
   b) Developing immunosuppressive techniques
   c) Performing surgeries without anesthesia
   d) Managing post-operative infections

4. Which of the following principles of bioethics was most directly impacted by Murray's emphasis on informed consent in transplantation?
   a) Nonmaleficence
   b) Autonomy
   c) Justice
   d) Beneficence

Clinical Vignette:

5. A patient with end-stage renal disease is evaluated for a kidney transplant. Reflecting on Joseph Murray's pioneering efforts, which ethical consideration must be prioritized when discussing the procedure with the patient?
   a) Ensuring the patient understands the risks and benefits of the transplant
   b) Minimizing the cost of the procedure
   c) Guaranteeing immediate availability of a donor organ
   d) Conducting the surgery without informing the patient to avoid stress

Basic Science Vignette:

6. A surgeon is preparing to perform a kidney transplant and is reviewing Joseph Murray's pioneering techniques to minimize organ rejection. What empirical approach, inspired by Murray, should the surgeon prioritize to ensure the best outcome for the patient?
   a) Rely solely on theoretical models of immune suppression
   b) Utilize evidence-based immunosuppressive protocols and monitor the patient's response closely
   c) Depend on historical success rates without further testing
   d) Use intuition and general medical knowledge without specific protocols

Philosophy Vignette:

7. A medical ethics committee is developing guidelines for patient consent in organ transplantation. Inspired by Joseph Murray's principles, what should be the main focus when obtaining informed consent from patients?
   a) Ensuring patients understand the theoretical benefits of transplantation
   b) Providing comprehensive and empirically verified information about the risks and benefits
   c) Using medical jargon to explain the procedure
   d) Minimizing the discussion of potential risks to avoid patient anxiety

Correct Answers:

1. False
2. True
3. b) Developing immunosuppressive techniques
4. b) Autonomy
5. a) Ensuring the patient understands the risks and benefits of the transplant
6. b) Utilize evidence-based immunosuppressive protocols and monitor the patient's response closely
7. b) Providing comprehensive and empirically verified information about the risks and benefits

Beyond The Chapter
Joseph Murray (1919-2012)

- *Surgery of the Soul: Reflections on a Curious Career* by Joseph E. Murray
- *Transplant: From Myth to Reality* by Nicholas L. Tilney
- *Joseph E. Murray and the Story of the First Human Kidney Transplant* by Franklin H. Epstein

***

Philosophy of Science

Medical Research

EPILOGUE

Epilogue

As we conclude our comprehensive exploration of "*Philosophy of Science*," we reflect on the profound journey through the epochs of human inquiry, scientific discovery, and ethical evolution. This book has meticulously traced the intricate tapestry woven by the threads of rationalism, empiricism, scientific method, medicine, and ethics. From the pioneering ideas of ancient philosophers to the cutting-edge innovations of modern scientists, we have witnessed the relentless pursuit of knowledge and the unwavering commitment to ethical responsibility.

Throughout the sixteen weeks, we have delved into the foundational concepts that underpin scientific rigor and ethical integrity. We began by establishing the principles of falsifiability, scientific methodology, and the critical distinction between correlation and causation, setting the stage for a nuanced appreciation of the scientific process. The exploration of paradigms and ethics highlighted the dynamic nature of knowledge and the frameworks that ensure scientific progress remains socially responsible.

The journey through various ethical frameworks, from deontology to utilitarianism, underscored the importance of integrity in scientific and healthcare practices. We examined the essential processes that uphold scientific rigor, from peer review to the ethical implications of AI and Big Data. The ethical dimensions of genetic advancements, precision medicine, and innovative technologies illuminated the balance between progress and stewardship.

Our exploration of stewardship, systems thinking, cognitive science, and the philosophies of key thinkers like Imre Lakatos and Paul Feyerabend challenged us to rethink the nature of knowledge and the ethical dimensions of scientific advancement. We traced the contributions of seminal thinkers from Plato to Copernicus, whose ideas shaped Western thought, scientific inquiry, and ethical frameworks.

The transformative period of the Scientific Revolution and the subsequent contributions of pioneers like William Harvey, René Descartes, and Isaac Newton highlighted the evolution of scientific thought and medical practice. The legacies of these visionaries underscored the power of rationalist and empirical approaches in advancing knowledge and human health.

The later weeks brought us to the modern era, where figures like Louis Pasteur, Alexander Fleming, and Albert Sabin revolutionized medical practices, emphasizing ethical principles of beneficence, nonmaleficence, and community protection. The contributions of ethical philosophers like Hans Jonas and clinicians like Henry K. Beecher reinforced the importance of informed consent and patient autonomy in clinical research and medical practice.

This journey has not only educated us on the historical and philosophical foundations of science but also inspired critical thinking and ethical considerations in contemporary scientific inquiry and medical practice. The enduring principles of autonomy, beneficence, nonmaleficence, and justice have been traced throughout the epochs, demonstrating their relevance and necessity in guiding scientific and medical advancements.

"*Philosophy of Science*" stands as a testament to the enduring quest for knowledge and the ethical imperatives that guide this pursuit. It serves as an essential resource for students, scholars, and professionals across multiple disciplines, providing a structured and thorough content that bridges historical and contemporary issues. As we move forward, let us carry the lessons learned and the inspirations garnered from this exploration, ensuring that our scientific endeavors are always accompanied by a steadfast commitment to ethical responsibility.

Final Reflections

In the grand tapestry of human history, the convergence of philosophy, science, and ethics has consistently propelled us toward greater understanding and improved well-being. This book, with its detailed examination of pivotal concepts, historical periods, and influential figures, aims to illuminate the path we have traveled and the ethical compass that must guide our future.

As we continue to push the boundaries of scientific knowledge and technological innovation, the lessons from our past remind us that progress is not merely a matter of discovery but also of responsibility. The intricate interplay between rationalism and empiricism, the integration of diverse knowledge systems, and the relentless pursuit of truth are the cornerstones of our intellectual heritage.

May this book inspire future generations to engage with science and ethics with curiosity, rigor, and integrity, ensuring that our collective journey continues to enhance human health, knowledge, and ethical responsibility. Philosophy of Science are not static fields but dynamic, ever-evolving disciplines that will continue to shape and be shaped by our quest for understanding and our commitment to ethical practice.

***

Philosophy of Science

Medical Research

AUTHOR

Jeffrey W. Bulger

Embarking on the literary journey of "*Philosophy of Science*" Dr. Jeffrey W. Bulger's academic background is both diverse and comprehensive. He holds a PhD in Philosophy from the University of Tennessee, with a specialization in Biomedical Ethics. His studies have encompassed Medical Ethics, Clinical Ethics, Philosophy of Science, Social Political Philosophy, Health Care Education, and Online Pedagogical Course Development. His studies have encompassed Medical Ethics, Clinical Ethics, Philosophy of Science, Social Political Philosophy, Health Care Education, and Online Pedagogical Course Development. Prior to this, he earned a Bachelor's degree in Geology, specializing in Petroleum Engineering, from the University of North Dakota, grounding his philosophical inquiries in the tangible realities of the natural world and its exploration.

Throughout his career, Dr. Bulger has dedicated himself to the development and delivery of a wide array of courses, mentoring over 50 students to present their research at the National Conference of Undergraduate Research (NCUR). This reflects his commitment to nurturing the academic and professional growth of the next generation of scholars. His efforts in teaching have been recognized through his inclusion in the Master Teacher Guild, a testament to the high regard in which he is held by both students and colleagues.

Beyond teaching, Dr. Bulger has engaged deeply with clinical education, research, and governance, working tirelessly to align administrative policies with ethical standards and educational mandates. His contributions to hospital ethics committees have been instrumental in enhancing decision-making processes and fostering collaboration among medical professionals, staff, and administrators.

His scholarly work spans authoring textbooks on biomedical ethics, leading online educational initiatives, and performing clinical ethics consultations. His research interests focus on shared decision-making and the ethical underpinnings of informed consent, contributing to the broader discourse on medical and scientific ethics.

Dr. Bulger is committed to inclusivity and understanding, valuing the diverse socio-economic, academic, cultural, and ethnic backgrounds of the educational, medical, and patient communities he serves. This commitment extends to accommodating individuals with diverse learning styles and abilities, ensuring his work is accessible to all.

With "*Philosophy of Science*," Dr. Bulger invites readers to explore the intricate interplay between science, philosophy, and ethics across the epochs of human inquiry. This book not only reflects his diverse academic background and professional experiences but also embodies his passion for lifelong learning and the pursuit of knowledge in the realms of biomedical and clinical research, offering a comprehensive and enriching exploration of the rich tapestry of human intellectual endeavor.

***

Philosophy of Science

Medical Research

Syllabus Content

Course Description, Objectives, Flipped Classroom, Weekly Schedule

Course Title:

Philosophy of Science:
Tracing Rationalism, Empiricism, Scientific Method, Medicine, and Ethics Through the Epochs of Human Inquiry

Instructor:

Name

Course Structure:

Duration: Adaptable for both a 10-week term and 16-week semester.

Class Frequency: Two one-hour classes per week or one two-hour class per week.

Course Description:

"Philosophy of Science: Medical Research" is an intellectually enriching course that delves into the evolution of scientific thought and its profound impact on modern medicine. This course, based on the text by Jeffrey W. Bulger, offers a comprehensive narrative that spans from the foundational ideas of early philosophers to contemporary advancements in science and medicine. It intricately intertwines the development of scientific paradigms with the ethical principles of autonomy, beneficence, nonmaleficence, and justice, providing students with an in-depth understanding of the philosophical and ethical underpinnings of biomedical research.

Course Objectives:

1. **Understand the Foundations of Scientific Inquiry:**
   - Comprehend the principles of falsifiability and the scientific method.
   - Analyze the historical development and evolution of the scientific method from ancient Greek philosophy to contemporary scientific practices.

2. **Analyze Historical Contributions:**
   - Identify and evaluate the contributions of pivotal thinkers from various historical periods, including Thales, Hippocrates, Newton, and Harvey.
   - Understand the impact of these contributions on the development of scientific, medical, and ethical thought.

3. **Investigate Contemporary Topics:**
   - Engage with modern challenges and advancements in science and medicine, including AI, bioinformatics, climate change, and precision medicine.
   - Analyze how contemporary issues are informed by historical and philosophical contexts.

4. **Develop Critical Thinking and Analytical Skills:**
   - Cultivate the ability to critically assess scientific theories and ethical dilemmas.
   - Enhance skills in logical reasoning, argumentation, and evidence-based analysis.

Course Content, Assessment, and Student Presentation(s):

Chapters: Each chapter is 3-4 pages, covering the five key elements: rationalism, empiricism, scientific method, medicine, and ethics.

Weekly Quizzes:

- Each chapter includes seven quiz questions: two true/false questions, two multiple-choice questions, one clinical vignette, one basic science vignette, and one philosophy vignette.
- Quizzes must be completed weekly on the LMS (e.g., D2L, Blackboard, Canvas) with a minimum score of 90%.

Exams:

- Online exams with random selections of quiz questions.
- Exams can only be taken once and only when fully caught up with weekly quizzes.

Flipped Classroom:

- Student Presenations
  - Students will give a 10-minute presentation on the assigned topics for the class.
  - Presentations will be followed by discussion, and/or a learning activity.
  - Topics and presentation dates will be assigned in advance.
- By engaging with these materials and activities, students will gain a thorough understanding of the historical and philosophical foundations of science and their contemporary implications.

Attendance:

- Mandatory for all sessions
- Attendance Make-Up Paper: All absences (excused or unexcused) will result in the requirement to author and submit an attendance make-up paper focusing on a topic(s) covered during the missed class period. The make-up paper is due within five days of the missed lecture and must include:
  - Topic Introduction
  - Discussion of the topic from a philosophical perspective.
  - Reference at least one scholarly citation to support the position on the specific topic.
  - The paper must be thoughtful, logical, organized, and well-developed.
  - 350-word minimum paper to be submitted on the LMS weekly module. A submission field will appear on the weekly attendance quiz when non-attendance is indicated. Student's name should be included in the text field when submitting their paper.

Weekly Schedule

WEEK ONE

Class Hour 1

- Syllabus
- The Big Picture

Class Hour 2

- Chapter 1: Karl Popper (1902-1994) — The Falsifiability Principle: Scientific Rigor
- Chapter 2: The Scientific Method — The Foundation of Inquiry: Empirical Rigor
- Chapter 3: Correlation and Causation — The Analytical Foundation: Accurate Inquiry

---

WEEK TWO

Class Hour 1

- Chapter 4: Thomas Kuhn (1922-1996) — The Paradigm Shifter: Scientific Revolutions
- Chapter 5: Epistemology — The Foundations of Knowledge: Inquiry and Understanding
- Chapter 6: Bioethics — The Moral Guide: Science and Ethics

Class 2

- Chapter 7: Deontology — The Moral Compass: Ethical Principles
- Chapter 8: Utilitarianism — The Greatest Good: Ethical Decision-Making
- Chapter 9: Virtue Ethics — The Moral Excellence: Character and Integrity

---

WEEK THREE

Class Hour 1

- Chapter 10: Universal Ethics — The Holistic Approach: Integrative Morality
- Chapter 11: Beauchamp and Childress — The Principlists: Ethical Framework in Bioethics
- Chapter 12: Peter Singer (1946-) — The Utilitarian Ethicist: Global Welfare and Animal Rights

Class Hour 2

- Chapter 13: Academic Freedom — The Catalyst for Truth: Unbounded Inquiry
- Chapter 14: Scholarship — The Beacon of Progress: Intellectual Rigor
- Chapter 15: History of Peer Review — The Evolution of Integrity: Critical Analysis

---

WEEK FOUR

Class Hour 1

- Chapter 16: Peer Review — The Pillar of Rigor: Scientific Integrity
- Chapter 17: P-Hacking — The Threat to Integrity: Ethical Research
- Chapter 18: Research Funding Source — The Ethical Foundation: Scientific Integrity

Class Hour 2

- Chapter 19: AI and Emergent Properties — The Technological Ethos: Rethinking Intelligence
- Chapter 20: Big Data — The Analytical Revolution: Precision and Ethics
- Chapter 21: Bioinformatics — The Computational Bridge: Biological Insights

EXAM I: 50Q - online

WEEK FIVE

Class Hour 1

- Chapter 22: CRISPR-Cas9 — The Genetic Precision: Ethical Innovation
- Chapter 23: Genetic Engineering and Biotechnology — The Transformative Frontier: Ethical Dilemmas
- Chapter 24: Human Genome Project — The Genetic Blueprint: Scientific and Ethical Insights

Class Hour 2

- Chapter 25: Large Language Models — The AI Frontier: Rethinking Knowledge
- Chapter 26: Nanotechnology — The Atomic Precision: Ethical Innovation
- Chapter 27: Pharmacogenomics — The Personalized Medicine: Genetic Insights

---

WEEK SIX

Class Hour 1

- Chapter 28: Precision Medicine — The Tailored Treatment: Efficacy and Safety
- Chapter 29: Quantum Computing — The Computational Revolution: Ethical Horizons
- Chapter 30: Digital Health and Telemedicine — The Accessible Healthcare: Modern Solutions

Class Hour 2

- Chapter 31: Virtual and Augmented Reality in Medicine — The Immersive Innovation: Transformative Healthcare
- Chapter 32: Richard Dawkins (1941-) — The Evolutionary Biologist: Gene-Centered Evolution
- Chapter 33: Climate Change — The Anthropocene Challenge: Ethical Imperatives

---

WEEK SEVEN

Class Hour 1

- Chapter 34: Environmentalism — The Stewardship Ethos: Planetary Preservation
- Chapter 35: Global Health Initiatives — The Equity Imperative: Health for All
- Chapter 36: Green Revolution — The Agricultural Paradigm: Balancing Innovation

Class Hour 2

- Chapter 37: Complexity Theory — The Holistic Framework: Understanding Systems
- Chapter 38: Epigenetics — The Dynamic Influence: Nature and Nurture
- Chapter 39: Microbiome — The Symbiotic Network: Health and Disease

EXAM II: 50Q - online

WEEK EIGHT

Class Hour 1

- Chapter 40: Neuroscience — The Cognitive Frontier: Understanding the Mind
- Chapter 41: Heraclitus' Ship of Theseus Paradox — The Identity Puzzle: Continuity and Change
- Chapter 42: Stem Cell Research — The Regenerative Potential: Ethical Frontiers

Class Hour 2

- Chapter 43: Systems Thinking — The Integrative Framework: Complex Problem-Solving
- Chapter 44: Imre Lakatos (1922-1976) — The Research Methodologist: Research Programs
- Chapter 45: Paul Feyerabend (1924-1994) — The Methodological Anarchist: Epistemological Pluralism

---

WEEK NINE

Class Hour 1

- Chapter 46: Daniel Dennett (1942-) — The Cognitive Philosopher: Interdisciplinary Dialogue
- Chapter 47: Consciousness — The Emergent Marvel: Beyond the Brain
- Chapter 48: Free Will and Determinism — The Dynamic Interplay: Freedom in Complexity

Class Hour 2

- Chapter 49: Medicine — The Dynamic Fusion: Science, Technology, Ethics
- Chapter 50: Hippocratic Oath — The Ethical Bedrock: Medical Integrity
- Chapter 51: Nuremberg Code — The Ethical Framework: Human Dignity

---

WEEK TEN

Class Hour 1

- Chapter 52: Declaration of Geneva — The Universal Guide: Medical Ethics
- Chapter 53: Declaration of Helsinki — The Ethical Compass: Biomedical Research
- Chapter 54: 45CFR46 — The Common Rule: Ethical Research

Class Hour 2

- Chapter 55: *Belmont Report* — The Ethical Blueprint: Human Research
- Chapter 56: IACUC — The Guardian of Ethics: Animal Research
- Chapter 57: 3Rs — The Ethical Triad: Humane Research

EXAM III: 50Q - online

WEEK ELEVEN

Class Hour 1

- Chapter 58: Thales (THAY-leez, 624-546 BCE) — The Water Philosopher: Transforming Myth to Science
- Chapter 59: Pythagoras (py-THAG-or-as, 570-495 BCE) — The Mathematical Metaphysicist: Numbers Governing Reality
- Chapter 60: Xenophanes (zeh-NOF-uh-neez, 570-475 BCE) — The Rational Critic: One God, Many Questions

Class Hour 2

- Chapter 61: Alcmaeon (Alk-MAY-on, 500-450 BCE) — The Anatomist: Brain as the Seat of Understanding
- Chapter 62: Hippocratic Corpus (460-unknown BCE) — The Medical Founders: Do No Harm
- Chapter 63: Socrates (469-399 BCE) — The Dialectical Philosopher: Unyielding Pursuit of Truth

---

WEEK TWELVE

Class Hour 1

- Chapter 64: Plato (427-347 BCE) — The Idealist: Forms and the Academy
- Chapter 65: Aristotle (384-322 BCE) — The Systematizer: Empiricism and Causality
- Chapter 66: Galen of Pergamum (GAY-len, 129-216 CE) — The Medical Synthesizer: Humoral Theory and Medical Ethics

Class Hour 2

- Chapter 67: Avicenna (Ibn Sina, 980-1037) — The Medical Synthesizer: The Canon of Medicine
- Chapter 68: Thomas Aquinas (1225-1274) — The Theological Synthesizer: Faith and Reason
- Chapter 69: Nicolaus Copernicus (1473-1543) — The Heliocentric Revolutionary: Shifting the Cosmic Paradigm

---

WEEK THIRTEEN

Class Hour 1

- Chapter 70: Girolamo Fracastoro (1478-1553) — The Germ Theorist: Seed-like Entities and Contagion
- Chapter 71: Paracelsus (1493-1541) — The Chemical Healer: Experimenting with Medicine
- Chapter 72: Ambroise Paré (1510-1590) — The Surgical Innovator: Compassionate Techniques

Class Hour 2

- Chapter 73: Andreas Vesalius (1514-1564) — The Anatomical Innovator: Revealing the Human Body
- Chapter 74: Francis Bacon (1561-1626) — The Empirical Philosopher: Method and Experimentatio
- Chapter 75: Galileo Galilei (1564-1642) — The Observational Pioneer: Telescopic Revelations

EXAM IV: 50Q - online

WEEK FOURTEEN

Class Hour 1

- Chapter 76: William Harvey (1578-1657) — The Circulatory Pioneer: Heart and Blood
- Chapter 77: René Descartes (1596-1650) — The Rationalist Thinker: Cogito, ergo sum
- Chapter 78: Antonie van Leeuwenhoek (1632-1723) — The Microscopic Discoverer: Unveiling the Invisible

Class Hour 2

- Chapter 79: Isaac Newton (1643-1727) — The Universal Theorist: Laws of Motion and Gravitation
- Chapter 80: David Hume (1711-1776) — The Empirical Skeptic: Questioning Causality
- Chapter 81: Immanuel Kant (1724-1804) — The Synthesizer: Rationalism and Empiricism

---

WEEK FIFTEEN

Class Hour 1

- Chapter 82: Edward Jenner (1749-1823) — The Vaccine Innovator: Conquering Smallpox
- Chapter 83: Thomas Malthus (1766-1834) — The Population Theorist: Limits of Growth
- Chapter 84: John Stuart Mill (1806-1873) — The Utilitarian Philosopher: Greatest Good for the Greatest Number

Class Hour 2

- Chapter 85: Charles Darwin (1809-1882) — The Evolutionary Theorist: Natural Selection
- Chapter 86: Ignaz Semmelweis (1818-1865) — The Hygiene Pioneer: Handwashing in Medicine
- Chapter 87: Gregor Mendel (1822-1884) — The Geneticist: Principles of Inheritance

---

WEEK SIXTEEN

Class Hour 1

- Chapter 88: Louis Pasteur (1822-1895) — The Germ Theory Pioneer: Microbial Revolution
- Chapter 89: Alexander Fleming (1881-1955) — The Antibiotic Pioneer: Penicillin
- Chapter 90: Albert Sabin (1906-1993) and Jonas Salk (1914-1995) — The Polio Warriors: Vaccine Pioneers

Class Hour 2

- Chapter 91: Hans Jonas (1903-1993) — The Ethical Philosopher: Responsibility in Science
- Chapter 92: Henry K. Beecher (1904-1976) — The Ethical Clinician: Informed Consent
- Chapter 93: Joseph Murray (1919-2012) — The Transplant Pioneer: Organ Transplants

EXAM V: 50Q - online

**

SYLLABUS
REVIEW QUESTIONS

True/False Questions:

1. The course "*Philosophy of Science: Medical Research*" spans from ancient Greek philosophy to contemporary scientific innovations.
   True or False?

2. The course is structured into Sixteen Weeks, each focusing on a different aspect of scientific and philosophical development.
   True or False?

3. Weekly quizzes for each chapter include questions such as True/False, Multiple-Choice, and Vignettes.
   True or False?

4. Students are required to attend all sessions and submit an attendance make-up paper for any absences, whether excused or unexcused.
   True or False?

5. The course covers contemporary topics such as AI, bioinformatics, climate change, and precision medicine, exploring their connections to historical and philosophical contexts.
   True or False?

Multiple-Choice Questions:

6. Which of the following periods is NOT explicitly mentioned in the course description as covered in the course?
   a) Early Classical
   b) Renaissance
   c) Industrial Revolution
   d) Contemporary

7. What are the foundational principles introduced at the beginning of the course?
   a) Falsifiability, scientific method, scientific revolutions, ethics, and emergent properties
   b) Rationalism, empiricism, consciousness, free will, and epistemology
   c) AI, bioinformatics, climate change, and precision medicine
   d) Logical reasoning, argumentation, and evidence-based analysis
8. Which of the following is NOT one of the course objectives?
   a) Understand the Foundations of Scientific Inquiry
   b) Analyze Historical Contributions
   c) Investigate Contemporary Topics
   d) Master Laboratory Techniques in Modern Science
9. In addition to true/false and multiple choice review questions, what other types of vignettes are included in the weekly quizzes?
   a) Historical, Pedigocial, and Discussion vignettes
   b) Clinical, Basic Science, and Philosophy vignettes
   c) Mathematical, Linguistical, and Sociological vignettes
   d) Literary, Contemporary, and Relational vignettes
10. What is the required minimum score for completing the weekly quizzes?
   a) 80%
   b) 85%
   c) 90%
   d) 95%

Correct Answers:

1. True
2. False
3. True
4. True
5. True
6. c) Industrial Revolution
7. a) Falsifiability, scientific method, scientific revolutions, ethics, and emergent properties
8. d) Master Laboratory Techniques in Modern Science
9. b) Clinical, Basic Science, and Philosophy vignettes
10. c) 90%

***

Made in the USA
Monee, IL
12 November 2026

dfb00cf5-f066-403f-aefd-1c1284a52eb5R01